MW01446009

THE BACKYARD

STARGAZER'S BIBLE

DISCOVER CONSTELLATIONS, GALAXIES, NEBULAE, METEORITES, AND MORE

Published in North America in 2024 by Abrams, an imprint of ABRAMS

First published in the United Kingdom in 2024 by William Collins, an imprint of HarperCollins*Publishers*

Text © HarperCollinsPublishers 2024
Images © individual copyright holders, see page 415 for details
Cover illustrations © Lynn Hatzius 2024

Written by Mary McIntyre, Ian Ridpath, and Rachel Federman
Foreword by Stephen Maran

The authors assert their moral right to be identified as the authors of this work. All rights reserved. No parts of this publication may be reproduced, stored in a retrieval system, or transmitted, in any form or by any means, electronic, mechanical, photocopying, recording, or otherwise, without the prior permission of the publishers

Library of Congress Control Number: 2024936190

ISBN 978-1-4197-7388-4
eISBN 979-8-88707-295-1

Original illustrations © Lynn Hatzius
Design by Eleanor Ridsdale Colussi
Picture research by Milena Harrison-Gray
Editor for HarperCollins: Caitlin Doyle
Editor for Abrams: Soyolmaa Lkhagvadorj

All reasonable efforts have been made by the author and publishers to trace the copyright owners of the material quoted in this book and of any images reproduced in this book. In the event that the author or publishers are notified of any mistakes or omissions by copyright owners after publication, the author and publishers will endeavor to rectify the position accordingly for any subsequent printing.

Printed and bound in Malaysia
10 9 8 7 6 5 4 3 2 1

Abrams books are available at special discounts when purchased in quantity for premiums and promotions as well as fundraising or educational use. Special editions can also be created to specification. For details, contact specialsales@abramsbooks.com or the address below.

Abrams® is a registered trademark of Harry N. Abrams, Inc.

DISCLAIMER
The publishers urge readers to be responsible stargazers and citizens of the Earth. Please see details on safe and sustainable stargazing. Additionally, please familiarize yourself with local restrictions and safety measures.

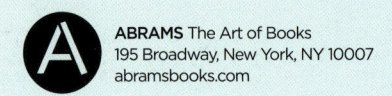

ABRAMS The Art of Books
195 Broadway, New York, NY 10007
abramsbooks.com

FSC
www.fsc.org
MIX
Paper
FSC™ C007454

THE BACKYARD STARGAZER'S BIBLE

THE HISTORY OF STARGAZING

LUNAR CYCLES, STAR CHARTS, PLANETS, AND ASTEROIDS

DISCOVER CONSTELLATIONS, GALAXIES, NEBULAE, METEORITES, AND MORE

NAKED-EYE ASTRONOMY, BINOCULARS, AND TELESCOPES

ASTROPHOTOGRAPHY, SOLAR AND LUNAR ART

ABRAMS, NEW YORK

CONTENTS

- 7 Foreword
- 15 How to use this book
- 16 How to stargaze safely and sustainably

19 Chapter One
Stargazing through the Ages
- 20 Stars, planets, and the celestial sphere
- 32 Our changing view of the universe
- 42 Notable astronomers
- 56 Charting the heavens

69 Chapter Two
Stargazing for Beginners
- 70 The star charts
- 72 The hemisphere charts
- 74 The monthly charts
- 98 The constellation charts
- 194 Stars, clusters, and nebulae
- 202 Double and multiple stars
- 204 Variable stars
- 208 The Milky Way and other galaxies
- 212 The Sun
- 218 The Moon
- 236 Planets and dwarf planets
- 268 Eclipses and occultations
- 272 Asteroids and comets
- 282 Meteors and meteorites

289 Chapter Three
Stargazing Tools and Techniques
- 290 Stargazing basics
- 304 Stargazing for free
- 310 Stargazing equipment
- 318 Urban and suburban skies
- 324 Dark sky sites
- 325 Astronomy clubs and organizations

329 Chapter Four
The Art of the Sky
- 330 The night sky in art
- 358 Astrophotography and processing
- 374 Timelapse sky photography tips
- 376 Solar art photography project
- 379 Lunar art photography project
- 384 Create your own astronomical art

- 389 Additional resources
- 392 Acknowledgements
- 394 Further reading and resources
- 396 Glossary
- 408 Index
- 415 Picture credits
- 416 Contributor biographies

Opposite Stargazing is a gateway to understanding our place in the universe. This stargazer is sitting beneath an epic Milky Way sky in Dartmoor National Park, U.K.

FOREWORD

Dr. Stephen P. Maran
Fellow of the American Association for the Advancement of Science
and Fellow of the Royal Astronomical Society

Astronomy is for everyone, everywhere. It's your entrée to visual delights in the heavens above, some spectacular and others subtle. Astronomy is also a gateway to understanding our place in the Universe. Beyond science, astronomical observations and insights inspire artistic expression: think Van Gogh's *Starry Night,* Blake's "To the Evening Star," or Holst's "Planets." A celestial effect such as moonlight may even foster romance, but that's another book.

You might casually glance up at the night sky on a pleasant evening, or perhaps look deliberately when prompted by news of an impending celestial event, such as an eclipse, a "supermoon," or a meteor shower. However, whether you get a good look at the phenomenon or not may well depend on where you are. Broadcast media, newspapers, or the internet can give fair warning of a sky happening, but you need a sky free of clouds and haze, and at night, darkness.

Urban and suburban residents have night views marred by interference from artificial lighting or even blocked by nearby buildings, so they will benefit from advice on how to spot stars, planets, and more. *The Backyard Stargazer's Bible* tells you how to begin making astronomical observations, cope with interfering glows ("light pollution"), and focus on what you want to see. It describes suitable equipment, basic concepts, and preferred observing locations.

The night sky was much more impressive in antiquity. Stars blazed, planets drew the observer's attention, and the Milky Way (depending on the season) was plain to see. Everyone got a good look, despite lacking lenses and telescopes, because there were also no electrical streetlights, illuminated billboards, or car headlights.

Ancient people looked at the heavens and wondered at what they found and

Opposite Stargazing can be done with the naked eye as well as with equipment such as binoculars or a telescope. These stargazers are using binoculars and a refractor telescope to study the sky. **Above** This stunning night sky showcases the Milky Way over Bruce Peninsula National Park in Ontario, Canada. This striking astrophotographic image has been enhanced to striking effect.

what it meant. They also systematically recorded observations, recognized recurring patterns in events, and noted one-off happenings. Importantly, they documented it all for posterity.

Today we view the sky from personal interest, as a hobby, or to conduct scientific research. Some of us speculate on whether there's truth in reports of UFOs, or if extraterrestrials have landed (although considering the state of the world, why would they?). But most people are too busy or have other pastimes. In contrast, the inhabitants of the ancient world thought sky events might foretell matters of life and death.

Records from early times reveal that many different civilizations were greatly concerned with the heavens. Looking up, they variously feared, worshiped, or studied celestial bodies. Based on what they saw, they adjusted their lives, deciding whether to wage war, or when to undergo a medical procedure, expect a prophet to appear, or plant crops for good results. They designed calendars based on movements of the Sun, Moon, Venus, or Mars. Some thought those objects were gods, bore messages, or were placed in the heavens and set in motion by God.

Sky phenomena that the ancients observed and recorded included the annual motions of the Sun, which rises at a different spot on the horizon every day and rises higher or lower in the sky with the various seasons. Also, the approximately monthly cycle of the Moon, varying from a thin crescent to a bright full Moon and back again.

Eclipses of the Sun and Moon

Today, schoolchildren learn about these phenomena and the physical nature of the Sun and Moon, but ancient people didn't understand what they were, or how and why they moved. But ingenuity prevailed. Making and analyzing careful observations of the Sun and Moon, some ancient viewers even learned to predict eclipses correctly.

Bright comets appeared unexpectedly, dazzling onlookers with their long tails. Scholars recorded visits from what we call Halley's Comet many times over the centuries, never realizing that the same comet had been seen before. The comet's visit in 1066 CE, 590 years before astronomer Edmond Halley was born, is depicted on a section of the Bayeux Tapestry.

Meteors flashed across the night sky. Sometimes one was so bright that it was witnessed in daylight. Very rarely, a rock was seen falling from the sky. Other meteorites, although not actually seen to land, were recognized later if they were made of iron and thus obviously different from the rock of the local area. If this happens nowadays, the finder must first eliminate the possibility that the metal hunk fell from a wayward satellite, launch vehicle, or airliner.

Hundreds of stars were cataloged, mapped, and named. Ancient Egyptian tombs and coffin lids were decorated with star maps. Many societies recognized patterns in the stars and named them as constellations (e.g., Ursa Major, the Great Bear) for what they might resemble, although the likeness is not always clear. The Chinese called their constellations "mansions."

As a bright planet (Mercury, Venus, Mars, Jupiter, or Saturn) moved from one constellation to the next, neared a bright star, or approached the Moon or another planet (think when "Jupiter aligns with Mars," as in the popular song), scholars predicted such events as victories, defeats, or the birth of a king.

Planets cross the background of stars and constellations at different rates. Some planets, notably Mars, even stop and (in American drivers' parlance) "hang a U-turn," so they now move in the direction opposite to the way that they had been going. These backward ("retrograde") motions puzzled thinkers from ancient times all the way up to Copernicus.

On very rare occasions, a bright "guest star" appeared that was new in the sky and

Above This detail from the Bayeux Tapestry depicts the appearance of what we now know to be Halley's Comet in 1066 CE. This was before the Norman Invasion of England in the 11th century and even 590 years before the birth of Edmond Halley, after whom it was named.

considered an omen. Eventually, it faded, never to be seen again. We believe that these were often exploding stars or supernovas. A guest star in 1054 CE was seen in China, Korea, the Middle East, and likely elsewhere at a position where we now find the Crab Nebula, the expanding, glowing remains of a supernova.

A bright "evening star" often was seen in the west after sunset, and a similar "morning star" sometimes appeared low in the east, hours before dawn. Some considered them to be two different objects, but the Sumerians and Maya knew better. They recognized the two "stars" as one and the same: Venus. They learned to predict when the planet would emerge as the morning star, and when it would disappear, then appear at dusk as the evening star. Sometimes the morning or evening star was not so bright; in this case it was Mercury.

Modern folks are not all so well informed. Astronomers frequently are asked, "What's that really bright star?", and the answer is usually "the planet Venus." When astronomically uninitiated individuals declare that they see a bright "UFO" moving rapidly past some clouds, it's almost always Venus, above and seen through clouds that are moving in the opposite direction of the supposed alien craft.

Few urban dwellers notice the Milky Way. However, it's a wonderful sight from a spot where sky conditions are clear and dark, especially when the Moon is below the horizon or just a crescent. No one knew that the Milky Way is composed of numerous faint stars that can't be distinguished individually, until Galileo pointed a telescope at it. Now you can visit dark sky locations to explore the Milky Way and penetrate the fog of stars with binoculars.

Opposite Few inhabitants of big cities tend to notice the Milky Way, especially with light pollution and inclement sky conditions, but it is there. On a clear night in the right conditions, it can be an incredible sight. Seen here is the bustling aerial skyline of Kuala Lumpur, in Malaysia.

Along or near the Milky Way, you'll find star clusters and glowing clouds (nebulas), especially with binoculars. You can spot a few galaxies, such as the Andromeda Galaxy in the northern sky, and the two Magellanic Clouds (visible from the southern hemisphere). The more you consult star charts or other suitable references, and the more you master the use of binoculars and a small telescope, the more you'll succeed in viewing.

Occasionally, two or more planets move close together in a "conjunction"—a striking sight that's an illusion that occurs when each planet is in roughly the same direction from Earth. They really are millions of miles apart, but you wouldn't know it from looking. Some experts think the biblical Star of Bethlehem was a conjunction, but it's still a mystery.

The constellations in view change with the time of night and season. There are 88 of them and you can spot many from your location, at one time of year or another. Some constellations, like Crux (the Southern Cross) can't be seen from most of the northern hemisphere. Others, like Ursa Minor (the Little Bear), are "out of sight" from the far south of the Earth.

A few other star patterns are well known, but technically not constellations, notably the Great Square of Pegasus, the Summer Triangle, and the Big Dipper (so-called in Canada and the United States, but known as the Plough in Ireland and the U.K.).

There are many more celestial treats that you'll witness, too. A random or "sporadic" meteor; you don't know when you'll see one until you do. Also, meteor showers, when many meteors are seen on a few adjacent nights. They occur at roughly the same dates, year after year, such as the famous Perseid meteors in August. The Leonid meteors fall in November, with much greater showers, "meteor storms," about every 33 years.

Every few years, a comet bright enough to see without binoculars or telescopes appears. It may have a long tail or two, and if you watch for a while, you may detect its motion across the background stars. Sometimes, astronomers have expected the comet, but occasionally it's a welcome surprise. The surprise comets are often much brighter than the ones we've been awaiting.

You can photograph the stars as well as observe them. Modern digital cameras available to hobby photographers are more capable than some of the equipment used at astrophysical observatories some few decades ago. If you take up this practice, you won't match the results from the Hubble and James Webb Space telescopes, but you will relish the products of your own work. Some nights you may spot Hubble as a small moving dot in the sky. It will be looking out into space, just like you are.

Astronomy is reflected widely in art, and astronomers have frequently employed their graphic talents to document observations. For centuries before the advent of photography, describing what was seen depended on writing and drawing. When Galileo discovered the moons of Jupiter, detected that the Sun rotates, and discovered that Venus has phases like those of the Moon, his careful sketches displayed the findings. Some amateur astronomers today prefer to draw telescopic impressions of the Moon or other targets rather than photograph them.

Many famous astronomers also engaged or now engage in the arts, especially music. Galileo was an accomplished lute player, while William Herschel (discoverer of planet Uranus) mastered several instruments, composed 24 symphonies, and was Director of Public Concerts in the English city of Bath. His sister Caroline Herschel, who discovered multiple comets and received the Gold Medal of the Royal Astronomical Society, performed as a soprano before concentrating on astronomy. Brian May, a contemporary British astrophysicist, is better known as lead guitarist and a songwriter in the rock band Queen. The Herschels would be envious of his royalties.

Celestial scenes, whether witnessed or imagined, have always graced the visual arts. One famous example is Giotto's *Adoration of the Magi,* which depicts the Star of Bethlehem as if it were a comet. The artist may have been inspired by Halley's Comet, which he viewed at its appearance of 1301 CE. One of Vincent van Gogh's greatest paintings, *The Starry Night,* is in the collection of the Museum of Modern Art in New York. Another glorious Van Gogh, *Road with Cypress and Star*, is a scene lit by the crescent Moon. Chesley Bonestell began painting space scenes in 1905. His paintings of then-imagined planets of stars beyond the Sun are eerily suggestive of "extrasolar" planetary systems discovered since the late 1990s.

Whether you intend to witness, photograph, serenade, or paint the Moon, stars, and planets, this book will be the launch pad. Take your time, but do come back to Earth.

On a clear night in an area with minimal light pollution, it is possible to see a wide range of constellations. **Above** Here, the constellation Orion is visible above Australian bushland. **Below** This enhanced image shows a night-lit forest with the constellation Gemini clearly visible, overlaid with a diagram to map out the stars.

HOW TO USE THIS BOOK

Welcome to the night sky. By beginning this celestial journey, you join millions of other star watchers around the world and take part in a history that began many thousands of years ago. With the Hubble Space Telescope, we can observe images of light from billions of years in the past. But just by stepping outside on a clear night in an area without too much artificial light, you can see several thousand years back in time.

Those with an abundance of hours to devote to a stargazing hobby may want to read straight through this book, from ancient times all the way through to tips for timelapse sky photography. You'll gain a vast understanding of star history, the constellations, and the Universe this way, along with guidelines for stargazing as well as an appreciation for astro art and photography.

Others may prefer to move more freely around the book, flipping to areas of particular interest first, or even searching for specific constellations. The at-a-glance constellations are organized by hemisphere. A bit further along, you'll find profiles of the 88 officially recognized constellations, from Andromeda to Vulpecula. The monthly star charts in Chapter 2 provide a useful blueprint for getting started. Find the month you're in and try to reserve a night near the New Moon for the best views. All stargazers would do well to review the safety and etiquette tips that follow this section before beginning any official exploration.

If you want to get a tad more sophisticated, with perhaps an eye toward equipment, choosing the right time of year, and other basic set-up questions first, start with Chapter 3. If, on the other hand, you're enchanted by the aesthetics of astronomy, perhaps you'd like to begin with Chapter 4, where you can dive into myriad artistic interpretations of all the night sky has to offer.

Those looking to go beyond the stars to other fascinating aspects of the Universe may be drawn to the later sections in Chapter 2. Here you'll find about galaxies, the Sun, the Moon, the planets and dwarf planets, eclipses, comets, meteorites, and more. There are double, multiple, and variable stars to be discovered here as well. And who could forget shooting stars, the stuff of myths and wishes?

You can be as dedicated or as dilettantish in this adventure as you desire. Those who become more committed can find out about clubs, organizations, conservation programs, and tips for taking care of your equipment at the end of Chapter 3. Perhaps you'll find yourself exploring an area you didn't expect to interest you—such as asteroids, the ancient origins of stargazing, or the orbit of a small moon around a distant dwarf planet.

The goal, of course, is to spend as much time looking up as possible. To that end, it's a good idea to study the at-a-glance constellations and constellation profiles in advance of any stargazing jaunts. You'll find it easier to orient yourself and get the most enriching experience this way. Use tabs to mark your favorite pages and go-to reference points. Study them while you have good light, but by all means take the book outside with you.

Here's hoping you have clear skies, a dark sky site, and time for your eyes to adjust to the magic above you.

Opposite The open ocean is an excellent showcase for the vastness of the night sky, allowing incredible vistas of sunsets and rises, constellations, planets, and other astronomical phenomena in panoramic view.

HOW TO STARGAZE SAFELY AND SUSTAINABLY

Stargazing is one of the easiest hobbies to begin. You simply need a clear night and a spot that isn't too bright. Then you'll get a chance to do something we don't do nearly enough these days—look up.

If you're serious about getting the best view and want to head to a deserted area, do a bit of research to make sure it's a safe one. Bring a blanket, a flashlight, and a friend. A reflective jacket is also a good idea if you'll be crossing dark roads. And if you think you'll need it, a compass.

The best time to stargaze in the northern hemisphere is from October through March—and from March to September in the southern hemisphere—when the nights are dark and the sky more likely to be clear. A night with a new moon, or just before or after, is the best. No matter when you go, wear warm clothes and pack an extra layer along with insect repellent. If you're in an area where there might be ticks, cover up, and perhaps even tuck your trousers into high socks.

Once you've found a good spot, give yourself time to adjust to a new landscape, one that begins more than four light years away. The stars we see are in fact snapshots of the past, as each of these years is the time it will take light to travel over that distance. It would take us tens of thousands of years to travel just one light year, which is more than nearly 6 trillion miles (9 trillion kilometers). The Sun, at about 93 million miles (150 million kilometers) from Earth, is of course our closest star.

There's a name for the process your eyes go through as they adjust. It's called "dark adaptation" and it takes about 30 minutes, so be patient as you wait. Beholding the night sky is always a neat lesson in perspective, isn't it? Here we are, so small, in a vast universe. Or as Carl Sagan put it, "Our planet is a lonely speck in the great enveloping cosmic dark." If you must use your cellphone, iPhones have a setting you can use that prevents the light from the phone interfering with your eyes adapting to the night. Altogether, with the naked eye, we can view approximately 2,000 stars on any given night.

It can be fun at first just to let yourself take in the full spectacle of the night sky, before you start trying to spot anything in particular. Depending on the season and your location, once your eyes calibrate, you may start to see familiar shapes. Perhaps a winged horse, a powerful hunter, or a vain queen. Many of us know the constellations by names that come from Greek mythology. (Pegasus is the winged horse, Orion the hunter, and Cassiopeia the queen.) The Plough, or Big Dipper, is a favorite, although some might be surprised to find out it's not officially a constellation. Rather, it is part of Ursa Major, the Great Bear, which makes it an asterism. Certain Indigenous tribes in North America thought of the Big Dipper as seven siblings who were chased into the sky. Some of the Inuit people see a group of caribou (Tukturjuit) in the same pattern.

If you want to see the stars in more detail, grab some binoculars. It's easy enough to pack a pair in your bag, or keep them in the car, and they'll allow you to level up stargazing with almost no effort.

If you're heading into the wilderness, or even just to a local park, the rules that apply to stargazing also apply to hiking and camping. Keep noise to a minimum, especially if people are living or camping nearby. (That means no blasting David Bowie's "Space Odyssey" through a portable speaker as you scope out a spot.) In summer, the sun doesn't set until well into the evening, so by the time you venture out, many may be already tucked up in bed.

Equally important during your nocturnal adventure is to remain a good steward of nature. Be respectful of wildlife and keep your distance—for your own safety and that of wild animals. And if you're bringing along a canine companion, make sure to pick up after your pup. Don't leave any debris, either. In the U.K., the Countryside Code offers official guidance for enjoying nature, encouraging visitors to "leave no trace" of their visit and to avoid harming or disturbing anything. (In Scotland, the Scottish Outdoor Access Code provides similar tips on protecting the environment.) Likewise in the U.S., the "Pack it in, pack it out" slogan commonly found on hiking trails and at national parks is a handy reminder. There are also various non-profit organizations around the world offering advice on how to protect and enjoy the outdoors. Leave No Trace, for example, has its headquarters in Boulder, Colorado. One of their principles, "Leave what you find," encourages outdoor explorers to "Leave rocks, plants, and other natural objects as you find them" and "Avoid introducing … non-native species."

Most of us know one of the most dangerous things you can leave behind is a cigarette butt. (This piece of advice falls under the "Pack it in, pack it out" adage, but given the unique risks involved, deserves a special mention.) Smoking may be prohibited—check the rules as needed—and, if you're planning to toast the Big Dipper, check the rules on alcohol as well.

If you're joining a designated "stargazing party," set up your spot so you're not right on top of someone else's. You'll likely come across a set of guidelines that may include arriving before dark so you can set up without any light interference, and parking with headlights facing away from the group. You may also be cautioned to take care walking around so as not to trip on telescopes or other star-viewing equipment. (Easier said than done when you can't use a flashlight.)

Once the sky darkens and the stargazing starts, it's common for stargazers to visit each other's sites and have a look from each other's telescopes. Of course, you should always ask permission before touching anyone else's equipment, and it's possible some people will be immersed in solo projects such as astrophotography. Typically, though, star-party attendees enjoy the camaraderie and the chance to poke around through someone else's scope.

When you've had enough gazing and you're ready to head home, it's customary to let people know you're leaving so they can be sure not to look at your car lights. And it goes without saying that you should drive slowly and use extreme caution in any area where people, and especially children, might be wandering around in the dark.

Above In any stargazing endeavor, be sure to be a good steward of nature. Be respectful of wildlife and other stargazers and avoid harming or disturbing your surroundings as much as possible.

CHAPTER ONE

STARGAZING THROUGH THE AGES

STARS, PLANETS, AND THE CELESTIAL SPHERE

Imagine that the Earth lies at the center of an enormous sphere that rotates around us once every day, carrying the stars and other celestial bodies with it. Thousands of years ago, this was thought to be the true arrangement of the heavens. Even though we now know better, astronomers still use this concept of an imaginary celestial sphere surrounding the Earth as a useful way of describing the positions and motions of celestial objects.

A casual glance at the celestial sphere shows that it's dotted with stars of different brightnesses. These remain fixed relative to each other as the sky rotates. In addition, a careful observer would notice five wandering stars that slowly change their position relative to the fixed stars from night to night. These wandering stars are the planets—a name that comes from the Greek word *planetes*, meaning "wanderer."

We now know that the planets orbit the Sun, but until the seventeenth century they too were thought to be carried around a central Earth on transparent spheres of their own, as were the Sun and Moon. The story of how we discovered the true structure of the Universe and our place within it is told in the section "Our changing view of the Universe," on pages 32–41.

The changing appearance of the sky

How much of the celestial sphere can be seen at a given time depends on three factors: the time of night, the time of year, and your latitude on Earth. Let's start by considering the effect of latitude.

To take an extreme example, at one of the Earth's poles (latitude 90°) the celestial pole would lie directly overhead—in other words, at the *zenith*—and the celestial equator would lie on the horizon. As the Earth turned, all stars would circle around the celestial pole without rising or setting. Nothing below the celestial equator would ever be visible.

At the other extreme, an observer stationed exactly on the Earth's equator, latitude 0°, would see the celestial equator directly overhead. The north and south celestial poles would lie on the north and south horizons respectively, and every part of the sky would be visible at one time or another. All stars would rise in the east and set in the west as the Earth rotated.

For most observers, the real sky appears somewhere between these two extremes: the celestial pole is at some intermediate altitude between horizon and zenith, and the stars closest to it circle around it without setting (they are said to be *circumpolar*), while the rest of the stars rise and set.

Above Celestial sphere: To understand celestial coordinates and motions, it helps to imagine that objects in the sky lie on a transparent sphere surrounding the Earth. The coordinates known as right ascension and declination can be visualized as circles on this sphere, along with the celestial equator and the ecliptic.

The exact angle of the celestial pole above the horizon depends on the observer's latitude. For someone at latitude 50° north, for instance, the north celestial pole is 50° above the northern horizon. As another example, if you were at latitude 30° south, the south celestial pole would be 30° above the southern horizon. In other words, the altitude of the celestial pole above the horizon is exactly equal to your latitude—a fact long recognized by navigators.

As the Earth turns, completing one 360° rotation every 24 hours, the stars march across the heavens at the rate of 15° per hour. Therefore the appearance of the sky changes with the time of night. An added complication is that the Earth is also orbiting the Sun each year, so the constellations on show change with the seasons.

For example, a constellation such as Orion, splendidly seen in December and January, will be in the daytime sky six months later and hence will then be invisible. The maps on page 74 will help you find out which stars are on view each month of the year, wherever you are on Earth.

Previous pages Only the faint site of a distant road gives away the place in history of this timeless image of the Milky Way at Balanced Rock, in Big Bend National Park, Texas. **Opposite** Similarly timeless is this photograph of a star-filled sky over Nuptse in Nepal.

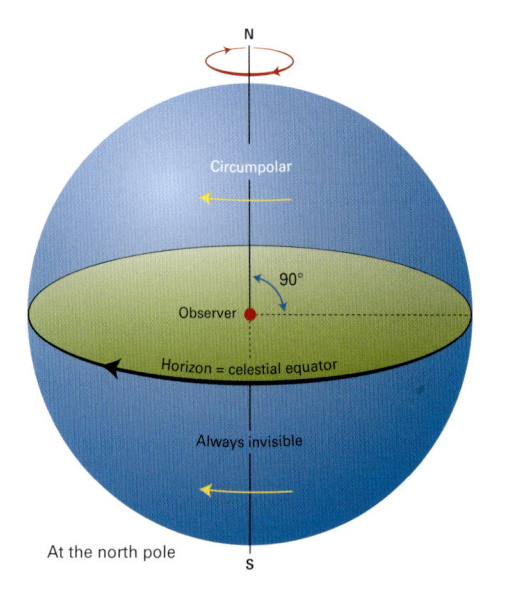

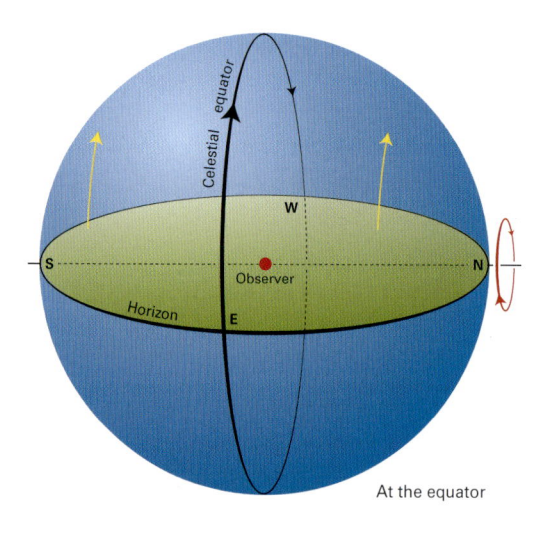

There is another circle on the celestial sphere that's of great importance: it is the *ecliptic*, the Sun's apparent path against the background stars as the Earth orbits it each year. This path is inclined at 23.5° to the celestial equator, the angle at which the Earth's axis is inclined to the vertical. There are four significant locations along the ecliptic, as we can see by looking at the celestial sphere. Two of these, known as the *equinoxes*, are the points at which the ecliptic intersects the celestial equator. They mark the start of astronomical spring and autumn. The other two are the most northerly and southerly points that the Sun reaches in the sky each year, when it lies 23.5° north and south of the celestial equator. These are called the *solstices*, and they mark the start of astronomical summer and winter.

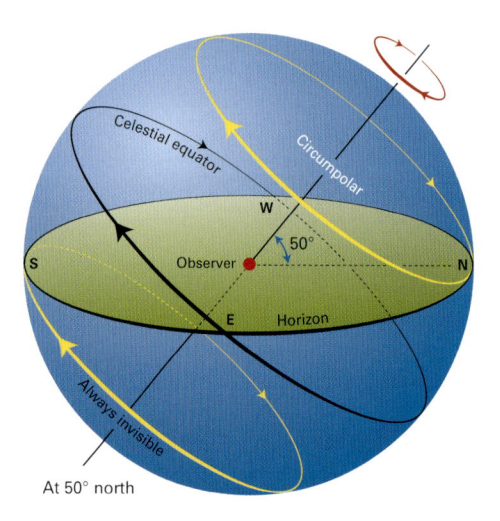

The changing appearance of the sky as seen from different latitudes on Earth. **Top left** For an observer at the Earth's pole, only one half of the sky is ever visible, with the other half permanently below the horizon. **Top right** At the equator all the sky is visible; as the Earth rotates, stars appear to rise in the east and set in the west. **Above** At intermediate latitudes the situation is between those two extremes. Part of the sky is always above the horizon (the part marked "Circumpolar"), but an equal part is always below the horizon, making it invisible. Stars between these two regions rise and set during the night.

22 THE BACKYARD STARGAZER'S BIBLE

Time and the seasons

Our main units of time, the year and the day, are based on the movements of the Earth. One orbit of the Sun is a year, while a day is one spin on its axis relative to the Sun. Unfortunately, there are not an exact number of days in a year: the Earth takes just under 365.25 days to orbit the Sun, and the odd fraction is accounted for by adding a day to the calendar every fourth year to create a leap year. By contrast, the division of the day into 24 hours, each of 60 minutes, is a human construct, unconnected with celestial movements.

Local time by the Sun varies with longitude around the globe, so clocks must be adjusted accordingly. To do so, the Earth is divided into 24 time zones, each offset by one hour from the next. The reference point is the zero line of longitude that runs through the Greenwich Observatory in London, which was adopted by international agreement as the Earth's prime meridian in 1884. Countries in each time zone keep time an exact number of hours—or in some cases, half hours—offset from Greenwich Mean Time (GMT), also commonly known as Universal Time (UT).

Time zones to the east of Greenwich are ahead of GMT, while those to the west are behind GMT. In the middle of the Pacific Ocean, 12 hours from Greenwich, is an imaginary line known as the International Date Line, where the date changes either one day back (if crossing from west to east), or one day forward (if going from east to west).

Years are divided into four seasons, and again there is an astronomical reason. Seasons result from the Earth's 23.5° axial tilt, which means that the north and south poles lean alternately toward and away from the Sun during the year.

Above Midnight Sun: Inside the Arctic and Antarctic Circles, the Sun does not set during the summer, but instead skims over the horizon around midnight. This sequence of photographic exposures of the Midnight Sun was taken from the shores of Lake Ozhogino in the far north of Russia.

On June 21 the north pole is tilted at its maximum toward the Sun, and the Sun lies overhead at noon on the Tropic of Cancer, 23.5° north of the equator. This date is the summer solstice in the northern hemisphere and the winter solstice in the south. Northern days are long and nights are short. Within the Arctic Circle the Sun never sets (the *Midnight Sun*), and within the Antarctic Circle it never rises (*polar night*).

Six months later, at the December solstice on December 21 or 22, the situation is reversed. The south pole is then tilted at its maximum toward the Sun and the Sun lies overhead at noon on the Tropic of Capricorn, 23.5° south of the equator. In between these two dates are the equinoxes, on March 20 and September 22 or 23, when the Sun lies overhead at noon on the equator; at the equinoxes, day and night are equal the world over—in fact the word equinox comes from the Latin meaning "equal night."

Astronomically speaking, the equinoxes and the solstices mark the start and finish of each season, but meteorologists use a different convention: they regard seasons as blocks of three full months, so that March to May is northern spring, June to August is northern summer, September to November is northern autumn, and December to February is northern winter.

Incidentally, if the Earth's axis were directly upright with respect to its orbit around the Sun, the celestial equator and ecliptic would then coincide. We would then have no seasons, for the Sun would always remain directly above the Earth's equator.

Stars and planets

In the night sky, stars appear to the naked eye as spiky, twinkling lights, and those stars near the horizon seem to flash and change color. The twinkling and flashing effects are due not to the stars themselves but to the Earth's atmosphere: turbulent air currents cause the stars' light to dance around. The steadiness of the atmosphere is referred to as the *seeing*. Steady air means good seeing. The spikiness of star images is due to optical effects in the observer's eye. In reality, stars are spheres of gas similar to our own Sun, emitting their own heat and light.

Stars come in various sizes, from giants to dwarfs, and in a range of colors according to their temperature. At first glance, all stars appear white, but more careful inspection reveals that certain stars are somewhat orange—notably Betelgeuse, Antares, Aldebaran, and Arcturus—while others, such as Rigel, Spica, and Vega, have a bluish tinge. Binoculars bring out the colors more readily than the naked eye does. Chapter 2 of this book explains more fully the different types of stars that exist.

By contrast, planets are cold bodies that shine by reflecting the Sun's light. The planets are constantly on the move as they orbit the Sun. Four of them can be easily seen with the naked eye: Venus, Mars, Jupiter, and Saturn. Venus, the brightest of all, appears as a dazzling object in the evening or morning sky. Mercury, the closest to the Sun, is more difficult to spot, always low down in twilight, but can be seen with the naked eye under favorable conditions. The planets are described in more detail in Chapter 2.

About 2,000 stars are visible to the naked eye on a clear, dark night, but you won't need to learn them all. Start by identifying the brightest stars and major constellations, and use these as signposts to the fainter, less prominent stars and constellations. Once you know the main features of the night sky, you'll never again be lost among the stars.

Constellations

The sky is divided into 88 sections known as constellations, which astronomers use as a convenient way of locating and naming celestial objects. Their names and boundaries are fixed by international agreement. Constellations have no physical reality; they are simply areas of sky. Most of the stars within them lie at vastly differing distances from Earth, and form a pattern simply by chance. When astronomers talk of an object being "in" a given constellation, they mean that it lies in that particular area of sky.

Some constellations are easier to recognize than others, such as the magnificent Orion, or the distinctively shaped Cassiopeia and Crux. Others are faint and obscure, such as Lynx and Telescopium. Whether large or small, bright or faint, each constellation is given a separate chart and description in this book.

The main constellations were devised at the dawn of history by Middle Eastern peoples who believed that they could see a likeness to certain fabled creatures and mythological heroes among the stars. Of particular importance were the 12 constellations of the zodiac, through which the Sun passes during its yearly path around the heavens. However, it should be realized that the astrological "signs" of the zodiac are not the same as the modern astronomical constellations, even though they share the same names. For the story of how we arrived at our modern system of constellations see pages 56–65.

As well as the officially recognized constellations, you can find other patterns among the stars called *asterisms*. An asterism can be composed of stars belonging to one or more constellations. Well-known examples are the Plough, or Big Dipper (part of Ursa Major), the Square of Pegasus, the Sickle of Leo, and the Teapot of Sagittarius.

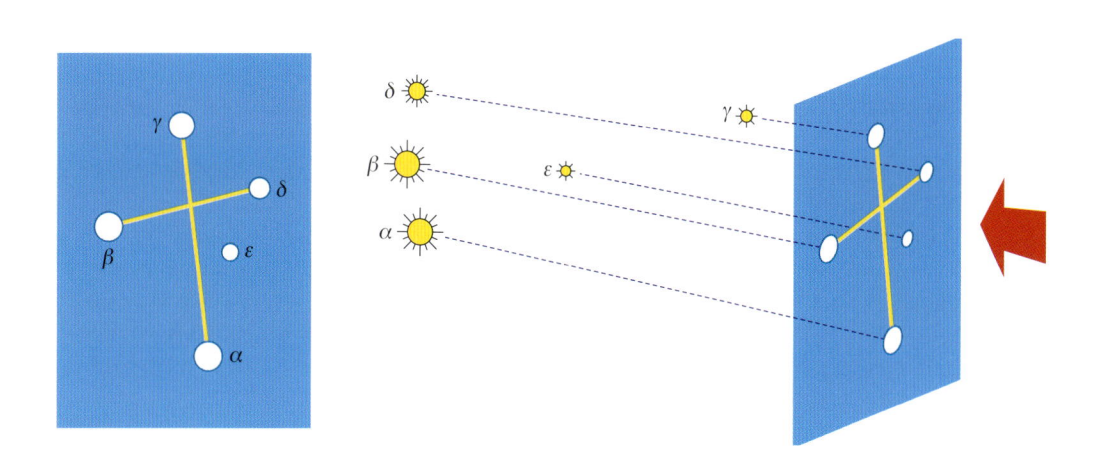

Above Stars in a constellation are usually unrelated to one another and just happen to lie in the same line of sight by chance. Here, the stars of Crux, the Southern Cross, are shown as they appear from Earth (left) and in a side-on view as they actually lie in space (right).

Naming the stars

The main stars in each constellation are labeled with a letter of the Greek alphabet, the brightest star usually (but not always) being termed α (alpha). Notable exceptions in which the stars marked β (beta) are in fact the brightest include the constellations Orion and Gemini. The entire Greek alphabet is given in the table on page 30.

Particularly confusing are the southern constellations Vela and Puppis, which were once joined with Carina to make the extensive figure of Argo Navis, the ship of the Argonauts. As a result of Argo's division into three parts, neither Vela nor Puppis possesses stars labeled α or β, and there are gaps in the sequence of Greek letters in Carina as well.

The system of labeling stars with Greek letters was introduced by the German astronomer Johann Bayer on his star atlas, called *Uranometria*, published in 1603, so these designations are often known as Bayer letters. In heavily populated constellations, where Greek letters ran out, fainter stars were assigned Roman letters, both lower case and capital, such as l Carinae, P Cygni, and L Puppis.

An additional system of identifying stars is that of Flamsteed numbers, originating from their order in a star catalog drawn up at Greenwich Observatory by the first Astronomer Royal of England, John Flamsteed (1646–1719). Examples are 61 Cygni and 70 Ophiuchi.

The genitive (possessive) case of the constellation's name is always used when referring to a star within it; hence Canis Major, for instance, becomes Canis Majoris, and the name α Canis Majoris means "the star α in Canis Major." All constellation names have standard three-letter abbreviations—for instance, the abbreviated form of Canis Major is CMa.

Prominent stars also have proper names by which they are commonly known. For example, α Canis Majoris, the brightest star in the sky, is better known as Sirius. Stars' proper names originate from several sources. Some, such as Sirius, Castor, and Arcturus, date back to ancient Greek times. Many others, such as Aldebaran, are of Arabic origin. Still others were added more recently by European astronomers who borrowed Arabic words in corrupted form; an example is Betelgeuse, which in its current form is meaningless in Arabic. The star names used in this book are those officially recognized by the International Astronomical Union.

Star clusters, nebulae, and galaxies have a different system of identification. The most prominent of them are given numbers prefixed by the letter M from a catalog compiled in the late eighteenth century by the French astronomer Charles Messier (1730–1817). For example, M1 is the Crab Nebula, and M31 the Andromeda Galaxy.

Messier's catalog contained 103 objects. A few more were added later by other astronomers, bringing the total to 110. A far more comprehensive listing, containing many thousands of objects, is the *New General Catalogue of Nebulae and Clusters of Stars* (NGC), compiled by J. L. E. Dreyer (1852–1926), with two supplements called the *Index Catalogues* (IC).

Messier numbers and NGC/IC numbers remain in use by astronomers, and both systems are used in this book. On the charts, such objects are labeled with their Messier number if they have one, or otherwise by their NGC number (without the "NGC" prefix) or IC number (prefixed by "I").

Star brightness

Stars appear of different brightnesses in the sky, for two reasons. Firstly, they give out different amounts of light. But also, and just as importantly, they lie at vastly differing distances. Hence, a modest star that's quite close to us can appear brighter than a tremendously powerful star that's a long way away.

Astronomers call a star's brightness its *magnitude*. The magnitude scale was introduced by the Greek astronomer Hipparchus in 129 BCE. Hipparchus divided the naked-eye stars into six classes of brightness, from 1st magnitude (the brightest stars) to 6th magnitude (the faintest visible to the naked eye). In his day there was no means of measuring star brightness precisely, so this rough classification sufficed. But with the coming of technology it became possible to measure a star's brightness to a fraction of a magnitude.

In 1856 the English astronomer Norman Pogson (1829–91) put the magnitude scale on a precise mathematical footing by defining a star of magnitude 1 as being exactly 100 times brighter than a star of magnitude 6. Since, on this scale, a difference of five magnitudes corresponds to a brightness difference of 100 times, a step of one magnitude is equal to a brightness difference of just over 2.5 times (the fifth root of 100).

Objects more than 250 times brighter than 6th magnitude are given negative (minus) magnitudes. For example, Sirius, the brightest star in the sky, is of magnitude −1.46. At the other end of the scale, stars fainter than magnitude 6 are given progressively larger positive magnitudes. The faintest objects detectable by telescopes on Earth are around magnitude 27.

Any object of magnitude 1.49 or brighter is said to be of 1st magnitude; objects from 1.50 to 2.49 are termed 2nd magnitude; and so on. The magnitude system may sound confusing at first, but it works well in practice and has the advantage that it can be extended indefinitely in both directions, to the very bright and the very faint.

When used without further qualification, the term "magnitude" refers to how bright a star appears to us in the sky; strictly, this is the star's apparent *magnitude*. But because the distance of a star affects how bright it appears, the apparent magnitude bears little relation to its actual light output, or *absolute magnitude*.

A star's absolute magnitude is defined as the brightness it would appear to have if it were at a standard distance from us of 10 parsecs (32.6 light years). The origin of the parsec is explained on page 28. The table shows the apparent and absolute magnitudes of the ten brightest stars visible in the night sky. Astronomers calculate the absolute magnitude from knowledge of the star's nature and its distance.

Absolute magnitude is a good way of comparing the intrinsic brightness of stars. For instance, our daytime star the Sun has an apparent magnitude of −26.7, but an absolute magnitude of 4.8 (when no sign is given, the magnitude is understood to be positive). Deneb (α Cygni) has an apparent magnitude of 1.2, but an absolute magnitude of −6.9. From a comparison of these absolute magnitudes we deduce that Deneb gives out about 50,000 times as much light as the Sun and hence is exceptionally luminous, even though there's nothing at first sight to mark it out as extraordinary.

A number of stars actually vary in their light output, for various reasons, and are a favorite subject for study by amateur astronomers. The nature of such so-called variable stars is discussed on pages 204–6.

Star distances

In the Universe, distances are so huge that astronomers have abandoned the puny kilometer (km) and have invented their own units. Most familiar of these is the *light year* (l.y.), the distance that a beam of light travels in one year. Light moves at the fastest known speed in the Universe, 299,792.5 kilometers per second. A light year is equivalent to 9.46 million million kilometers.

On average, stars are several light years apart. For instance, the closest star to the Sun, Proxima Centauri (actually a member of the α Centauri triple system), is 4.3 light years away. Sirius is 8.6 l.y. away, and Deneb 1,400 l.y. away.

The distance of the nearest stars can be found directly in the following way. A star's position is measured accurately when the Earth is on one side of the Sun, and then remeasured six months later when the Earth has moved around its orbit to the other side of the Sun. When viewed from two widely differing points in space in this way, a nearby star will appear to have shifted slightly in position with respect to more distant stars (see diagram below).

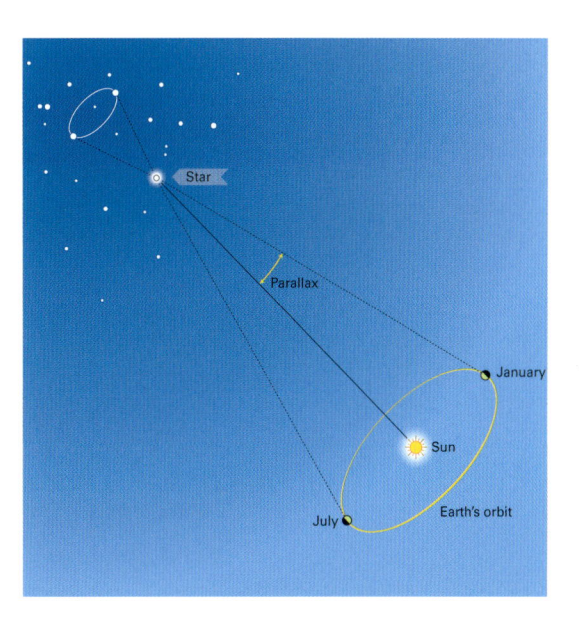

Above Parallax: As the Earth moves around its orbit, nearby stars will appear to change in position slightly against the celestial background, as seen through a telescope. This shift in position is known as the star's "parallax." The nearer the star is to us, the greater its parallax. Here, the amount of parallax is exaggerated for clarity.

This effect is known as parallax, and applies to any object viewed from two vantage points against a fixed background, such as a tree against the horizon. A star's parallax shift is so small that under normal circumstances it's unnoticeable—in the case of Proxima Centauri, which has the greatest parallax shift of any star, the amount is about the same as the width of a small coin seen at a distance of 1.4 miles (2 kilometers). Once the star's parallax shift has been measured, a simple calculation reveals how far away it is.

An object close enough to us to show a parallax shift of 1" (1 second of arc) would, in the jargon of astronomers, be said to lie at a distance of 1 *parsec*, equivalent to 3.26 light years. In practice, no star is this close; the parallax of Proxima Centauri is 0".77. Astronomers frequently use parsecs in preference to light years because of the ease of converting parallax into distance: a star's distance in parsecs is simply the inverse of its parallax in seconds of arc. For example, a star 2 parsecs away has a parallax of 0".5, at 4 parsecs away its parallax is 0".25, and so on.

The farther away a star is, the smaller its parallax. Beyond about 50 light years a star's parallax becomes too small to be measured accurately by telescopes on Earth. Before the launch of the European Space Agency's astrometry satellite Hipparcos in 1989, astronomers had been able to establish reliable parallaxes for fewer than 1,000 stars from Earth; from space, Hipparcos increased the number of reliable parallaxes to over 100,000. A newer spacecraft, called Gaia, launched in 2013, has extended the number to more than a billion, and with greater accuracy.

Before the days of accurate parallaxes, astronomers had to use an indirect method of finding star distances. First they estimated the star's absolute magnitude by studying the spectrum of its light. They then compared this estimated absolute magnitude with the observed apparent magnitude to determine the star's distance. The distance obtained in this way was open to considerable error, and the values quoted in various books and catalogs often differed widely as a result. Most of the distances of stars given in this book are accurate to better than 10 percent, with the uncertainties tending to become greater for the more distant stars.

Star positions

To determine positions of objects in the sky, astronomers use a system of coordinates similar to latitude and longitude on Earth. The celestial equivalent of latitude is called *declination* and the equivalent of longitude is called *right ascension*. Declination is the easier to understand as, like latitude, it's measured in degrees, minutes, and seconds of arc (abbreviated °, ', and ") from 0° on the celestial equator to 90° at the celestial poles.

Right ascension seems more puzzling at first. It's expressed not in degrees but in hours, minutes, and seconds (abbreviated h, m, and s), from 0h to 24h. This is because of the connection between the Earth's rotation and time—the Earth rotates through 15° in one hour, so one hour of right ascension is equivalent to an angle of 15°. As with terrestrial longitude, right ascension is taken to start from an agreed zero meridian. In this case, the 0h line of right ascension passes through the point where the Sun crosses the celestial equator on its way north each year. Technically, this point is known as the *vernal* (or spring) *equinox*. Right ascension runs from west to east, which is the direction the Earth turns.

One additional effect that becomes important over long periods of time is that the Earth is wobbling slowly on its axis, like a spinning top. The axis remains inclined at an angle of 23.5°, but the position in the sky to which the north and south poles of the Earth are pointing moves slowly. This wobbling of the Earth in space is termed *precession*. As a result of precession, the Earth's north and south poles describe a large circle on the sky, taking 26,000 years to return to their starting places. Hence the positions of the celestial poles are always changing, albeit very slowly, as are the two points at which the Sun's path (the ecliptic) cuts the celestial equator.

As an example of the effects caused by precession, Polaris will not always be the Pole Star. Although Polaris currently lies less than 1° from the celestial pole, that's just a matter of chance. In 11,000 years' time the north celestial pole will lie near Vega in the constellation Lyra, having moved through Cepheus and Cygnus in the interim. Similarly, the vernal equinox, which lay in Aries between 1865 and 67 BCE, now lies in Pisces and in 2597 CE will have reached Aquarius.

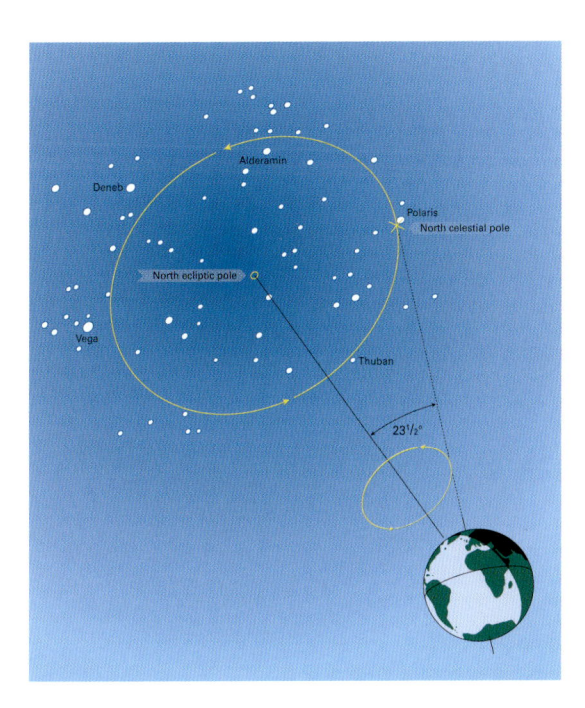

The effect of precession means that the coordinates of all celestial objects—the cataloged positions of stars, galaxies, and even constellation boundaries—are continually drifting. Astronomers draw up catalogs and star charts for a standard reference date, or epoch, commonly chosen to be the start or middle of a century. The epoch of the star positions in this book is the year 2000. For most general purposes, precession does not introduce a noticeable error until after about 50 years, so the charts in this book will be usable without amendment until halfway through the twenty-first century.

Proper motions

All the stars visible in the sky are members of a vast wheeling mass of stars called the Galaxy. Those stars visible to the naked eye are among the nearest to us in the Galaxy. More distant stars in the Galaxy crowd together in a hazy band called the Milky Way, which can be seen arching across the sky on dark nights.

The Sun and the other stars are all orbiting the center of the Galaxy; the Sun takes about 250 million years to complete one orbit. Other stars move at different speeds, like cars in different lanes on a highway. As a result, stars are all very slowly changing their positions relative to one another.

Such stellar movement, termed *proper motion*, is so slight that it's undetectable to the naked eye, even over a human lifetime, but it can be measured through telescopes. As with many other aspects of stellar positions, our knowledge of proper motions has been radically improved by the Hipparcos and Gaia satellites.

If ancient Greek astronomers could be transported forward to the present day, they would notice little difference in the sky, with the exception of Arcturus, a fast-moving bright star that has drifted more than two Moon diameters from its position 2,000 years ago. Over very long periods of time, the proper motions of stars distort the shapes of all constellations considerably. The diagrams on page 30 show some examples of how proper motions will alter some familiar patterns.

An additional long-term effect of stellar motions is to change the apparent magnitudes of stars as they move toward us or away. For example, Sirius will brighten by 20 percent over the next 60,000 years as its distance shrinks by 0.8 light years. Then, as it moves away again, it will be superseded as the brightest star in the sky by Vega, which will peak at magnitude −0.8 nearly 300,000 years from now.

Above Precession: The Earth is very slowly wobbling in space like a tilted spinning top, a movement known as "precession." As a consequence, the celestial poles trace out a complete circle on the sky every 26,000 years. Only the path of the north celestial pole is shown in this diagram, but the effect applies to the south pole as well.

Constellations

URSA MAJOR (The Plough)

In 100,000 years

CYGNUS

In 100,000 years

LEO

In 100,000 years

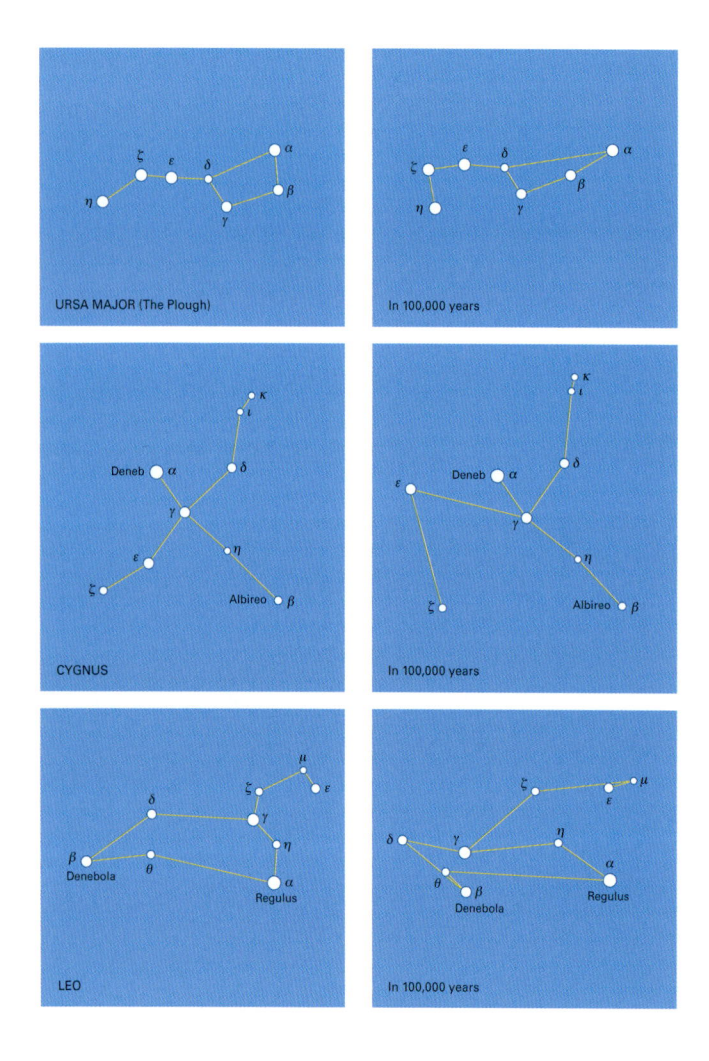

Above Proper motions: The shapes of the constellations are gradually changing due to the motions of the stars over long periods of time, known as "proper motions." Here, the familiar star patterns of the Plough, Cygnus, and Leo are shown as they appear today and as they will appear in 100,000 years' time.

The Greek alphabet

α	alpha
β	beta
γ	gamma
δ	delta
ε	epsilon
ζ	zeta
η	eta
θ	theta
ι	iota
κ	kappa
λ	lambda
μ	mu
ν	nu
ξ	xi
ο	omicron
π	pi
ρ	rho
σ	sigma
τ	tau
υ	upsilon
φ or f	phi
χ	chi
ψ	psi
ω	omega

The ten brightest stars as seen from Earth

Star name	Constellation	Apparent magnitude	Absolute magnitude	Distance (l.y.)
Sirius	Canis Major	−1.46	+1.43	8.60
Canopus	Carina	−0.74	−5.62	309
Rigil Kentaurus	Centaurus	−0.27	+4.12	4.32
Arcturus	Boötes	−0.05	−0.31	36.7
Vega	Lyra	+0.03	+0.60	25.0
Capella	Auriga	+0.08	−0.51	42.8
Rigel	Orion	+0.13	−6.98	863
Procyon	Canis Minor	+0.37	+2.64	11.5
Betelgeuse	Orion	+0.4 (variable)	−5.52	498
Achernar	Eridanus	+0.46	−2.69	139

Magnitude difference converted to brightness difference

Difference in magnitude	Difference in brightness
0.5	1.6
1.0	2.5
1.5	4.0
2.0	6.3
2.5	10
3.0	16
3.5	25
4.0	40
5.0	100
6.0	250
7.5	1,000
10	10,000
12.5	100,000
15	1,000,000

OUR CHANGING VIEW OF THE UNIVERSE

Today, we know that the Earth is simply a small, rocky planet orbiting an average star within an enormous spiral-shaped galaxy of billions of other stars. But this has been only a relatively recent recognition.

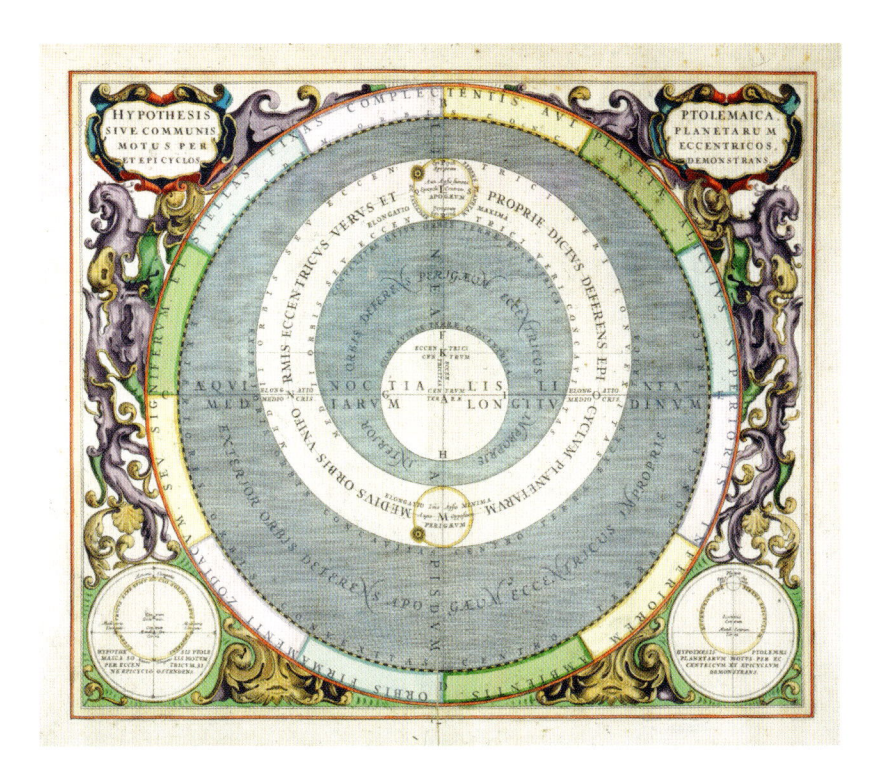

Until the sixteenth century, astronomers continued to accept the ancient Greek view that the Earth was at the center of the Universe and that all else rotated around it on transparent spheres nested like the layers of an onion. The innermost sphere, closest to Earth, carried the Moon. Then came the planets Mercury and Venus, with the Sun on the fourth sphere out. Beyond the Sun came Mars, Jupiter, Saturn, and, finally, the background stars. All celestial motions were to be accounted for by the movement of these heavenly spheres, while the central (and spherical) Earth remained immobile.

The most advanced version of this Earth-centered, or geocentric, view of the Universe was developed around 150 CE by the Greek astronomer Claudius Ptolemy (c.100–c.170), who worked at Alexandria in northern Egypt, at that time part of the Roman Empire. As a result it's usually known as the Ptolemaic system.

Opposite This enhanced NASA image of a solar eclipse over planet Earth demonstrates just how far we've come in our changing view of the Universe. **Above** Ptolemaic view: A diagram of the ancient Greek view of the Universe, explaining how the planets were thought to orbit the Earth on combinations of circles. This diagram comes from an atlas called *Harmonia Macrocosmica* by the Dutch-German cartographer Andreas Cellarius, published in 1660.

The Ptolemaic system

Underlying all Greek cosmological models was the fundamental assumption, dating back to the philosopher Plato around 400 BCE, that the heavenly bodies must travel at uniform speed along circular paths, since it was thought that only the perfection of a circle would be acceptable in the heavens. Unfortunately, this assumption was seriously flawed, for the motions of the planets are not smooth and uniform.

Most noticeably, the outer planets Mars, Jupiter, and Saturn are seen at times to slow down and perform a backward, or *retrograde*, loop against the stars before resuming their onward travel. In addition, planets change in brightness during the year, an indication that their distances from Earth change, too. Even the Sun and Moon appear to move faster or slower as they move along their orbits.

Over the centuries several Greek astronomers and mathematicians—notably Plato's pupil Eudoxus, and Ptolemy's predecessor Hipparchus—had made extensive modifications to the basic geocentric theory in an attempt to account for these irregularities. Ptolemy's version, published in his book in around 150 CE, was the culmination of these efforts.

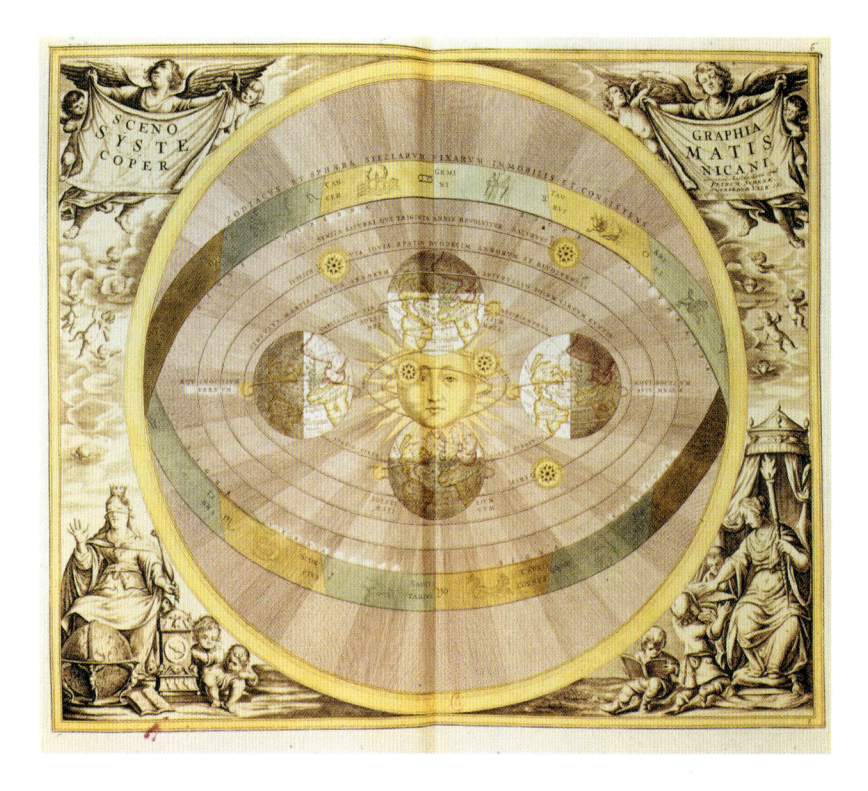

Three main techniques were used to modify the basic model, while retaining the essential principle of circular motion. Most important was the *epicycle*, a smaller circle described by each planet as it moved on its orbit around the Earth. This, in particular, would explain the retrograde motions of the outer planets. In a second refinement, the center of the orbit could be offset as necessary from the center of the Earth. More complicated, and controversial, was Ptolemy's introduction of the equant, a point offset from the center of an orbit by the same amount as the Earth, but on the opposite side; the epicycle would, according to Ptolemy's theory, have uniform motion with respect to the deferent, although not to the Earth.

All these devices could be adjusted at will to explain the observations of planetary positions and brightnesses. Yet for all the complexity of the Ptolemaic system (or perhaps because of it), perfect accuracy was still not attainable.

The Copernican system

Despite its flaws, the Ptolemaic theory reigned supreme for 1,400 years, until a rival emerged in the hands of a Polish clergyman and astronomer called Nicolaus Copernicus (1473–1543). Dissatisfied with the shortcomings of the Ptolemaic system, Copernicus realized that a better solution could be obtained by placing the Sun at the center of the Universe, relegating the Earth to the status of a planet, and setting it spinning daily. He was not the first to consider the idea of a heliocentric (Sun-centered) Universe, but he was the first to develop a practical version.

Copernicus published his revolutionary new model in 1543, the year of his death, in a book called *De revolutionibus orbium coelestium* ("On the Revolutions of the Heavenly Spheres"). In the Copernican system, the planets were arranged around the central Sun in the sequence we know them today, with the stars

Above Copernican system: A diagram of the heliocentric view of the Universe, put forward by Nicolaus Copernicus, with the Earth shown in four positions as it orbits the Sun, from Andreas Cellarius's *Harmonia Macrocosmica* published in 1660.

(the true nature of which was still not known) on an enclosing outer sphere, as in Ptolemy's model.

This arrangement provided natural explanations for retrograde motion (it was due to the Earth catching up and overtaking the outer planets on its smaller orbit), the changing brightness of planets, and the increase in the orbital period of the planets with distance from the Sun. However, Copernicus remained tied to the ancient axiom of perfect circular motion. Hence his system was still burdened by the need for contrivances such as epicycles and, overall, its accuracy in predicting planetary positions was little better than Ptolemy's system.

Orbits become elliptical

The new Copernican model soon had its adherents, but most astronomers were not persuaded. For it to become generally accepted, one further breakthrough was needed. That was provided half a century later by the German mathematician Johannes Kepler (1571–1630), who worked with the great Danish observer Tycho Brahe (1546–1601).

Tycho had spent two decades observing the stars and planets with the most accurate instruments of the day, designed and built by himself, from an observatory on the island of Hven, off the coast of Denmark. When Tycho died, he left behind a great cache of observations of stellar and planetary positions, from which Kepler was able to calculate the true shape of the planetary orbits.

Kepler found that they were not combinations of circles at all, but ellipses, i.e. squashed circles. Planets moved faster along the part of the orbit that was closest to the Sun, and more slowly farther out. With this one finding, published in 1609, Kepler swept away two millennia of misguided assumptions about spheres and epicycles and replaced them with a simple geometrical curve.

Further observational evidence to support the heliocentric theory was soon forthcoming with the invention of the telescope, which occurred in the same year that Kepler published his findings. Until then, all positional measurements had been made with simple sighting instruments without optics, and it had not been possible to learn anything about the physical nature of the Moon and planets beyond what could be seen with the naked eye.

Through the telescope

The invention of the telescope is usually attributed to a Dutch optician called Hans Lipperhey (*c.*1570–1619), who in 1608 placed two spectacle lenses at either end of a tube about 20 inches (50 centimeters) long to produce a device that magnified about three times. The principle was easy enough to grasp, and when the Italian astronomer Galileo Galilei (1564–1642) heard about the invention, he set about making his own more powerful versions with powers up to 30 times. Galileo turned his telescopes on the sky in the autumn of 1609, and the wonders he beheld heralded one of the greatest revolutions in the history of astronomy.

Galileo saw that the surface of the Moon was not smooth and perfect as the ancient Greek philosophers had presumed, but had mountains and other topographic features like those of the Earth. In January 1610 he discovered four moons orbiting Jupiter, just like Copernicus had said the planets orbited the Sun. Later that year, Galileo noted that Venus goes through a cycle of phases, which could only be possible if it orbited the Sun. And everywhere he looked in the sky he saw faint stars beyond reach of the naked eye, hinting at a vastly larger Universe than had hitherto been suspected. Galileo's observations, published in a 60-page book called *Sidereus Nuncius* ("Starry Messenger") in 1610, proved devastating to the geocentric hypothesis.

Yet serious questions remained. What was the force that kept objects in orbit around the Sun? If we lived on a spinning Earth, why did objects not fly off? Answers were provided by one of the greatest geniuses in the history of science, the English physicist and mathematician Isaac Newton (1642–1727).

Top left Two of Galileo's original telescopes which he used to observe the Moon and planets in 1609-10, preserved at the Museo Galileo in Florence, Italy. **Top right** Galileo's sketches of the Moon at different phases as seen through his telescope, showing craters and mountains on its surface, as published in his book *Sidereus Nuncius* in 1610. **Right** William Herschel built this giant telescope at his home in Slough, England, in 1789. With a mirror 48 inches (1.2 meters) in diameter, a tube 40 feet (12 meters) long, and supported by a complex wooden framework, it was the largest telescope in the world at the time.

A matter of gravity

Having seen an apple fall from a tree in his garden in Lincolnshire, England, so the popular story goes, Newton set to thinking about the force that governed the motions of bodies, all the way from apples out to the Moon and planets. We now call that force gravity.

Newton formulated a series of laws governing the motions of bodies and the gravitational attraction between them. According to these laws, all bodies attract each other with a force that depends on their masses, and which falls off with distance. Gravity acts as though a body's mass is concentrated at its center, so the gravitational force acts along a line joining the centers of the bodies.

Newton's laws were published in 1687 in a book called the *Philosophiae naturalis principia mathematica* "Mathematical Principles of Natural Philosophy"), usually known simply as the *Principia*—one of the most important scientific books ever written. The laws explained why the planets moved around the Sun in elliptical orbits, and provided a physical basis for the heliocentric model of the Universe.

Newton's slightly younger contemporary Edmond Halley (1656–1742) used Newton's laws to calculate that a comet seen in 1531, 1607, and 1682 was not three separate comets but the same comet orbiting the Sun every 75 or 76 years, and he predicted that it would return in or around 1758. Its subsequent reappearance as Halley had predicted was a spectacular confirmation of Newton's laws, and it was named Halley's Comet in his honor.

Above An apple falling from this tree in Isaac Newton's garden at Woolsthorpe Manor, Lincolnshire, England, is said to have set him thinking about the force of gravity. A piece of wood from this tree was taken into orbit aboard a Space Shuttle in 2010.

Sizing up the Universe

Having established the layout of the Solar System, astronomers began to turn their attention to the size and structure of the Universe beyond. It was now realized that stars were similar to the Sun, but much farther away. The question was: how far?

Several astronomers made attempts to estimate the distance of the stars by comparing their brightness with that of either the Sun or one of the outer planets. Newton himself made one such estimate, concluding that Sirius, the brightest star in the night sky, lay about 15 light years away. This was in fact an overestimate—the true distance is 8.6 light years—but was a clear indication of the immense distances of the stars.

It was not until 1838 that the first truly accurate distance of a star was obtained, by the German astronomer Friedrich Wilhelm Bessel (1784–1846). Bessel succeeded in measuring the very slight annual shift in position of the star 61 Cygni as the Earth orbited the Sun, known as a parallax. From the amount of change in the star's position Bessel calculated that it lay 10.4 light years away. This was 1 light year short of the true distance, but was impressively accurate for its time.

In the 1780s William Herschel (1738–1822), an energetic English (but German-born) musician-turned-astronomer, began to make estimates of the number of stars visible in different parts of the sky. Herschel had already made his name in 1781 by discovering an unknown planet, Uranus, which lay twice as far from the Sun as Saturn. After this discovery of a new planet he was funded by the King of England, George III, which enabled him to build larger and more powerful telescopes at his home in Slough, England.

From his surveys, Herschel deduced that we live in a lens-shaped galaxy of stars, the densest part of which we see as the Milky Way, with the Sun somewhere near the center. He depicted the arrangement in a diagram published in 1785. As far as was known, this galaxy constituted the entire Universe. In fact, Herschel's assumption that we were centrally placed in the galaxy turned out to be as misguided as the ancient geocentric model, but it would be another century and a half before this was realized.

Telescope technology had moved on considerably by Herschel's time. Large telescopes used mirrors instead of lenses and as a result could be made in much bigger sizes, although the mirrors were cast from shiny metal rather than made of silver-coated glass as today. Herschel, who was a master telescope maker as well as an assiduous observer, built telescopes with mirrors up to 48 inches (120 centimeters) across, the largest in the world at that time.

With his giant reflectors, Herschel began to peer far into the depths of space, cataloging around 2,500 faint, hazy objects, the nature of which became a long-standing matter of speculation. Around 100 of these had previously been cataloged by the French comet hunter Charles Messier (1730–1817), and consequently bear catalog numbers prefixed by an M for his name. Herschel was able to resolve some of these into individual clusters of stars, but others remained fuzzy. Were these also clusters that were too far away to be resolved with existing telescopes, or were they luminous clouds of gas? Or were they, as the German philosopher Immanuel Kant (1724–1804) had speculated, actually other galaxies?

This question intrigued the Irish astronomer Lord Rosse (1800–67). In 1845, Rosse built what was then the world's largest telescope, a reflector with a mirror 72 inches (1.8 meters) in diameter, at his home at Birr Castle in central Ireland. Through this giant telescope he saw that the object cataloged by Messier as M51 had a spiral structure, and he went on to discover spiral structures in other nebulae, too. M51 is now known appropriately as the Whirlpool. The full implications of this discovery would not become apparent until the twentieth century, when it was realized that these so-called spiral nebulae were indeed other galaxies, far outside our own.

Astronomy becomes physical

During the nineteenth century, two new techniques emerged that were to radically change the face of astronomy and give birth to a new branch of science, called astrophysics. The first of these techniques was spectroscopy, the ability to analyze the chemical composition of a star or other object from details in its spectrum, obtained by passing its light through a prism in a device called a spectroscope.

In 1814, the German physicist Joseph von Fraunhofer (1787–1826) examined the Sun's spectrum through such a device and found that it was crossed by hundreds of dark lines of varying thickness, now called Fraunhofer lines, although he did not know what they were. He also found similar lines in the spectra of bright stars such as Sirius.

The explanation was provided by Fraunhofer's countrymen Gustav Kirchhoff (1824–87) and Robert Bunsen (1811–99), who discovered in the 1850s that the lines were due to cooler gas in the Sun's outer layers absorbing light at particular wavelengths from the hotter regions below. Since each chemical element had its own signature of lines that could be measured in the laboratory, it was therefore possible to deduce the composition of the Sun and other stars from the lines in their spectra.

Armed with this new insight, astronomers began to study and classify stellar spectra in the hope of learning more about the compositions of stars and how they evolved (see also Chapter 2, page 197). But gathering spectral data at the telescope was a slow and inaccurate process. What was needed was a quick and simple way of recording the lines so they could be measured carefully later. That recording system was the second of the new techniques to emerge: photography.

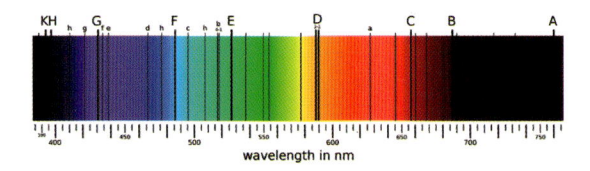

Top left A drawing of M51, the Whirlpool nebula, made by Lord Rosse through his giant telescope, which revealed its spiral shape. Such spiral nebulae were later revealed to be separate galaxies far outside our own Milky Way. This drawing was published in 1850. **Above** Solar spectrum: The Sun's spectrum is crossed by dark lines that reveal the elements of which it is composed. The lines are called Fraunhofer lines after their discoverer, the German physicist Joseph von Fraunhofer. Other stars also display similar lines, which allow us to determine their composition as well.

Photographing the stars

Photography transformed astronomy in the latter half of the nineteenth century. Not only was a photograph more accurate than a hand-drawn sketch, long exposures could bring into view objects that were too faint for the eye to see, thereby opening up a completely new view of the Universe.

Progress was rapid as photographic plates became more sensitive. The first, very crude, photograph of the Moon was taken in 1840 by the American scientist John William Draper (1811–82), although photographic plates were then so slow that it required an exposure time of 20 minutes. He then turned to photographing the Sun's spectrum. His son Henry Draper (1837–82) went several steps further, obtaining the first photographic spectrum of a star, Vega, in 1872, and in 1880, the first photograph of the brightest central part of the Orion Nebula.

One of the most accomplished of the early astrophotographers was a British amateur, Isaac Roberts (1829–1904), who photographed the sky through a 20-inch (50-centimeter) reflector at his observatory near Liverpool, England. His most celebrated images, taken in 1887 and 1888, revealed for the first time that the great nebula in Andromeda, M31, was spiral in shape, just like M51 and all the other spiral nebulae previously discovered by Lord Rosse.

Spectroscopic studies in the 1860s by the English scientist William Huggins (1824–1910) had settled the long-standing dispute over the nature of the nebulae by showing that there were two distinct types: the spiral nebulae were composed of distant stars, whereas the irregular nebulae, such as the one in Orion, were simply clouds of gas. Yet there was still no way of telling how far away any of these nebulae actually were.

A larger Galaxy

A major advance in understanding the true extent of the Universe was made in 1912 at Harvard College Observatory, Massachusetts, by the American astronomer Henrietta Leavitt (1868–1921), who discovered an important relationship between the inherent brightness of Cepheid variable stars and their period of variation. Cepheids are pulsating variables, named after their prototype, Delta Cephei, and they are highly luminous, so they can be seen over great distances.

While measuring the brightness changes of Cepheids on photographic plates of the Small Magellanic Cloud, a dense patch of stars in the southern hemisphere, Leavitt noticed that the time a Cepheid took to go through a cycle of variations was directly linked to its brightness, with the slowest pulsators being the brightest. Hence, by measuring the pulsation period of a Cepheid variable star, astronomers could find how inherently bright it was (its *absolute magnitude*); comparing that with how bright it appears in our sky (its *apparent magnitude*) they could then work out its distance. This was the vital yardstick that astronomers needed to gauge the true scale of space, and it's still in use today.

On the other side of the United States, at Mount Wilson Observatory in California, Harlow Shapley (1885–1972) used Leavitt's discovery to make the first accurate measurement of the size of our Galaxy. At Mount Wilson he had access to the Observatory's 60-inch (1.5-meter) reflector, at that time the largest operational telescope in the world. With it he estimated the distance of the many globular clusters of stars scattered around our Galaxy from the Cepheid variables that lay within them.

Above M31, the great nebula in Andromeda, photographed by Isaac Roberts in December 1888 with an exposure of four hours. Roberts's photographs of M31 showed for the first time that it was spiral in form. Although it was then termed a "nebula," we now know that it is actually a spiral galaxy.

Shapley noted that the clusters were not distributed symmetrically around the sky but were concentrated on one side, toward Sagittarius. From this he deduced that the Sun is not centrally placed in the Galaxy, as had been thought since the time of William Herschel, but lies about two-thirds of the way to the edge, a revelation comparable to the dethroning of the Earth from the center of the Solar System three centuries before.

From his calculations of the clusters' distances, Shapley concluded in 1918 that the Galaxy was about ten times larger than previously supposed, and that the Sun lay about 50,000 light years from the center. These figures have since been revised downward to a distance of about 30,000 light years from the center, and an overall diameter for our Galaxy of about 100,000 light years, but the overall picture that Shapley outlined was correct.

Shapley was wrong about one thing, though. He still thought that the enlarged Galaxy was alone and surrounded by empty space—in effect, it was the entire Universe—and that the spiral nebulae were either part of it or, at the very most, just beyond it. The power of a new telescope on Mount Wilson would soon prove otherwise.

Nebulae become galaxies

Along with Galileo's original spyglass and the Hubble Space Telescope, the 100-inch (2.5-meter) reflector on Mount Wilson is one of the most influential telescopes in history, for it finally cracked the problem of the spiral nebulae and demonstrated that we live in an expanding Universe of unfathomable size.

With the extra light grasp of this powerful new eye on the sky, the American astronomer Edwin Hubble (1889–1953) was able to photograph individual stars in the outer regions of M31, the great spiral in Andromeda. Among these stars were Cepheid variables, the first of which he detected in 1923. He eventually found 40 of them, from which he calculated the distance to M31.

His result was 900,000 light years away (the modern figure, based on better understanding of Cepheids, is 2.5 million light years). This placed it well beyond the limits of even Shapley's enlarged Galaxy, and finally demonstrated that the spiral "nebulae" were separate galaxies. The Universe had suddenly become much bigger.

The expanding Universe

Hubble went on to make an even more significant discovery with the Mount Wilson telescope: the galaxies are moving apart from one another, as though the Universe is expanding like a balloon being inflated. This astounding conclusion, announced in 1929, came from a study of the spectra of distant galaxies, which revealed that their light was being lengthened in wavelength as a result of high-speed recession. Such a lengthening of wavelength is termed a *redshift*, because the light from the galaxy is moved toward the red (longer-wavelength) end of the spectrum.

Hubble's observations demonstrated that the amount of redshift in the light from galaxies increases with their distance, a relationship that became known as Hubble's Law. Hence measuring the redshift of a distant galaxy gives astronomers an indication of how far away it is, right to the limits of the observable Universe.

Since the Universe is expanding, it's logical to conclude that it was once smaller and more densely packed than it is now. In 1931 a Belgian mathematician called Georges Lemaître (1894–1966) proposed that the Universe had originated in the explosion of a highly compressed, super-dense blob of matter some 10 billion years ago, and the galaxies are

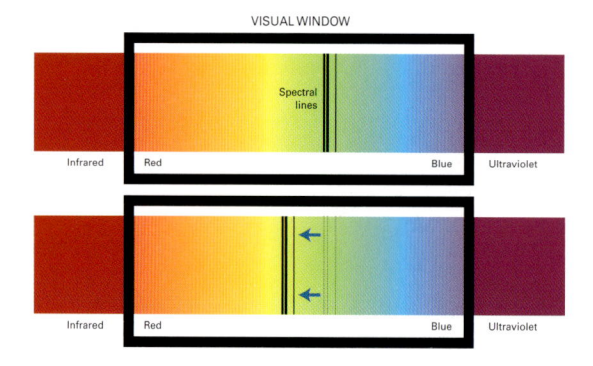

Above The redshift in the light from a distant galaxy is measured by the change in position of dark lines in its spectrum. Here, the change in position is shown in the range of wavelengths to which the human eye is sensitive (the visual window). All light from the galaxy is shifted by the same amount, so while some of the galaxy's light moves out of sight beyond the red end of the visual window, other wavelengths move into the visible window from the ultraviolet.

the fragments from that explosion, still flying apart as the space between them expands. This explosive event was later dubbed the Big Bang and it's now the generally accepted theory of cosmology. Astronomy had progressed from Copernicus to the Big Bang in less than 400 years.

Direct evidence for the Big Bang came in 1965, when two American radio engineers, Arno Penzias (b. 1933) and Robert Wilson (b. 1936), working at Bell Telephone Laboratories in New Jersey, detected very faint radio noise at short wavelengths coming from all over the sky. Known as the cosmic microwave background, this radio noise is thought to be energy left over from the Big Bang. For this discovery, Penzias and Wilson were awarded the Nobel Prize in Physics in 1978.

According to the best current estimates, the Big Bang took place 13,800 million (13.8 billion) years ago—that's the age of the Universe as we know it. As far as anyone can tell, the Universe will continue to expand for ever.

Top The 100-inch (2.5-meter) reflector on Mount Wilson, California, U.S.A, was used by the American astronomer Edwin Hubble in the 1920s to prove that spiral nebulae were actually separate galaxies and to measure their distances. Further observations with this telescope led to the discovery that the Universe is expanding. **Left** In 1965, this horn-shaped antenna at Bell Telephone Laboratories in Holmdel, New Jersey, U.S.A, detected radiation from space at short radio wavelengths. This radiation is thought to be energy left over from the Big Bang and is now known as the "cosmic microwave background."

NOTABLE ASTRONOMERS

Since the beginning of time, the night sky has fascinated us. Awash with blazing stars, distant planets, and the splendor of astronomical phenomena, the sky has always been an incredible sight to behold. Ancient people systematically recorded observations and recurring events, and artists captured the majesty. All as we continue to do today. In the pages that follow, the discoveries and hypotheses of incredible astronomers ancient and contemporary are explored.

Ptolemy
(Around 150 CE)

Ptolemy, also known as Claudius Ptolemaeus, was a Greek astronomer and geographer who worked at Alexandria in northern Egypt in the second century CE. He is regarded as the greatest astronomical authority of ancient times. Around 150 CE he wrote a compendium of Greek astronomy called the *Mathematike Syntaxis,* better known by its later Arabic title *Almagest*, meaning "the greatest." This contained a catalog of over 1,000 stars divided into 48 constellations, based on the work of an earlier Greek astronomer, Hipparchus (*c.*190–*c.*120 BCE). These 48 constellations, still recognized by astronomers today, form the basis of our modern constellation system. In addition, Ptolemy developed a geocentric (Earth-centered) model of the Universe that was accepted until the seventeenth century, when it was overthrown by the heliocentric (Sun-centered) model.

Opposite Claudius Ptolemaeus, better known simply as Ptolemy, compiled a catalogue of over 1,000 stars arranged into 48 constellations around 150 CE.

Ulugh Beg
(1394–1449)

Ulugh Beg (1394–1449) was a Mongol sultan and astronomer who made the first major update to Ptolemy's catalog of stars from his observatory in Samarkand, in present-day Uzbekistan. In the 1420s, Ulugh Beg built an enormous stone sextant with a radius of 130 feet (40 meters), to make highly accurate observations of the altitudes of celestial objects as they crossed the meridian. With this and other instruments, Ulugh Beg and his fellow astronomers measured the length of the year to an accuracy of 25 seconds, the tilt of the Earth's axis to within a fraction of a degree, and re-observed the stars in Ptolemy's to produce the most accurate catalog of its time, published in 1437. His observatory was demolished after his death, although the lower part of it, which extended underground, still exists.

Top Statue of Ulugh Beg at the site of the observatory he built in the 1420s in Samarkand, Uzbekistan. It is now a World Heritage Site.

Nicolaus Copernicus
(1473–1543)

Tycho Brahe
(1546–1601)

Nicolaus Copernicus (1473–1543) was a Polish theologian and astronomer who devised a heliocentric (Sun-centered) model of the Universe that challenged the ancient Greek geocentric (Earth-centered) view put forward by Ptolemy. Copernicus developed his ideas over many years but was reluctant to publish them for fear of opposition from the Church. His revolutionary theory finally appeared in 1543, shortly before his death, in a book called *De revolutionibus orbium coelestium* ("On the Revolutions of the Heavenly Spheres"). In it, he proposed that all the planets, including the Earth, revolved around a central Sun on transparent spheres, with an outermost sphere carrying the fixed stars, the true nature of which was not then known. However, Copernicus's model was still hampered by the traditional dogma that all planetary motions must be explained by circles or combinations of circles. As a result, the heliocentric view did not become fully accepted until the discovery by Johannes Kepler that planetary orbits are actually elliptical in shape.

Top left Nicolaus Copernicus published the heliocentric view of cosmology in 1543. He holds a celestial sphere with the Sun at the center in this monument in his hometown of Torun, Poland. **Top right** Tycho Brahe made the most accurate observations of the stars and planets in the pre-telescopic era from his two observatories on the Danish island of Hven.

Tycho Brahe (1546–1601) was a Danish astronomer, now considered the greatest observer in the days before the telescope. In 1572 he saw a brilliant new star appear in Cassiopeia that remained visible for over a year; his observations showed that it lay beyond the Moon, thereby demonstrating that the heavens were not perfect and unchanging as the ancient Greek philosophers had maintained. His doubts about the ancient Greek dogmas were reinforced when his observations of a comet in 1577 demonstrated that it moved between the planets. With support from the King of Denmark, Tycho set up two observatories on the Danish island of Hven: Uraniborg, which incorporated living quarters, and Stjerneborg, in which instruments were placed partly underground for protection. Tycho made great improvements in the design and construction of his instruments, which enabled him to make far more accurate observations of planetary motions than ever before, and to produce the most accurate star catalog of the pre-telescopic era. After the Danish king withdrew his financial support Tycho moved to Prague, in 1599, where Johannes Kepler (see opposite) became his assistant. Kepler used Tycho's planetary observations to draw up the three laws of planetary motion that confirmed the heliocentric theory of the Solar System.

Galileo Galilei
(1564–1642)

Johannes Kepler
(1571–1630)

Galileo Galilei (1564–1642) was an Italian astronomer and physicist who in 1610 published the first observations made with a telescope. Galileo had heard of the invention of the telescope in the Netherlands in 1609 and set out to make his own improved versions. With these simple refractors he made a series of significant astronomical discoveries, including mountains on the Moon, the four main satellites of Jupiter (now known as the Galilean satellites), the phases of Venus, and that the Milky Way was made up of countless faint stars. He reported his discoveries in his book of 1610 called *Sidereus Nuncius* ("Starry Messenger") and later used them to argue in favor of the heliocentric model of the Universe developed by Nicolaus Copernicus (see opposite). His other notable scientific work included experiments on falling bodies, in which he demonstrated that objects of different mass fall at the same speed, not at different speeds according to their mass, as the ancient Greeks had maintained.

Top Galileo Galilei published the first astronomical observations made with a telescope, in 1610, including the discovery of mountain and craters on the Moon and the four main moons of Jupiter.

Johannes Kepler (1571–1630) was a German mathematician and astronomer who in 1600 became assistant to Tycho Brahe (see opposite) in Prague, where he began calculating tables of planetary motion from Tycho's accurate observations. During this work he realized that the orbits of the planets were elliptical in shape (not circular as had hitherto been believed), and that the speed of each planet varies along its orbit, being fastest when closest to the Sun. These were his first two laws of planetary motion, which he published in 1609, and they helped to establish the heliocentric (Sun-centered) model of the Solar System. Kepler went on to discover a third law, which linked the period of a planet's orbit to its average distance from the Sun. After Tycho's death in 1601, Kepler prepared his catalog of 1,000 stars for publication. This appeared in 1627 as part of the *Rudolphine Tables*, a set of tables for calculating planetary positions. Kepler's other work included a book about the supernova of 1604, popularly known as Kepler's Star, and a book about optics and telescope lenses called *Dioptrice*.

Top Johannes Kepler discovered the laws of planetary motion, demonstrating that the planets, including the Earth, orbit the Sun on elliptical paths.

Isaac Newton
(1642–1727)

Ole Rømer
(1644–1710)

Isaac Newton (1642–1727) was an English physicist and mathematician who revolutionized physics and astronomy with the discovery of the laws of motion and gravitation. He published these laws in 1687 in a book called the *Philosophiae naturalis principia mathematica* ("Mathematical Principles of Natural Philosophy"), better known as the *Principia*, which is widely regarded as the greatest scientific book ever written. His three laws of motion state that a body continues in a state of rest or uniform motion in a straight line unless acted upon by an external force; that the acceleration produced by a force is directly proportional to the force and takes place in the direction in which the force acts; and to every action there is an equal and opposite reaction. His law of gravitation states that any two bodies attract each other with a force that depends on the product of their masses divided by the square of the distance between them. Newton also studied optics, discovering that white light is made up a range of colors (the spectrum), and in 1668 made the first working reflecting telescope of a design that is now known as a Newtonian.

Top Isaac Newton discovered the laws governing the gravitational attraction between objects, which explained why smaller bodies, such as planets, orbit larger bodies, such as the Sun.

Ole Rømer (1644–1710) was a Danish astronomer who in 1676 made the first measurement of the speed of light. Rømer found that the times of the eclipses of Jupiter's innermost known moon, Io, as it orbited the planet varied by up to 11 minutes from predictions during the course of a year. He realized that this must be due to the changing distance of Jupiter from the Earth, which would affect the time light took to reach us from Jupiter. From the offset between the predicted and observed times of the eclipses, other astronomers calculated that the speed of light was around 200,000 km/s. This is two-thirds of the correct value, the error being due in part to uncertainties in the size of the Solar System, but it was a good first attempt and demonstrated that the speed of light was not infinitely fast, as had been widely believed until then.

Top Ole Rømer made the first measurement of the speed of light, demonstrating that it was not infinitely fast as had been believed.

Edmond Halley
(1656–1742)

Nicolas Louis de Lacaille
(1713–62)

Edmond Halley (1656–1742) was an English astronomer and mathematician who discovered that comets orbit the Sun on highly elliptical paths. In 1677, while still a student, he traveled to the island of St. Helena in the South Atlantic Ocean from where he made the first catalog of the southern stars with the aid of a telescope. Halley used Isaac Newton's law of gravity to calculate that a comet he had observed in 1682 was the same as those seen in 1531 and 1607 and predicted that it would return again in or about 1758. It did so, although he was no longer alive to see it, and it was subsequently named Halley's Comet. In 1718 he announced that certain bright stars had changed in position since ancient Greek times, an effect known as "proper motion," although his results are now thought to have mostly been due to observational errors. In 1720 Halley became England's second Astronomer Royal and spent the next 18 years studying the orbital motion of the Moon around the Earth.

Top Edmond Halley used Isaac Newton's laws of gravity to calculate that comets, like planets, orbit the Sun on elliptical paths and predicted the return of the comet that now bears his name.

Nicolas Louis de Lacaille (1713–62) was a French astronomer who made the first major survey of the southern sky. In 1751–52 he cataloged nearly 10,000 stars from the Cape of Good Hope in South Africa with a small telescope. From these stars he created 14 new constellations representing instruments from the arts and sciences—all of which are still recognized—and adjusted some existing constellations to make way for them. Lacaille's figures filled in the southern sky and were the last new constellations to be accepted by astronomers. His final catalog and chart of the southern sky were published posthumously in 1763.

Top Nicolas Louis de Lacaille made the first major survey of the southern skies, cataloging nearly 10,000 stars and introducing 14 new southern constellations.

Charles Messier
(1730–1817)

William Herschel
(1738–1822)

Charles Messier (1730–1817) was a French astronomer who made a famous list of hazy-looking objects that might be mistaken for comets, which were his main interest. Messier discovered over a dozen comets (plus some joint discoveries), but he is best remembered for his catalog of nebulous objects; these consist of not just nebulae but also star clusters and galaxies. Objects in Messier's catalog are given M numbers, such as M1 (the Crab Nebula), the first in the list; M31, the Andromeda Galaxy; and M42, the Orion Nebula. Messier's first list, containing 45 objects, was published in 1774. An expanded version with 23 more objects appeared in 1780 followed by his final catalog of 103 objects in 1781. Later astronomers have extended the Messier catalog to 110 objects based on his notes.

Top left Charles Messier compiled a catalog of over one hundred star clusters and nebulae. These are now known as Messier, or M, objects.
Top right William Herschel discovered the planet Uranus and cataloged thousands of faint nebulae, many of which later turned out to be distant galaxies, using large telescopes of his own construction.

William Herschel (1738–1822) was an English astronomer who discovered the planet Uranus and made comprehensive surveys of the night sky with large telescopes of his own construction. Herschel was a musician by profession but became more interested in astronomy. On March 13, 1781 he discovered what he thought was a comet but turned out to be a new planet, subsequently named Uranus. It was the first planet ever discovered with the aid of a telescope. With his sister Caroline Herschel (1750–1848) he undertook multiple surveys of the entire sky visible from England, cataloging numerous double stars and about 2,500 nebulae, realizing that some of them were actually clusters of faint stars. His observations of double stars revealed that many are in mutual orbit around each other, i.e. they are genuine binaries. Caroline herself discovered eight comets and over a dozen nebulae. With the financial support of King George III, William built what was the largest telescope in the world, with a mirror 48 inches (1.2 meters) in diameter, at his home in Slough, England. From his mapping of the stars he deduced that our Galaxy was shaped like a lens, its thickest part forming the Milky Way, and that we lie near the center, although this latter deduction was later proved wrong. Among his other advances, William discovered the existence of infrared radiation. William's son John Herschel (see opposite) extended the family's survey of the sky to the southern hemisphere.

Friedrich Bessel
(1784–1846)

John Frederick William Herschel
(1792–1871)

Friedrich Bessel (1784–1846) was a German astronomer and mathematician who made the first accurate measurement of the distance of a star. His long-term measurements of star positions revealed several that showed a steady movement over time, known as "proper motion," indicating that they were relatively close to the Sun. In 1837–38 he carefully monitored the one with the largest known motion, called 61 Cygni, and found an annual shift in position known as a "parallax," from which its distance could be calculated. According to Bessel's measurements, 61 Cygni lay 10.4 light years away, 1 light year short of the true distance. In 1844 he announced that the bright stars Sirius and Procyon showed oscillations in their proper motions, which indicated the presence of unseen companions; these companions were later detected visually and turned out to be white dwarfs.

Top Friedrich Bessel made the first accurate determination of the distance of a star other than the Sun.

John Frederick William Herschel (1792–1871) was an English astronomer and physicist, the son of William Herschel (see opposite). In 1834 he took his father's 20-foot- (6-meter-) long reflector to the Cape of Good Hope in South Africa to survey the southern part of the sky that was below the horizon from England. During the next four years he cataloged over 2,100 double stars, plus 1,700 nebulae and clusters. He combined these with William Herschel's discoveries to produce the *General Catalogue of Nebulae and Clusters of Stars*, published in 1864. This formed the basis for the New General Catalogue (NGC) still used by astronomers. Among his other scientific work, John Herschel studied the local flora at the Cape, was a pioneer of photography, and invented the blueprint.

Top John Frederick William Herschel, son of William Herschel, continued his father's work of cataloging clusters and nebulae by surveying the southern skies from the Cape of Good Hope.

Lord Rosse
(1800–67)

William Huggins
(1824–1910)

Lord Rosse (1800–67), full name William Parsons, 3rd Earl of Rosse, was an Irish astronomer who in 1845 built what was then the world's largest telescope, a reflector with a mirror 72 inches (1.8 meters) in diameter, at his home at Birr Castle. The instrument was commonly known as the Leviathan on account of its size. With this telescope he discovered the spiral structure of what were later realized to be galaxies, starting with the Whirlpool Galaxy, M51. This was the first step to establishing that our Galaxy is only one of countless others through the Universe. Rosse also gave the Crab Nebula its name because he thought it appeared to have legs like those of a crab.

Top Lord Rosse, also known as William Parsons, 3rd Earl of Rosse, built the world's largest telescope at the time, at his home in Birr Castle.

William Huggins (1824–1910) was an English astronomer who pioneered the study of celestial bodies with a spectroscope. Working initially with the English chemist William Allen Miller (1817–70), he discovered that stars consist of incandescent gas, similar to the Sun, and that nebulae such as the one in Orion are also gaseous but cool. Analysis of the light from the so-called "nebula" in Andromeda, on the other hand, showed that it was composed of stars, and was thus a separate galaxy. Huggins also measured the first radial velocity of a star, Sirius. Following his marriage in 1875, he and his wife, Margaret Lindsay Huggins (1848–1915), began to record spectra photographically, producing an atlas of stellar spectra in 1899. Their work helped found the new branch of astronomy known as astrophysics.

Top Willian Huggins pioneered the use of spectroscopy to analyze the composition of stars and nebulae.

Henrietta Leavitt
(1868–1921)

Albert Einstein
(1879–1955)

Henrietta Leavitt (1868–1921) was an American astronomer who devised a method for measuring distances in the Universe. While studying Cepheid variable stars in the Small Magellanic Cloud, she noticed that the time a Cepheid took to go through its cycle of variations was directly linked to its brightness, with the brightest ones having the longest periods. This relationship, published in 1912, is now known as the "period-luminosity relation," or the Leavitt law. Leavitt's discovery meant that by measuring the period of a Cepheid in a distant galaxy, astronomers could identify its inherent brightness (its absolute magnitude); comparing that brightness with how bright it actually appeared (its apparent magnitude) then allowed the distance to be calculated. This method helped astronomers to estimate the scale of the Universe and is still used today.

Top Henrietta Leavitt discovered the law that links the brightness of Cepheid variable stars to their period of variation, providing a valuable distance indicator for astronomers.

Albert Einstein (1879–1955) was a German-born American physicist and mathematician who developed theories of time, space, and gravitation that form the basis of our modern understanding of the Universe. In 1905 he published the special theory of relativity which included the famous equation $E = mc^2$; this explains that matter can be turned into energy, which is what happens in the atomic fusion processes that power stars. It also stated that the speed of light is the maximum possible speed in the Universe. In 1915 Einstein announced the general theory of relativity. This described gravitation as a curvature of space and predicted the existence of gravitational waves, both of which have been confirmed by observations.

Top Albert Einstein developed the special and general theories of relativity that underpin our modern understanding of the physical Universe.

Arthur Eddington
(1882–1944)

Harlow Shapley
(1885–1972)

Arthur Eddington (1882–1944) was an English astrophysicist who made the first experimental confirmation of Albert Einstein's theory of general relativity. At a total eclipse of the Sun in 1919, Eddington photographed the position of background stars around the Sun and found that they had moved slightly from their normal positions. This was due to the bending of light passing through the Sun's gravitational field, as predicted by Einstein. Eddington also studied the internal structure of stars, realizing that they must be powered by the release of atomic energy in their cores. He discovered that the luminosity of a star is governed by its mass, the most massive stars being the most luminous, and vice versa.

Top Arthur Eddington made the first experimental confirmation of Einstein's theory of relativity, as well as developing mathematical models that explained the internal structure of stars.

Harlow Shapley (1885–1972) was an American astronomer who discovered that the Sun does not lie near the center of the Galaxy, as had previously been believed, but instead is in its outer regions. Shapley measured the distances to globular clusters using the period-luminosity law for Cepheid variable stars discovered by Henrietta Leavitt (see page 51), finding that our Galaxy was much larger than previously supposed, although he overestimated its size. From the distribution of globular clusters, which mostly lay in one half of the sky, Shapley deduced that the center of our Galaxy lies in Sagittarius and that the Sun is about two-thirds of the way to the edge. This was another step in the process of dethroning the Sun from the center of the Universe that had been started by Copernicus.

Top Harlow Shapley discovered that our galaxy is much larger than previously thought and that the Solar System lies in its outer regions.

Edwin Hubble
(1889–1953)

Georges Lemaître
(1894–1966)

Edwin Hubble (1889–1953) was an American astronomer who discovered that the Universe is expanding, a finding that was the foundation of modern cosmology. In 1923, using the new 100-inch (2.5-meter) reflector at Mount Wilson Observatory, Hubble resolved individual stars in the Andromeda "nebula," M31, which proved that it was in fact a separate galaxy of stars outside our own. Hubble went on to show that other so-called nebulae were also separate galaxies at vast distances from us, and in 1925 he developed the now-standard classification of galaxies into spirals, barred spirals, ellipticals, and irregulars. Hubble measured the velocities of galaxies through space from redshifts in their spectra and concluded in 1929 that galaxies are receding from us with speeds that increase with their distance, thereby demonstrating that the Universe is expanding; the actual rate of expansion is known as the Hubble constant. The Hubble Space Telescope is named in his honor.

Top Edwin Hubble demonstrated that spiral nebulae are actually separate galaxies and discovered the expansion of the Universe.

Georges Lemaître (1894–1966) was a Belgian mathematician and cosmologist who originated the Big Bang model of the Universe, which is the currently accepted theory. In 1927 he used Einstein's theory of general relativity to predict that the Universe should be expanding. His prediction was confirmed two years later by Edwin Hubble (see left), who discovered that galaxies are moving apart at speeds that increase with their distance, a relationship that is now known as the Hubble–Lemaître law. In 1931 he proposed that the Universe had originated in the explosion of a so-called primeval atom some 10 billion years ago, from which it has been expanding ever since; this explosive event was later dubbed the Big Bang by the English cosmologist Fred Hoyle.

Top Georges Lemaître proposed that the Universe had its origin in an immense explosion now known as the Big Bang.

Cecilia Payne-Gaposchkin
(1900–79)

Cecilia Payne-Gaposchkin (1900–79), born Cecilia Payne, was an English-born American astronomer who discovered that stars consist mostly of hydrogen and helium, the two simplest gases. She made her discovery by studying the spectra of stars. Her conclusion, published in 1925, was not widely accepted at first since stars were then thought to be of a similar composition to the Earth, albeit very much hotter. The discovery also demonstrated that the Universe itself must be composed mostly of hydrogen and helium, which was fundamental to our understanding of the origin and evolution of the Universe. In 1934 Payne married the Russian–American astronomer Sergei Gaposchkin (1898–1984) and together they made extensive studies of variable stars.

Top Cecilia Payne-Gaposchkin discovered that stars are mostly made of the lightest gases in the Universe, hydrogen and helium.

Karl Jansky
(1905–50)

Karl Jansky (1905–50) was an American communications engineer who made the first detection of radio noise from space, thereby marking the birth of radio astronomy. His discovery came in 1931–32 while searching for sources of interference on long-distance telephony at Bell Telephone Laboratories in New Jersey. Jansky found that one source of radio noise lay in the direction of the center of our galaxy in Sagittarius. He theorized that the noise came from gas between the distant stars of the Milky Way, which we now know to be the case. The unit of radio flux is named the "jansky" in his honor. Jansky never followed up these observations and the next steps were left to an American radio amateur, Grote Reber (1911–2002), who built his own radio telescope in 1937, with which he made the first radio maps of the sky.

Top Karl Jansky discovered the existence of radio waves coming from space, founding the science of radio astronomy.

Hans Bethe
(1906–2005)

Hans Bethe (1906–2005) was a German-born American theoretical physicist who discovered the atomic processes that generate energy in stars. In 1938, he and the American physicist Charles Critchfield (1910–94) published a paper that explained how a chain of nuclear reactions that fuse four hydrogen nuclei (protons) to form one nucleus of helium, known as the "proton-proton chain," would account for energy generation inside low-mass stars. The following year Bethe discovered a second reaction, called the "carbon-nitrogen cycle," which would explain the energy generation inside stars of higher mass than the Sun. This was the first detailed explanation of energy generation in stars, for which Bethe was awarded the Nobel Prize in Physics in 1967.

Top Hans Bethe won the Nobel Prize in Physics for explaining the atomic processes that generate energy inside stars. **Right** Located at Griffith Observatory in Los Angeles, California, Astronomers Monument pays tribute to the incredible work of Galileo, Copernicus, Herschel, Hipparchus, Kepler, and Newton.

CHARTING THE HEAVENS

Modern astronomers recognize a total of 88 constellations, filling the sky like pieces of a jigsaw from the north celestial pole to the south, as laid down by the International Astronomical Union (IAU), astronomy's governing body. But until the sixteenth century, only 48 constellations were known to Western astronomers. These were the figures listed by the Greek astronomer Ptolemy in his summary of Greek astronomical knowledge popularly called the *Algamest,* written around 150 CE.

Many of these constellation figures originated with earlier civilizations in the Middle East, such as the Babylonians—over 3,000 years ago—and were subsequently adopted by the Greeks, who changed their names and added their own myths and legends to them. Through these myths, as recounted by Greek and Roman writers such as Aratus, Hyginus, and Ovid, we can still imagine the legendary characters among the stars, such as Perseus holding the severed head of Medusa the Gorgon; Orion, the Hunter, with raised club and shield; Cygnus, the Swan, flying along the Milky Way; and Taurus, the Bull, snorting the night air.

The *Almagest* contained the positions and brightnesses of over 1,000 stars, about half of those easily visible to the naked eye. But there were no illustrations of the constellations, just written descriptions of where the stars lay within each imaginary figure—for example, "the star on the end of the tail" in Ursa Minor, or "the bright, reddish star on the right shoulder" in Orion. Later artists and astronomers were left to create their own depictions based on Ptolemy's descriptions.

Our first view of the constellation figures as imagined by the ancient Greeks is on a sculpture called the Farnese Atlas, after Cardinal Alessandro Farnese, who acquired it in the sixteenth century. In this sculpture, the kneeling figure of Atlas carries on his shoulders a globe of the heavens, and on the globe we can see representations of the constellations known to Ptolemy. This is the oldest known celestial globe, made in Rome sometime in the second century CE and probably a copy of a much older Greek original. It would be several centuries before any better representations of the sky emerged, and they did so not in Europe but the Middle East.

Opposite Albrect Durer's chart of the Celestial Map, created in woodcut in 1515. **Top left** A table of stars from the first printed edition of Ptolemy's *Almagest* which appeared in 1515, around 1,400 years after the original was written. The columns contain a verbal description of the star's position in the constellation, its coordinates on the celestial sphere, and a figure indicating its brightness. **Top right** This sculpture of Atlas holds a globe of the heavens containing images of the 48 constellations known to the ancient Greeks. The statue is known as the Farnese Atlas, because it was once owned by the Italian cardinal Alessandro Farnese, and is thought to date from the second century CE.

STARGAZING THROUGH THE AGES 57

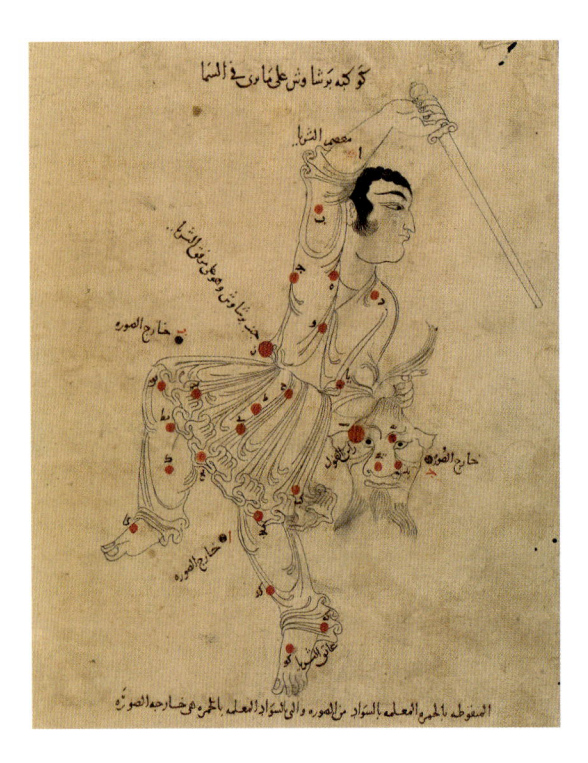

Enter the Arabs

After Ptolemy, Greek science went into permanent decline. By the eighth century CE, the center of intellectual inquiry had moved east to Baghdad, where the Greek works were translated into Arabic. It was then that Ptolemy's book received its Arabic name the *Almagest*, meaning "the greatest," by which we still know it.

One of the greatest Arabic astronomers of that era was Abd al-Rahman al-Sūfī (903–86), who produced a revised and updated version of Ptolemy's star catalog in the *Book of the Fixed Stars* around 964 CE. Al-Sūfī adopted the same 48 constellations as listed by Ptolemy but, unlike the *Almagest*, al-Sūfī's version included illustrations of each constellation, although they were drawn in an Islamic style, not a Greek style. Each constellation was shown in two orientations—one as seen in the sky and the other in mirror image, as on a celestial globe.

Since this was still long before the invention of printing, all books were hand-written and each manuscript had different illustrations. The example shown here is from an early copy of al-Sūfī's book, made around 1010 CE, which shows an Arabic artist's impression of Perseus in flowing robes, holding the severed head of Medusa the Gorgon, here drawn as a bearded man.

Greek scientific knowledge had long been lost in the West, but that was about to change. From the tenth century onwards, the works of Ptolemy and other Greek scientists were reintroduced into Europe through the Arab conquest of Spain. There they were translated from Arabic into Latin, the scientific language of the day. It was through this roundabout route of Ptolemy's catalog being translated into Arabic and then re-translated into Latin that we ended up with Greek constellations with Latin names and containing stars with Arabic titles.

Albrecht Dürer's star charts

With the development of the printing press in the fifteenth century, it became possible to produce large numbers of books and charts at will, and astronomical works were prominent among them. The first printed copy of the *Almagest* appeared in 1515, based on a translation from Arabic into Latin that was made in Toledo, Spain, around 1175. Also in 1515, the first European printed charts of the sky appeared in the form of a pair of woodcuts drawn by the German artist Albrecht Dürer (1471–1528), with the technical help of Johannes Stabius, an Austrian mathematician, and Conrad Heinfogel, a German astronomer.

Dürer's charts were based on Ptolemy's star catalog in the *Almagest*, which was still the most authoritative source. The 48 Ptolemaic constellations were depicted on two hemispheres. One hemisphere showed the 12 figures of the zodiac and all constellations north of it, while the other contained all constellations south of the zodiac that were known to Ptolemy. Stars in each constellation were numbered in the order they appeared in Ptolemy's catalog so that they could easily be identified. Dürer drew the constellations in mirror image, as they would appear on a celestial globe, since at that time astronomers were accustomed to using globes rather than charts.

Dürer's visualizations brought Ptolemy's constellation descriptions to life, and set an artistic style that was to be followed by many later cartographers. But the area around the south celestial pole remained empty of stars, the celestial equivalent of *terra incognita*, because it had been permanently below the horizon to the Greeks. Charting this blank area of the celestial sphere was the next task for astronomers.

Opposite An illustration of Perseus from a manuscript of the *Book of the Fixed Stars* by the Arabic astronomer Abd al-Rahman al-Sūfī written around 964 CE. This manuscript was a revised and updated version of Ptolemy's *Almagest* and contained the same 48 constellations as recognised by the Greeks. **Above** Chart of the northern hemisphere constellations drawn by Albrecht Dürer in 1515, depicting the constellations as described by Ptolemy in the *Almagest*. The constellations of the zodiac are arranged around the rim.

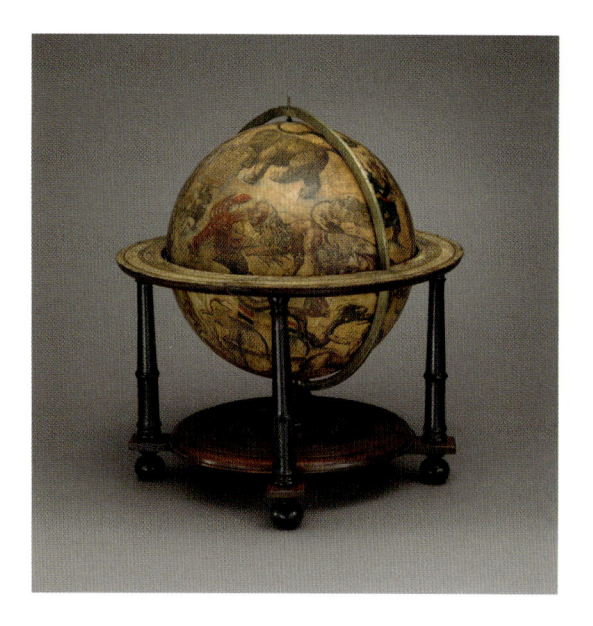

Exploring the southern sky

Petrus Plancius (1552–1622), a distinguished Dutch cartographer, was determined to fill in the blank area of sky around the south celestial pole. To help him, he instructed several members of the first Dutch trading expedition to the East Indies in 1595 to observe the southern stars during their voyage. Foremost among these was the chief navigator, Pieter Dirkszoon Keyser (c.1540–96).

Keyser died during the voyage but his list of over 100 southern stars was delivered to Plancius when the fleet returned to Holland in August 1597. These stars appeared on a celestial globe made by Plancius in 1598, divided into 12 new constellations representing some of the wonderful things that the explorers had seen on their travels, such as a peacock, a flying fish, and a bird of paradise. Plancius himself invented three constellations that appeared on his globes, namely Camelopardalis, the Giraffe; Columba, the Dove; and Monoceros, the Unicorn, and was also the first to show the Southern Cross as a separate figure. These constellations were the first significant additions to the sky since Ptolemy's time.

Another member of that first expedition, Frederick de Houtman (1571–1627), made a second foray to the far east, during which he expanded Keyser's list of southern stars. De Houtman published his catalog on his return to Holland in 1603, and his observations, divided into the same 12 constellations as those of Keyser, first appeared that same year on a celestial globe by the Dutch cartographer Willem Janszoon Blaeu (1571–1638), a rival of Plancius.

Johann Bayer's *Uranometria*

At the same time as the far southern heavens were being mapped, a major new catalog of the northern sky was nearing completion. It was compiled by a Danish astronomer, Tycho Brahe (1546–1601), who is regarded as the finest observer of pre-telescopic times. Tycho's observations, ten times more accurate than any before him, formed the basis of the first great star atlas, the *Uranometria*, published in 1603 by Johann Bayer (1572–1625), a lawyer and keen amateur astronomer in Augsburg, Germany.

Bayer's atlas devoted a page to each of the 48 Ptolemaic constellations, with an imaginative engraving of each figure. He supplemented star positions and brightnesses from Tycho's catalog with his own observations, although he had to fall back on the *Almagest* for the more southerly stars that could not be seen from northern Europe. The *Uranometria* also included a plate depicting the 12 new southern constellations invented by the Dutch, thereby making it the first star atlas to cover the entire sky.

Bayer's atlas was also notable for another reason: it introduced the system of labeling the main stars in each constellation with Greek letters, a system that rapidly caught on and is still in use today. For more about stellar nomenclature, see "Naming the Stars" on page 26.

Johannes Hevelius and seven new constellations

As instrumentation improved, so did the accuracy with which star positions could be measured. Johannes Hevelius (1611–87), a wealthy Polish brewer, decided to create a new catalog, larger and more accurate than Tycho's, for which he built new and highly accurate sighting instruments. From a rooftop observatory above his three adjoining houses in Danzig (present-day Gdańsk), Hevelius recorded the positions and brightnesses of over 1,500 stars, 50 percent more than Ptolemy or Tycho.

Top This celestial globe by the Dutch cartographer Willem Janszoon Blaeu, first issued in 1603, included the 12 new southern constellations invented by the Dutch navigators who first charted the southern sky at the end of the 16th century.

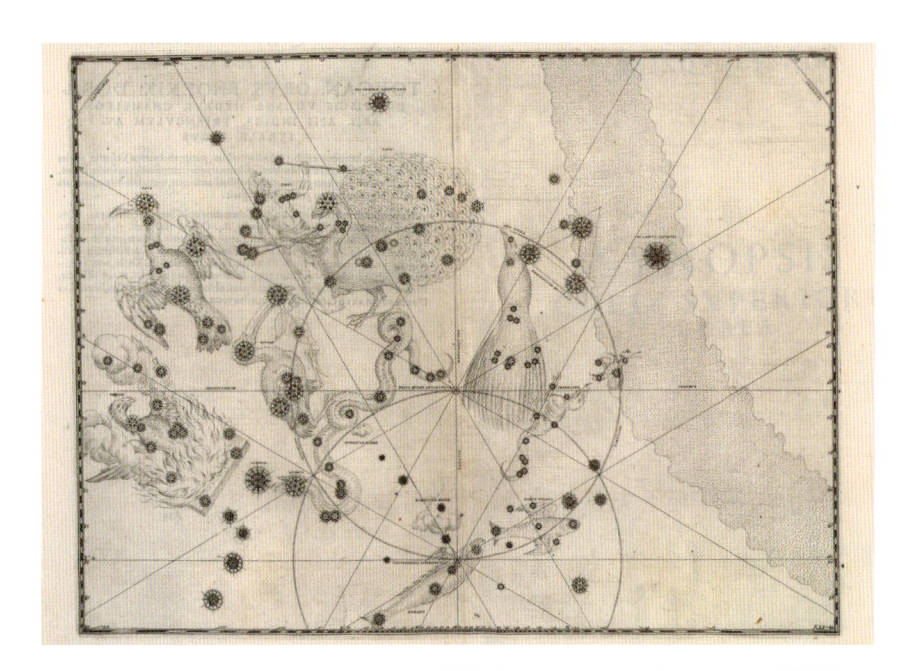

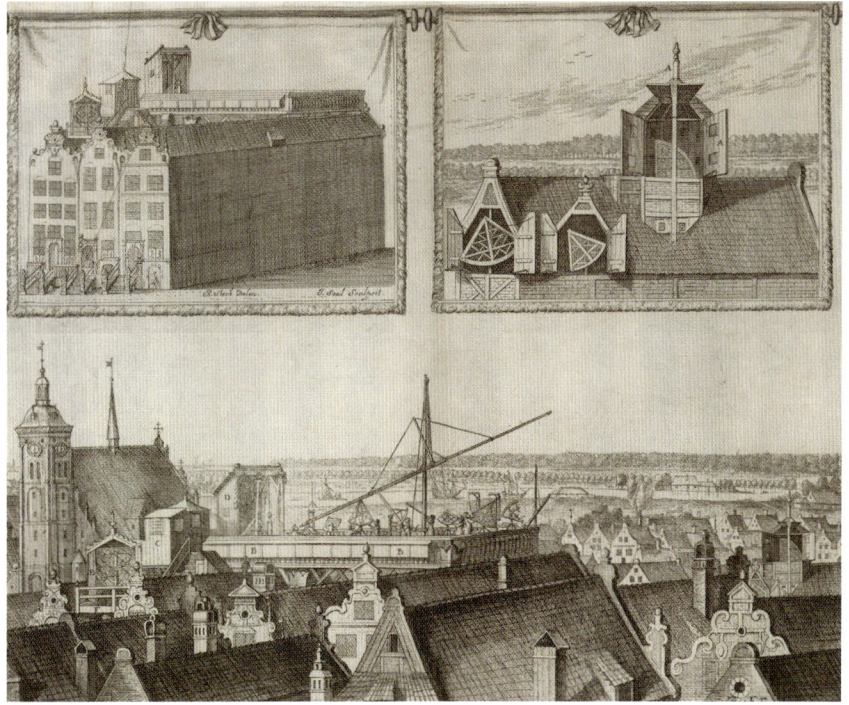

Top Johann Bayer's *Uranometria*, of 1603, was the first great star atlas. It included this chart of the 12 new southern constellations depicting exotic creatures seen by Dutch navigators during their voyages of exploration to the East Indies. **Bottom** Johannes Hevelius, a 17th-century Polish astronomer, built a rooftop observatory over his three adjacent houses. From here, he mapped the Moon and compiled a major catalog of over 1,500 stars.

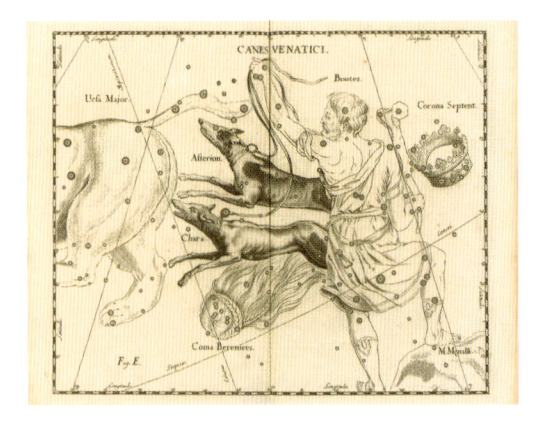

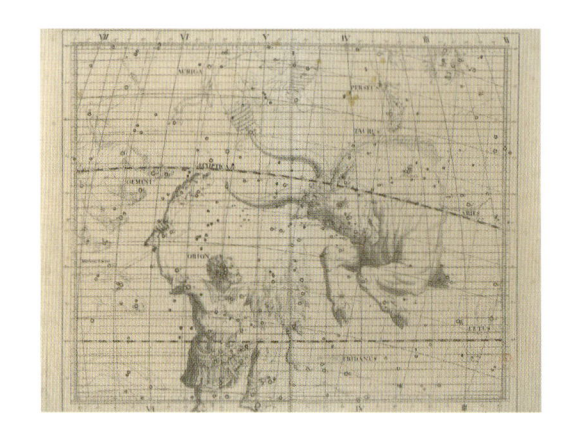

Hevelius owned various telescopes, which he used for observing the Moon, Sun, and planets. But for measuring star positions he preferred to follow Tycho by using naked-eye sights on quadrants and sextants. Because of the quality of these instruments, and his skill as an observer, Hevelius's star positions were a significant improvement on those of Tycho.

Hevelius's immense star catalog, *Catalogus Stellarum Fixarum*, the product of some 30 years of labor, was published posthumously in 1690, accompanied by an atlas with plates of each constellation, *Firmamentum Sobiescianum*. In his atlas and catalog, Hevelius introduced ten new constellations to fill in the gaps between the existing Ptolemaic figures. Three of these were not adopted by other astronomers, namely Cerberus (the triple-headed monster that guarded the gates of Hades), Mons Maenalus (a mountain beneath the feet of Boötes), and Triangulum Minus (a small companion to Triangulum proper), but the remaining seven were and now form part of the officially recognized pantheon of 88 (see table on page 66).

In his atlas, Hevelius drew the constellations as they appear on a globe, as Dürer had done, so they appear back to front by comparison with the way we see them from Earth. This made it difficult for observers to match up the figures with the real sky. Another drawback of his atlas was that Hevelius did not label the stars in any way, such as with Bayer letters, so it can be difficult to identify many of the stars at a glance. All major star atlases after Hevelius turned the figures face-on.

John Flamsteed's *Atlas Coelestis*

The next advance in charting the sky was made in Britain. In 1675, King Charles II of England appointed John Flamsteed (1646–1719) as the first Astronomer Royal at the new national observatory at Greenwich, near London. His main task was to create a new star catalog to assist navigation at sea. The result, published posthumously in 1725, was the first major star catalog produced with the aid of a telescope. It contained the positions of nearly 3,000 stars visible from Greenwich down to 8th magnitude, measured with unprecedented precision. Accompanying the catalog was a set of 25 elegantly engraved charts called the *Atlas Coelestis*.

Whereas Bayer and Hevelius had devoted a page of their atlases to a single constellation, Flamsteed's atlas introduced the modern style of dividing the sky into sections, each containing more than one constellation. As a result, some constellations overlap between charts—see, for example, the tableau of Orion facing Taurus. As well as the Ptolemaic constellations visible from Greenwich, Flamsteed's atlas included six of the new figures invented by Hevelius (he left out Scutum, which was later reinstated by Johann Bode.)

Flamsteed's catalog and atlas rapidly became standard reference works. As was the usual practice at the time, he listed his stars on a constellation-by-

Top left Johannes Hevelius introduced ten new constellations on his star atlas, *Firmamentum Sobiescianum*, published posthumously in 1690. Seven of his inventions are still recognized today, including this one, Canes Venatici, representing a pair of hunting dogs. **Top right** John Flamsteed's *Atlas Coelestis*, published posthumously in 1729, was the first star atlas compiled with the aid of a telescope, plotting many more stars than previous atlases and with greater accuracy. This chart depicts Orion and Taurus apparently engaged in combat. **Opposite** Nicolas Louis de Lacaille spent a year observing the southern skies from the Cape of Good Hope in 1751–2. On his return, he presented the French Royal Academy of Sciences with a star chart from which the painting shown here was made. It includes 14 new southern constellations that he invented.

constellation basis, with the stars in each constellation being ordered by right ascension, the celestial equivalent of longitude on Earth. In a French edition of Flamsteed's catalog published in 1783, the French astronomer Joseph Jérôme de Lalande (1732–1807) inserted a column in which he allocated numbers to the stars within each constellation. These numbers were subsequently adopted as a convenient means of identifying stars that did not have Bayer letters, and they are now referred to as Flamsteed numbers, even though they were not actually introduced by Flamsteed himself

Hevelius had filled in the gaps in the northern sky, but knowledge of the southern sky remained sketchy. To remedy this, a French astronomer named Nicolas Louis de Lacaille (1713–62) sailed to South Africa in 1750. He set up a small observatory near the famous Table Mountain at the Cape of Good Hope, which housed a tiny telescope of only half an inch (13.5-millimeter) aperture, mounted on a quadrant. With this simple instrument he recorded the positions and brightnesses of nearly 10,000 stars over the course of a year, an astounding total in that short time.

Lacaille created 14 new southern constellations out of these stars, which first appeared on a painted chart he presented to the French Royal Academy of Sciences in 1754. Whereas the Dutch navigators had named most of their new constellations after exotic animals, Lacaille's were named after instruments from science and the arts such as Antlia, the Air Pump; Fornax, the Chemical Furnace; and Pictor, the Painter's Easel. The exception was Mensa, commemorating the Table Mountain under which he had carried out his observations. Lacaille also divided up the unwieldy Greek constellation of Argo Navis, the Ship of the Argonauts, into the subsections Carina, the Keel or Hull; Puppis, the Stern; and Vela, the Sails. These are still recognized as separate figures today, as are all Lacaille's other inventions (see table on page 66).

Johann Bode's *Uranographia*

Celestial mapping reached new heights with an atlas called *Uranographia* (not to be confused with Johann Bayer's earlier *Uranometria*) published in 1801 by the German astronomer Johann Elert Bode (1747–1826), director of the Berlin Observatory. Over 17,000 stars down to 8th magnitude are plotted, taken from the observations of astronomers including Flamsteed, Lacaille, Lalande, and Bode himself.

Bode's *Uranographia* covered the sky in 20 beautifully engraved plates and is arguably the finest pictorial star atlas ever produced. As well as depicting the established constellations of Ptolemy, Plancius, Hevelius, and Lacaille, Bode included more recent inventions by several other astronomers. Three of them he devised himself: Officina Typographica, the Printing Shop, commemorating the invention of printing; Machina Electrica, the Electrostatic Generator; and Lochium Funis, the Ship's Log and Line. Two others, Globus Aerostaticus, the Hot Air Balloon, and Felis the Cat, were suggested by the Frenchman Lalande. However, none of these constellations were officially adopted and Lacaille's 14 southern constellations remain the last recognized additions to the sky.

One notable innovation on Bode's atlas was the introduction of constellation borders, represented by snaking lines. Until then, there had been nothing on charts to indicate where one constellation might end and the next one begin. Other cartographers adopted Bode's idea, but the freehand borders they drew were ill-defined and varied from atlas to atlas.

An additional complication was that certain stars had traditionally been shared between constellations. For example, according to Ptolemy's description, the star at the tip of the bull's horn in Taurus also served to mark the right foot of Auriga, the Charioteer. Bayer labeled it twice, as Beta Tauri on his chart of Taurus, and Gamma Aurigae on his chart of Auriga. As another example, Alpha Andromedae, the head of

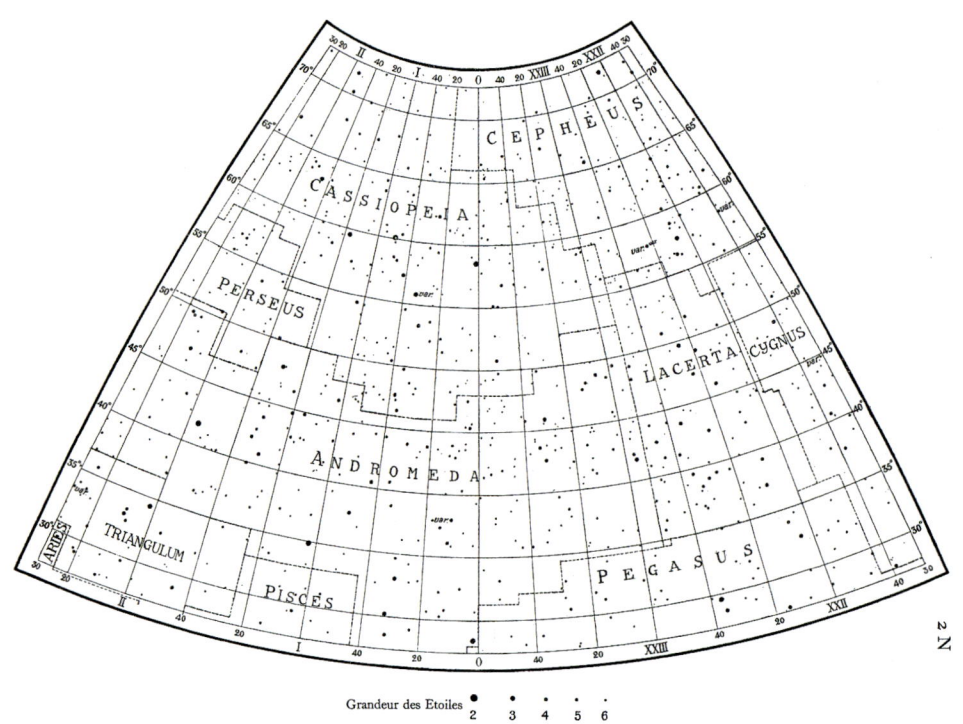

Andromeda, was also part of the body of Pegasus, where it was labeled by Bayer as Delta Pegasi.

Properly defined boundaries were called for, along with a list of officially agreed constellations, and that was one of the first tasks of the International Astronomical Union (IAU), astronomy's governing body, after its formation in 1920.

The final 88

At its first General Assembly in 1922, the IAU adopted the list of 88 constellations that we still use today, with officially approved names and three-letter abbreviations (see table on page 66). No longer could an enterprising cartographer tuck a new constellation into an overlooked corner of the sky in the hope of achieving celestial immortality. Devising boundaries for these 88 figures, though, required more thought.

The IAU entrusted the job to Eugène Delporte (1882–1955), a Belgian astronomer at the Royal Observatory in Brussels. Since there are no natural geographical features in the sky like those on Earth that can be used as borders, Delporte decided to divide up the sky along coordinate lines of right ascension and declination, resulting in the zig-zag outlines that we know today. Delporte's results, published by the IAU in 1930 in a book called *Délimitation scientifique des constellations*, amount to an international treaty on the demarcation of the sky.

Sadly, the pictorial figures that decorated the old star atlases have vanished, although a trace of them can still be found in the lines that link the main stars of a constellation to indicate its overall shape. However, using our imagination on a clear, dark night, we can still visualize the beasts and heroes of Greek legend that once populated the sky.

The Chinese sky

Other civilizations had different constellations from those in the West, and the best-known are those from China, where the skies were closely monitored for astrological purposes. By the end of the third century CE, Chinese astronomers had an elaborate system of 283 constellations, depicting facets of everyday Chinese life rather than the myths of the West. Chinese constellations were small, and usually consisted of far fewer stars than Western ones.

The example shown here is the north polar region of the sky drawn on a paper scroll that was found in the early twentieth century in caves at Dunhuang, on the Silk Road trade route in China. Dating from the seventh century CE, it is the oldest surviving paper star map in the world.

The familiar shape of the Plough, or Big Dipper, can be made out below center, since the Chinese saw a similar pattern here to what Western astronomers saw, but little else is recognizable. The dots on Chinese maps were not graded in size according to the brightness of the star, which adds to the difficulty of identification.

Jesuit missionaries introduced Western constellations to China in the seventeenth century, after which the indigenous Chinese constellation system died out.

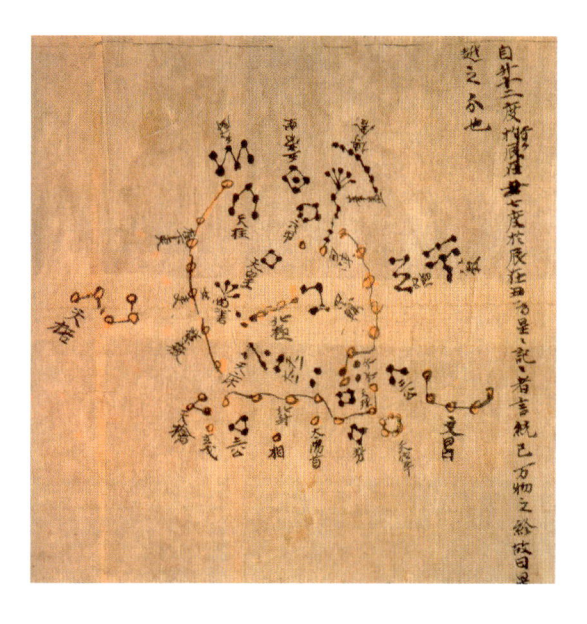

Opposite top Plate XI from Johann Bode's great star atlas, called *Uranographia*, published in 1801. Among the constellations on this chart are Aries, the ram, and Pisces, the fishes. **Opposite bottom** Official boundaries to the constellations were fixed in 1930 by a Belgian astronomer, Eugène Delporte, acting on behalf of the International Astronomical Union. This is his chart for part of the northern sky, including Cassiopeia and Andromeda. The constellation boundaries follow lines of right ascension and declination. **Right** Ancient Chinese constellations differed markedly from Western constellations, being usually much smaller and including fainter stars. This chart of the north polar region of the sky dates from the 7th century CE and was found in caves at Dunhuang, China, along the Silk Road. Among the star patterns shown here only the familiar shape of the Plough, or Big Dipper.

THE 88 CONSTELLATIONS

Name	Genitive	Abbrevn.	Area (square degs.)	Order of size	Origin*
Andromeda	Andromedae	And	722	19	1
Antlia	Antliae	Ant	239	62	6
Apus	Apodis	Aps	206	67	3
Aquarius	Aquarii	Aqr	980	10	1
Aquila	Aquilae	Aql	652	22	1
Ara	Arae	Ara	237	63	1
Aries	Arietis	Ari	441	39	1
Auriga	Aurigae	Aur	657	21	1
Boötes	Boötis	Boo	907	13	1
Caelum	Caeli	Cae	125	81	6
Camelopardalis	Camelopardalis	Cam	757	18	4
Cancer	Cancri	Cnc	506	31	1
Canes Venatici	Canum Venaticorum	CVn	465	38	5
Canis Major	Canis Majoris	CMa	380	43	1
Canis Minor	Canis Minoris	CMi	183	71	1
Capricornus	Capricorni	Cap	414	40	1
Carina	Carinae	Car	494	34	6
Cassiopeia	Cassiopeiae	Cas	598	25	1
Centaurus	Centauri	Cen	1060	9	1
Cepheus	Cephei	Cep	588	27	1
Cetus	Ceti	Cet	1231	4	1
Chamaeleon	Chamaeleontis	Cha	132	79	3
Circinus	Circini	Cir	93	85	6
Columba	Columbae	Col	270	54	4
Coma Berenices	Comae Berenices	Com	386	42	2
Corona Australis	Coronae Australis	CrA	128	80	1
Corona Borealis	Coronae Borealis	CrB	179	73	1
Corvus	Corvi	Crv	184	70	1
Crater	Crateris	Crt	282	53	1
Crux	Crucis	Cru	68	88	4
Cygnus	Cygni	Cyg	804	16	1
Delphinus	Delphini	Del	189	69	1
Dorado	Doradus	Dor	179	72	3
Draco	Draconis	Dra	1083	8	1
Equuleus	Equulei	Equ	72	87	1
Eridanus	Eridani	Eri	1138	6	1
Fornax	Fornacis	For	398	41	6
Gemini	Geminorum	Gem	514	30	1
Grus	Gruis	Gru	366	45	3
Hercules	Herculis	Her	1225	5	1
Horologium	Horologii	Hor	249	58	6
Hydra	Hydrae	Hya	1303	1	1
Hydrus	Hydri	Hyi	243	61	3
Indus	Indi	Ind	294	49	3
Lacerta	Lacertae	Lac	201	68	5
Leo	Leonis	Leo	947	12	1
Leo Minor	Leonis Minoris	LMi	232	64	5
Lepus	Leporis	Lep	290	51	1

Name	Genitive	Abbrevn.	Area (square degs.)	Order of size	Origin*
Libra	Librae	Lib	538	29	1
Lupus	Lupi	Lup	334	46	1
Lynx	Lyncis	Lyn	545	28	5
Lyra	Lyrae	Lyr	286	52	1
Mensa	Mensae	Men	153	75	6
Microscopium	Microscopii	Mic	210	66	6
Monoceros	Monocerotis	Mon	482	35	4
Musca	Muscae	Mus	138	77	3
Norma	Normae	Nor	165	74	6
Octans	Octantis	Oct	291	50	6
Ophiuchus	Ophiuchi	Oph	948	11	1
Orion	Orionis	Ori	594	26	1
Pavo	Pavonis	Pav	378	44	3
Pegasus	Pegasi	Peg	1121	7	1
Perseus	Persei	Per	615	24	1
Phoenix	Phoenicis	Phe	469	37	3
Pictor	Pictoris	Pic	247	59	6
Pisces	Piscium	Psc	889	14	1
Piscis Austrinus	Piscis Austrini	PsA	245	60	1
Puppis	Puppis	Pup	673	20	6
Pyxis	Pyxidis	Pyx	221	65	6
Reticulum	Reticuli	Ret	114	82	6
Sagitta	Sagittae	Sge	80	86	1
Sagittarius	Sagittarii	Sgr	867	15	1
Scorpius	Scorpii	Sco	497	33	1
Sculptor	Sculptoris	Scl	475	36	6
Scutum	Scuti	Sct	109	84	5
Serpens	Serpentis	Ser	637	23	1
Sextans	Sextantis	Sex	314	47	5
Taurus	Tauri	Tau	797	17	1
Telescopium	Telescopii	Tel	252	57	6
Triangulum	Trianguli	Tri	132	78	1
Triangulum Australe	Trianguli Australis	TrA	110	83	3
Tucana	Tucanae	Tuc	295	48	3
Ursa Major	Ursae Majoris	UMa	1280	3	1
Ursa Minor	Ursae Minoris	UMi	256	56	1
Vela	Velorum	Vel	500	32	6
Virgo	Virginis	Vir	1294	2	1
Volans	Volantis	Vol	141	76	3
Vulpecula	Vulpeculae	Vul	268	55	5

*Origin:

1. One of the original 48 Greek constellations listed by Ptolemy. The Greek figure of Argo Navis has since been divided into Carina, Puppis, and Vela.
2. Considered by the Greeks as part of Leo; made separate by Caspar Vopel in 1536.
3. The 12 southern constellations of Pieter Dirkszoon Keyser and Frederick de Houtman, c. 1600.
4. Four constellations added by Petrus Plancius.
5. Seven constellations of Johannes Hevelius.
6. The 14 southern constellations of Nicolas Louis de Lacaille, who also divided the Greeks' Argo Navis into Carina, Puppis, and Vela.

CHAPTER TWO

STARGAZING FOR BEGINNERS

THE STAR CHARTS

Star charts have been created throughout history and across cultures to indicate the location of stars and other astronomical objects in the night sky and have been used for navigation, data collection, and stargazing. Modern charts offer a celestial map of these objects laid out on a grid system for easy reference. The pages that follow detail charts showing the complete northern and southern hemispheres and monthly star charts.

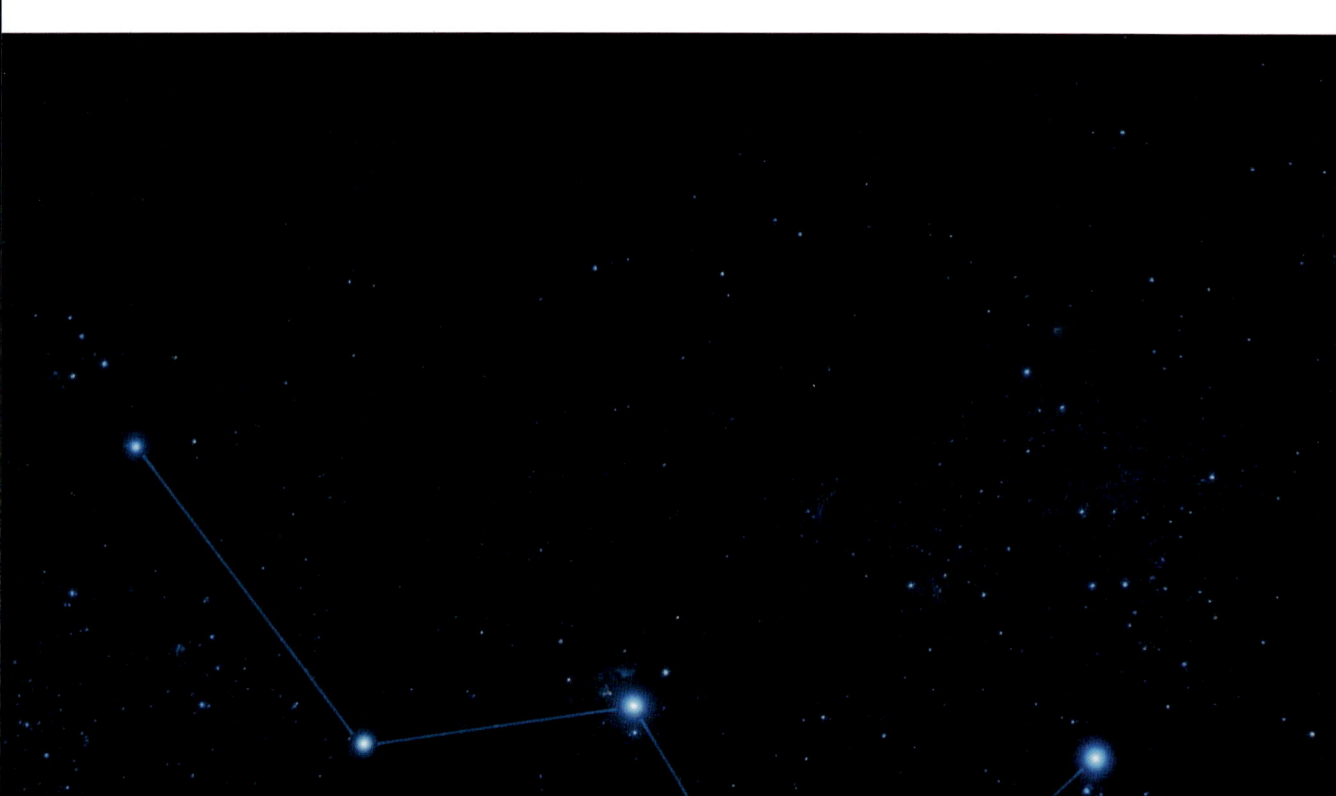

The hemisphere charts

On the following two pages are charts showing the complete northern and southern hemispheres of the sky. As well as the main stars of each hemisphere, these charts depict the hazy band of the Milky Way; the dashed red line is the ecliptic, the Sun's path in the heavens. When planets are visible, they will be found near the ecliptic.

Around the rim of each chart are listed the months of the year, to help you find which constellations are best placed at about 10 P.M. local time each month, or 11 P.M. when daylight-saving time (DST, or summer time) is in operation.

Observers in mid-northern latitudes should take the northern hemisphere chart and turn it so that the month of observation is at the bottom. The chart will show the sky that's visible when you face due south that evening. Rotate the chart 15° anticlockwise for each hour after 10 P.M., and turn it clockwise for each hour before 10 P.M.

Observers in mid-southern latitudes should take the southern hemisphere chart and turn it so that the month of observation is at the bottom. The chart will then show the stars as they appear when you are facing due north. Turn the chart 15° clockwise for each hour after 10 P.M., and 15° anticlockwise for each hour before.

In all cases above, for 10 P.M. read 11 P.M. when DST is in operation. Up-to-date information on the application of DST worldwide can be found on the following websites:

webexhibits.org/daylightsaving/
www.timeanddate.com/time/dst/

The monthly charts

Next comes a series of maps showing the sky as it appears when facing north or south at 10 P.M. (11 P.M. DST) in mid-month from various latitudes. The first set of maps is for use in the northern hemisphere from latitudes 60° to 10° north; the second set is for use from the equator to 50° south. (They will also be usable for about 10° either side of this range without significant error.) Curved lines on each map depict the horizon for each latitude. Taking these monthly charts in conjunction with those of the complete celestial hemispheres, you should be able to identify the stars in the sky no matter where you are on Earth.

The constellation charts

Starting on page 72, individual charts of each constellation can be found, accompanied by descriptions of the brightest stars and main objects of interest. All stars within each constellation down to magnitude 6.0 are shown, plus the most prominent examples of what are termed *deep-sky objects* (star clusters, nebulae, and galaxies). The total number of stars shown is about 5,000.

All constellation maps are to the same scale, with the exception of the rambling Hydra, the largest constellation of all, which is drawn to a significantly smaller scale to fit the page. Areas of particular interest in certain constellations, such as the Hyades and Pleiades clusters in Taurus, and the Orion Nebula, are shown to a larger scale in special detail charts.

We hope that the charts and descriptions in this book will serve as trusty companions for many nights of exploration under the stars. Good stargazing!

Previous pages Stargazing is an incredible opportunity to glimpse the universe from almost anywhere on Earth. Here, a stunning vista of the Milky Way can be seen above the Wadden Flats of Germany, along the Wadden Sea. **Opposite** Constellation charts show stars' location in the night sky, with a grid overlaid to show the constellation itself. Seen here is the constellation Cassiopeia, with its distinctive W-shape of its five brightest stars.

THE HEMISPHERE CHARTS

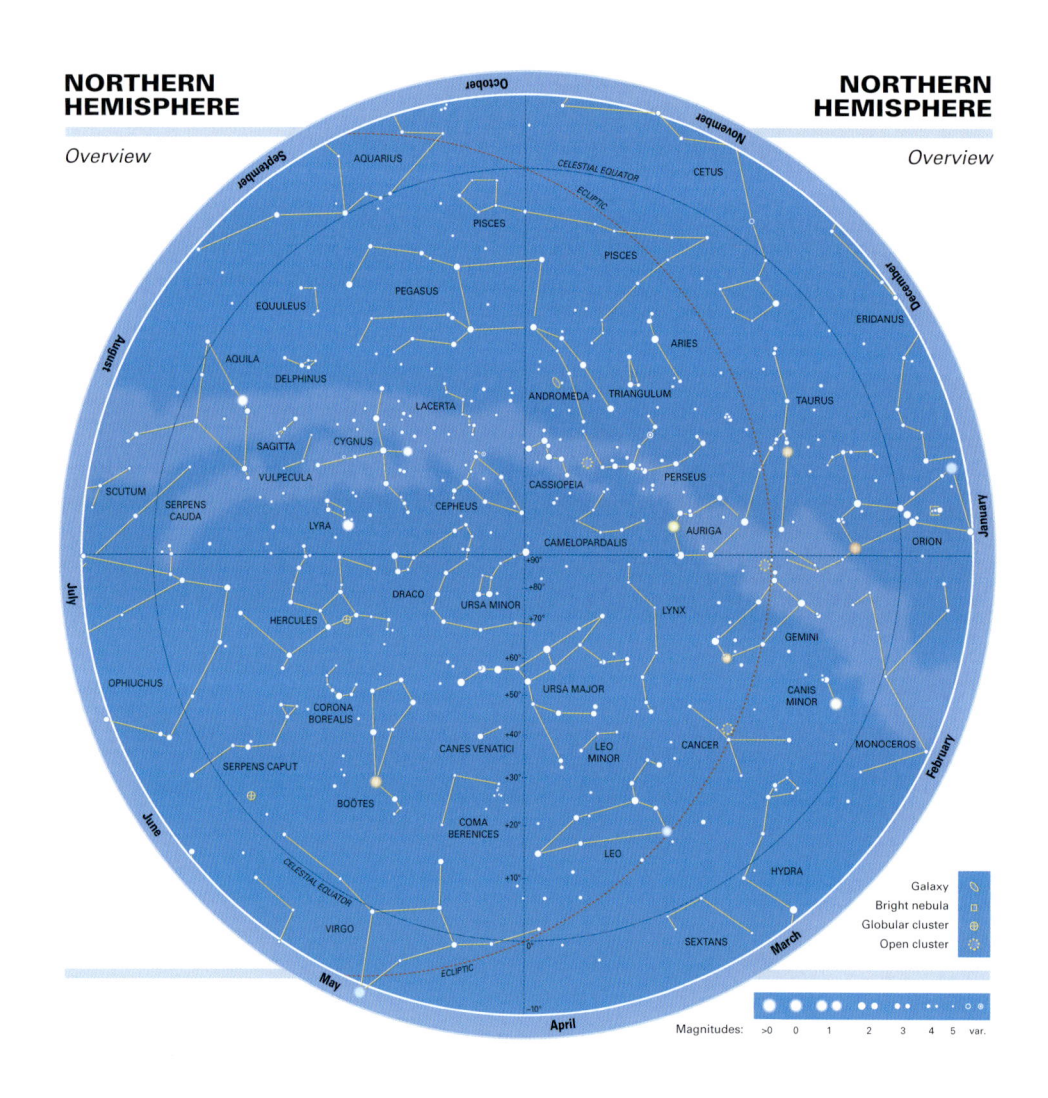

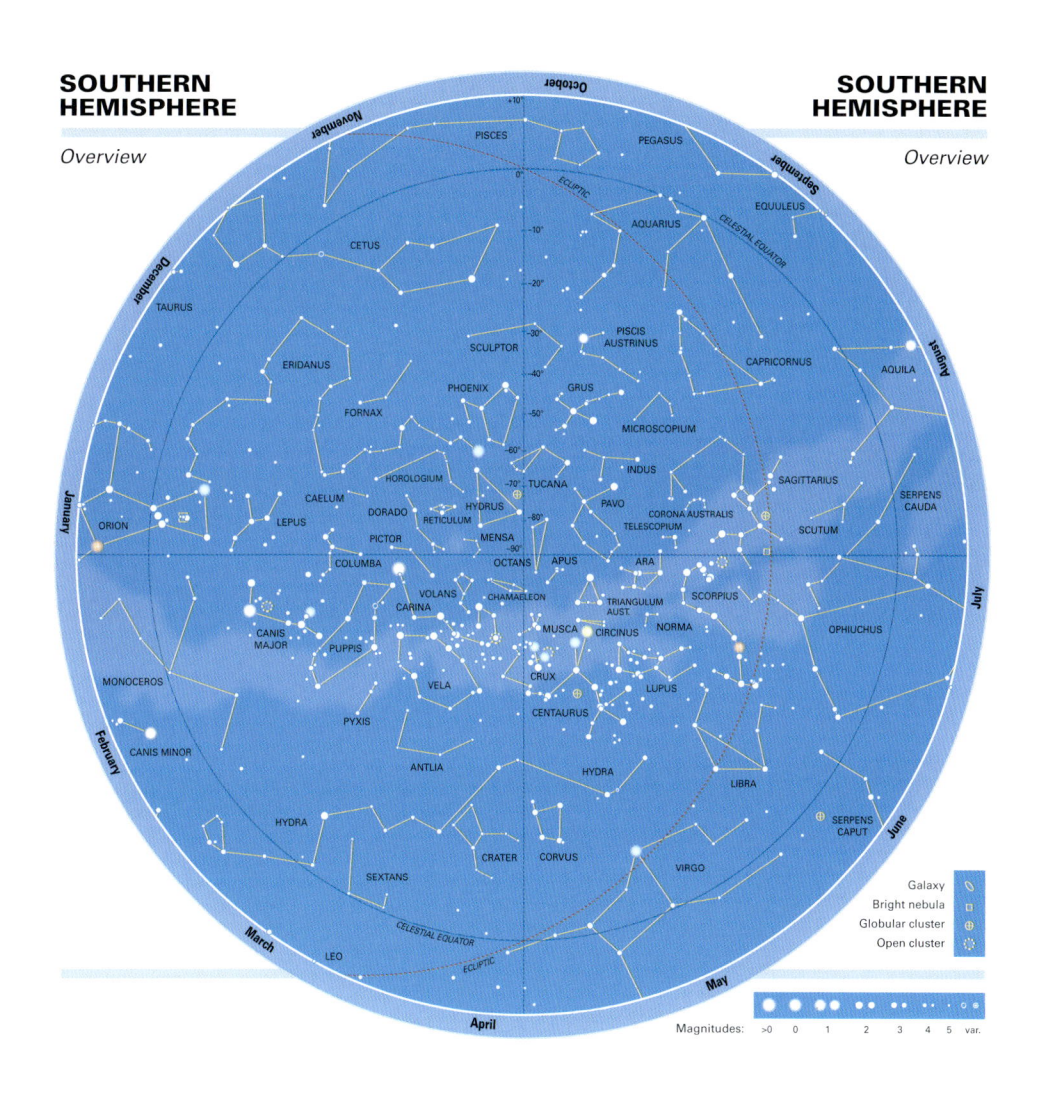

SOUTHERN HEMISPHERE
Overview

SOUTHERN HEMISPHERE
Overview

THE MONTHLY CHARTS

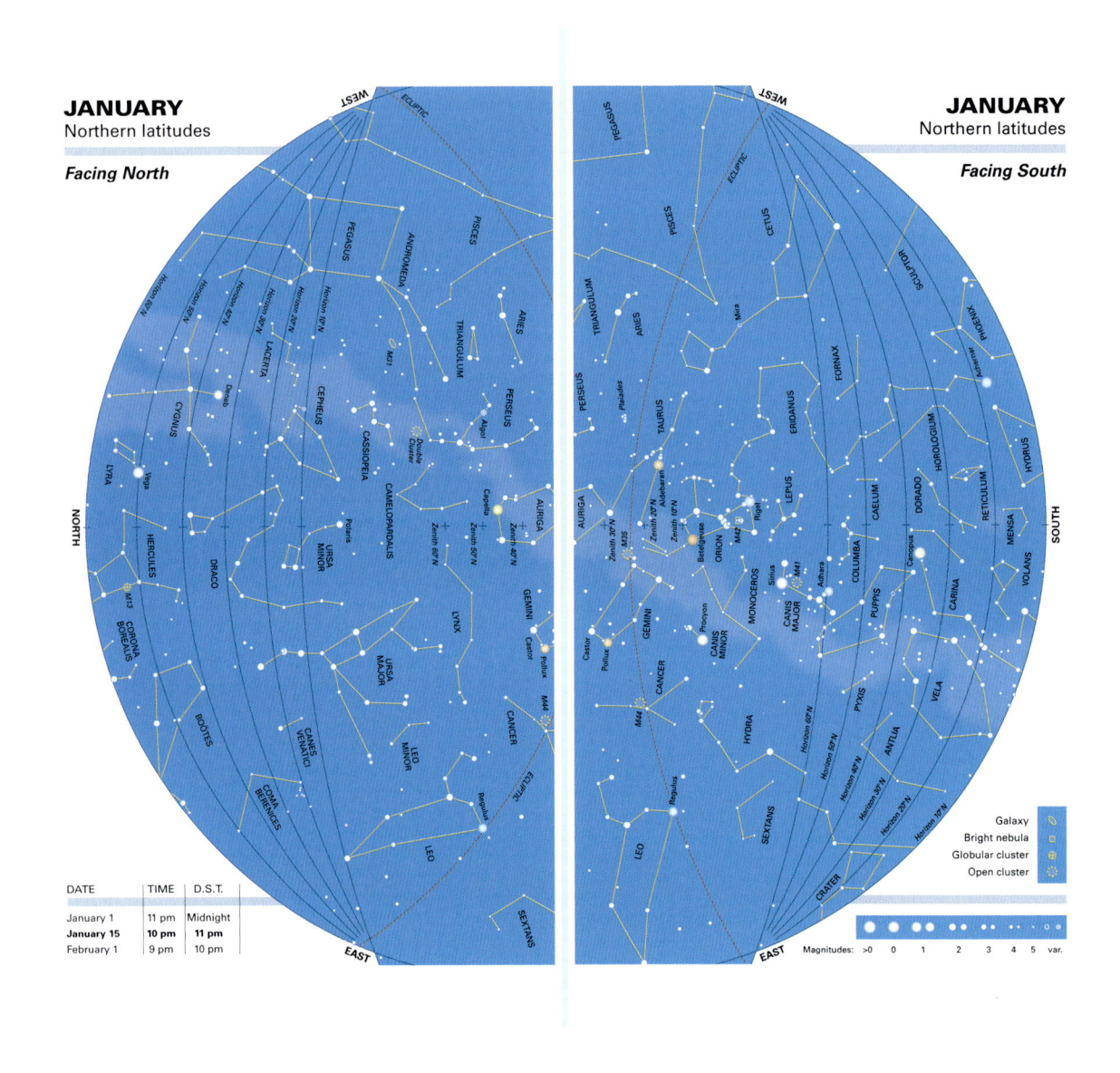

JANUARY
Northern latitudes

Facing North

JANUARY
Northern latitudes

Facing South

DATE	TIME	D.S.T.
January 1	11 pm	Midnight
January 15	**10 pm**	**11 pm**
February 1	9 pm	10 pm

Galaxy
Bright nebula
Globular cluster
Open cluster

Magnitudes: >0 0 1 2 3 4 5 var.

JANUARY
Southern latitudes

Facing North

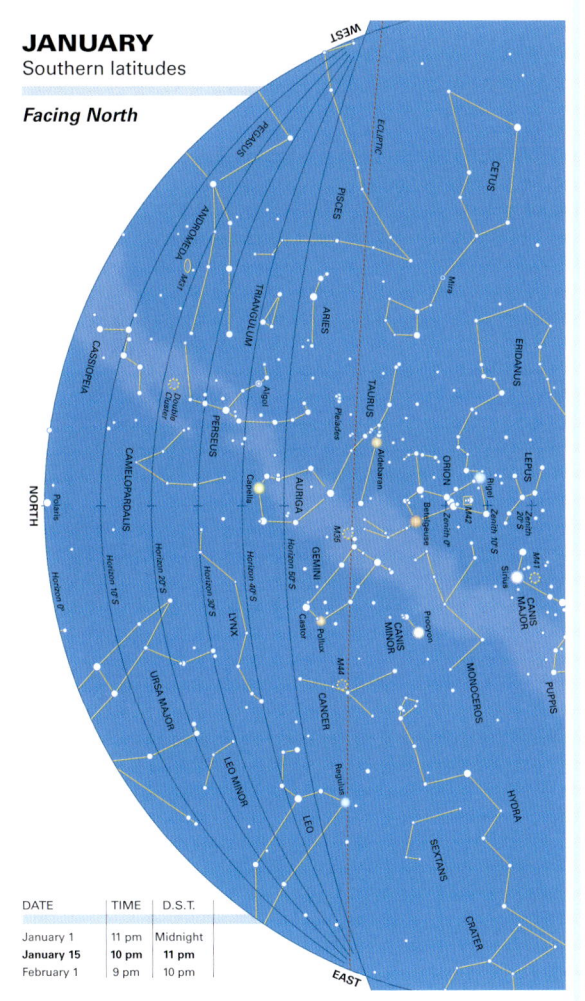

JANUARY
Southern latitudes

Facing South

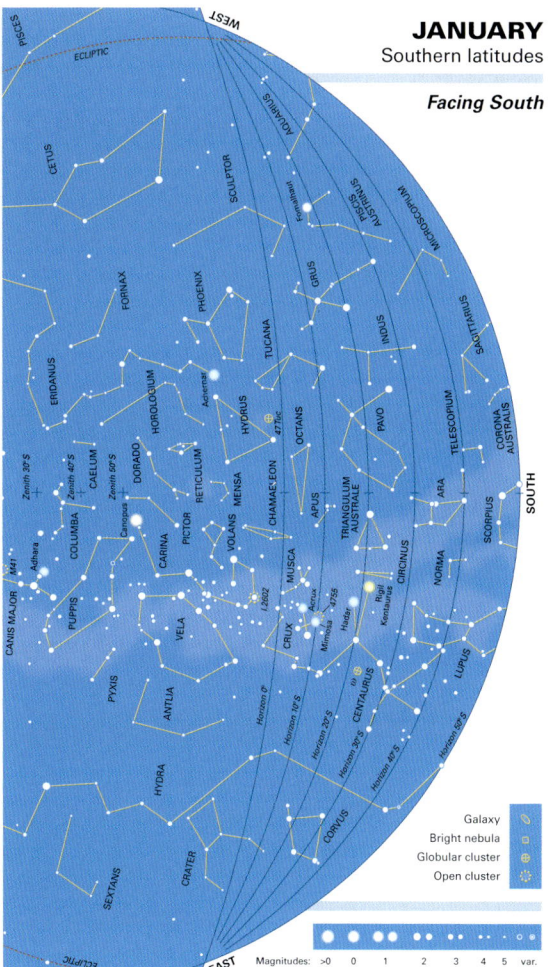

DATE	TIME	D.S.T.
January 1	11 pm	Midnight
January 15	**10 pm**	**11 pm**
February 1	9 pm	10 pm

Galaxy
Bright nebula
Globular cluster
Open cluster

Magnitudes: >0 0 1 2 3 4 5 var.

STARGAZING FOR BEGINNERS 75

FEBRUARY
Northern latitudes

Facing North

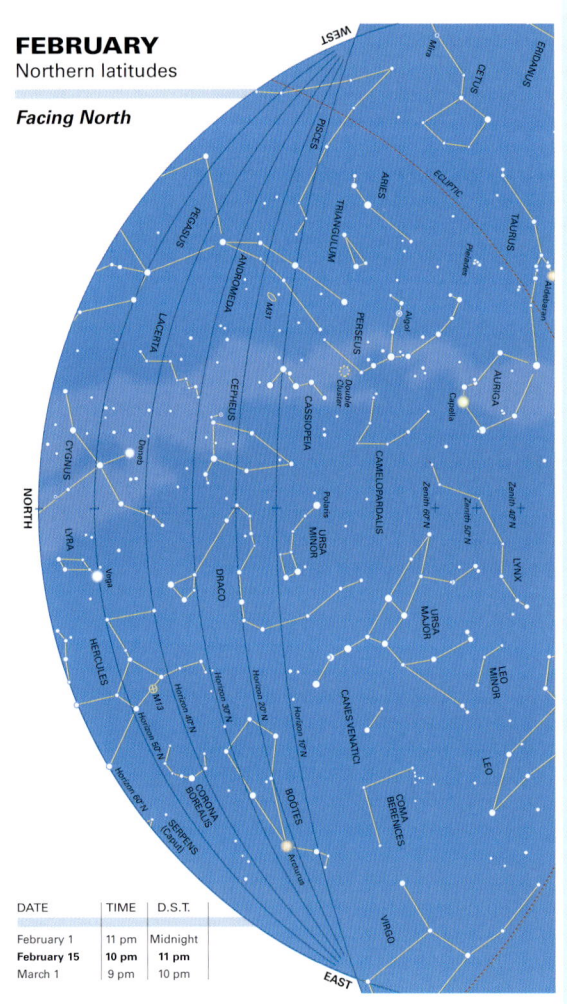

FEBRUARY
Northern latitudes

Facing South

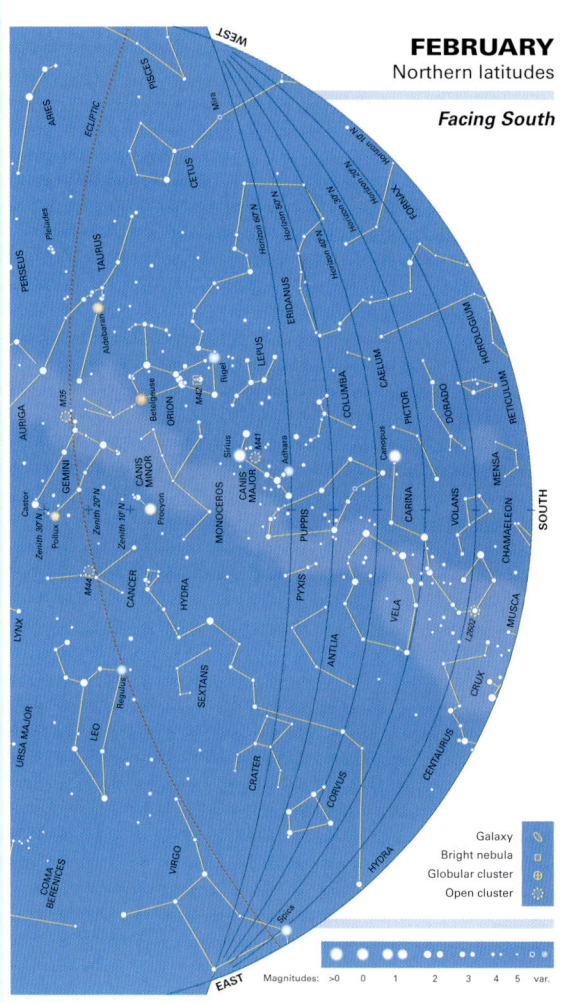

DATE	TIME	D.S.T.
February 1	11 pm	Midnight
February 15	**10 pm**	**11 pm**
March 1	9 pm	10 pm

Galaxy
Bright nebula
Globular cluster
Open cluster

Magnitudes: >0 0 1 2 3 4 5 var.

FEBRUARY
Southern latitudes

Facing North

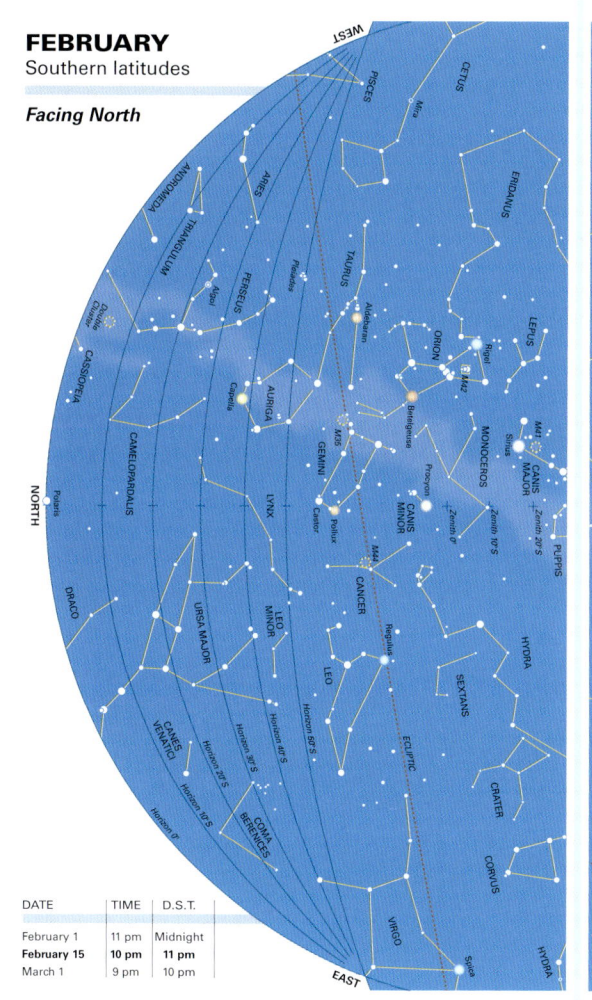

FEBRUARY
Southern latitudes

Facing South

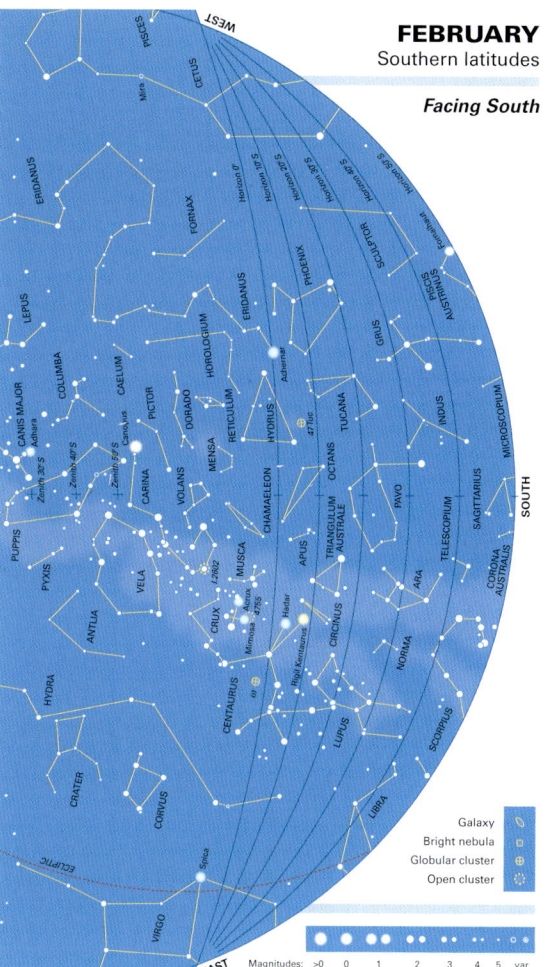

DATE	TIME	D.S.T.
February 1	11 pm	Midnight
February 15	**10 pm**	**11 pm**
March 1	9 pm	10 pm

Galaxy
Bright nebula
Globular cluster
Open cluster

Magnitudes: >0 0 1 2 3 4 5 var.

STARGAZING FOR BEGINNERS 77

MARCH
Northern latitudes

Facing North

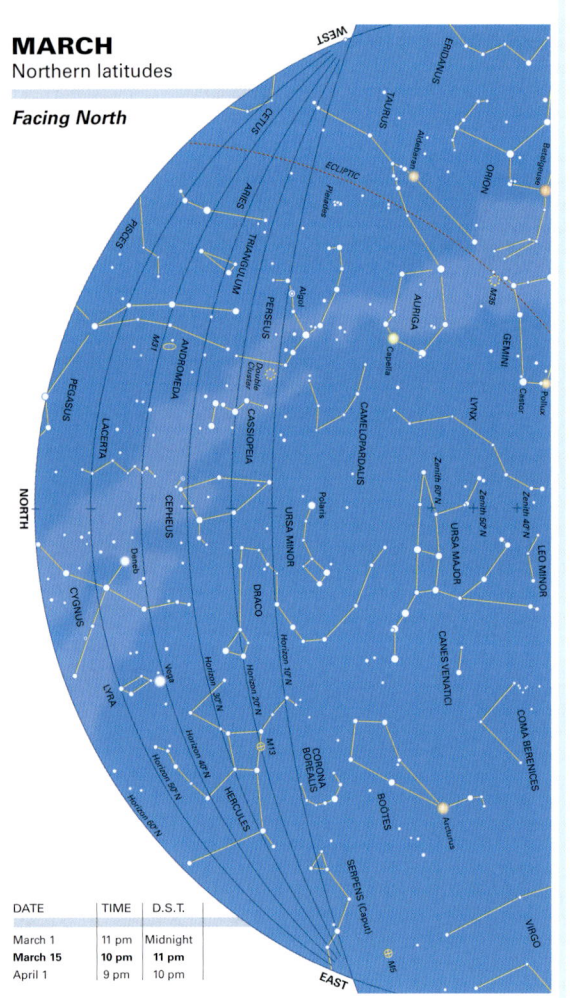

MARCH
Northern latitudes

Facing South

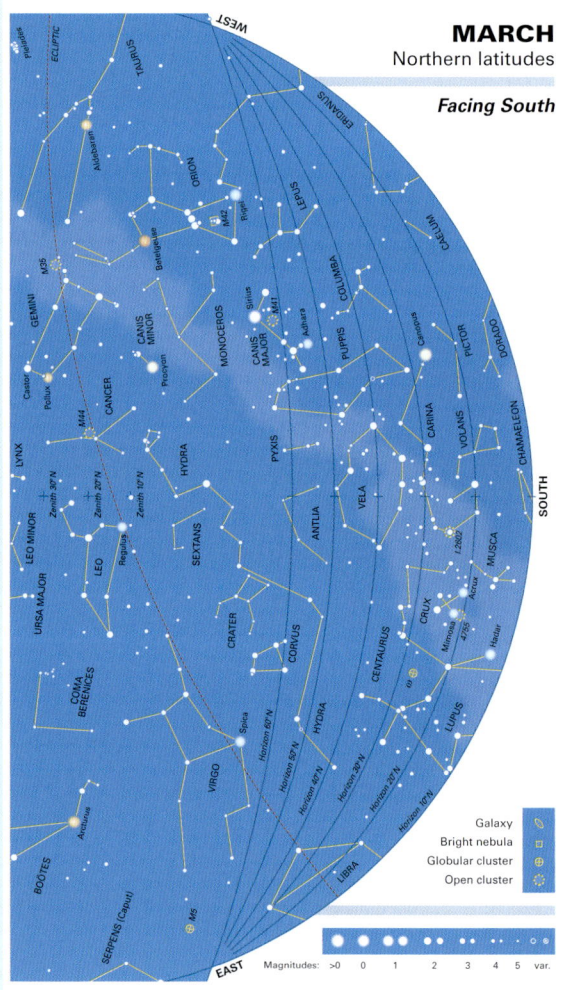

DATE	TIME	D.S.T.
March 1	11 pm	Midnight
March 15	10 pm	11 pm
April 1	9 pm	10 pm

Galaxy
Bright nebula
Globular cluster
Open cluster

Magnitudes: >0 0 1 2 3 4 5 var.

78 THE BACKYARD STARGAZER'S BIBLE

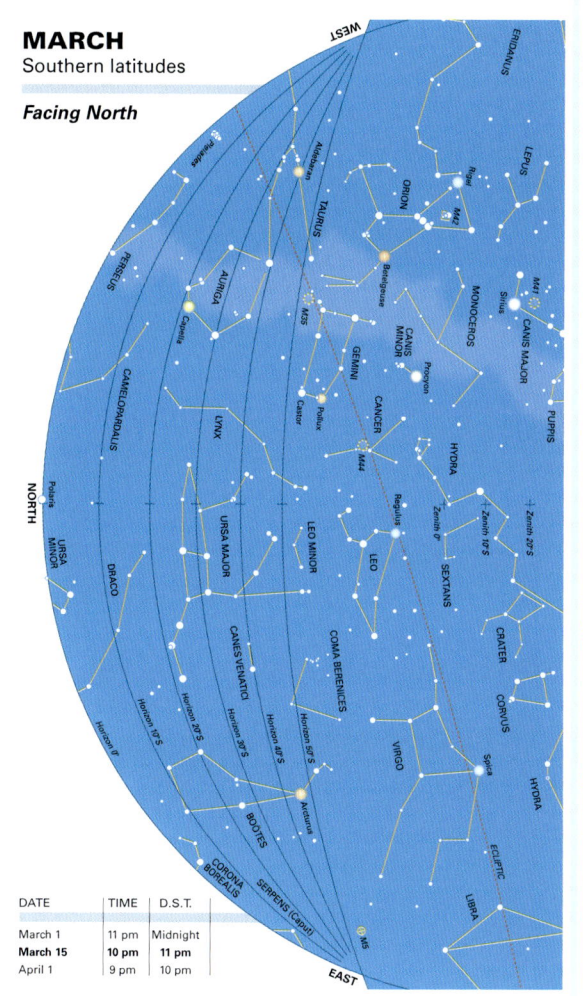

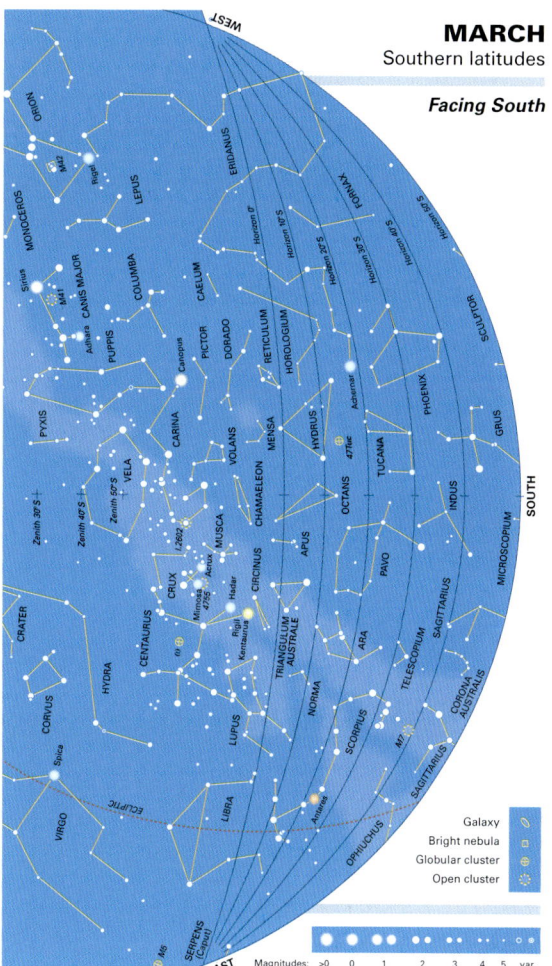

STARGAZING FOR BEGINNERS 79

APRIL
Northern latitudes

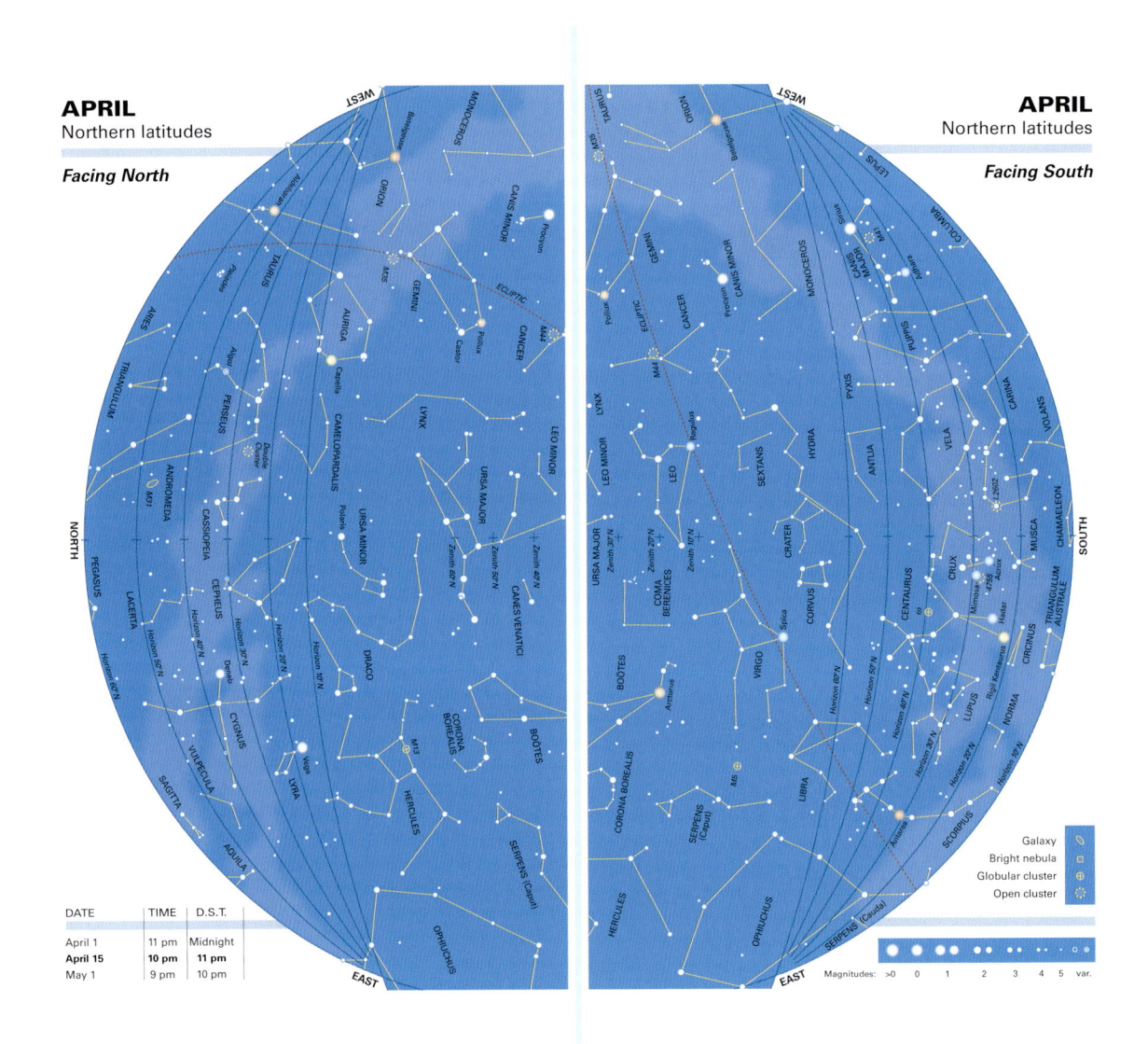

Facing North

APRIL
Northern latitudes

Facing South

DATE	TIME	D.S.T.
April 1	11 pm	Midnight
April 15	**10 pm**	**11 pm**
May 1	9 pm	10 pm

Galaxy
Bright nebula
Globular cluster
Open cluster

Magnitudes: >0 0 1 2 3 4 5 var.

80 THE BACKYARD STARGAZER'S BIBLE

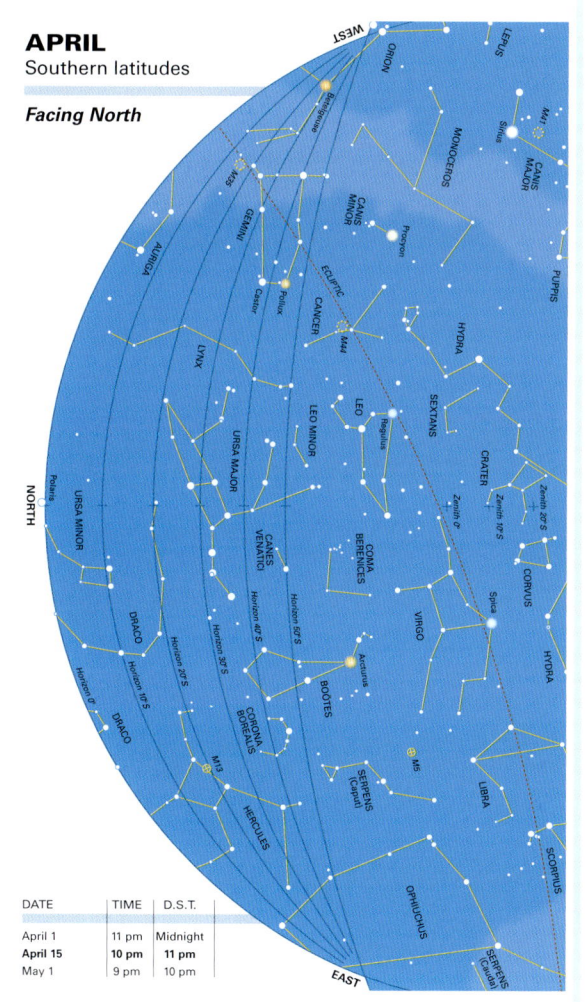

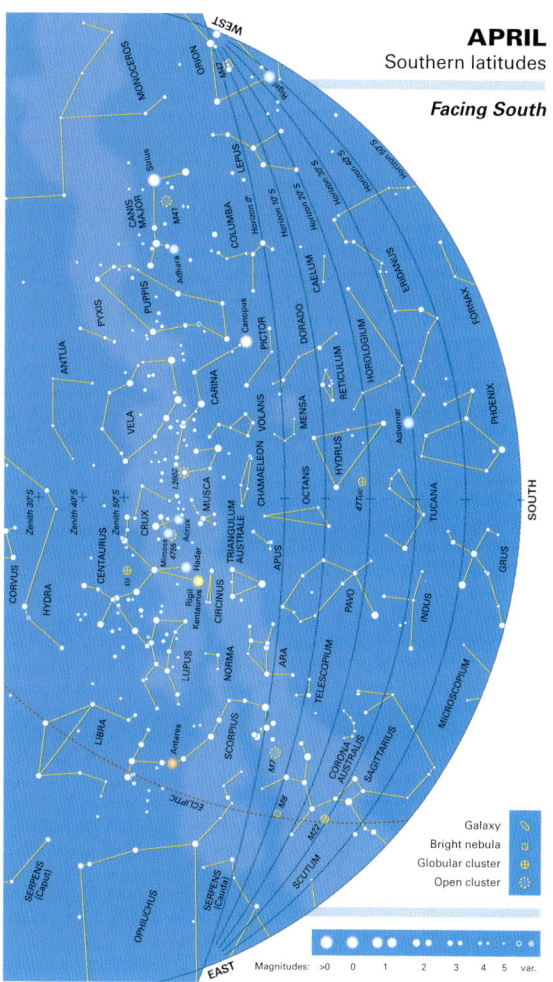

STARGAZING FOR BEGINNERS 81

MAY
Northern latitudes

Facing North

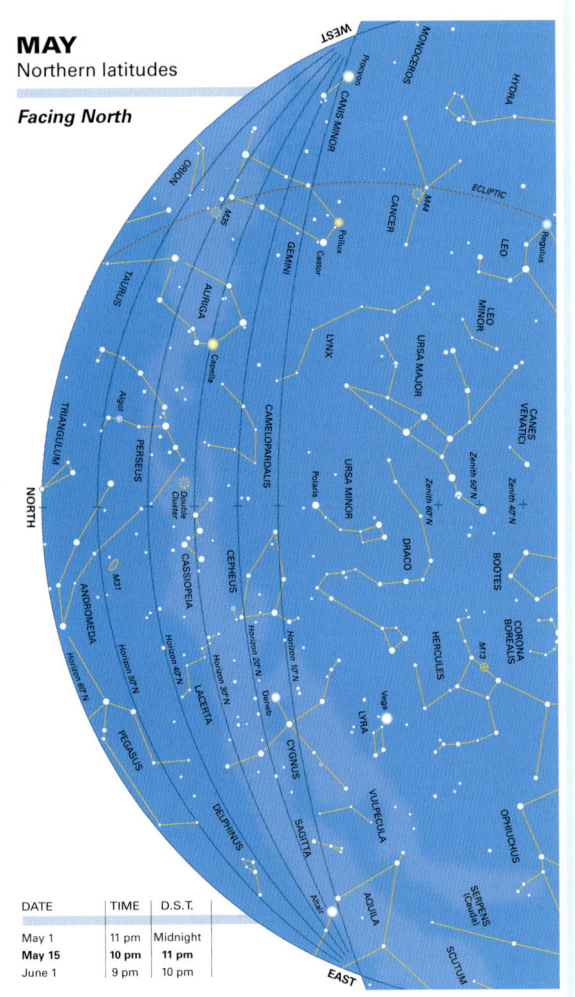

MAY
Northern latitudes

Facing South

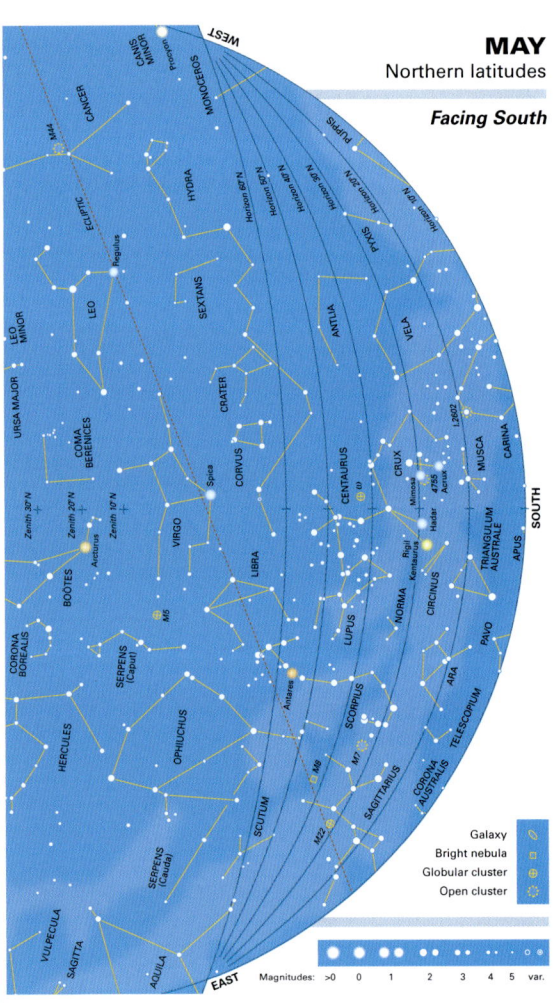

DATE	TIME	D.S.T.
May 1	11 pm	Midnight
May 15	10 pm	11 pm
June 1	9 pm	10 pm

Galaxy
Bright nebula
Globular cluster
Open cluster

Magnitudes: >0 0 1 2 3 4 5 var.

82 THE BACKYARD STARGAZER'S BIBLE

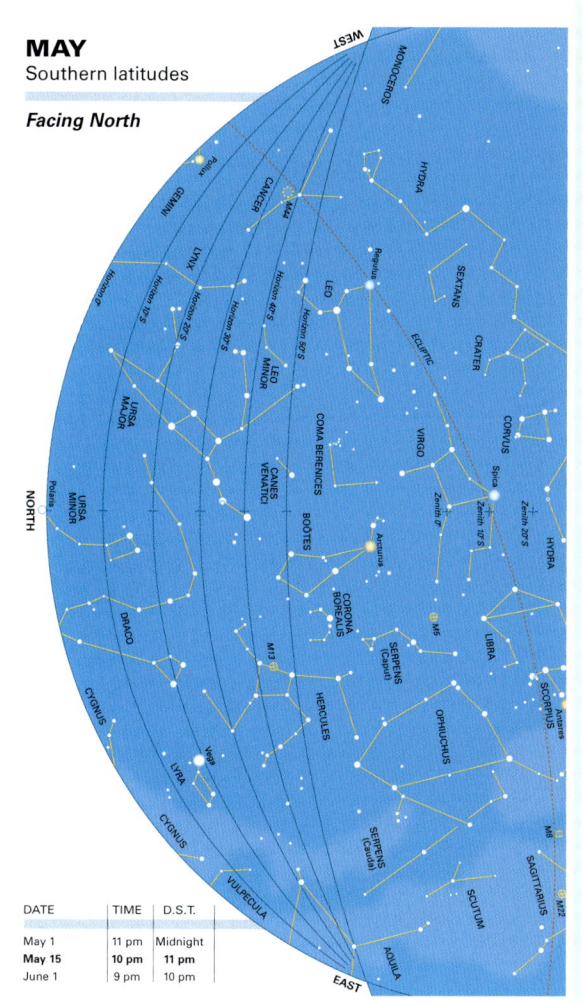

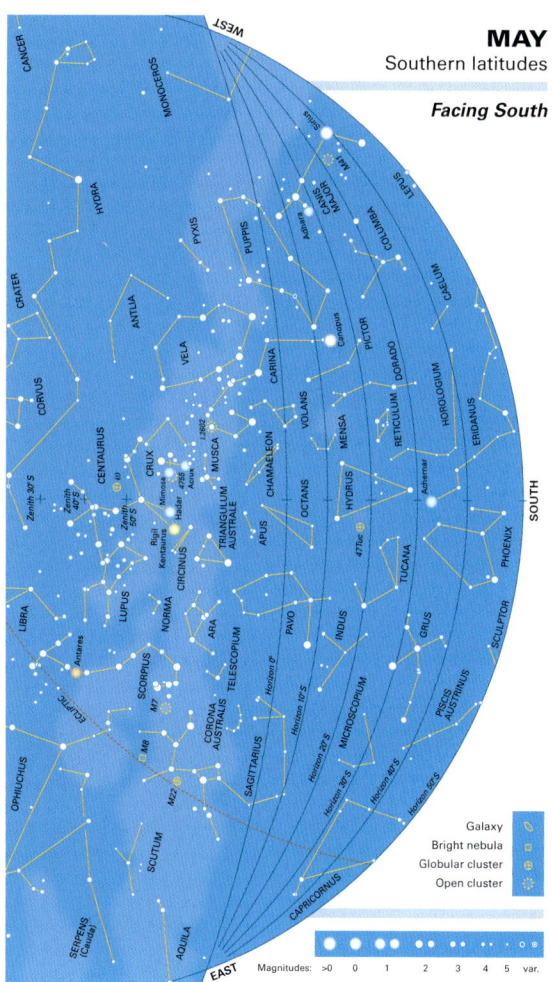

STARGAZING FOR BEGINNERS 83

JUNE
Northern latitudes

Facing North

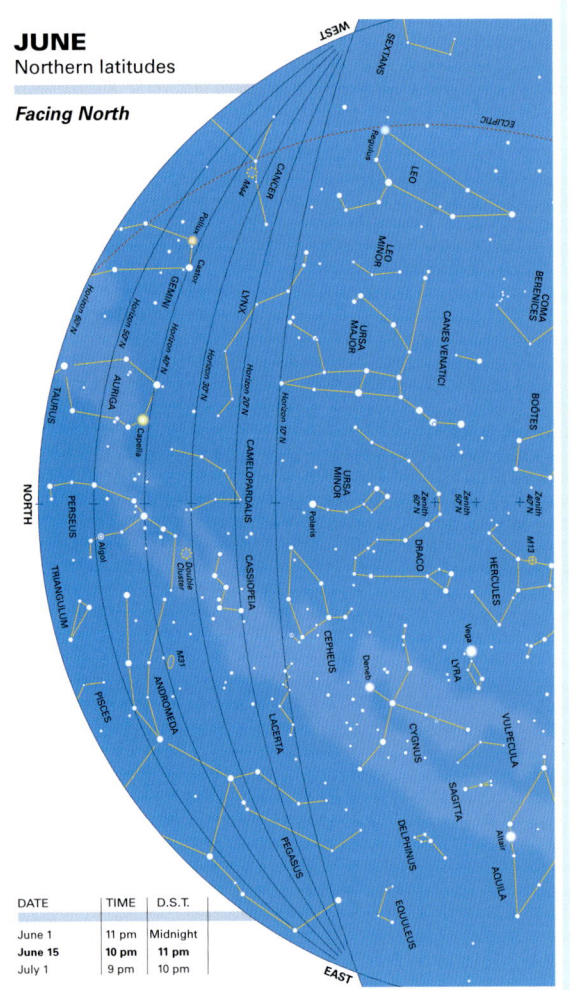

Facing South

JUNE
Northern latitudes

DATE	TIME	D.S.T.
June 1	11 pm	Midnight
June 15	**10 pm**	**11 pm**
July 1	9 pm	10 pm

Galaxy
Bright nebula
Globular cluster
Open cluster

Magnitudes: >0 0 1 2 3 4 5 var.

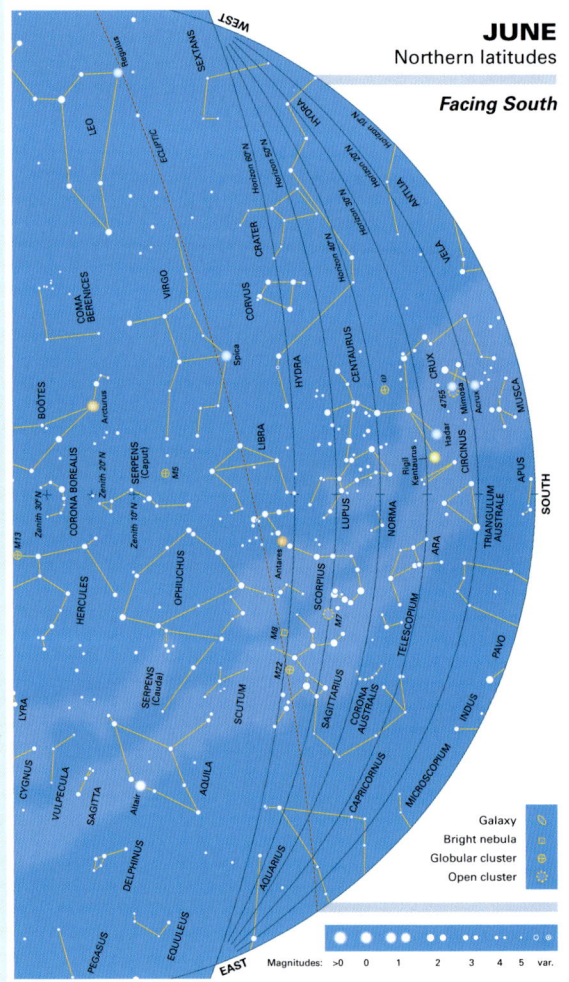

JUNE
Southern latitudes

Facing North

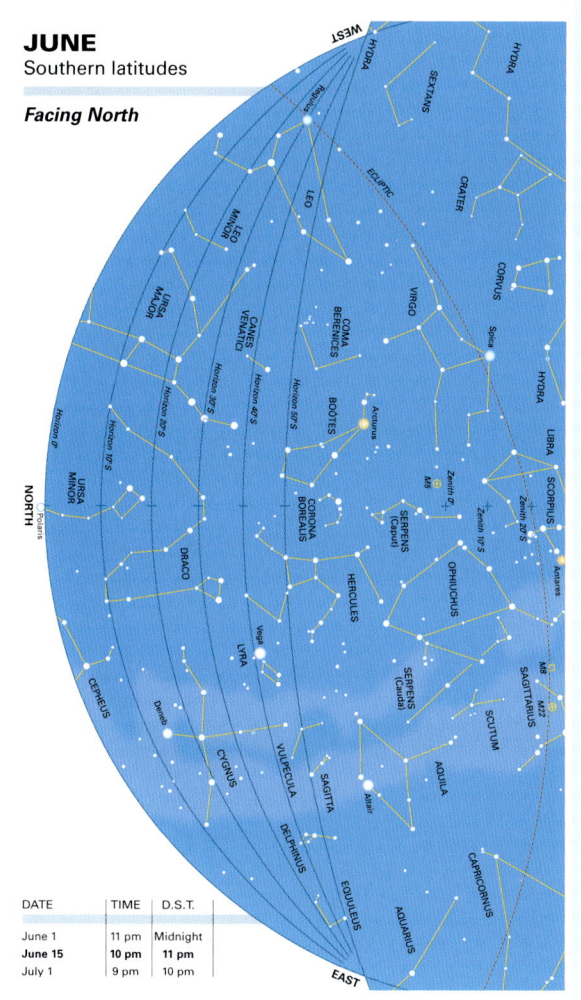

JUNE
Southern latitudes

Facing South

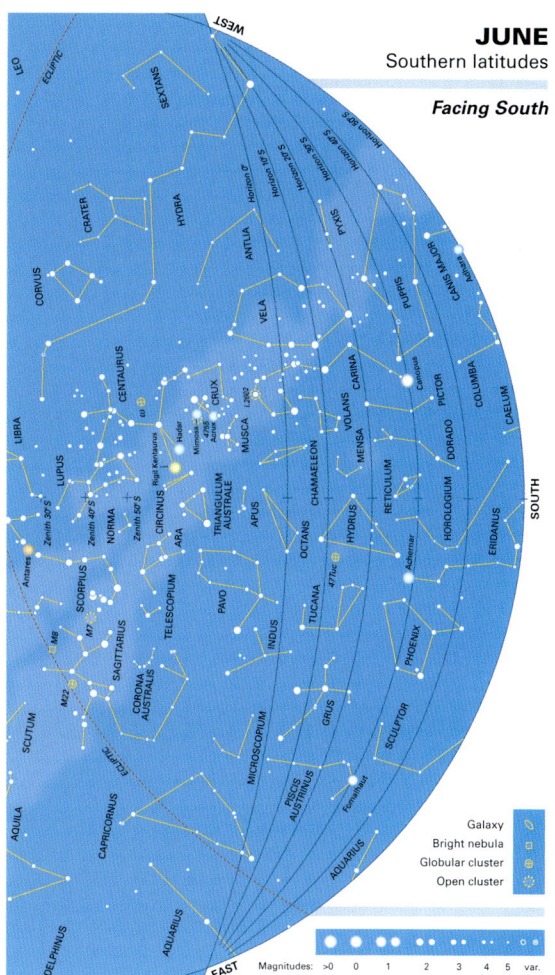

DATE	TIME	D.S.T.
June 1	11 pm	Midnight
June 15	**10 pm**	**11 pm**
July 1	9 pm	10 pm

Galaxy
Bright nebula
Globular cluster
Open cluster

Magnitudes: >0 0 1 2 3 4 5 var.

STARGAZING FOR BEGINNERS

86 THE BACKYARD STARGAZER'S BIBLE

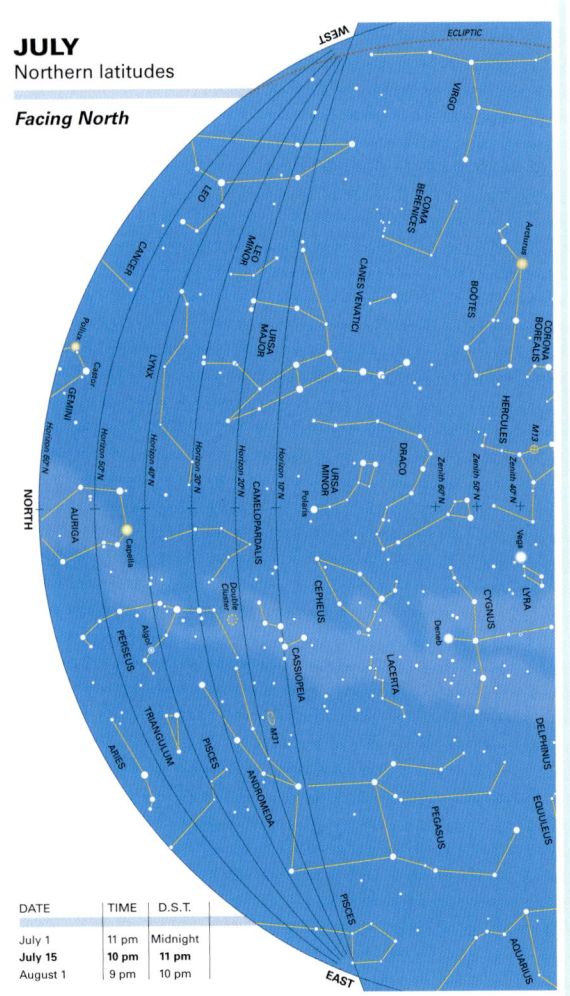

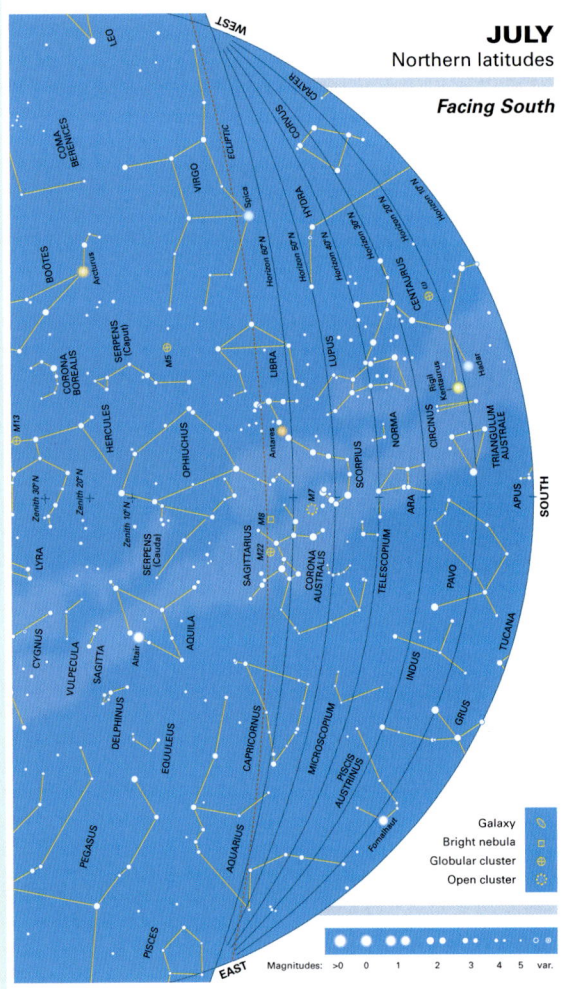

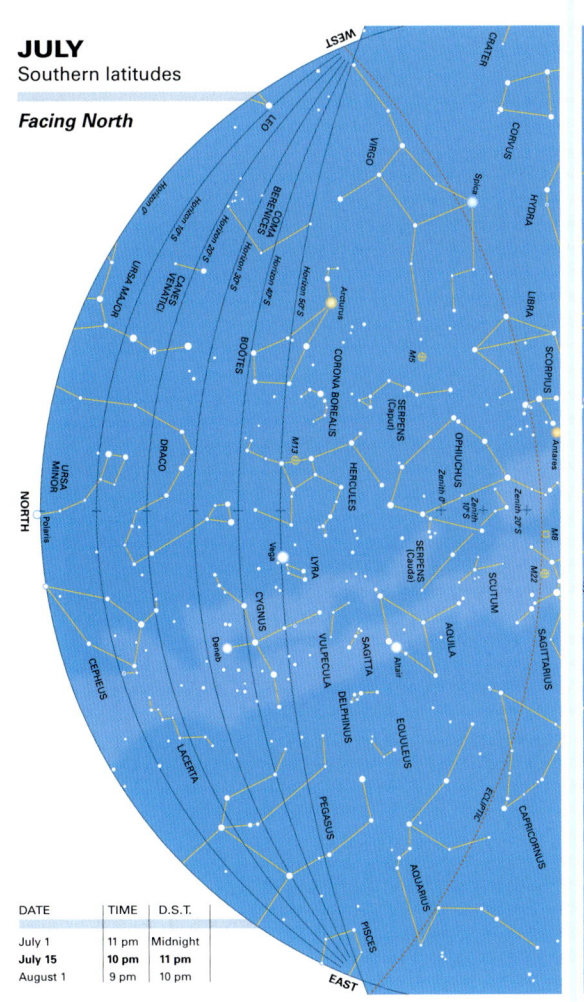

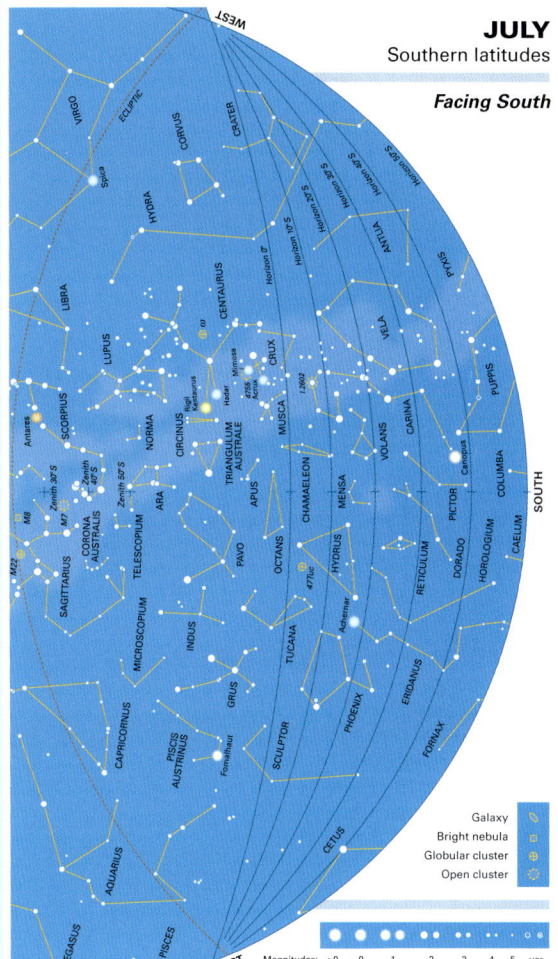

STARGAZING FOR BEGINNERS 87

AUGUST
Northern latitudes

Facing North

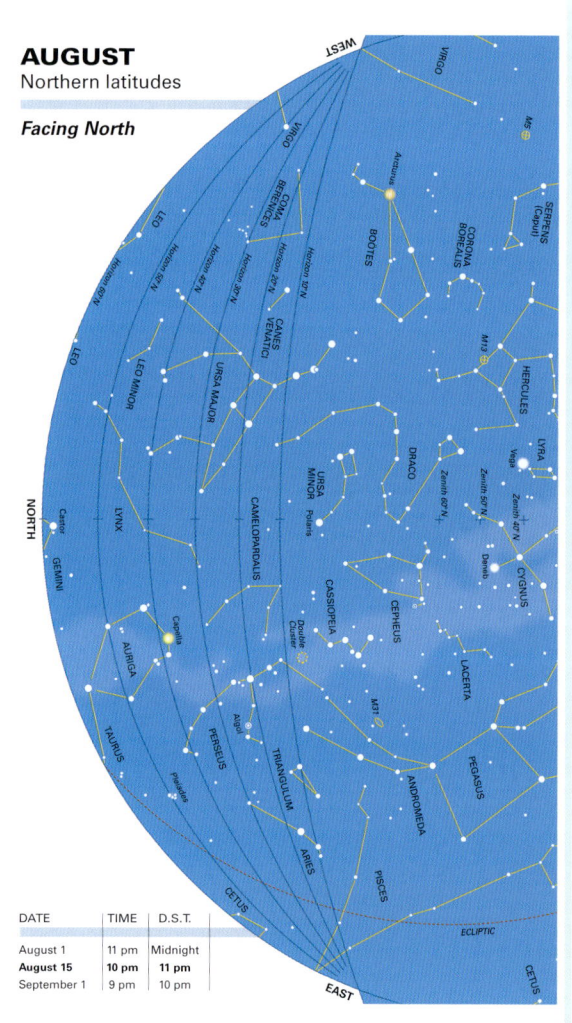

AUGUST
Northern latitudes

Facing South

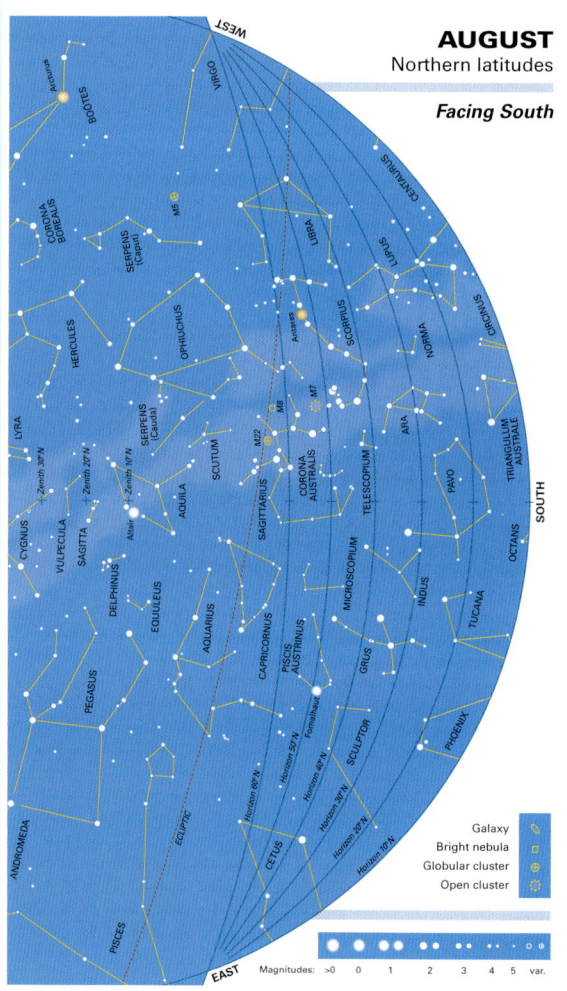

DATE	TIME	D.S.T.
August 1	11 pm	Midnight
August 15	**10 pm**	**11 pm**
September 1	9 pm	10 pm

Galaxy
Bright nebula
Globular cluster
Open cluster

Magnitudes: >0 0 1 2 3 4 5 var.

AUGUST
Southern latitudes

Facing North

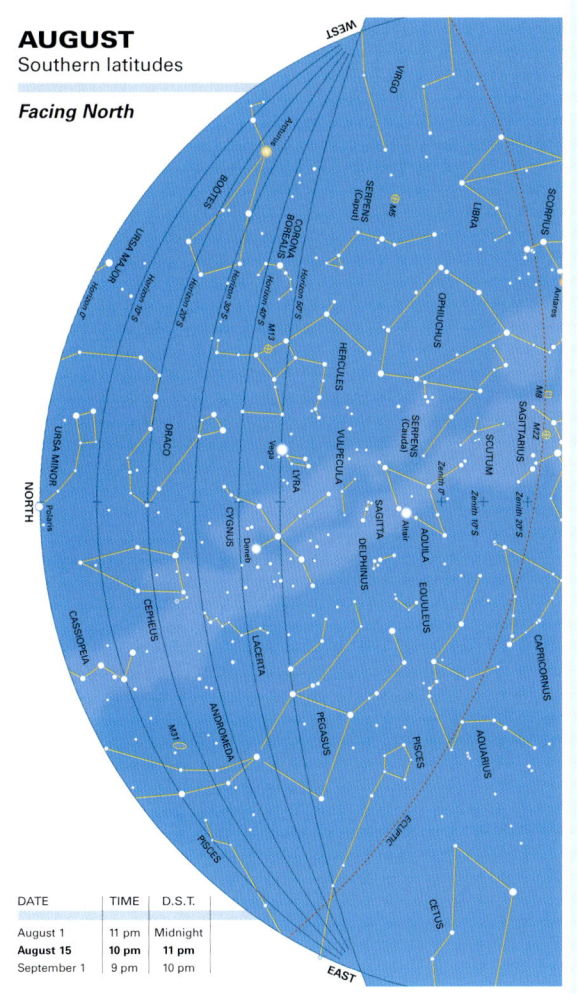

AUGUST
Southern latitudes

Facing South

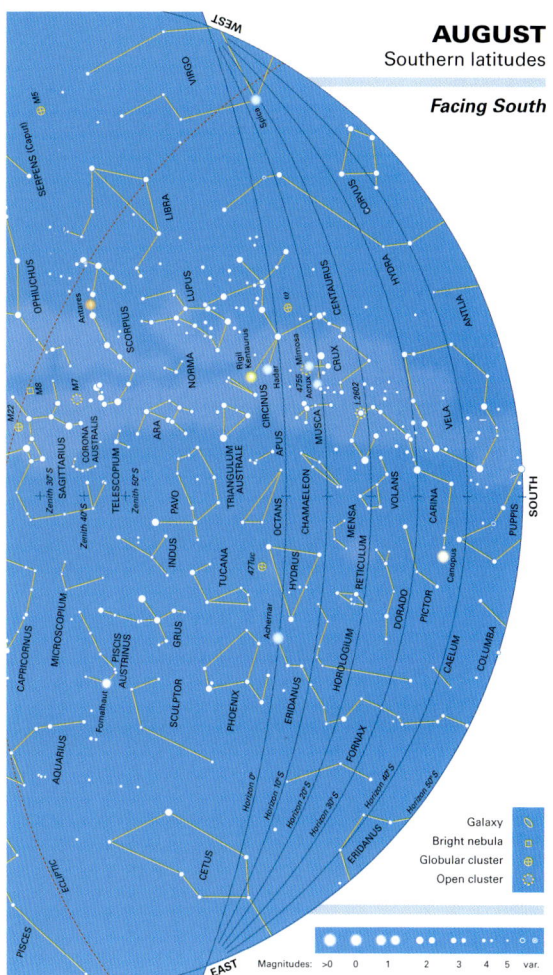

DATE	TIME	D.S.T.
August 1	11 pm	Midnight
August 15	**10 pm**	**11 pm**
September 1	9 pm	10 pm

Galaxy
Bright nebula
Globular cluster
Open cluster

Magnitudes: >0 0 1 2 3 4 5 var.

STARGAZING FOR BEGINNERS 89

SEPTEMBER
Northern latitudes

Facing North

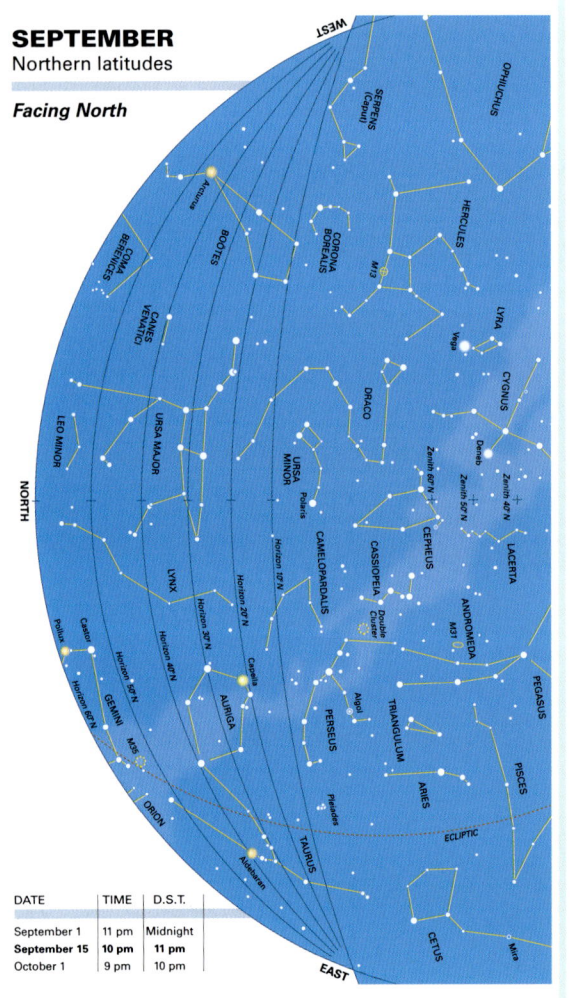

SEPTEMBER
Northern latitudes

Facing South

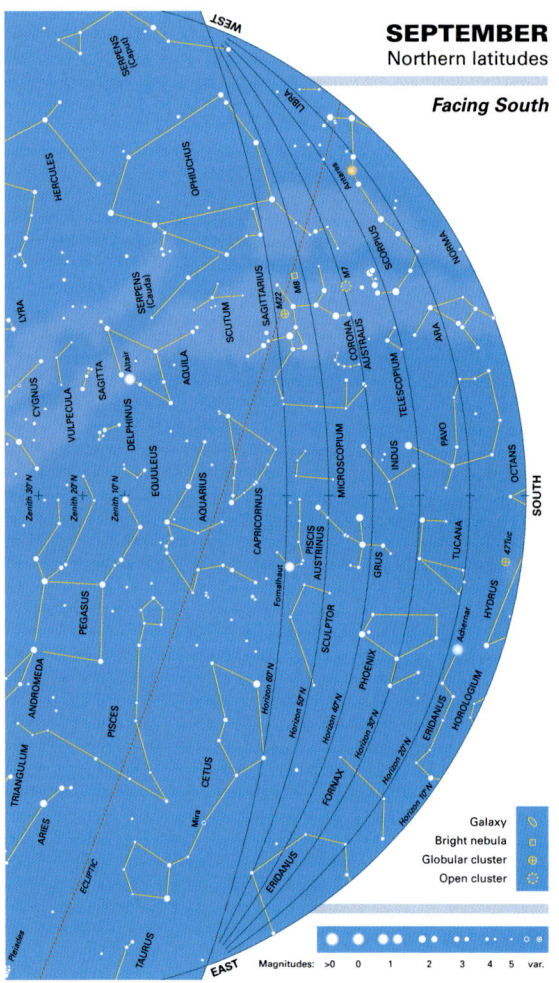

DATE	TIME	D.S.T.
September 1	11 pm	Midnight
September 15	**10 pm**	**11 pm**
October 1	9 pm	10 pm

Galaxy
Bright nebula
Globular cluster
Open cluster

Magnitudes: >0 0 1 2 3 4 5 var.

SEPTEMBER
Southern latitudes

Facing North

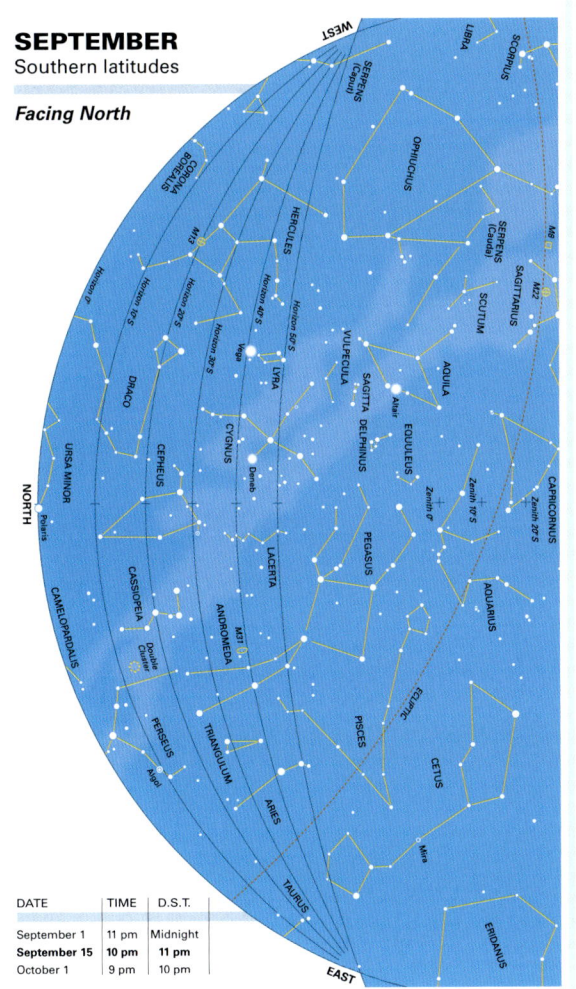

SEPTEMBER
Southern latitudes

Facing South

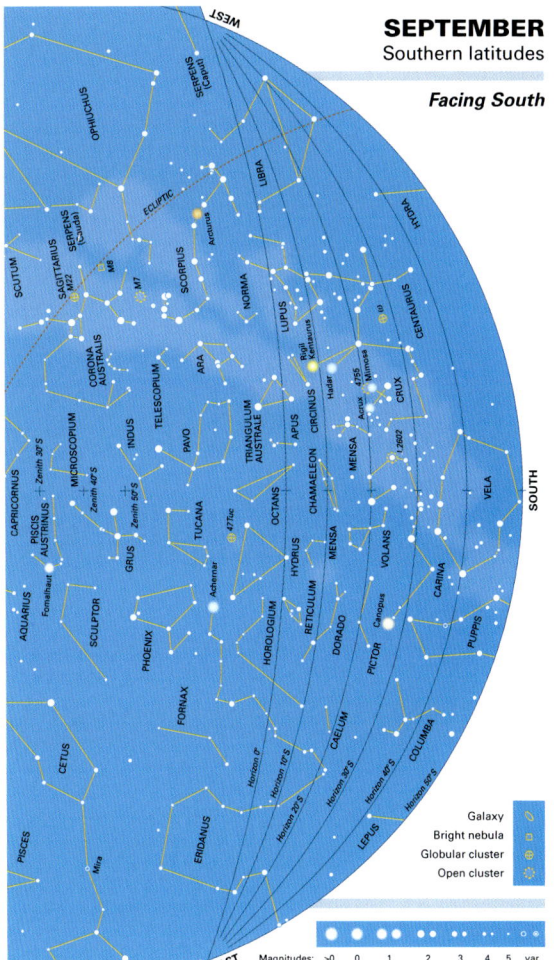

DATE	TIME	D.S.T.
September 1	11 pm	Midnight
September 15	**10 pm**	**11 pm**
October 1	9 pm	10 pm

STARGAZING FOR BEGINNERS 91

OCTOBER
Northern latitudes

Facing North

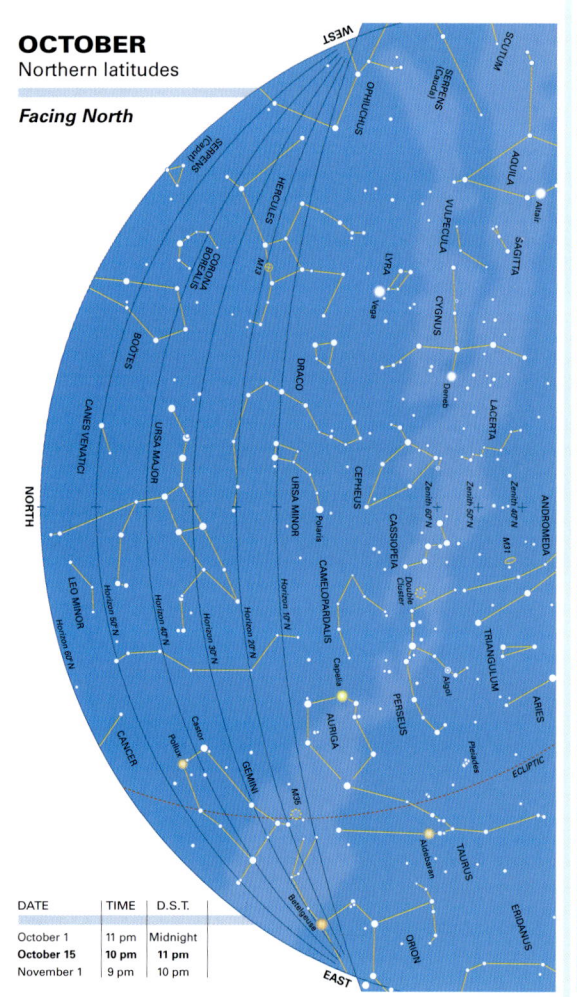

OCTOBER
Northern latitudes

Facing South

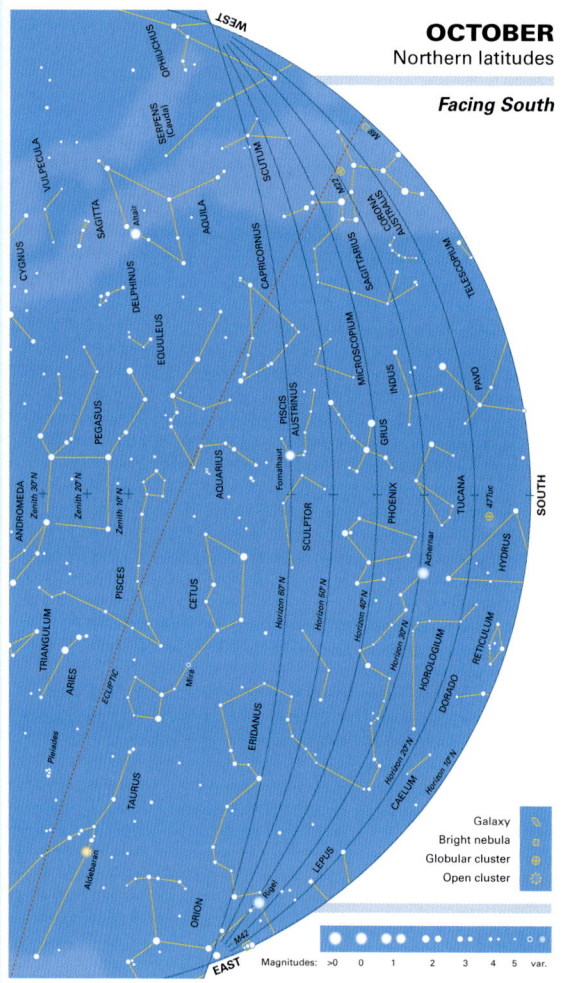

DATE	TIME	D.S.T.
October 1	11 pm	Midnight
October 15	**10 pm**	**11 pm**
November 1	9 pm	10 pm

OCTOBER
Southern latitudes

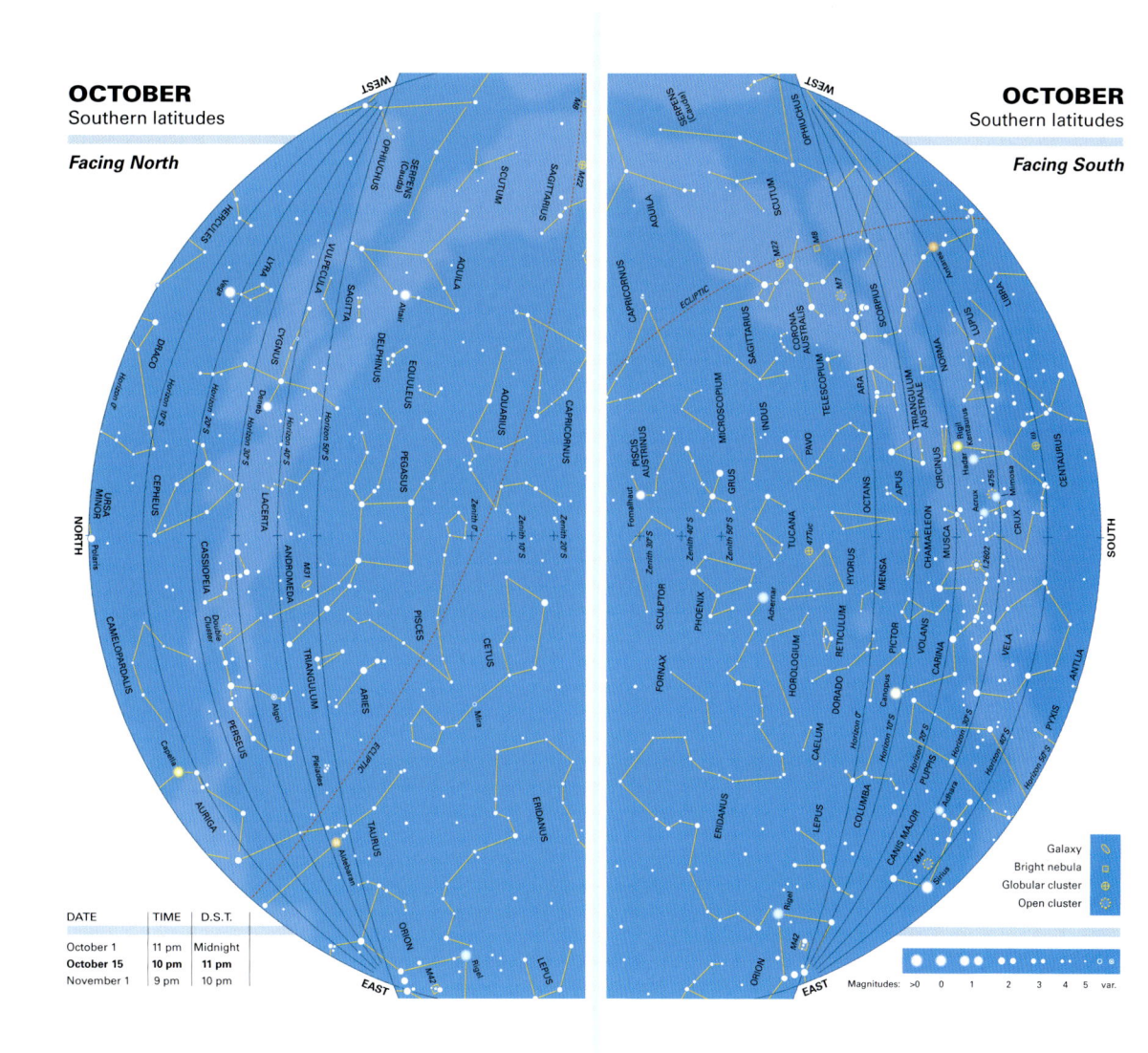

Facing North

DATE	TIME	D.S.T.
October 1	11 pm	Midnight
October 15	**10 pm**	**11 pm**
November 1	9 pm	10 pm

OCTOBER
Southern latitudes

Facing South

Galaxy
Bright nebula
Globular cluster
Open cluster

Magnitudes: >0 0 1 2 3 4 5 var.

STARGAZING FOR BEGINNERS 93

NOVEMBER
Northern latitudes

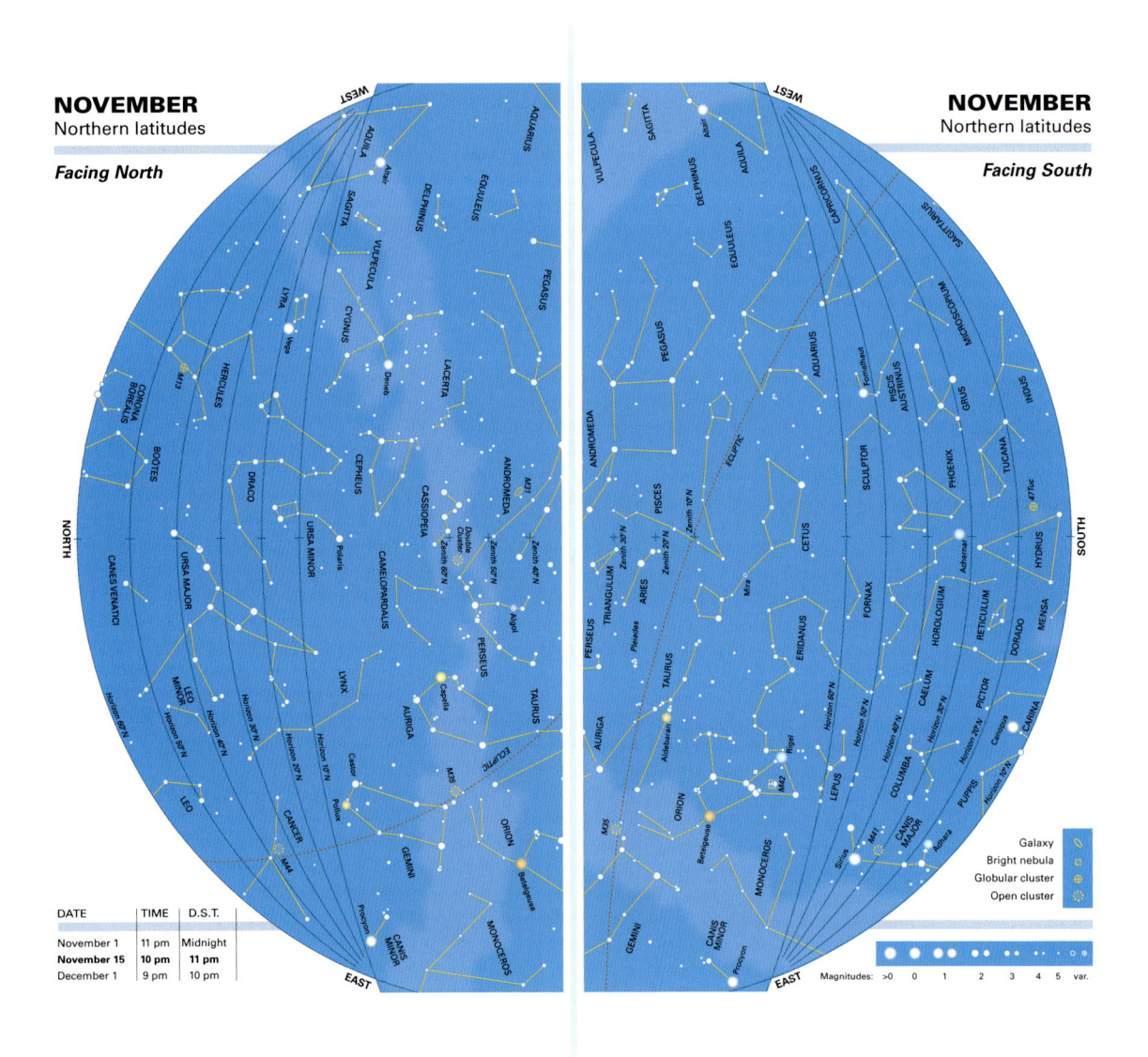

Facing North

NOVEMBER
Northern latitudes

Facing South

DATE	TIME	D.S.T.
November 1	11 pm	Midnight
November 15	**10 pm**	**11 pm**
December 1	9 pm	10 pm

Galaxy
Bright nebula
Globular cluster
Open cluster

Magnitudes: >0 0 1 2 3 4 5 var.

NOVEMBER
Southern latitudes

Facing North

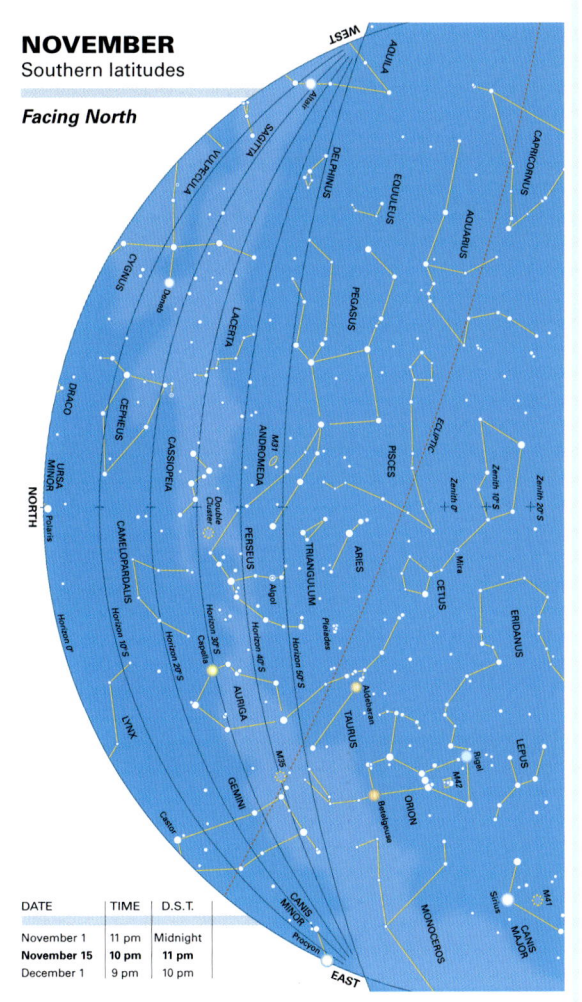

DATE	TIME	D.S.T.
November 1	11 pm	Midnight
November 15	**10 pm**	**11 pm**
December 1	9 pm	10 pm

NOVEMBER
Southern latitudes

Facing South

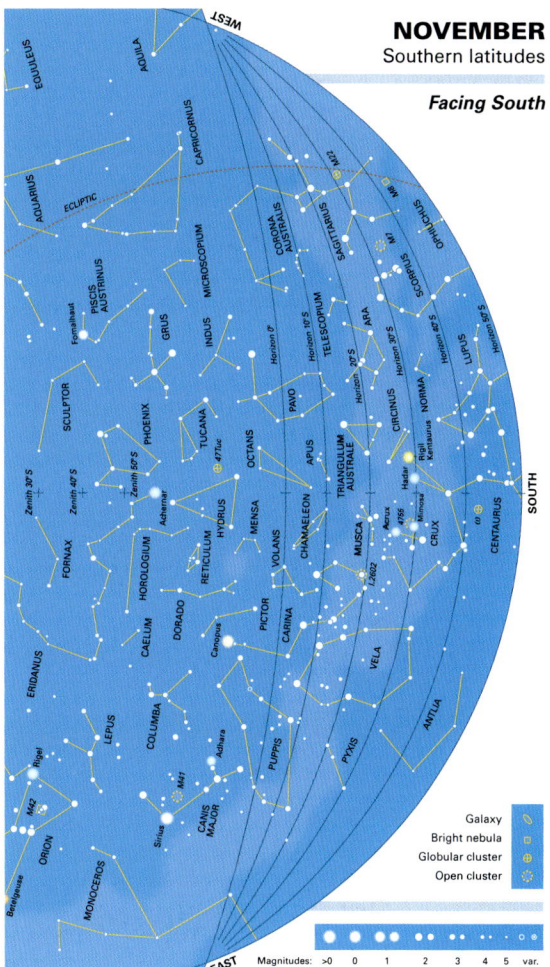

Galaxy
Bright nebula
Globular cluster
Open cluster

Magnitudes: >0 0 1 2 3 4 5 var.

STARGAZING FOR BEGINNERS 95

DECEMBER
Northern latitudes

Facing North

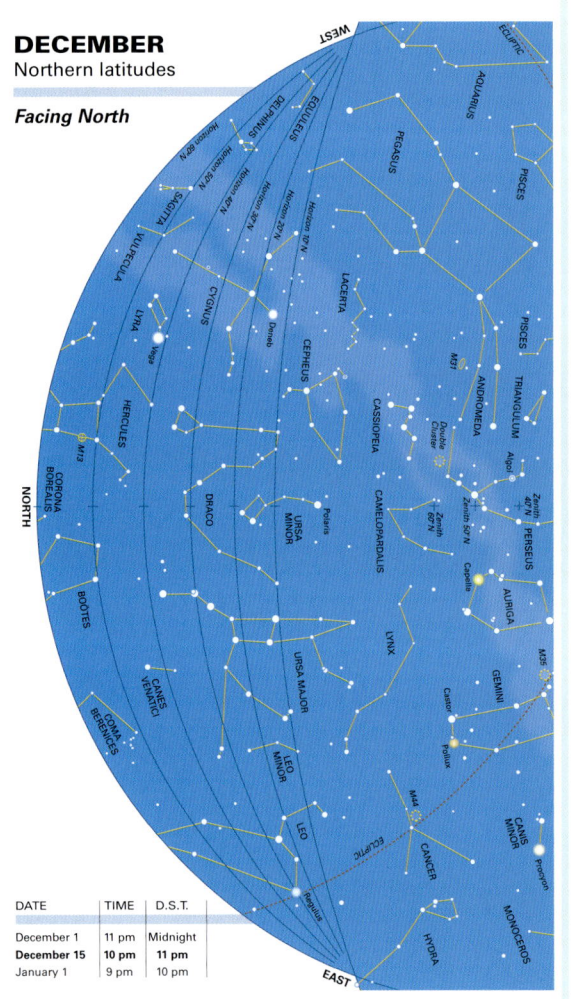

DECEMBER
Northern latitudes

Facing South

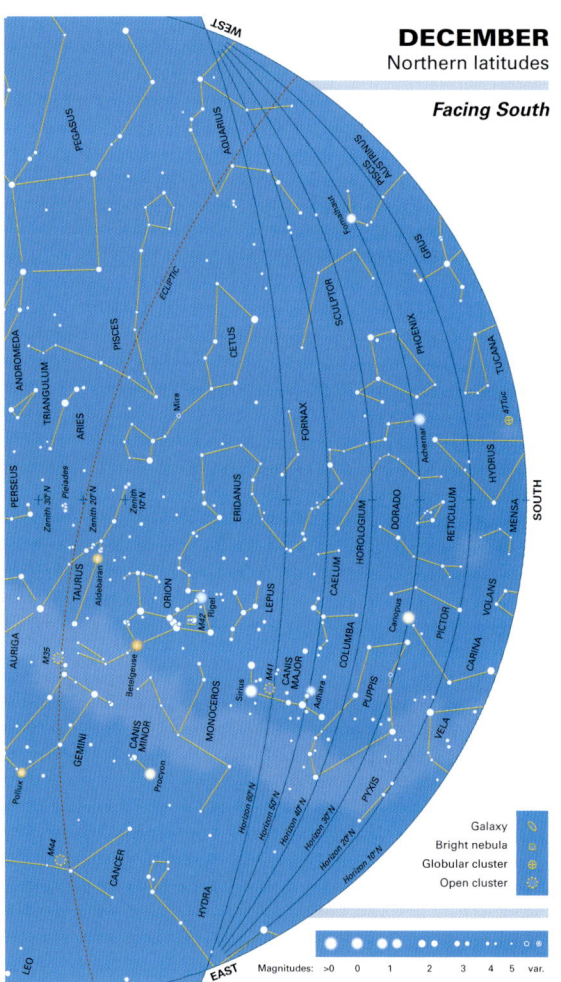

DATE	TIME	D.S.T.
December 1	11 pm	Midnight
December 15	**10 pm**	**11 pm**
January 1	9 pm	10 pm

Galaxy
Bright nebula
Globular cluster
Open cluster

Magnitudes: >0 0 1 2 3 4 5 var.

96 THE BACKYARD STARGAZER'S BIBLE

DECEMBER
Southern latitudes

Facing North

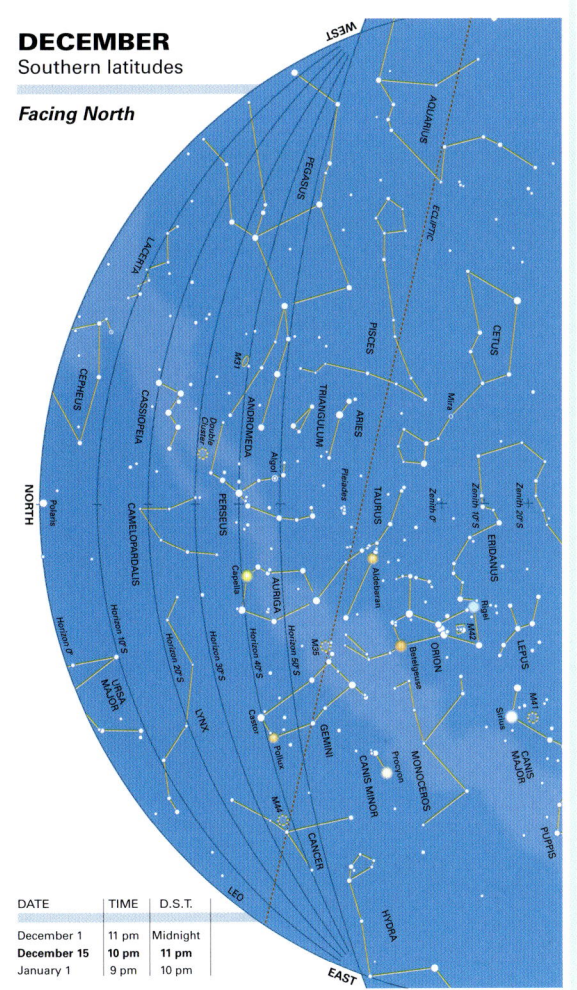

DECEMBER
Southern latitudes

Facing South

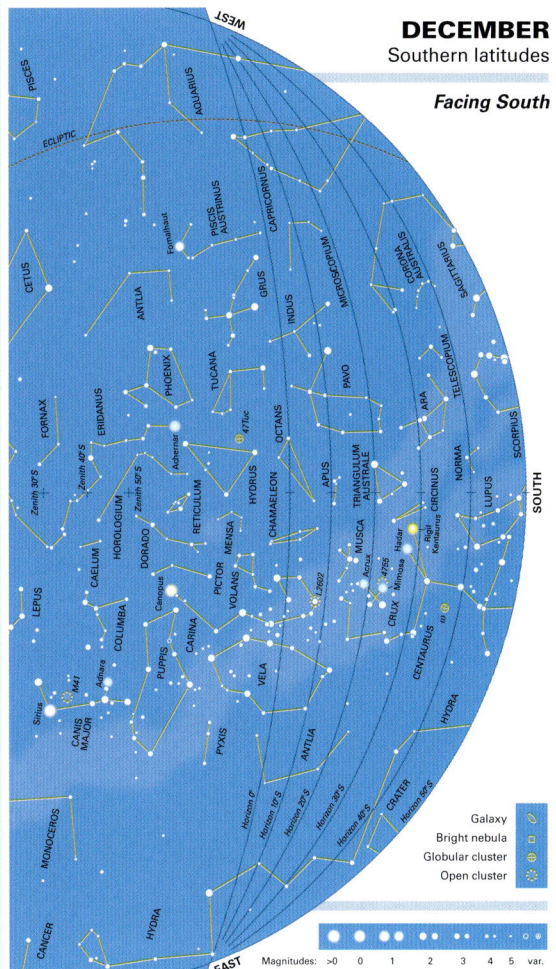

DATE	TIME	D.S.T.
December 1	11 pm	Midnight
December 15	**10 pm**	**11 pm**
January 1	9 pm	10 pm

Galaxy
Bright nebula
Globular cluster
Open cluster

Magnitudes: >0 0 1 2 3 4 5 var.

STARGAZING FOR BEGINNERS

THE CONSTELLATION CHARTS

This section details the 88 constellations in the night sky as shown on grid charts. Each constellation profile outlines the constellation's location, magnitude, type of star, and the history of its discovery. In the constellation charts that follow, each constellation is shown with an image of its location in the sky. Due to the size and location of the constellations, some images are shown in portrait mode to allow for full viewing of the constellation on the page.

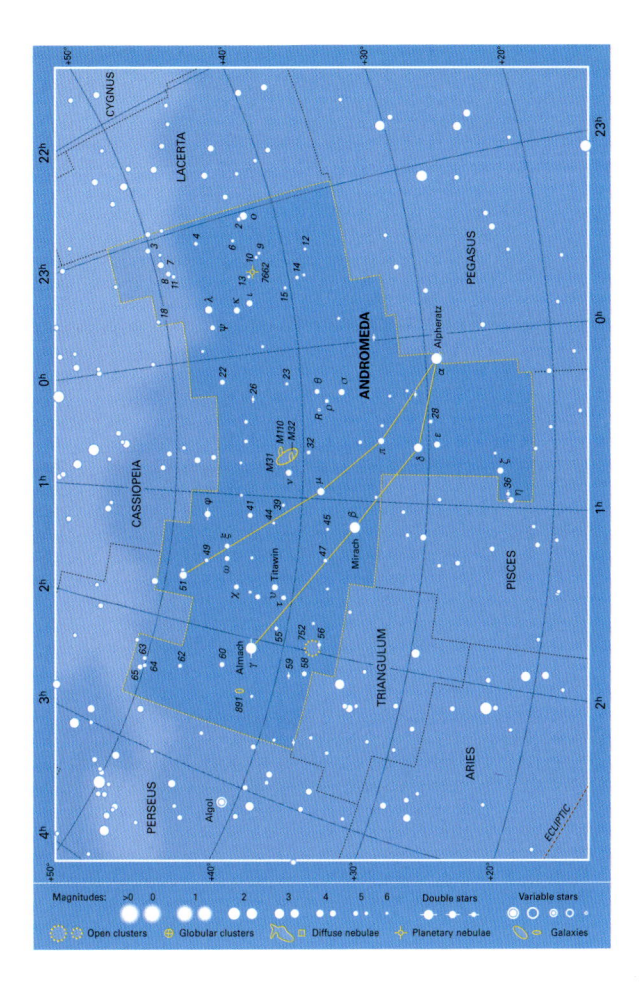

Andromeda

Andromeda represents the daughter of Queen Cassiopeia, who was chained to a rock as a sacrifice to the sea monster Cetus until saved by Perseus, whom she subsequently married. The constellation was one of the 48 listed by Ptolemy. Despite its fame, Andromeda isn't particularly striking, as its brightest stars are only 2nd magnitude. Its most prominent feature is a crooked line of four stars extending from the Great Square of Pegasus. The first of these stars marks a corner of the Square, although it is actually part of Andromeda. This star, known as Alpheratz, marks the head of the chained Andromeda; another star in the line, Mirach, represents her waist, and a third, Almach, is her chained foot. The most celebrated object in the constellation is the Andromeda Galaxy (M31), a spiral galaxy like our own Milky Way; it is the most distant object visible to the naked eye. Two stars leading from Mirach, or β (beta) Andromedae, act as a guide to it.

α (alpha) Andromedae, 0h 08m +29°.1, (Alpheratz), mag. 2.1, is a blue-white subgiant 97 l.y. away.

β (beta) And, 1h 10m +35°.6, (Mirach), mag. 2.1, is a red giant 197 l.y. away.

γ (gamma) And, 2h 04m +42°.3, (Almach), 390 l.y. away, is an outstanding triple star. Its two brightest components, of mags. 2.2 and 4.8, form one of the finest pairs for small telescopes: their colors are orange and blue. The fainter, blue star also has a close 6th-mag. blue-white companion that orbits it every 63 years. This fainter pair will be at their widest separation around 2048 and should be divisible in 200-mm apertures for several years either side of this date.

δ (delta) And, 0h 39m +30°.9, mag. 3.3, is an orange giant 103 l.y. away.

μ (mu) And, 0h 57m +38°.5, mag. 3.9, is a white main-sequence star 122 l.y. away.

π (pi) And, 0h 37m +33°.7, 600 l.y. away, is a blue-white main-sequence star of mag. 4.4 with a mag. 8.6 companion visible in small telescopes.

υ (upsilon) And, 1h 37m, +41°.4, (Titawin), mag. 4.1, is a yellow-white main-sequence star 44 l.y. away. Astronomers have found that it is accompanied by four planets, the first known multi-planet system around a star other than our own Sun.

56 And, 1h 56m +37°.3, is a yellow giant star of mag. 5.7, 327 l.y. away, with an orange giant companion, mag. 5.9 and 1,160 l.y. away, easily split in binoculars. The two stars are found near the star cluster NGC 752 but are actually closer to us.

M31 (NGC 224), 0h 43m +41°.3, the Andromeda Galaxy, is a spiral galaxy 2.5 million l.y. away. It is visible to the naked eye as an elliptical fuzzy patch, and becomes more prominent when seen through binoculars or a telescope with low magnification (too high a power reduces the contrast and renders the fainter parts of the galaxy less visible). Dark lanes can be seen in the spiral arms surrounding the nucleus, but the full extent of the galaxy only becomes apparent in long-exposure photographs (visual observers see just its brightest, central portion). If the entire Andromeda Galaxy were bright enough to be seen by the naked eye, it would appear five or six times the diameter of the full Moon, as demonstrated in the composite image on page 74. M31 is accompanied by two small satellite galaxies, the equivalents of our Magellanic Clouds but both elliptical rather than irregular in shape. The brighter of these, M32 (also known as NGC 221), is visible in small telescopes as a fuzzy, 8th-mag. star-like glow 0.5° south of M31's core. The second companion, M110 (NGC 205), is larger but visually more elusive, and over 1° northwest of M31.

NGC 752, 1h 58m +37°.7, is a widely spread cluster of about 60 stars of 9th mag. and fainter, 1,500 l.y. away, visible in binoculars and easily resolved in telescopes.

NGC 7662, 23h 26m +42°.6, is one of the brightest and easiest planetary nebulae to see with a small telescope. At low powers it appears as a fuzzy, 9th-mag. blue-green star popularly called the Blue Snowball, but magnifications of ×100 or so reveal its slightly elliptical disk. Larger apertures show a central hole; the central star is a difficult object for amateur telescopes. NGC 7662 lies about 5,700 l.y. away.

Opposite The Andromeda galaxy, M31, captured in a beautiful image from an observatory in Germany.

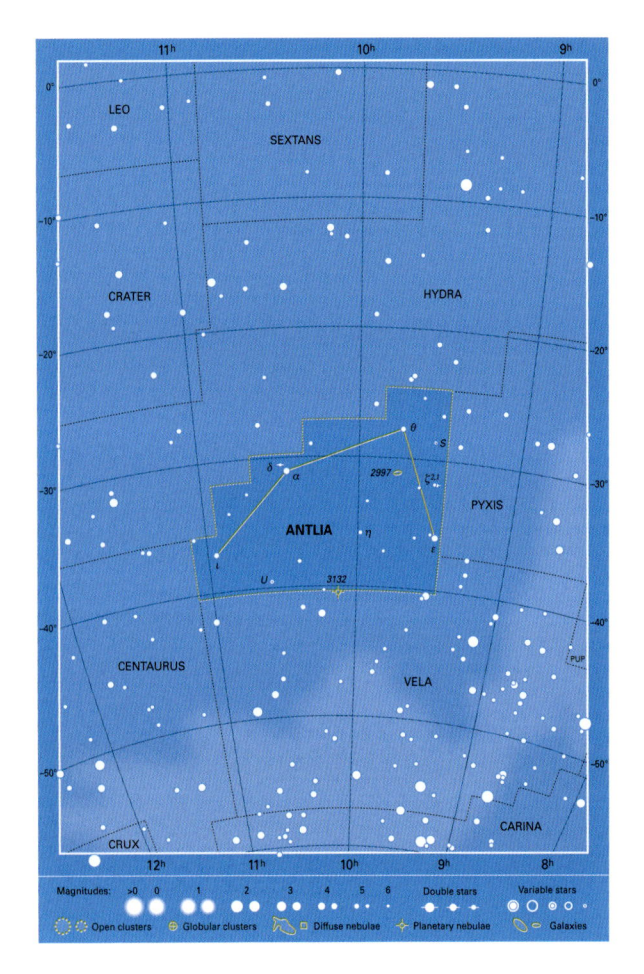

Antlia, the Air Pump

An obscure southern constellation introduced on a map published in 1756 by the French astronomer Nicolas Louis de Lacaille, and named to commemorate the air pump invented by the French physicist Denis Papin. Lacaille, the first person to map the naked-eye stars in the southern skies completely (see page 47), introduced 14 new constellations to fill gaps between existing ones. Most of these new figures are unremarkable, as is Antlia.

α (alpha) Antliae, 10h 27m −31°.1, mag. 4.3, the constellation's brightest star, is an orange giant 366 l.y. distant.

δ (delta) Ant, 10h 30m −30°.6, 465 l.y. away, is a blue-white star of mag. 5.6, with a mag. 9.8 companion visible in small telescopes.

ζ1 ζ2 (zeta1 zeta2) Ant, 9h 31m −31°.9, is a wide pair of stars of mags. 5.7 and 5.9, 340 and 370 l.y. away, visible in binoculars. Small telescopes show that ζ1 Ant is itself double, with components of mags. 6.2 and 6.8.

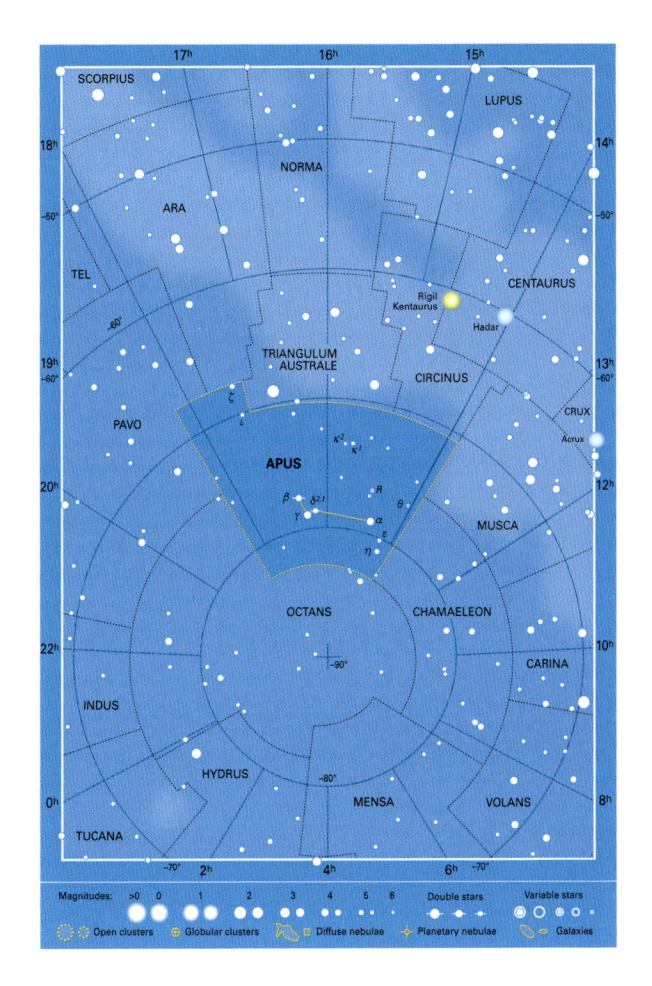

Apus, the Bird of Paradise

A faint constellation near the south celestial pole—one of the 12 new figures introduced in the 1590s by the Dutch navigators Pieter Dirkszoon Keyser and Frederick de Houtman during voyages to the southern hemisphere. This one represents the exotic bird of paradise, native to New Guinea.

α (alpha) Apodis, 14h 48m −79°.0, mag. 3.8, is an orange giant star 447 l.y. away.

β (beta) Aps, 16h 43m −77°.5, mag. 4.2, is an orange giant 156 l.y. away.

γ (gamma) Aps, 16h 33m −78°.9, mag. 3.9, is an orange giant 153 l.y. away.

δ1 δ2 (delta1 delta2) Aps, 16h 20m −78°.7, is a naked-eye or binocular pair of unrelated red and orange giant stars of mags. 4.7 and 5.2, 762 and 563 l.y. away.

θ (theta) Aps, 14h 05m −76°.8, 389 l.y. away, is a red giant that varies semi-regularly between 5th and 6th mags. every 4 months or so.

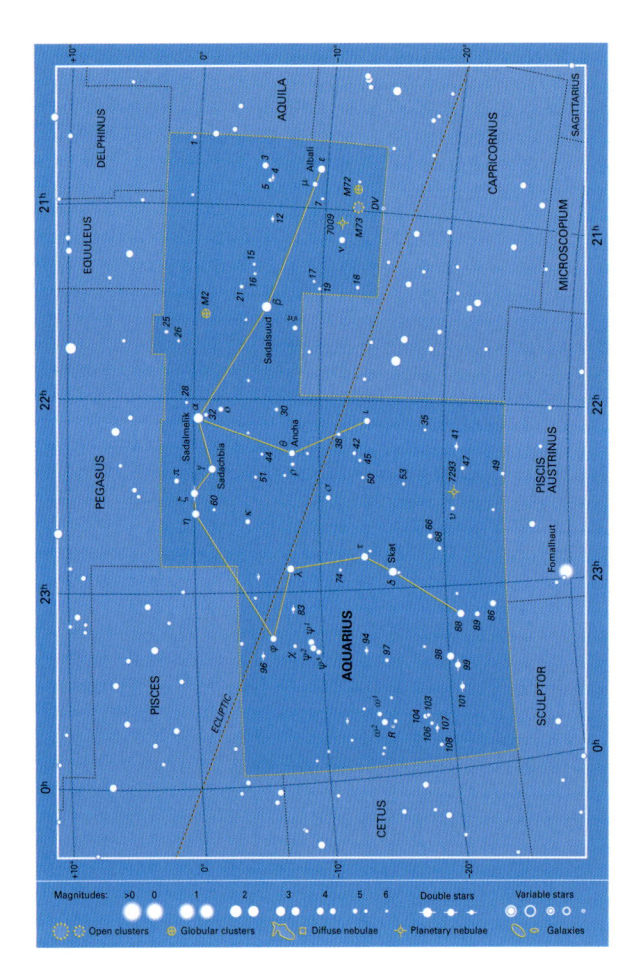

Aquarius, the Water Carrier

Aquarius is one of the most ancient constellations. The Babylonians saw in this area of sky the figure of a man pouring water from a jar. In Greek mythology, the constellation represents Ganymede, a shepherd boy carried off by Zeus to Mount Olympus, where he became wine-waiter to the gods.

The most prominent part of Aquarius is the Y-shaped asterism of four stars representing the Water Jar itself, centered on the star ζ (zeta) Aquarii. The flow of water from the jar is represented by a cascade of stars running southwards from κ (kappa) Aquarii to Fomalhaut in the mouth of the southern fish, Piscis Austrinus. Aquarius lies in an area of "watery" constellations that includes Pisces, Cetus, and Capricornus.

The Sun, on its annual passage around the sky, is within the boundaries of Aquarius from late February to early March. Aquarius will one day contain the vernal equinox, the point at which the Sun crosses into the northern celestial hemisphere each year. This astronomically important point, from which the coordinate of right ascension is measured, will move into Aquarius from neighboring Pisces in 2597 CE because of the effect of precession. Hence the so-called Age of Aquarius is a long way off yet.

Two main meteor showers radiate from Aquarius each year. The first, the Eta Aquariids, is the richer, reaching a maximum of 40 meteors per hour around May 5–6. The Delta Aquariids produce about 15 meteors per hour around July 29. Each shower is named after the bright star closest to its radiant.

α (alpha) Aquarii, 22h 06m −0°.3, (Sadalmelik, from the Arabic for "the lucky stars of the king"), mag. 2.9, is a yellow supergiant 524 l.y. away.

β (beta) Aqr, 21h 32m −5°.6, (Sadalsuud, from the Arabic for "luckiest of the lucky stars"), mag. 2.9, is a yellow supergiant 537 l.y. away.

γ (gamma) Aqr, 22h 22m −1°.4, (Sadachbia), mag. 3.8, is a blue-white main-sequence star 126 l.y. away.

δ (delta) Aqr, 22h 55m −15°.8, (Skat), mag. 3.3, is a blue-white main-sequence star 141 l.y. away.

ε (epsilon) Aqr, 20h 48m −9°.5, (Albali), mag. 3.8, is a blue-white main-sequence star 244 l.y. away.

ζ (zeta) Aqr, 22h 29m −0°.0, 95 l.y. away, is a celebrated binary consisting of twin white stars of mags. 4.3 and 4.5, orbiting each other every 430 years or so. The two stars are gradually moving apart as seen from Earth and hence are becoming increasingly easy to divide in small telescopes.

M2 (NGC 7089), 21h 34m −0°.8, is a mag. 6.5 globular cluster easily visible in binoculars or small telescopes, but requiring 100 mm aperture to resolve the brightest individual stars. This rich and highly concentrated globular lies 40,000 l.y. away.

M72 (NGC 6981), 20h 54m −12°.5, is a 9th-mag. globular cluster 40,000 l.y. away, much smaller and less impressive than M2. NGC 7009, 21h 04m −11°.4, is a famous planetary nebula, 4,000 l.y. away, known as the Saturn Nebula because of its resemblance to that planet when seen in large telescopes (see the photograph on page 80). But in most amateur telescopes, of 75 mm aperture or more, it appears as merely an 8th-mag. blue-green ellipse of similar apparent size to the globe of Saturn. Its central star is of mag. 11.5.

NGC 7293, 22h 30m −20°.8, is the nearest planetary nebula to the Sun, about 650 l.y. away, and is commonly known as the Helix Nebula. It is the largest planetary nebula in apparent size, covering 0.25° of sky, half the apparent size of the Moon. Its popular name arises because in early photographs it appeared to consist of two overlapping turns of a spiral, although that structure is less apparent in modern color images, which instead emphasize its ring shape (see the photograph above). Despite its size, the Helix Nebula is quite faint and is best found with binoculars or very low power on a telescope, when it appears as a circular misty patch, and not as impressive as its large size would suggest.

102 THE BACKYARD STARGAZER'S BIBLE

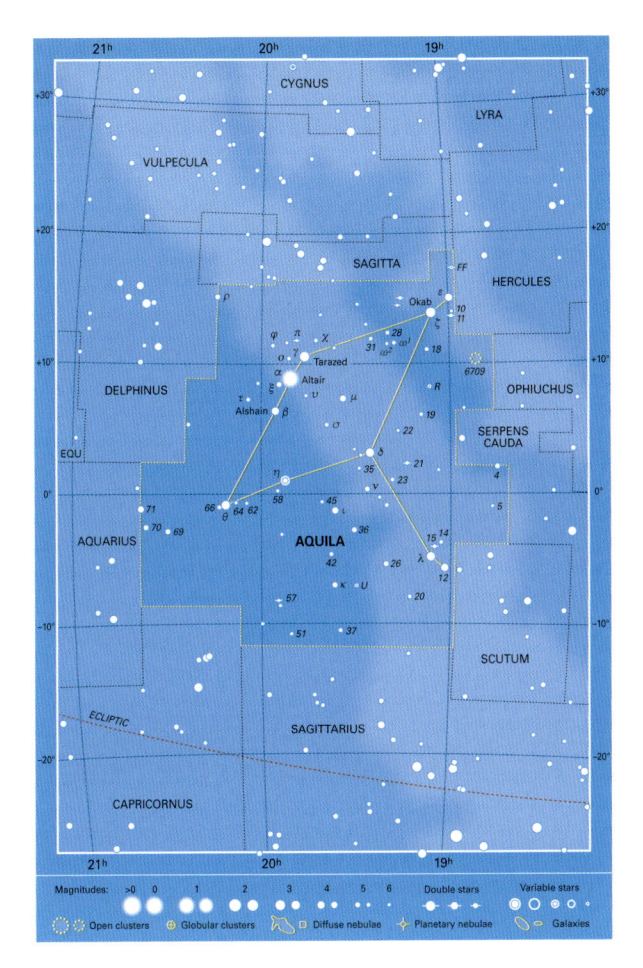

Aquila, the Eagle

A constellation dating from ancient times, representing the bird that in Greek mythology carried the thunderbolts of Zeus. According to Greek myth, Zeus sent an eagle (or, in a variant of the tale, turned himself into an eagle) to abduct the shepherd boy Ganymede, represented in the sky by the neighboring constellation of Aquarius.

Aquila's brightest star, Altair, forms one corner of a giant stellar triangle (known in the northern hemisphere as the Summer Triangle) that is completed by Deneb in Cygnus and Vega in Lyra. The name Altair comes from the Arabic *al-nasr al-ta'ir*, meaning "the flying eagle." Two fainter stars, β (beta) and γ (gamma) Aquilae, stand like sentinels either side of it; these are called Alshain and Tarazed, both from the Persian *shahin-i tarazu*, a translation of an Arabic name meaning "the balance," which referred jointly to these two stars and Altair. Aquila lies in the Milky Way and contains rich starfields, particularly toward Scutum in the south. It is an abundant area for novae.

α (alpha) Aquilae, 19h 51m +8°.9, (Altair, "the flying eagle"), mag. 0.76, is a white star 16.7 l.y. away, among the closest naked-eye stars to us.

β (beta) Aql, 19h 55m +6°.4, (Alshain), mag. 3.7, is a yellow subgiant 44 l.y. away.

γ (gamma) Aql, 19h 46m +10°.6, (Tarazed), mag. 2.7, is an orange giant star 395 l.y. away.

ζ (zeta) Aql, 19h 05m +13°.9, (Okab), mag. 3.0, is a blue-white star 83 l.y. away.

η (eta) Aql, 19h 52m +1°.0, a yellow-white supergiant 890 l.y. away, is one of the brightest Cepheid variable stars. Its brightness ranges from mag. 3.5 to 4.3 with a period of 7.2 days.

15 Aql, 19h 05m −4°.0, is an orange giant star of mag. 5.4, 314 l.y. away, with a purplish mag. 6.8 companion, 640 l.y. away, easily visible in small telescopes.

57 Aql, 19h 55m −8°.2, is an easy double for small telescopes, consisting of a bluish main-sequence star of mag. 5.7 with a mag. 6.4 companion, both about 450 l.y. away.

R Aql, 19h 06m +8°.2, 690 l.y. away, is a red giant variable of Mira type, about 400 times the diameter of the Sun, ranging from 6th to 12th mag. every 9 months or so. It lies about 760 l.y. away.

FF Aql, 18h 58m +17°.4, is a yellow-white supergiant Cepheid variable that ranges from mag. 5.2 to 5.5 every 4.5 days. It lies about 1,700 l.y. away.

NGC 6709, 18h 52m +10°.3, is a loosely scattered cluster of some 40 stars of mags. 9 to 11, about 3,600 l.y. away.

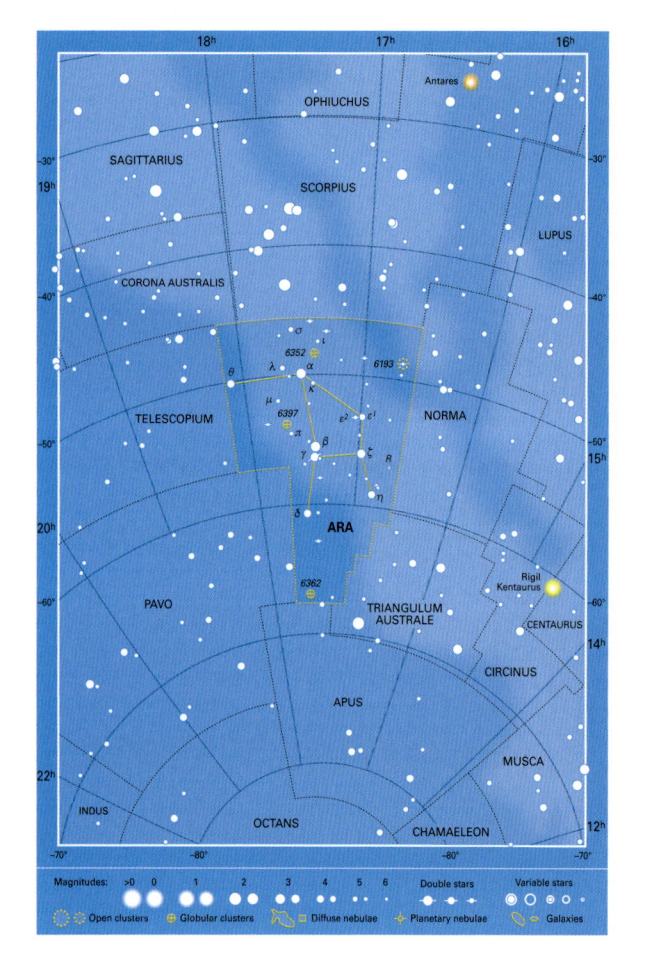

Ara, the Altar

This constellation, although faint and relatively little known, originated with the Greeks, who visualized it as the altar on which the gods of Olympus swore an oath of allegiance before their war against the Titans, who at that time ruled the Universe. After their victory, the Olympian gods divided up the Universe between them. Zeus, the greatest god of all, was allotted the sky and placed the altar of the gods among the stars in lasting memory of their victory. Old star atlases depict Ara as the altar on which Centaurus, the centaur, is about to sacrifice Lupus, the wolf. Ara lies in a rich part of the Milky Way, south of Scorpius.

α (alpha) Arae, 17h 32m −49°.9, mag. 2.9, is a blue-white star 267 l.y. away.

β (beta) Ara, 17h 25m −55°.5, mag. 2.8, is an orange supergiant 715 l.y. away.

γ (gamma) Ara, 17h 25m −56°.4, mag. 3.3, is a blue supergiant 1,100 l.y. away.

δ (delta) Ara, 17h 31m −60°.7, mag. 3.6, is a blue-white star 225 l.y. away.

ζ (zeta) Ara, 16h 59m −56°.0, mag. 3.1, is an orange giant star 486 l.y. away.

NGC 6193, 16h 41m −48°.8, is a 5th-mag. open cluster of about 30 stars 4,000 l.y. away, about half the apparent size of the full Moon. The brightest member is a blue-white star of mag. 5.6, which small telescopes show has a companion of mag. 6.9. Around the cluster is an irregularly shaped patch of faint nebulosity, NGC 6188, which shows up well only in photographs.

NGC 6397, 17h 41m −53°.7, is a 6th-mag. globular cluster that appears as a fuzzy star through binoculars; under good conditions it is visible to the naked eye. Small telescopes resolve the brightest stars in its outer regions, which extend across at least half the apparent diameter of the full Moon. It is the second-closest globular to us, 7,800 l.y. away (M4 in Scorpius is the closest).

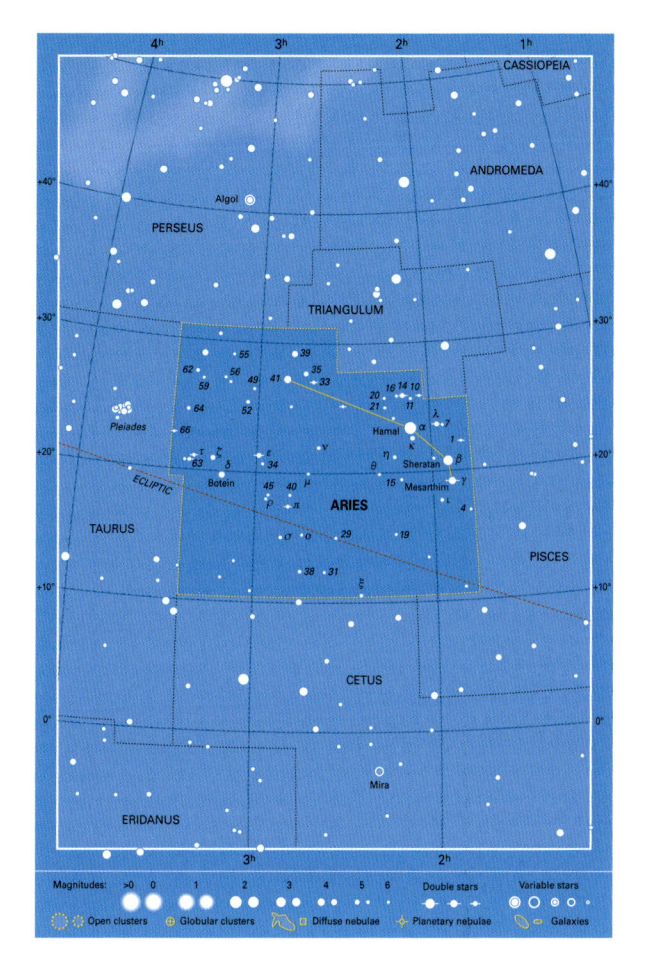

Aries, the Ram

A constellation lying between Taurus and Andromeda, the origin of which dates back to ancient times. Aries represents the ram of Greek legend whose golden fleece was sought by Jason and the Argonauts. Despite its faintness, Aries has assumed great importance in astronomy because, over 2,000 years ago, it contained the point where the Sun passes from south to north across the celestial equator each year. This point, the vernal equinox, is still sometimes known as the First Point of Aries, although it no longer lies in Aries, having moved into neighboring Pisces in 67 BCE as a result of the slow wobble of the Earth in space known as precession. It is now headed for Aquarius (see diagram left). Currently, the Sun passes through Aries from late April to mid-May.

α (alpha) Arietis, 2h 07m +23°.5, (Hamal, from the Arabic for "lamb"), mag. 2.0, is an orange giant star 66 l.y. away.

β (beta) Ari, 1h 55m +20°.8, (Sheratan, from the Arabic for "two"), mag. 2.6, is a blue-white star 59 l.y. away.

γ (gamma) Ari, 1h 54m +19°.3, (Mesarthim), is a striking double consisting of twin white stars of mags. 4.5 and 4.6, about 170 l.y. away, clearly visible through small telescopes even under low magnification. To the eye they appear as a single star of magnitude 3.8.

ε (epsilon) Ari, 2h 59m +21°.3, 330 l.y. away, is a challenging double star for apertures of 100 mm or more. High magnification reveals a tight pair of white stars of mags. 5.2 and 5.6.

λ (lambda) Ari, 1h 58m +23°.6, 132 l.y. away, is a white main-sequence star of mag. 4.8 with a yellow mag. 6.7 companion visible in small telescopes or even good binoculars.

π (pi) Ari, 2h 49m +17°.5, 780 l.y. away, is a blue-white main-sequence star of mag. 5.3, with a close mag. 7.9 companion, difficult to distinguish in the smallest telescopes.

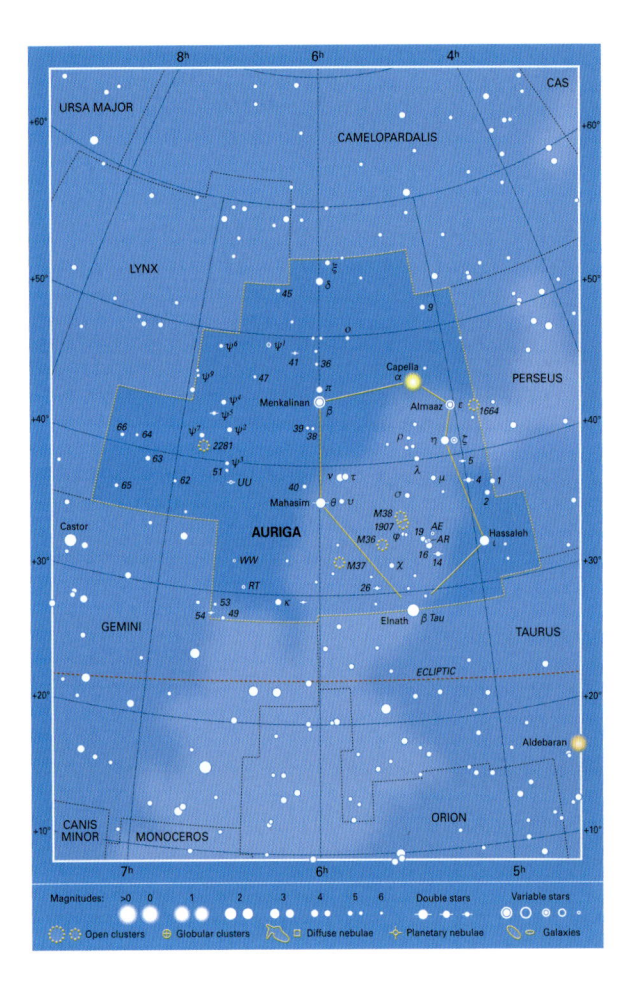

Auriga, the Charioteer

A large and prominent constellation, usually identified by the ancient Greeks as Erichthonius, a legendary king of Athens and a skilled charioteer. Auriga's leading star is Capella, sixth-brightest in the whole sky and the most northerly first-magnitude star. In legend this star represented the she-goat Amaltheia, who suckled the infant Zeus; the stars ζ (zeta) and η (eta) Aurigae are supposedly her kids and were visualized as being carried in the charioteer's arms. The star marking the charioteer's foot was originally shared with Taurus the Bull, but it is now assigned exclusively to Taurus as β (beta) Tauri. However, it once bore the alternative designation of γ (gamma) Aurigae and is shown on the chart here to complete the classical figure.

α (alpha) Aurigae, 5h 17m +46°.0, (Capella, "she-goat"), mag. 0.08, lies 43 l.y. away. It is a spectroscopic binary, consisting of two yellow giant stars orbiting every 104 days, although they do not eclipse each other.

β (beta) Aur, 6h 00m +44°.9, (Menkalinan, "shoulder of the charioteer"), 81 l.y. away, is an eclipsing variable of mag. 1.9, consisting of two blue-white stars that orbit every 3.96 days, causing two dips in brightness of 0.1 mag. on each orbit.

ε (epsilon) Aur, 5h 02m +43°.6, (Almaaz), a white supergiant about 3,000 l.y. away, is an eclipsing binary of exceptionally long period. Normally it appears at mag. 3.0, but every 27 years it sinks to mag. 3.8 as it is eclipsed by a dark companion, remaining at minimum for a year. One theory is that ε Aurigae's companion is a binary star enveloped in a disk of matter. Its next eclipse starts in late 2036.

ζ (zeta) Aur, 5h 02m +41°.1, 790 l.y. away, is a famous eclipsing binary of contrasting stars: an orange giant orbited every 972 days by a smaller blue companion. During eclipses, ζ Aurigae's brightness drops from mag. 3.7 to 4.0.

θ (theta) Aur, 6h 00m +37°.2, (Mahasim), 166 l.y. away, is a blue-white main-sequence star of mag. 2.6. It has a yellowish companion of mag. 7.2 that, because of its closeness and relative faintness, needs at least 100 mm aperture and high magnification to distinguish. This is a tough double for steady nights.

ψ1 (psi1) Aur, 6h 25m +49°.3, is a variable orange supergiant that ranges between mags. 4.7 and 5.7 with no set period. It lies roughly 7,000 l.y. away.

4 Aur, 4h 59m +37°.9, 161 l.y. away, is a double star of mags. 5.0 and 8.2, divisible through small telescopes.

14 Aur, 5h 15m +32°.7, 286 l.y. away, is a white main-sequence star of mag. 5.0, with a mag. 7.3 companion visible in small telescopes.

RT Aurigae, 6h 29m +30°.5, a yellow-white supergiant 1,800 l.y. away, is a Cepheid variable ranging between mag. 5.0 and 5.8 every 3.7 days. It is also known as 48 Aurigae.

UU Aur, 6h 37m +38°.4, is a semi-regular variable, ranging between 5th and 7th mags. with a rough period of 235 days. It is a giant star appearing deep red in color. It lies about 1,600 l.y. away.

M36 (NGC 1960), 5h 36m +34°.1, is a small, bright open cluster of about 60 stars, visible in binoculars and resolvable into stars in small telescopes (see the photograph on the previous page). It appears about one-third the diameter of the full Moon and is the most prominent of Auriga's three Messier clusters. M36 lies 3,900 l.y. away.

M37 (NGC 2099), 5h 52m +33°.5, is the largest and richest of the three Messier clusters in Auriga, containing about 150 stars. In binoculars the cluster appears as a hazy, unresolved patch about half the diameter of the full Moon, but a 100-mm telescope resolves it into a sparkling field of faint stardust, with a brighter orange star at the center. It lies 4,900 l.y. away.

M38 (NGC 1912), 5h 29m +35°.8, is a large, scattered cluster of about 100 faint stars, visible in binoculars, with a noticeable cross-shape when seen through a telescope. Its distance is 3,700 l.y. Half a degree south of it lies the small fuzzy blob of NGC 1907, a much smaller and fainter cluster about 5,300 l.y. away.

NGC 2281, 6h 49m +41°.1, is a binocular cluster of about 30 stars, 1,700 l.y. away. Through a telescope, the stars appear to be arranged in a crescent, with four brighter stars forming a diamond shape.

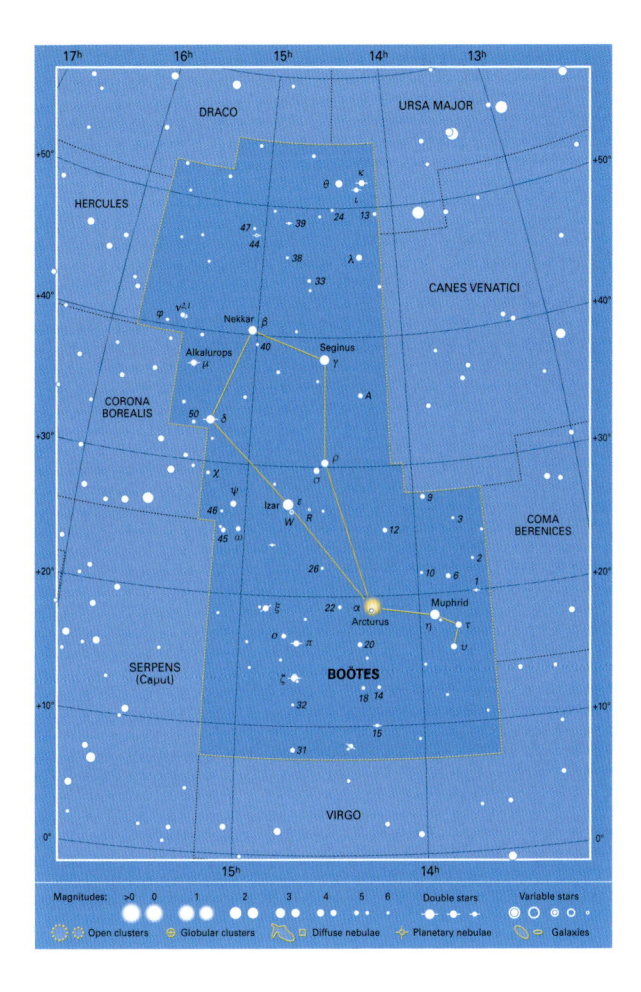

Boötes, the Herdsman

An ancient constellation, representing a herdsman driving a bear (Ursa Major) around the sky; the herdsman is often depicted holding the leash of the hunting dogs, Canes Venatici. The name of the constellation's brightest star, Arcturus, actually means "bear-keeper" in Greek. In Greek mythology, Boötes represents Arcas, the son of Zeus and the nymph Callisto—neighboring Ursa Major represents Callisto herself, who was turned into a bear by Zeus's jealous wife Hera.

Arcturus is the brightest star in the northern half of the sky, and is easily identified: the curving handle of the Big Dipper or Plough points to it. Arcturus forms the base of a large "Y" shape completed by ε (epsilon) Boötis, γ (gamma) Boötis, and α (alpha) Coronae Borealis. The other stars of the constellation are much fainter than Arcturus, but include many doubles of interest.

One of the year's most abundant meteor showers, the Quadrantids, radiates from the northern part of Boötes, an area of sky that was once occupied by the now-abandoned constellation of Quadrans Muralis, the Mural Quadrant (hence the shower's name). The Quadrantid meteors reach a peak of about 100 per hour on January 3–4 each year, although they are not as bright as other rich showers, such as the Perseids and Geminids.

α (alpha) Boötis, 14h 16m +19°.2, (Arcturus), mag. −0.05, is the fourth-brightest star in the entire sky. It is an orange giant, about 27 times the diameter of the Sun, lying 37 l.y. away; its ruddy color is noticeable to the naked eye, and is more striking with optical aid. Arcturus has a mass similar to our Sun's, and it is believed that the Sun will swell up to become a red giant like Arcturus a few billion years from now.

β (beta) Boo, 15h 02m +40°.4, (Nekkar, corrupted from the Arabic for "ox-driver," referring to the whole constellation), mag. 3.5, is a yellow giant 235 l.y. away.

γ (gamma) Boo, 14h 32m +38°.3, (Seginus), mag. 3.0, is a white star 87 l.y. away.

δ (delta) Boo, 15h 16m +33°.3, mag. 3.5, is a yellow giant or subgiant 120 l.y. away. It has a wide binocular companion of mag. 7.9.

ε (epsilon) Boo, 14h 45m +27°.1, (Izar, Arabic for "girdle" or "loincloth"), 220 l.y. away, is a celebrated double star: an orange giant primary of mag. 2.5 with a blue main-sequence companion of mag. 4.8. This close double of contrasting colors requires a telescope of at least 75 mm at ×100 power or more, because the bright primary tends to overwhelm its fainter companion, but its appearance when split has led to the alternative name Pulcherrima, meaning "most beautiful."

ι (iota) Boo, 14h 16m +51°.4, 96 l.y. away, is a wide double star of mags. 4.8 and 7.4.

κ (kappa) Boo, 14h 13m +51°.8, is an easy double star for small telescopes, consisting of unrelated components of mags. 4.5 and 6.7, at distances of 162 and 154 l.y.

μ (mu) Boo, 15h 24m +37°.4, (Alkalurops, from the Greek for "club" or "staff"), 120 l.y. distant, is an attractive triple star. To the naked eye it appears as a blue-white star of mag. 4.3, but binoculars reveal a wide companion of mag. 6.5. Telescopes of 75-mm aperture with high magnification show that this companion actually consists of two close stars of mags. 7.0 and 7.6; they orbit each other every 256 years.

ν1 ν2 (nu1 nu2) Boo, 15h 31m +40°.8, is an unrelated binocular duo: ν1 is an orange giant of mag. 5.0, 970 l.y. distant, while ν2 is a white main-sequence star, also of mag. 5.0, 440 l.y. away.

π (pi) Boo, 14h 41m +16°.4, 310 l.y. away, is a double star with blue-white components of mags. 4.9 and 5.8, visible in small telescopes.

ξ (xi) Boo, 14h 51m +19°.1, 22 l.y. away, is a showpiece double for small telescopes, consisting of yellow and orange stars of mags. 4.7 and 6.8, orbiting each other every 153 years.

44 Boo, 15h 04m +47°.7, 42 l.y. away, also known as i Boo, is a complex double–variable star. To the naked eye it appears as a yellow star of mag. 4.8. In fact, it is a binary of mags. 5.1 and 6.0, orbiting every 215 years. At their closest, around the year 2021, the stars were indivisible in most amateur telescopes, but are now opening out. By 2026 they will be divisible in 150-mm apertures, in 100 mm from 2028, and in 75 mm after 2030. The fainter star is itself an eclipsing binary with a period of 6.4 hours and a range of 0.6 mag.

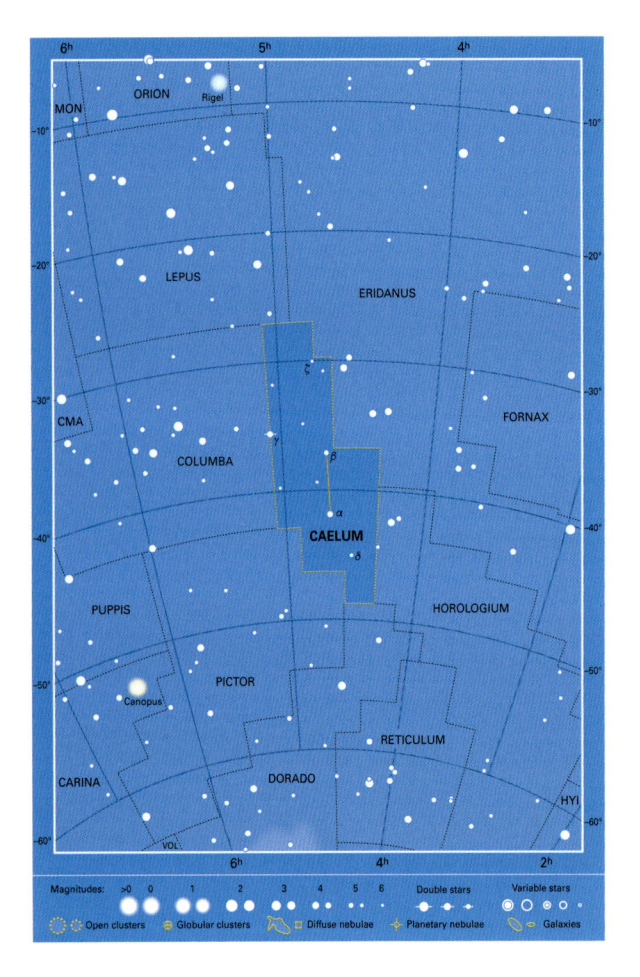

Caelum, the Chisel

An obscure, almost irrelevant constellation at the foot of Eridanus, representing a pair of engraving tools, or burins. It was one of the figures representing instruments of the arts and sciences introduced in the 1750s by the French astronomer Nicolas-Louis de Lacaille during his mapping of the southern sky.

α (alpha) Caeli, 4h 41m −41°.9, mag. 4.4, is a white main-sequence star 66 l.y. away.

β (beta) Cae, 4h 42m −37°.1, mag. 5.0, is a white subgiant 94 l.y. away.

γ (gamma) Cae, 5h 04m −35°.5, mag. 4.6, is an orange giant 185 l.y. away. It has a close 8th-mag. companion, difficult to see in the smallest apertures because of the brightness contrast.

δ (delta) Cae, 4h 31m −45°.0, mag. 5.1, is a blue-white star 680 l.y. away.

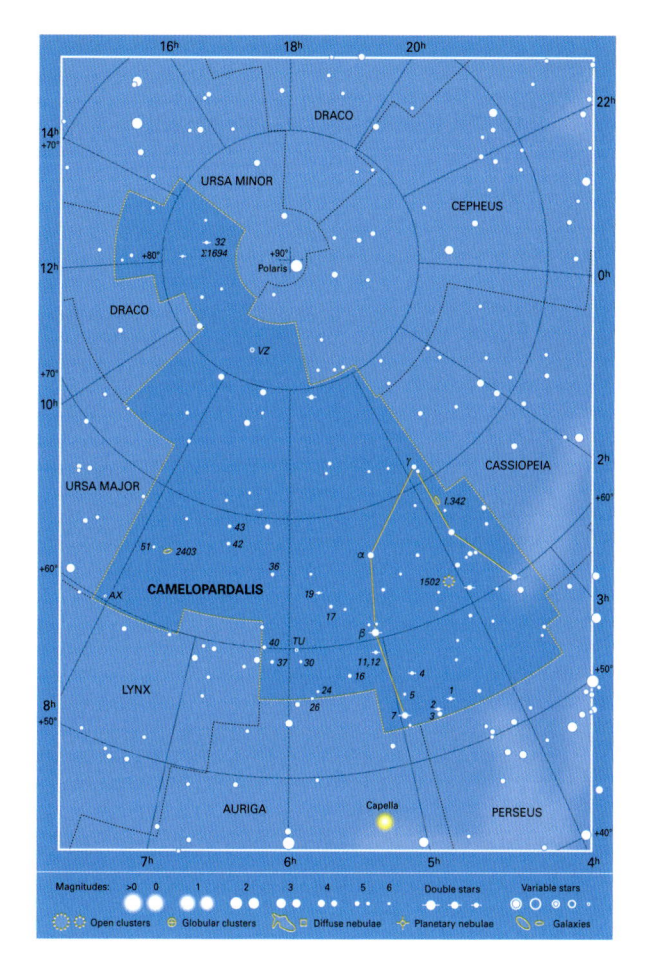

Camelopardalis, the Giraffe

A faint and obscure constellation of the north polar region of the sky, occupying an area that was left blank by the ancient Greeks. It was invented in 1612 by the Dutch theologian and astronomer Petrus Plancius. The German astronomer Jakob Bartsch wrote in 1624 that it represented the camel on which Rebecca rode into Canaan for her marriage to Isaac. But Camelopardalis is a giraffe not a camel, so its true meaning remains unclear.

α (alpha) Camelopardalis, 4h 54m +66°.3, mag. 4.3, is a highly luminous blue supergiant star approximately 6,000 l.y. away, exceptionally distant for a naked-eye star.

β (beta) Cam, 5h 03m +60°.4, at mag. 4.0 the brightest star in the constellation, is a yellow supergiant 870 l.y. distant. It has a wide mag. 7.4 companion star visible in small telescopes or even good binoculars.

11 Cam, 5h 06m +59°.0, mag. 5.1, forms an easy binocular pairing with 12 Cam, mag. 6.1. Both lie at a similar distance from us, about 700 l.y., but are too widely separated to be a genuine double.

Σ 1694 (Struve 1694), 12h 49m +83°.4, is a pair of unrelated blue-white stars, mags. 5.3 and 5.8, easily split in small telescopes. In some old catalogs the star was listed as 32 Cam.

NGC 1502, 4h 08m +62°.3, is a small 6th-mag. open star cluster with about 45 members visible in binoculars or a small telescope, somewhat triangular in shape and with an easy 7th-mag. double star at its center. It lies about 3,500 l.y. away. Note a chain of 15 or more stars called Kemble's Cascade, which runs for 2.5° from NGC 1502 towards Cassiopeia, parallel to the Milky Way (see illustration left). The brightest star in the chain is of 5th mag., about halfway along.

NGC 2403, 7h 37m +65°.6, is an 8th-mag. spiral galaxy, 0.25° long and visible as an elliptical glow in a 100-mm telescope. It lies about 12 million l.y. away.

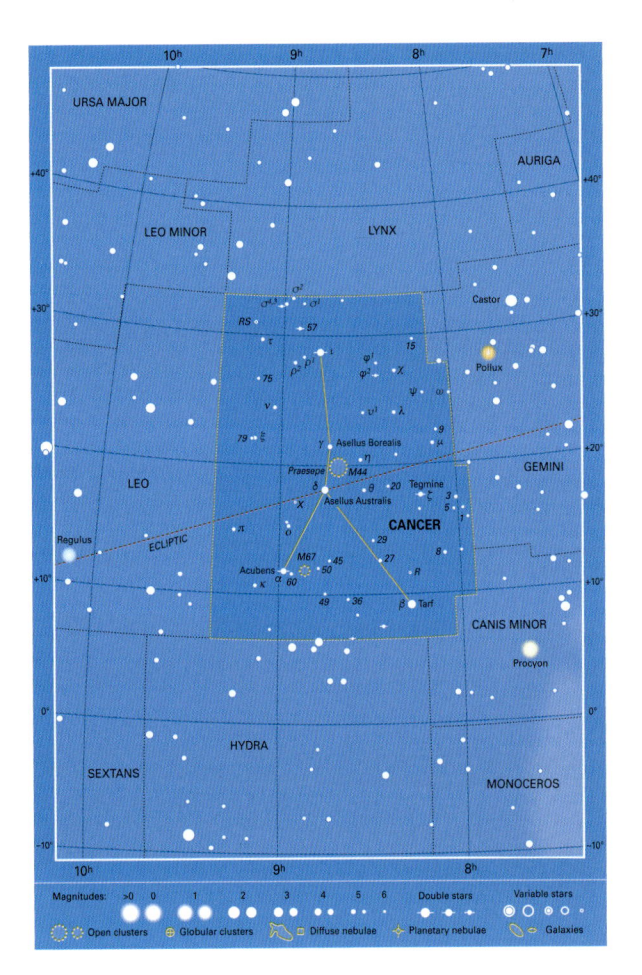

Cancer, the Crab

Cancer represents the crab that attacked Hercules when he was fighting the multiheaded Hydra; the luckless crab was crushed underfoot by mighty Hercules but was subsequently elevated to the heavens.

In ancient times, the Sun reached its most northerly point in the sky each year while it was in Cancer. The date on which the Sun is farthest north of the Earth's equator, on or around 21 June, is the northern summer solstice; on this day the Sun appears overhead at noon at latitude 23.5° north on Earth. This latitude came to be known as the Tropic of Cancer, a name it retains today even though, because of the effect of precession, the Sun now lies in the constellation of Taurus on that date. Currently the Sun is within the borders of Cancer from around July 21 to August 10 each year.

With only two stars brighter than mag. 4.0, Cancer is the faintest of the 12 constellations of the zodiac, but it nevertheless contains much of interest, notably the star cluster Praesepe, the Manger. Praesepe is flanked by a pair of 4th- and 5th-mag. stars, Asellus Borealis and Asellus Australis, the northern and southern donkeys, visualized as feeding at the stellar manger.

α (alpha) Cancri, 8h 58m +11°.9, (Acubens, "the claw"), mag. 4.2, is a white star 178 l.y. away. It has a 12th-mag. companion visible with telescopes of 75 mm aperture and over.

β (beta) Cnc, 8h 17m +9°.2, (Tarf), mag. 3.5, an orange giant 320 l.y. away, is the brightest star in the constellation.

γ (gamma) Cnc, 8h 43m +21°.5, (Asellus Borealis ,"northern donkey"), mag. 4.7, is a white star 175 l.y. away.

δ (delta) Cnc, 8h 45m +18°.2, (Asellus Australis, "southern donkey"), mag. 3.9, is an orange giant star 134 l.y. away.

ζ (zeta) Cnc, 8h 12m +17°.6, (Tegmine), 80 l.y. away, is an interesting multiple star. To the naked eye it appears of mag. 4.6, but a small telescope reveals two yellow stars of mags. 4.9 and 6.2; they form a genuine binary with an estimated orbital period of about 700 years. Larger telescopes split the brighter component into a tight binary of mags. 5.4 and 6.0, orbital period 59.4 years. When at their closest, around the year 2048, at least 200 mm will be needed to separate this pair, but at their widest, around 2078, 100 mm should be sufficient.

ι (iota) Cnc, 8h 47m +28°.8, 346 l.y. away, is a yellow giant of mag. 4.0 with a blue-white mag.-6.6 companion just visible in binoculars, or easily seen through a small telescope.

M44 (NGC 2632), 8h 40m +20°.0, Praesepe ("manger"), commonly called the Beehive Cluster, is a swarm of about 50 stars of 6th mag. and fainter, visible as a misty patch to the naked eye and best seen through binoculars. The brightest member of this open cluster, ε (epsilon) Cancri, is mag. 6.3. Praesepe sprawls over 1.5° of sky, three times the apparent diameter of the Moon. The distance to the cluster's center is just over 600 l.y.

M67 (NGC 2682), 8h 50m +11°.8, is a smaller and denser open cluster than M44, visible as a Moon-sized misty ellipse in binoculars or small telescopes. Apertures of at least 75 mm are needed to resolve the brightest of its 200 or so individual stars, which are of 10th mag. and fainter. It lies 2,900 l.y. away.

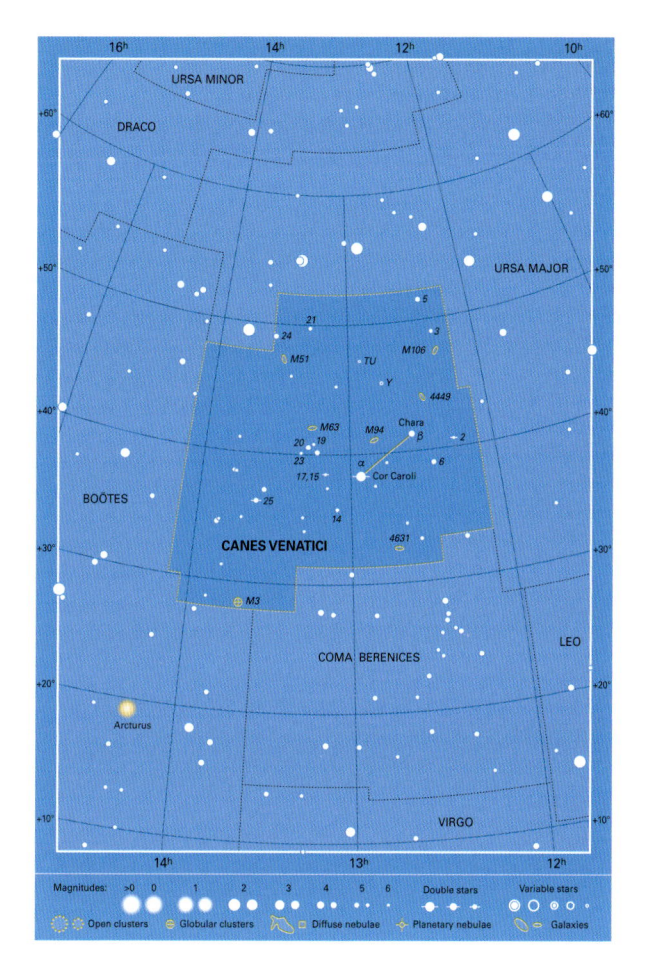

Canes Venatici, the Hunting Dogs

A constellation introduced in 1687 by the Polish astronomer Johannes Hevelius, consisting of a sprinkling of faint stars below Ursa Major. It represents two dogs, Asterion ("starry") and Chara ("dear"), held on a leash by neighboring Boötes as they pursue the Great Bear around the pole. Canes Venatici contains numerous galaxies, the most famous being M51, the Whirlpool, a beautiful face-on spiral. It was the first galaxy in which spiral form was detected, by Lord Rosse in 1845 with his 72-inch (1.8-meter) reflector at Birr Castle, Ireland.

α (alpha) Canum Venaticorum, 12h 56m +38°.3, is popularly called Cor Caroli, meaning "Charles's heart," a reference to the executed King Charles I of England. The name was given in the 1670s by Sir Charles Scarborough, physician to King Charles I's son, Charles II. It is a double star of mags. 2.9 and 5.6, easily split in small telescopes. Both stars are white, but various observers have reported subtle shades of color in them when viewed through a telescope. The brighter star, α2, is the standard example of a rare class of stars with strong and variable magnetic fields; its brightness fluctuates slightly but not enough to be noticeable to the eye. Cor Caroli is 105 l.y. away.

β (beta) CVn, 12h 34m +41°.4, (Chara, "dear"), mag. 4.3, is the only other star of any prominence in the constellation. It is a yellow main-sequence star similar to the Sun, 28 l.y. away.

Y CVn, 12h 45m +45°.4, is a semi-regular variable supergiant of deep red color popularly known as La Superba. Its range is about mag. 4.9 to 5.9 and its period is approximately 270 days. It lies about 1,000 l.y. away.

M3 (NGC 5272), 13h 42m +28°.4, is a rich globular cluster located midway between Cor Caroli and Arcturus, regarded as one of the finest globulars in the northern sky. At 6th mag. it is on the naked-eye limit, but can be picked up easily as a hazy star in binoculars or a small telescope; a 6th-mag. star nearby acts as a guide. In small telescopes, the cluster appears as a condensed ball of light with a faint outer halo. Apertures of 100 mm or more are needed to resolve individual stars in its outer regions. M3 is about 30,000 l.y. away.

M51 (NGC 5194), 13h 30m +47°.2, the Whirlpool Galaxy, is an 8th-mag. spiral galaxy about 26 million l.y. away, with a smaller satellite galaxy, NGC 5195, apparently lying at the end of one of its arms; in reality, this companion lies slightly behind M51, having brushed past it some time in the past 100 million years or so. The Whirlpool can be seen in binoculars, appearing elongated. It is disappointing in small telescopes, which show a faint milky radiance around the star-like nuclei of the galaxy and its satellite; apertures of at least 250 mm are needed to see the spiral arms well. Nevertheless, M51 is well worth hunting for on clear, dark nights.

M63 (NGC 5055), 13h 16m +42°.0, is a 9th-mag. spiral galaxy visible in small telescopes as an elliptical haze with a somewhat mottled texture. It is popularly known as the Sunflower Galaxy because of its appearance in large instruments.

M94 (NGC 4736), 12h 51m +41°.1, is a compact spiral galaxy presented nearly face-on. In amateur telescopes it looks like an 8th-mag. comet, with a fuzzy star-like nucleus surrounded by an elliptical halo. M94 is about 15 million l.y. away.

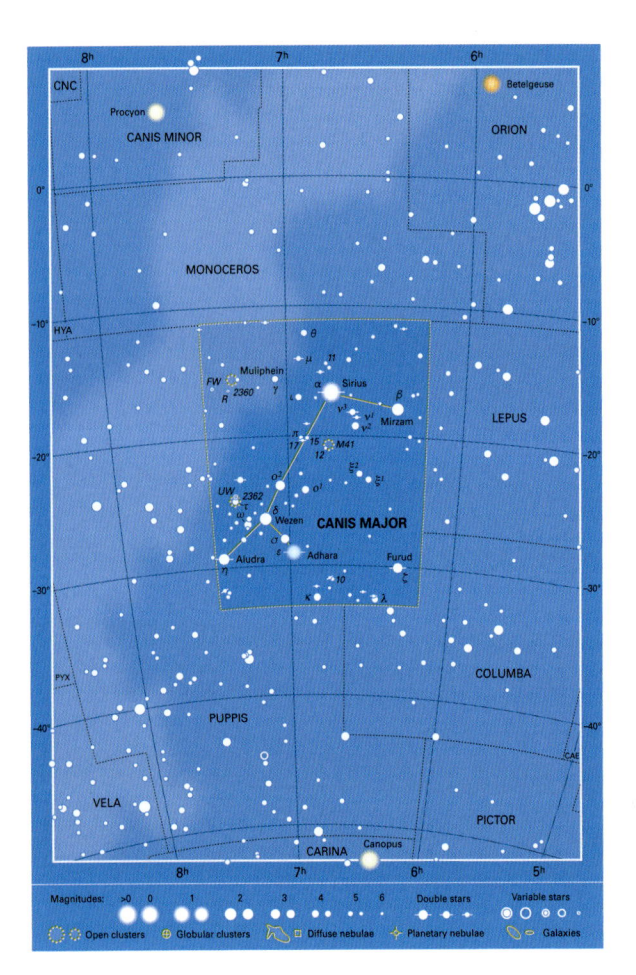

Canis Major, the Greater Dog

An ancient constellation, representing one of the two dogs (the other being Canis Minor) following at the heels of Orion. In other Greek legends it was identified as Laelaps, a dog so swift that no prey could outrun it. Laelaps was sent in pursuit of a fox that could never be caught. The chase would have continued for eternity had not Zeus plucked up the dog and placed it in the sky as Canis Major, although without the fox.

Canis Major contains many brilliant stars, making it one of the most prominent constellations; its leading star, Sirius, is the brightest star in the entire night sky as seen from Earth. Sirius features in many legends, and the ancient Egyptians based their calendar on its yearly motion around the sky.

α (alpha) Canis Majoris, 6h 45m −16°.7, (Sirius, from the Greek meaning "searing" or "scorching"), mag. −1.46, is a brilliant white main-sequence star 8.6 l.y. away, one of the Sun's closest neighbors. It has a white dwarf companion of mag. 8.4 that orbits it every 50.1 years. The brilliance of Sirius overpowers this white dwarf so that even when the two stars are at their greatest separation, as between the years 2020 and 2025, telescopes of 200 mm aperture or more and steady atmospheric conditions are required to see it. The two will next be at their closest together in 2043, when they will be inseparable in all but the largest apertures.

β (beta) CMa, 6h 23m −18°.0, (Mirzam, "the announcer," i.e. of Sirius), mag. 2.0, is a blue giant 490 l.y. away. It is a pulsating star whose variations, of a few hundredths of a magnitude every 6 hours, are undetectable to the naked eye.

δ (delta) CMa, 7h 08m −26°.4, (Wezen, "the weight"), mag. 1.8, is a white supergiant star about 1,600 l.y. away.

ε (epsilon) CMa, 6h 59m −29°.0, (Adhara, "the virgins"), mag. 1.5, is a blue giant 405 l.y. away. It has a mag.-7.5 companion that is difficult to see in small telescopes because of the glare from the primary.

η (eta) CMa, 7h 24m −29°.3, (Aludra), mag. 2.4, is a blue supergiant about 2,000 l.y. away.

μ (mu) CMa, 6h 56m −14°.0, mag. 5.1, is an orange giant about 950 l.y. away, with a close 7th-mag. blue-white companion difficult to pick up in the smallest apertures because of the magnitude contrast.

ν1 (nu1) CMa, 6h 36m −18°.7, mag. 5.7, is a yellow giant 280 l.y. away, with a mag.-7.4 companion, visible in small telescopes.

W CMa, 7h 19m −24°.6, is an eclipsing binary of β (beta) Lyrae type that varies between mags. 4.8 and 5.3 in 4.4 days. It lies about 3,800 l.y. away.

M41 (NGC 2287), 6h 47m −20°.7, is a large and bright open cluster of about 80 stars, easily visible through binoculars or a small telescope and, with a total magnitude of 4.5, detectable by the naked eye under good conditions—it was known to the ancient Greeks. A low-power view in a small telescope shows the individual stars grouped in bunches and curves, covering an area of sky equivalent to the apparent diameter of the Moon. The brightest stars in the cluster are 7th-mag. orange giants. M41 is 2,300 l.y. away.

NGC 2362, 7h 19m −25°.0, is a compact cluster surrounding the mag. 4.4 blue giant star τ (tau) Canis Majoris, which is a genuine member. Small telescopes show about 60 stars in the cluster, which lies some 4,800 l.y. away.

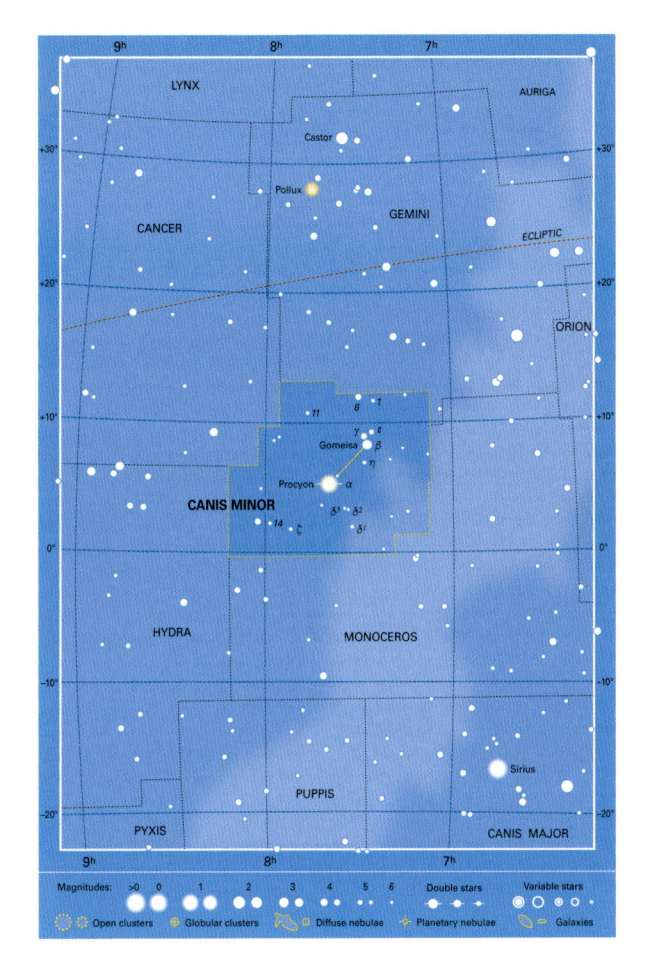

Canis Minor, the Lesser Dog

The second of the two dogs of Orion, the other being Canis Major. In one Greek legend, the constellation represents Maera, dog of Icarius, the man who was first taught to make wine by the god Dionysus. When some shepherds drank the wine they became intoxicated. Suspecting that Icarius had poisoned them they killed him. Maera the dog ran howling to Icarius's daughter Erigone and led her to her father's body. Both Erigone and the dog took their own lives where Icarius lay. Zeus placed their images among the stars as a reminder of the tragic affair. In this story, Icarius is identified with the constellation Boötes, Erigone is Virgo, and Maera is Canis Minor.

Apart from its leading star Procyon, which is the eighth-brightest star in the sky, there are few objects of importance in Canis Minor. Procyon forms a prominent equilateral triangle with the bright stars Sirius (in Canis Major) and Betelgeuse (in Orion).

α (alpha) Canis Minoris, 7h 39m +5°.2, (Procyon, from the Greek meaning "preceding the dog," referring to its rising before Sirius), mag. 0.37, is a yellow-white subgiant 11.5 l.y. away, and therefore among the nearest stars to the Sun. Like Sirius, Procyon has a white dwarf companion, but this star, of mag. 10.8, is even more difficult to see than the companion of Sirius, requiring the use of large professional telescopes. Procyon's companion orbits it every 41 years.

β (beta) CMi, 7h 27m +8°.3, (Gomeisa), mag. 2.9, is a blue-white main-sequence star 162 l.y. away.

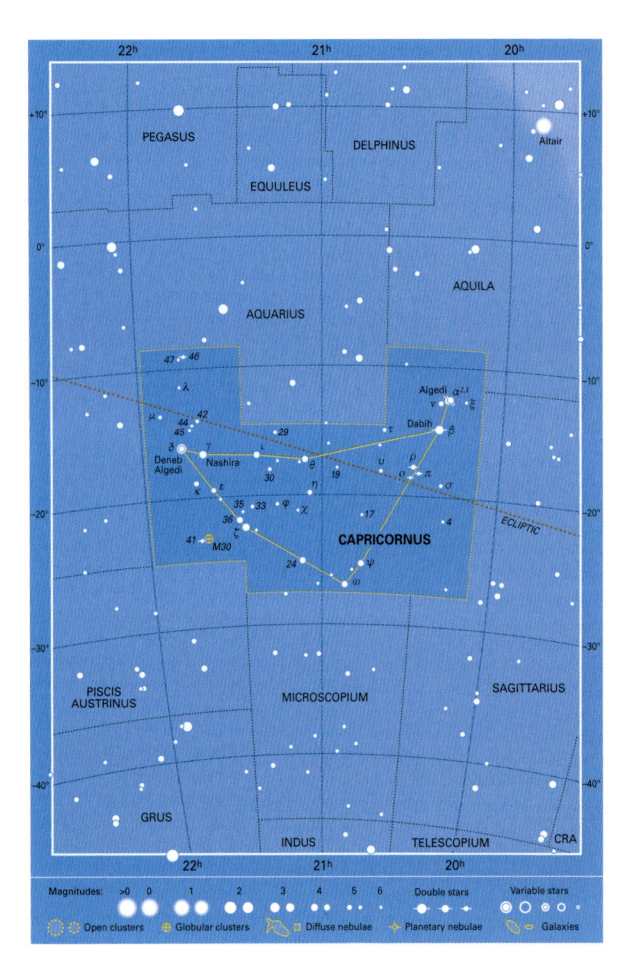

Capricornus, the Sea Goat

Capricornus is depicted as a goat with a fish's tail. Amphibious creatures feature prominently in ancient legends, and the origin of Capricornus certainly dates back to ancient times. In Greek legend, the constellation represented the goat-headed god Pan, who jumped into a river to escape the approach of the monster Typhon, turning his lower half into a fish.

Before about 130 BCE the Sun lay in Capricornus when it reached its farthest point south of the equator each year; this point, known as the northern winter solstice, currently occurs on December 21 or 22 (the date can vary from year to year). The latitude on Earth at which the Sun appears overhead at noon on that date, 23.5° south, became known as the Tropic of Capricorn. Because of precession, the northern winter solstice has since moved from Capricornus into the neighboring constellation Sagittarius (and will reach Ophiuchus in the year 2269), but the Tropic of Capricorn retains its name.

Capricornus is the smallest constellation of the zodiac. The Sun is within its boundaries from late January to mid-February.

$α1$ $α2$ (alpha1 alpha2) Capricorni, 20h 18m −12°.5, (Algedi, from the Arabic meaning "the kid," in reference to the constellation as a whole) is a multiple star, consisting of a yellow supergiant of mag. 4.3 and an unrelated yellow giant of mag. 3.6, lying 810 and 109 l.y. away respectively; the two stars are visible separately with the naked eye or binoculars. Telescopes reveal that each star is itself double. The fainter of the pair, $α1$, has a wide mag.-9.6 companion visible in small telescopes; $α2$ has its own companion of mag. 11. Telescopes of at least 100 mm aperture show that this faint companion is itself composed of two 11th-mag. stars. $α$ Capricorni is therefore a fascinating hybrid system.

$β$ (beta) Cap, 20h 21m −14°.8, (Dabih, from the Arabic meaning "the lucky stars of the slaughterer"), about 350 l.y. away, is a golden-yellow giant star of mag. 3.1, with a wide, blue-white companion of mag. 6.1, visible through binoculars or small telescopes.

$γ$ (gamma) Cap, 21h 40m −16°.7, (Nashira), mag. 3.7, is a white star 170 l.y. away.

$δ$ (delta) Cap, 21h 47m −16°.1, (Deneb Algedi), mag. 2.8, is the brightest star in the constellation. It is an eclipsing binary of $β$ (beta) Lyrae type, varying by a barely perceptible 0.2 mag. over 24.5 hours. It lies 39 l.y. away.

$π$ (pi) Cap, 20h 27m −18°.2, 650 l.y. away, is a blue-white main-sequence star of mag. 5.1 with a close mag. 8.5 companion visible with a small telescope.

M30 (NGC 7099), 21h 40m −23°.2, is a 7th-mag. globular cluster 24,000 l.y. away, visible in small telescopes and resolvable in 100 mm aperture, notably centrally condensed with fingerlike chains of stars extending northward. It lies next to 41 Capricorni, a 5th-mag. foreground star.

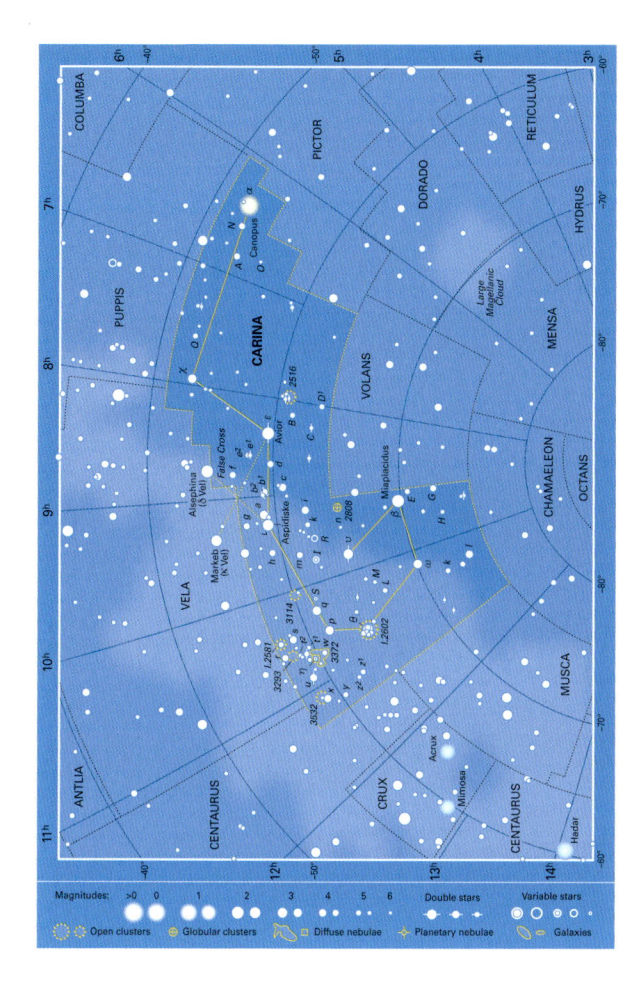

Carina, the Keel

This constellation was originally part of the extensive Argo Navis, the ship of the Argonauts, until that figure was subdivided into three parts in 1763 by the French celestial surveyor Nicolas Louis de Lacaille. The other two parts are Puppis and Vela. Carina, the smallest of the three, represents the ship's keel or hull. As a part of Argo Navis, Carina originated in ancient Greek times and is associated with the legend of Jason and the Argonauts and their quest for the Golden Fleece. Carina lies in the Milky Way, providing rich starfields and clusters for binoculars. The stars ι (iota) and ε (epsilon) Carinae, together with κ (kappa) and δ (delta) Velorum, form a shape known as the False Cross, sometimes confused with the real Southern Cross.

α (alpha) Carinae, 6h 24m −52°.7, (Canopus), mag. −0.74, the second-brightest star in the sky, is a white giant 309 l.y. away. It is named after the helmsman of the Greek king Menelaus, and appropriately enough is now used by spacecraft as a guide for navigation.

β (beta) Car, 9h 13m −69°.7, (Miaplacidus), mag. 1.7, is a blue-white giant star 113 l.y. away.

ε (epsilon) Car, 8h 23m −59°.5, (Avior), mag. 1.9, is an orange giant star about 600 l.y. away.

η (eta) Car, 10h 45m −59°.7, 7,500 l.y. away, is a peculiar nova-like variable star embedded in the nebula NGC 3372. In the past, η Carinae has fluctuated erratically in brightness, reaching a maximum of mag. −1 in 1843, when it was temporarily the second-brightest star in the sky; it subsequently settled at around 6th mag., but brightened to 5th mag. in 1998 and reached 4th mag. in 2015. The star is estimated to be over 100 times more massive and 4 million times brighter than the Sun, with an unseen companion orbiting it every 5.5 years. The binary pair is surrounded by a shell of dust and gas thrown off in the 1843 outburst.

θ (theta) Car, 10h 43m −64°.4, mag. 2.8, is a blue-white main-sequence star 456 l.y. away, a member of the sparkling open cluster IC 2602.

ι (iota) Car, 9h 17m −59°.3, (Aspidiske), mag. 2.3, is a white supergiant 766 l.y. away.

υ (upsilon) Car, 9h 47m −65°.1, is a double star divisible in small telescopes, consisting of a white giant of mag. 3.0 about 1,400 l.y. away, with a probably unrelated white companion of mag. 6.0.

l Car, 9h 45m −62°.5, a yellow supergiant 1,640 l.y. away, is the brightest Cepheid, whose variations are large enough to be obvious to the naked eye. It rises and falls between mags. 3.3 and 4.1 every 35.6 days.

R Car, 9h 32m −62°.8, 590 l.y. away, is a red giant variable star of Mira type that ranges between 4th and 10th mags. with a period of 306 days.

S Car, 10h 09m −61°.5, about 1,500 l.y. away, is a red giant variable similar to nearby R Car, varying from 5th to 10th mag. over 150 days.

NGC 2516, 7h 58m −60°.9, is a naked-eye open cluster of some 80 stars, 1,350 l.y. away, as large as the full Moon. It is a sparkling sight in binoculars, which show it to be cross-shaped. Its brightest member is a mag.-5.2 red giant.

NGC 3114, 10h 03m −60°.1, is a widely scattered open cluster the same apparent size as the full Moon, containing stars of 6th mag. and fainter, around 3,400 l.y. away. It is best seen in binoculars and small telescopes with low power.

NGC 3372, 10h 44m −59°.9, is a celebrated diffuse nebula easily visible to the naked eye as a brilliant patch of the Milky Way four Moon diameters wide, surrounding the erratic variable star η (eta) Carinae. The nebula shines from the light of brilliant young stars born within it. Binoculars and small telescopes show jeweled star clusters and swirls of glowing gas alternating with dark lanes. Most famous is a dark notch, called the Keyhole because of its distinctive shape, silhouetted against the nebula's brightest central portion near η Carinae itself. The whole NGC 3372 nebula lies about 7,500 l.y. from us, the same distance as η Carinae.

NGC 3532, 11h 06m −58°.7, 1,580 l.y. away, is an outstanding open cluster, visible to the naked eye as a brighter patch nearly 1° wide among rich Milky Way starfields and glorious in binoculars. It contains 150 or so stars of 7th mag. and fainter, including several orange giants, arranged in an elliptical shape with a star-free lane across its center. The mag.-3.8 yellow supergiant x Carinae (V382 Car) at the cluster's edge is not a member but a background object four times farther off.

IC 2602, 10h 43m −64°.4, is a large and brilliant open cluster containing a few dozen stars, popularly known as the Southern Pleiades, centered just under 500 l.y. away. Its brightest members are visible to the naked eye, notably the blue-white star θ Carinae, mag. 2.8. The whole cluster appears twice as wide as the full Moon.

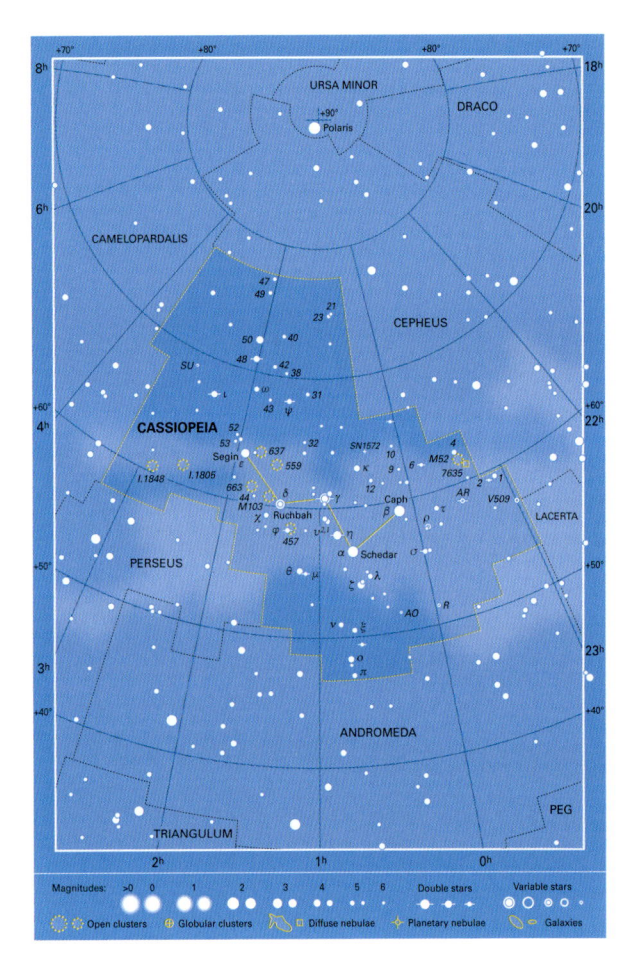

Cassiopeia

In Greek legend, Cassiopeia was the beautiful but boastful queen of Ethiopia, wife of King Cepheus and mother of Andromeda. In the sky she is depicted sitting in a chair. The constellation is easily identifiable by the distinctive W-shape of its five brightest stars. Cassiopeia lies on the opposite side of the Pole Star from Ursa Major, in a rich part of the Milky Way.

The brilliant supernova outburst of 1572, which reached mag. −4, comparable to Venus, occurred near the star κ (kappa) Cassiopeiae, at 0h 25.3m, +64° 09'. It is now known as Tycho's Star because it was observed by the great Danish astronomer Tycho Brahe. The remains of another supernova, which erupted around 1660 but which went unseen at the time, form the strongest radio source in the sky, Cassiopeia A; it lies at 23h 23.4m, +58° 50', and is some 10,000 l.y. away.

α (alpha) Cassiopeiae, 0h 41m +56°.5, (Schedar, "the breast"), mag. 2.2, is an orange giant star 228 l.y. away. It has a wide 9th-mag. companion, which is unrelated.

β (beta) Cas, 0h 09m +59°.1, (Caph), mag. 2.3, is a white giant star 55 l.y. away.

γ (gamma) Cas, 0h 57m +60°.7, 550 l.y. away, is a remarkable blue-white variable star that rotates so rapidly that it is unstable, throwing off rings of gas from its equatorial region at irregular intervals. The ejections of gas cause it to vary unpredictably between mags. 3.0 and 1.6. Currently it hovers around mag. 2.2.

δ (delta) Cas, 1h 26m +60°.2, (Ruchbah, "the knee"), mag. 2.7, is a blue-white subgiant star 100 l.y. away. It is an eclipsing binary of the Algol type that varies by about 0.1 mag. with the relatively long period of 2 years and 1 month.

ε (epsilon) Cas, 1h 54m +63°.7, (Segin), mag. 3.4, is a blue-white star 465 l.y. away.

η (eta) Cas, 0h 49m +57°.8, 19 l.y. away, is a beautiful double star with yellow and red components of mags. 3.5 and 7.4, visible in small telescopes. They form a true binary with a period of 480 years.

ι (iota) Cas, 2h 29m +67°.4, 148 l.y. away, is a mag.-4.6 white star with a wide mag.-8.5 companion visible through a 60-mm telescope. With an aperture of 100 mm and high magnification, the brighter star is seen to have a closer mag.-6.9 yellow companion of its own, making this an impressive triple.

ρ (rho) Cas, 23h 54m +57°.5, is a yellow-white supergiant, one of the most luminous stars known, giving out as much light as half a million Suns. It is a semi-regular pulsating variable, ranging between mag. 4.1 and 6.2 every 2.25 years or so. Its distance is not known accurately, but is probably over 3,000 l.y.

σ (sigma) Cas, 23h 59m +55°.8, around 1,300 l.y. away, is a close pair of mags. 5.0 and 7.2 that appear green and blue, in striking contrast to the warmer hues of η Cas. An aperture of 75 mm and high power will split the pair.

ψ (psi) Cas, 1h 26m +68°.1, 200 l.y. away, is a mag.-4.7 orange giant with a wide 9th-mag. companion visible in a small telescope. High powers reveal that this companion is itself a close binary.

M52 (NGC 7654), 23h 24m +61°.6, is an open cluster of about 100 stars, 5,400 l.y. away, visible as a misty patch in binoculars. It is somewhat kidney-shaped, with an 8th-mag. orange star embedded at one edge, like a poorer version of the Wild Duck Cluster (M11 in Scutum). M52 can be resolved into stars with 75 mm aperture.

M103 (NGC 581), 1h 33m +60°.7, is a small, elongated group of about 25 faint stars, 8,700 l.y. away. What appears to be the brightest member, a double of 7th and 10th mags. near the cluster's northern tip, is in fact a foreground object.

NGC 457, 1h 19m +58°.3, is a loose open cluster of about 80 stars, 10,000 l.y. away, seemingly arranged in chains. It has been given various imaginative names inspired by its shape, such as the Dragonfly Cluster, the ET Cluster, and the Owl Cluster. The mag.-5.0 white supergiant φ (phi) Cas on its southern outskirts is probably a true member.

NGC 663, 1h 46m +61°.2, is a prominent binocular cluster of about 80 stars appearing half the size of the full Moon, 10,000 l.y. away.

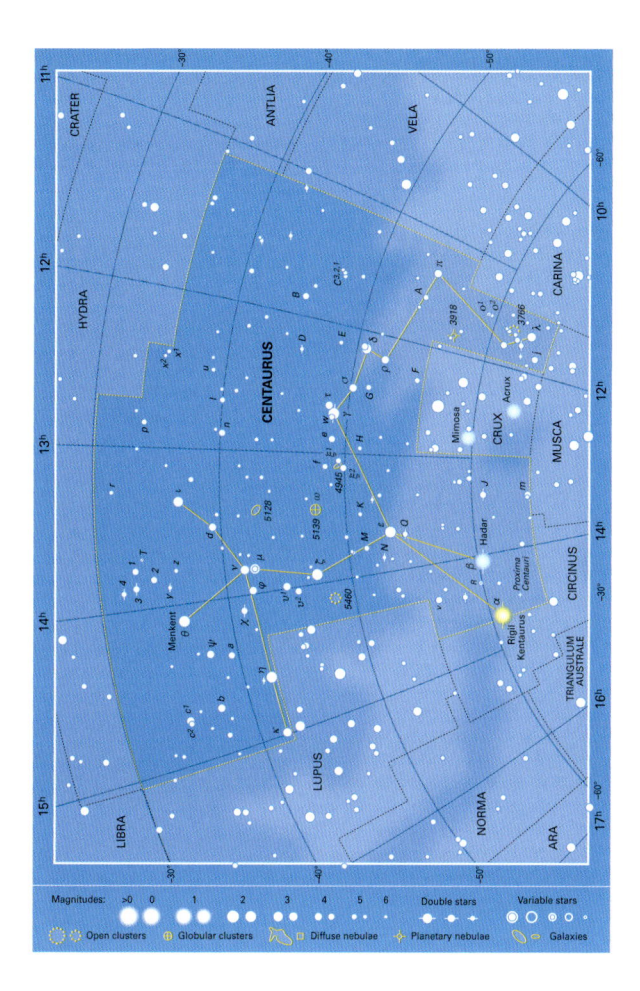

Centaurus, the Centaur

A large and rich constellation representing a centaur, the mythical beast that was half man, half horse. Reputedly, Centaurus depicts the scholarly centaur Chiron, the tutor of many Greek gods and heroes, who was raised to the sky after being accidentally struck by a poisoned arrow from Hercules.

A line from α (alpha) through β (beta) Centauri points to Crux, the Southern Cross. Currently α and β Centauri are about 4.5° apart, but the rapid proper motion of α Cen is taking it toward β Cen, and about 4,000 years from now the two will be little more than 0.5° (one Moon diameter) apart in our skies, making a stunning naked-eye pairing. Of particular note in Centaurus is the closest star to the Sun, Proxima Centauri, an 11th-mag. red dwarf, which is an outlying companion of α Centauri.

One of the strongest radio sources in the sky, Centaurus A, is associated with the unusual galaxy NGC 5128. Centaurus lies in a prominent part of the Milky Way and contains more naked-eye stars than any other constellation: 281 brighter than magnitude 6.5, according to the Hipparcos Catalogue.

α (alpha) Centauri, 14h 40m −60°.8, (Rigil Kentaurus, sometimes abbreviated as Rigil Kent, "foot of the centaur") lies 4.3 l.y. away. To the naked eye it shines at mag. −0.27, the third-brightest star in the sky. But the smallest of telescopes reveals that it consists of twin yellow stars of mags. 0.01 and 1.33; the brighter of these is very similar to the Sun. They orbit each other every 80 years and are always divisible in amateur telescopes, although at their closest, around the years 2037–8, they will need 75 mm aperture. Also associated with α Centauri is an 11th-mag. red dwarf called Proxima Centauri, lying 2° away and therefore not even in the same telescopic field of view (see the finder chart above). This star is estimated to take as long as a million years to orbit its two brilliant companions. At present, Proxima Centauri is about 0.1 l.y. closer to us than the two other members of α Centauri. Proxima Centauri is a flare star, suddenly brightening by as much as one magnitude for several minutes at a time.

β (beta) Cen, 14h 04m −60°.4, (Hadar), mag. 0.6, is a blue giant 390 l.y. away. It is in fact a very close double with a 4th-mag. companion, divisible only in large apertures.

γ (gamma) Cen, 12h 42m −49°.0, 130 l.y. away, is a close double with blue-white components of mag. 2.8 and 2.9, orbiting each other every 84 years. Together they shine as a star of mag. 2.2. When at their closest, as between 2010 and 2020, amateur telescopes are unable to split them, but they are currently moving apart and are divisible in 150 mm from 2026. They are widest apart around 2053–6, when 75 mm should divide them.

3 Cen, 13h 52m −33°.0, 330 l.y. away, is a blue-white giant star of mag. 4.5 with a main-sequence companion of mag. 6.0, forming a striking pair in small telescopes.

R Cen, 14h 17m −59°.9, is a red giant variable of Mira type, which varies between mags. 5.3 and 11.8 in about 17 months. It lies nearly 3,000 l.y. away.

ω (omega) Cen (NGC 5139), 13h 27m −47°.5, is the largest and brightest globular cluster in the sky — so prominent that it was labeled as a star on early charts. It appears as a hazy star of mag. 3.7 to the naked eye, noticeably elliptical in shape and as large as the full Moon. It is inherently the most luminous of all globulars, with the light output of a million Suns. Small telescopes or even binoculars begin to resolve its outer regions into stars, and it is a showpiece for all apertures. Its brilliance and large apparent size are due in part to its relative closeness, 17,000 l.y., which places it among the nearest globular clusters to us.

NGC 3766, 11h 36m −61°.6, is a naked-eye open cluster of about 100 stars of mag. 7 and fainter, resolvable in binoculars, 7,100 l.y. away.

NGC 3918, 11h 50m −57°.2, is an 8th-mag. planetary nebula about 15,000 l.y. away, discovered in the 19th century by John Herschel and called by him the Blue Planetary. It is similar in appearance to the planet Uranus, but three times the apparent diameter. Its central star, of 11th mag., should be detectable in modest amateur instruments.

NGC 5128, 13h 25m −43°.0, is a peculiar 7th-mag. galaxy, visually the brightest galaxy outside the Local Group, known to radio astronomers as Centaurus A. In long-exposure photographs such as the one on page 295 it appears as a giant elliptical galaxy with an encircling band of dust. It is thought to result from a merger between an elliptical and a spiral galaxy. Under good skies it is visible in binoculars as a rounded glow, but at least 100 mm aperture is necessary to trace its outline and the dark bisecting lane of dust. NGC 5128 is 12 million l.y. away.

NGC 5460, 14h 08m −48°.3, is a large, 6th-mag. open cluster of about 40 stars visible in binoculars or small telescopes. It lies 2,400 l.y. away.

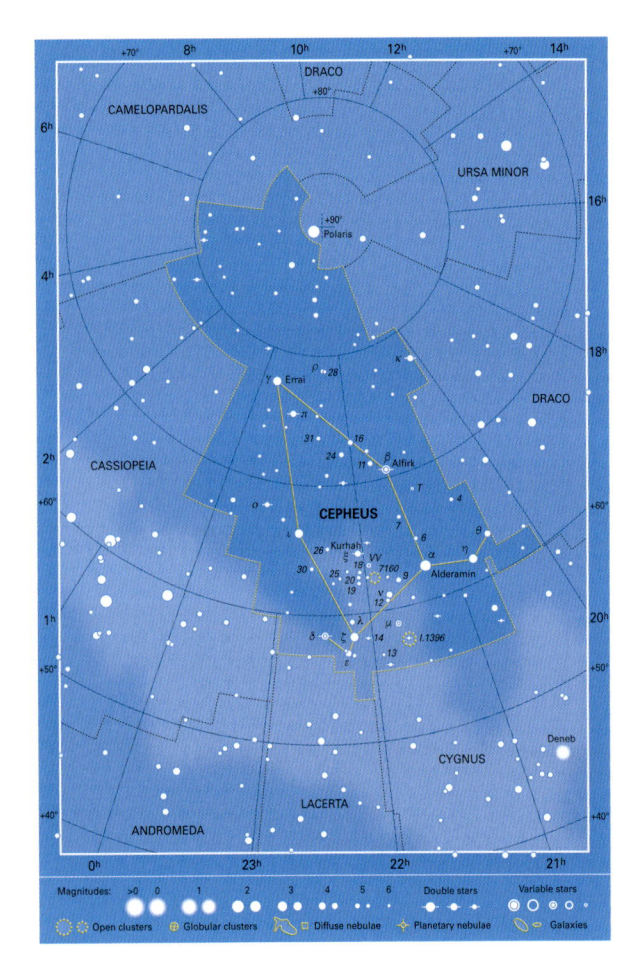

Cepheus

An ancient constellation representing the mythological King Cepheus of Ethiopia, the husband of Cassiopeia and father of Andromeda, who are themselves depicted by nearby constellations. Cepheus is replete with double and variable stars, including the celebrated δ (delta) Cephei, prototype of the Cepheid variables, which are used as "standard candles" for distance-finding in space. This star's fluctuations in light output were discovered in 1784 by the English amateur astronomer John Goodricke.

α (alpha) Cephei, 21h 19m +62°.6, (Alderamin), mag. 2.5, is a white main-sequence star 49 l.y. away.

β (beta) Cep, 21h 29m +70°.6, (Alfirk, meaning "flock," i.e. of sheep), 685 l.y. away, is both a double and a variable star. A small telescope shows that this blue giant, of mag. 3.2, has a mag.-8.6 companion. β Cephei is the prototype of a class of pulsating variable stars (also known as β Canis Majoris stars) with periods of a few hours and tiny brightness fluctuations. Over a period of 4.6 hours or so, β Cephei varies by 0.1 mag., an amount indistinguishable to the naked eye but detectable by sensitive instruments.

γ (gamma) Cep, 23h 39m +77°.6, (Errai, "the shepherd"), mag. 3.2, is an orange giant or subgiant 45 l.y. away.

δ (delta) Cep, 22h 29m +58°.4, 920 l.y. away, is a famous pulsating variable star, the prototype of the classic Cepheid variables. This yellow supergiant varies between mags. 3.5 and 4.4 in 5 days 9 hours, changing in size between about 40 and 46 times the Sun's diameter as it does so. Less well known is that δ Cephei is also an attractive double star for binoculars or the smallest telescopes, with a wide, bluish mag.-6.1 companion.

μ (mu) Cep, 21h 44m +58°.8, is a famous red star, called the Garnet Star by William Herschel because of its striking tint, which is notable in binoculars. μ Cephei is a red supergiant, a prominent example of a class of variable stars known as semi-regular variables. It varies between mags. 3.4 and 5.1 with a period of just over 2 years. The distance is uncertain, but is probably several thousand light years. Its diameter is estimated to be over one thousand times that of the Sun, making it one of the largest stars known, a distinction it coincidentally shares with VV Cephei in the same constellation (see below).

ξ (xi) Cep, 22h 04m +64°.6, (Kurhah), 101 l.y. away, is a double star of mags. 4.4 and 6.3 visible in small telescopes. The components are blue-white and yellow and form a true binary with an estimated orbital period of around 2,500 years.

ο (omicron) Cep, 23h 19m +68°.1, 200 l.y. away, is an orange giant of mag. 4.9 with a close mag.-7.3 companion for telescopes of 60 mm aperture and above. They orbit each other about every 2,200 years.

T Cep, 21h 10m +68°.5, 600 l.y. away, is a red giant variable of the Mira type, about 500 times the diameter of the Sun, ranging between mags. 5.2 and 11.3 in around 13 months.

VV Cep, 21h 57m +63°.6, is an enormous red supergiant that varies semi-regularly between mags. 4.8 and 5.4. It is also an eclipsing binary with the unusually long period of 20.5 years, but the light dip during eclipse of its 8th-mag. blue companion is too slight to be noticeable to the naked eye. The distance of VV Cephei is around 3,250 l.y. and its diameter is probably over a thousand Suns, placing it among the largest stars known, a distinction it shares with another red supergiant in the constellation, Mu Cephei.

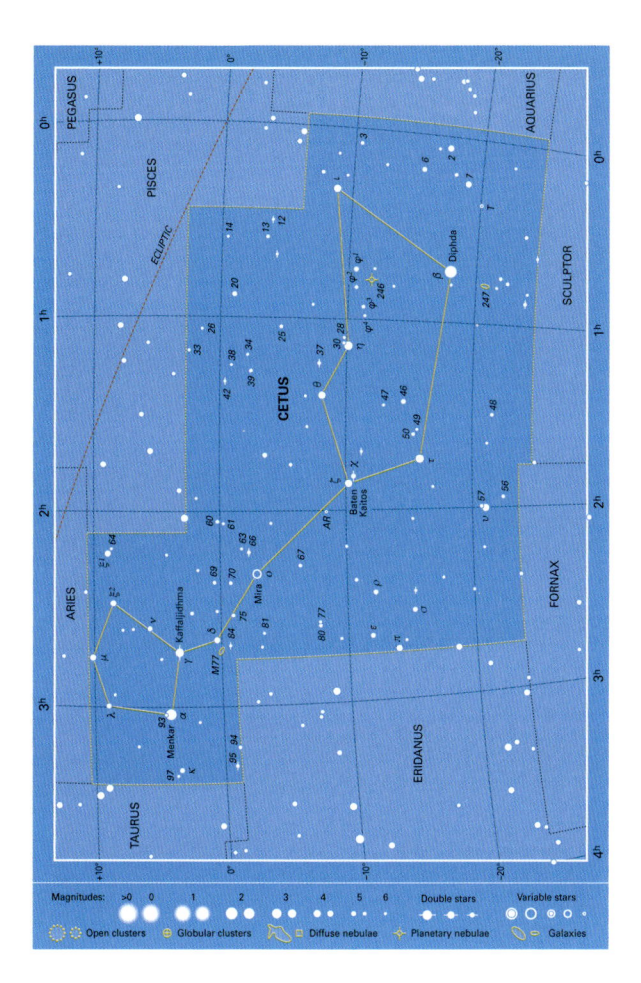

Cetus, the Whale

An ancient constellation depicting the sea monster that was about to devour Andromeda before she was rescued by Perseus. In the sky, Cetus is found basking on the banks of Eridanus, the River. The constellation is large but not prominent; nevertheless it contains several stars of particular interest, notably o (omicron) Ceti and τ (tau) Ceti.

One faint but famous star is UV Ceti, position 1h 38.8m, −17° 57', which consists of a pair of 13th-mag. red dwarfs 8.7 l.y. away. One of these is the prototype of a class of erratic variables known as "flare stars"; these are red dwarfs that undergo sudden increases in light output lasting only a few minutes. The outbursts of the flare star component of UV Ceti can take it from its normal level of 13th mag. to as bright as 7th mag.

α (alpha) Ceti, 3h 02m +4°.1, (Menkar, "nose"), mag. 2.5, is a red giant star 250 l.y. away. Binoculars show a wide mag.-5.6 blue-white apparent companion, 93 Ceti, which is actually unrelated to it, lying nearly twice as far away.

β (beta) Cet, 0h 44m −18°.0, (Diphda), mag. 2.0, the brightest star of the constellation, is an orange giant 96 l.y. away.

γ (gamma) Cet, 2h 43m +3°.2, (Kaffaljidhma), 75 l.y. away, is a close double star needing telescopes of at least 60 mm aperture and high power to split. The stars are of mags. 3.5 and 6.1, the colors bluish and yellow.

o (omicron) Cet, 2h 19m −3°.0, (Mira, "the amazing one"), 300 l.y. away, is the prototype of a famous class of red giant long-period variable stars. Mira itself varies between about 3rd and 9th mags. (although it can become as bright as 2nd mag.) in an average of 332 days, changing in diameter from about 400 to 500 times the size of the Sun as it does so. Mira's light variations were first noted in 1596 by the Dutch astronomer David Fabricius, making it the first variable star (other than novae) to be discovered. The finder chart on page 116 shows various comparison stars down to mag 9.5, from which Mira's brightness can be estimated.

τ (tau) Cet, 1h 44m −15°.9, mag. 3.5, a yellow main-sequence star, is one of the nearest stars to us, lying 11.9 l.y. away. Its main claim to fame is that, of all the nearby single stars, it is the one most like the Sun. At least four planets have been discovered in orbit around it, although we do not yet know whether any of these could be habitable.

M77 (NGC 1068), 2h 43m −0°.0, is a small, softly glowing 9th-mag. face-on spiral galaxy with a 10th-mag. star-like nucleus. Apertures of 100 mm show the brighter patches in the spiral arms. M77 is the brightest example of a Seyfert galaxy, a close relative of the quasars, and it is a radio source. It lies roughly 50 million l.y. away.

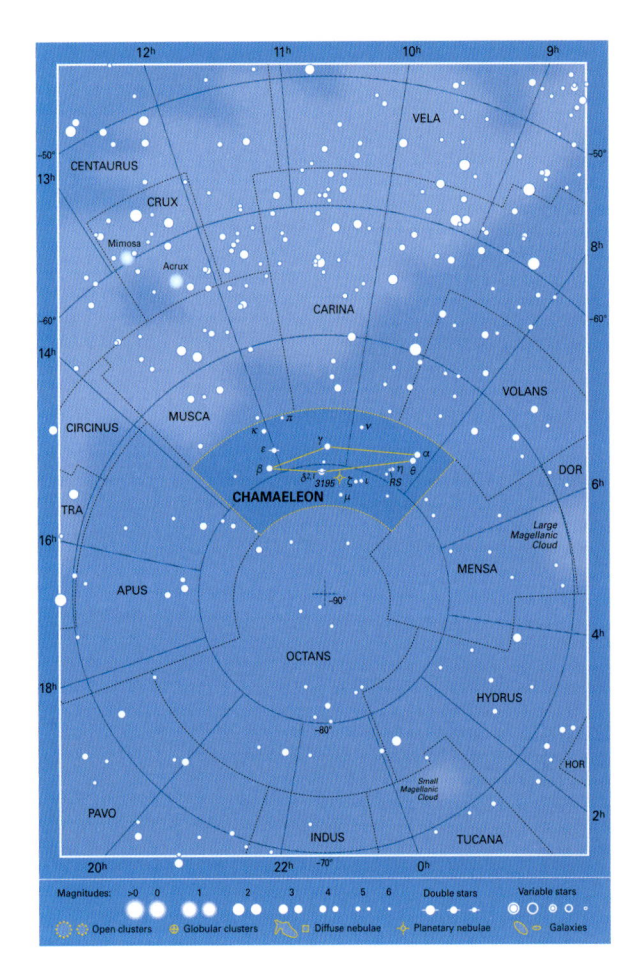

Chamaeleon, the Chameleon

A faint and unremarkable constellation between Carina and the south celestial pole, representing the color-changing lizard. It was one of the constellations representing exotic animals introduced by the Dutch navigators Pieter Dirkszoon Keyser and Frederick de Houtman at the end of the sixteenth century.

α (alpha) Chamaeleontis, 8h 19m −76°.9, mag. 4.0, is a white star 64 l.y. away.

β (beta) Cha, 12h 18m −79°.3, mag. 4.2, is a blue-white main-sequence star 304 l.y. away.

γ (gamma) Cha, 10h 35m −78°.6, mag. 4.1, is an orange giant 375 l.y. away.

δ1 δ2 (delta1 delta2) Cha, 10h 45m −80°.5, consists of a wide pair of stars clearly seen in binoculars: δ1 Cha, mag. 5.5, is an orange giant about 350 light years away, and δ2 Cha, mag. 4.4, is a blue main-sequence star 375 l.y. away. They both have different proper motions, so are not related.

NGC 3195, 10h 09m −80°.9, is a faint planetary nebula of similar apparent size to the planet Jupiter, needing at least 100 mm aperture to be seen well.

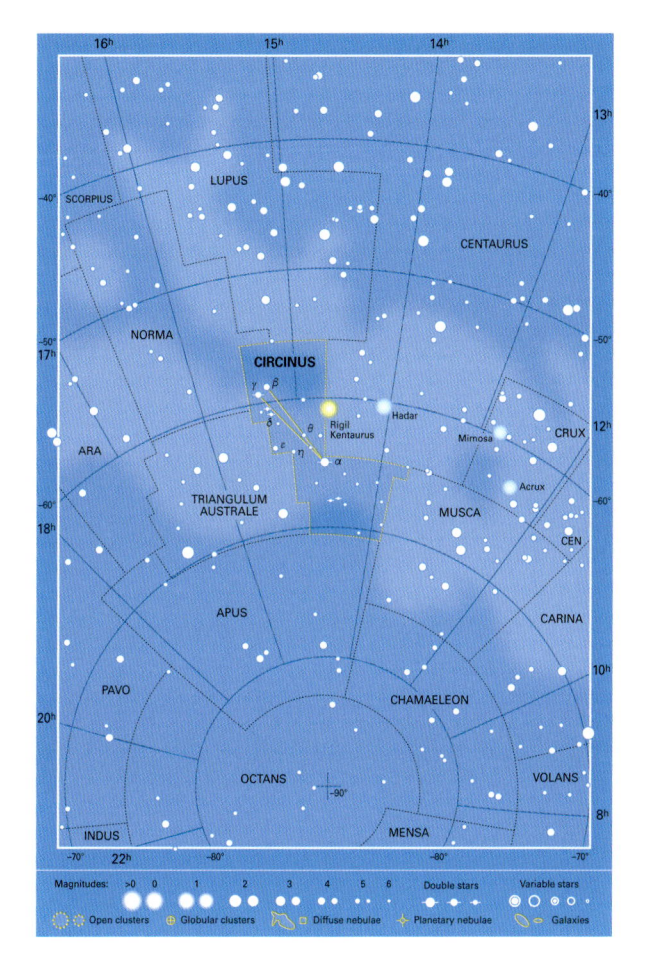

Circinus, the Compasses

Another of the small and obscure southern constellations introduced in 1756 by the French astronomer Nicolas Louis de Lacaille. It represents a pair of dividing compasses as used by draftspeople and surveyors and is appropriately placed in the sky next to Norma, the Set Square. It is overshadowed by the brilliance of neighboring Centaurus.

α (alpha) Circini, 14h 43m −65°.0, 54 l.y. away, is a white main-sequence star of mag. 3.2 with a companion of mag. 8.5 easily visible in small telescopes.

γ (gamma) Cir, 15h 23m −59°.3, 450 l.y. away, consists of a very close binary pair of blue and yellow stars of mags. 4.9 and 5.6, requiring an aperture of at least 150 mm and high magnification to see separately. Their orbital period is estimated to be 600 years.

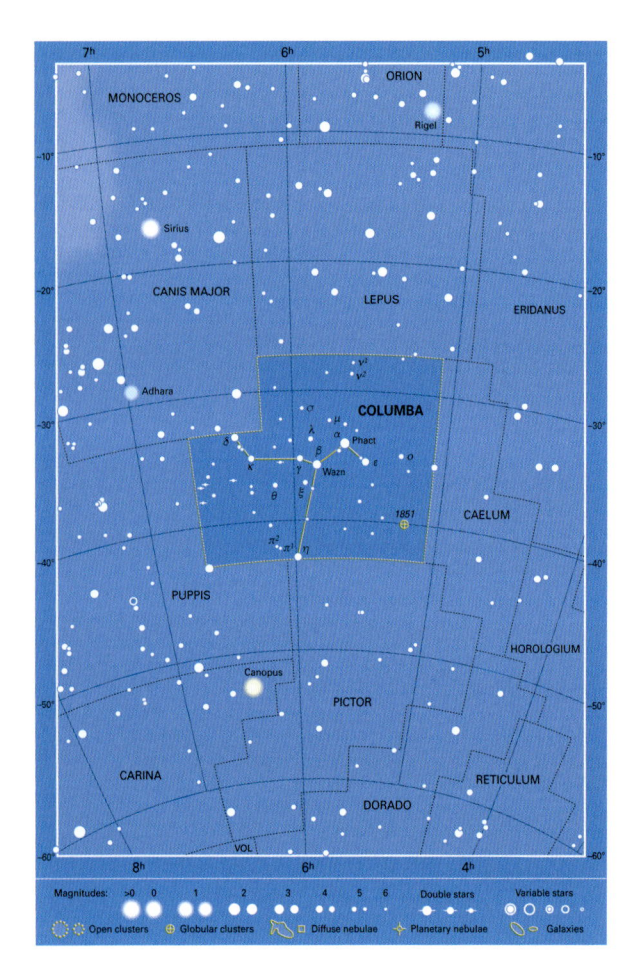

Columba, the Dove

A constellation representing the dove sent out from Noah's Ark to find dry land after the Biblical Flood. The Dutchman Petrus Plancius formed it in 1592 from some stars adjacent to Canis Major that had not previously been part of any constellation. Columba is placed next to Puppis, the stern of the ship Argo, so it might alternatively be imagined as the dove that the Argonauts sent ahead to help them pass safely between the Clashing Rocks (the Symplegades) at the mouth of the Black Sea. Columba contains little of interest for amateur telescopes.

α (alpha) Columbae, 5h 40m −34°.1, (Phact, "ring dove"), mag. 2.6, is a blue-white star around 270 l.y. away.

β (beta) Col, 5h 51m −35°.8, (Wazn), mag. 3.1, is an orange giant 87 l.y. away.

NGC 1851, 5h 14m −40°.1, is a 7th-mag. globular cluster, visible in small telescopes but requiring moderate apertures to resolve its brightest stars. It lies 37,000 l.y. away.

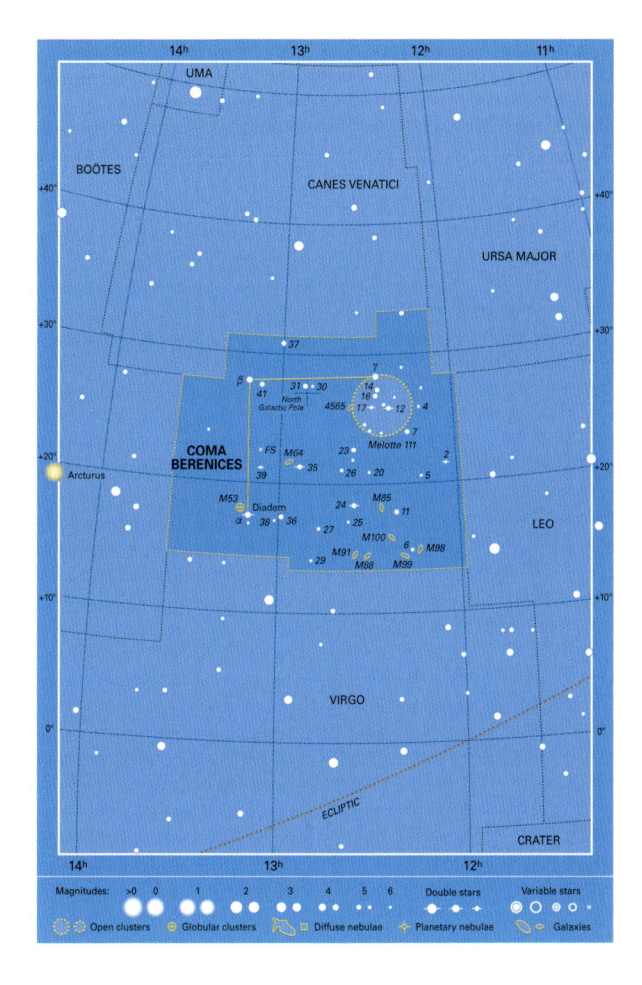

Coma Berenices, Berenice's Hair

This faint constellation represents the flowing locks of Queen Berenice of Egypt, who cut off her hair in gratitude to the gods for the safe return of her husband Ptolemy III Euergetes from battle. Although the legend dates from Greek times, this group of stars was regarded as part of Leo until 1536 when the German cartographer Caspar Vopel made them into a separate constellation. The main part of the queen's severed tresses is represented by the extensive Coma Star Cluster (Melotte 111). Coma Berenices also contains another type of cluster—a cluster of galaxies. The Coma Cluster of galaxies lies about 280 million l.y. away, so its members are too faint for all but the largest amateur telescopes. But the constellation also contains some brighter galaxies, members of the nearer Virgo Cluster, the brightest of which are visible in amateur telescopes. The north pole of our Galaxy lies in Coma Berenices.

α (alpha) Comae Berenices, 13h 10m +17°.5, (Diadem), mag. 4.3, is a tight binary 58 l.y. away, consisting of twin yellow-white stars of mags. 4.8 and 5.5 that orbit each other every 26 years. Even at their widest, around 2035–6, they will be at the limit of resolution of a 200-mm telescope.

β (beta) Com, 13h 12m +27°.9, mag. 4.3, is a yellow main-sequence star 30 l.y. away.

γ (gamma) Com, 12h 27m +28°.3, mag. 4.3, is an orange giant 164 l.y. away, which appears to be a member of the Coma Star Cluster but is actually a foreground star.

24 Com, 12h 35m +18°.4, 365 l.y. away, is a beautiful colored double star for small telescopes, consisting of an orange giant of mag. 5.0 with a blue-white mag. 6.3 companion.

35 Com, 12h 53m +21°.2, 270 l.y. away, is a tight binary star with yellow and white components of mags. 5.1 and 7.1 orbiting every 490 years or so and divisible in 150-mm apertures. Small telescopes show a wider 9th-mag. companion.

FS Com, 13h 06m +22°.6, 600 l.y. away, is a red giant that varies semi-regularly between mags. 5.3 and 6.1 every two months or so.

Coma Star Cluster (Melotte 111), 12h 25m +26°, is a scattered group of about 50 stars best seen in binoculars. The cluster's brightest members, of 5th mag., form a noticeable V-shape extending for several degrees south of γ (gamma) Com, which is not actually a member but a foreground star. The brightest true member seems to be 12 Com, mag. 4.8. The cluster lies about 280 l.y. away.

M53 (NGC 5024), 13h 13m +18°.2, is an 8th-mag. globular cluster 50,000 l.y. away, visible in small telescopes as a rounded, hazy patch.

M64 (NGC 4826), 12h 57m +21°.7, is a famous spiral galaxy, called the Black Eye Galaxy because of a dark cloud of dust silhouetted against its nucleus. This dark dust lane shows up well in apertures above 150 mm; observers with smaller instruments must content themselves simply with locating this 9th-mag. galaxy some 20 million l.y. away, closer than the Virgo Cluster and not a member of it.

M85 (NGC 4382), 12h 25m +18°.2, is a 9th-mag. elliptical galaxy in the Virgo Cluster, 55 million l.y. away. Small telescopes show a brighter, star-like center.

M88 (NGC 4501), 12h 32m +14°.4, is a 10th-mag. spiral galaxy in the Virgo Cluster, 55 million l.y. away. It is presented at an angle to us, so that it appears elliptical.

M99 (NGC 4254), 12h 19m +14°.6, is a 10th-mag. spiral galaxy 55 million l.y. away in the Virgo Cluster, presented face-on so that it appears almost circular.

M100 (NGC 4321), 12h 23m +15°.8, is a 9th-mag. Virgo Cluster spiral seen face-on, similar to M99 but larger. The Hubble Space Telescope has measured its distance accurately at 52.5 million l.y.

NGC 4565, 12h 36m +26°.0, is a 10th-mag. spiral galaxy seen edge-on. It is the most famous of the edge-on spirals, and is pictured [above]. Apertures of 100 mm show its cigar-shaped body with a central bulge and star-like core, but larger instruments are needed to trace the dark band (actually a dust lane) that splits it lengthwise. NGC 4565 is not a member of the Virgo Cluster but only about half as far, around 30 million l.y. away.

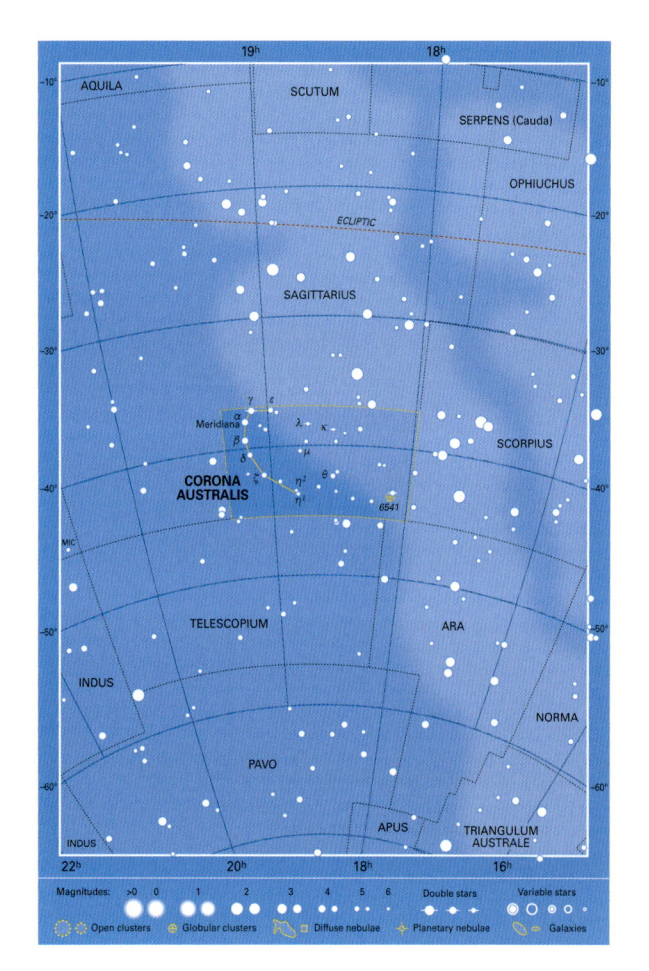

Corona Australis, the Southern Crown

The southern counterpart of the Northern Crown (Corona Borealis), Corona Australis has been known since the time of the Greek astronomer Ptolemy in the second century CE who visualized it not as a crown but as a wreath. In one legend it represents the crown placed in the sky by Bacchus when he rescued his dead mother from the Underworld; alternatively, it could simply have slipped from the head of the centaur Sagittarius, at whose feet it lies. Although faint, it is a distinctive figure and is situated on the edge of the Milky Way.

α (alpha) Coronae Australis, 19h 09m −37°.9, (Meridiana), mag. 4.1, is a blue-white main-sequence star 120 l.y. away.

β (beta) CrA, 19h 10m −39°.3, mag. 4.1, is an orange giant 440 l.y. away.

γ (gamma) CrA, 19h 06m −37°.1, 56 l.y. away, consists of a pair of yellow-white stars, mags. 4.5 and 6.4, orbiting every 122 years, forming a neat double for small telescopes. The stars were closest together during the 1990s, but as they move apart again they have become progressively easier to split with 100-mm aperture.

κ (kappa) CrA, 18h 33m −38°.6, is a pair of blue-white stars of mags. 5.6 and 6.2, easily divisible in small telescopes. They both lie about 700 l.y. away but seemingly do not form a true binary.

λ (lambda) CrA, 18h 44m −38°.3, 200 l.y. away, is a mag.-5.1 blue-white star with a wide 10th-mag. companion, visible in small telescopes.

NGC 6541, 18h 08m −43°.7, is a 6th-mag. globular cluster 22,000 l.y. away, visible in binoculars or small telescopes.

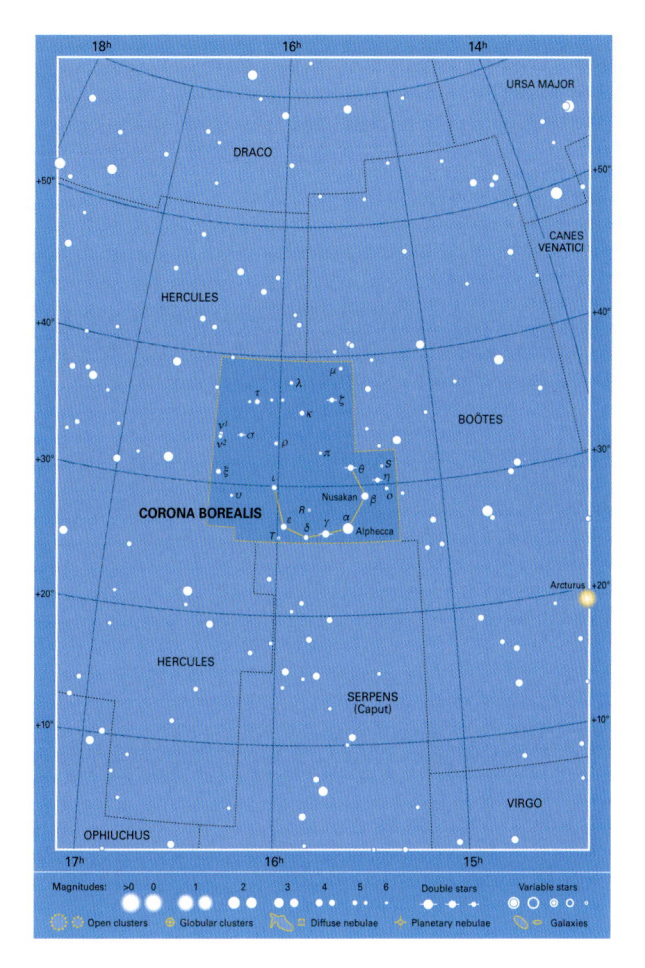

Corona Borealis, the Northern Crown

An ancient constellation, representing the jeweled crown worn by Princess Ariadne of Crete when she married Dionysus (known to the Romans as Bacchus) and cast by him into the sky to mark the happy event. It consists of an arc of seven stars, all but one of 4th magnitude, the exception being Alphecca (a word derived from the Arabic name for the constellation), which is of 2nd magnitude. Alphecca is set in the crown like a central gemstone. Corona Borealis contains a famous cluster of about 400 galaxies more than 1 billion light years away. Being so very distant, the galaxies are no brighter than 16th magnitude and are thus far beyond the reach of amateur telescopes.

α (alpha) Coronae Borealis, 15h 35m +26°.7, (Alphecca), mag. 2.2, is a blue-white subgiant 75 l.y. distant. It is an eclipsing binary of the Algol type, but its variation every 17.4 days is only 0.1 mag., too slight to be noticeable to the naked eye.

β (beta) CrB, 15h 28m +29°.1, (Nusakan), mag. 3.7, is a white main-sequence star with peculiar chemical composition 112 l.y. away.

ζ (zeta) CrB, 15h 39m +36°.6, is a pair of blue-white main-sequence stars, mags. 5.0 and 5.9, visible in small telescopes, lying just over 500 l.y. away.

ν1 ν2 (nu1 nu2) CrB, 16h 22m +33°.8, is a wide binocular pair of red and orange giants of mags. 5.2 and 5.4. They lie about 650 and 570 l.y. from us and are moving in different directions so do not constitute a true binary.

σ (sigma) CrB, 16h 15m +33°.9, 74 l.y. away, is a pair of yellow stars for small telescopes, of mags. 5.6 and 6.4. They form a genuine binary with an estimated orbital period of about 650 years.

R CrB, 15h 49m +28°.2, is a remarkable yellow supergiant star lying within the arc of the Crown, halfway between the stars α (alpha) and ι (iota). It usually appears about 6th mag., but occasionally and unpredictably drops in a matter of weeks to as faint as mag. 15, from where it may take many months or even years to regain its former brightness. Recent catastrophic declines in the star's brilliance were seen in 1983, 1995, 1999, and 2007, the last one being particularly deep and lasting for several years. It recovered briefly before dropping again in 2015. Less extreme fades are more frequent and can occur at any time. These sudden dips in the light of R Coronae Borealis are believed to be due to the accumulation of carbon particles (i.e. soot) in its atmosphere. R Coronae Borealis is estimated to lie just over 2,000 l.y. away.

S CrB, 15h 21m +31°.4, is a variable of Mira type, ranging from 6th to 14th mag. in just under a year. Its distance is about 1,250 l.y.

T CrB, 16h 00m +25°.9, is another spectacular variable star, known as the Blaze Star, which performs in almost the opposite way to R Coronae Borealis. It is a recurrent nova, usually slumbering at around mag. 11 but which can suddenly and unpredictably brighten to mag. 2. Its last recorded outburst was in 1946, and the previous one was 80 years before that. It is not known when it may erupt again. It is around 3,000 l.y. away.

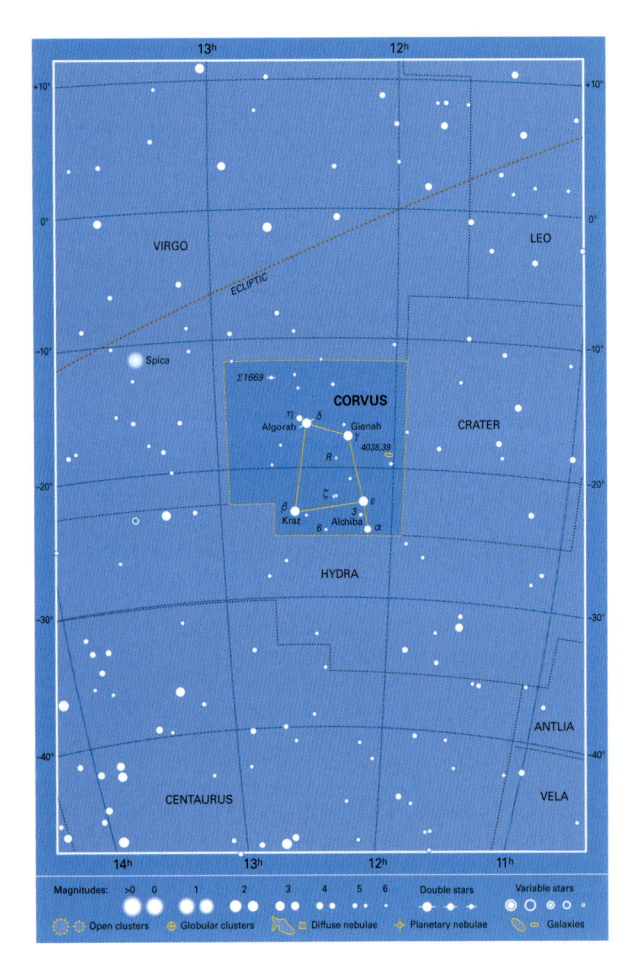

Corvus, the Crow

In Greek legend, Corvus is associated with the neighboring constellations Crater, the Cup, and Hydra, the Water Snake. The crow is said to have been sent by Apollo to fetch water in a cup, but along the way it dallied to eat figs. When the crow returned to Apollo, it carried the water snake in its claws, claiming that this creature had been blocking the spring and was the cause of its delay. Apollo, seeing through the lie, banished the trio to the sky, where the Crow and the Cup lie on the back of Hydra. For its misdeed the crow was condemned to suffer from eternal thirst, which is why crows croak so harshly; in the sky, the cup is just out of the thirsty crow's reach. In another legend, a snow-white crow brought Apollo the bad news that his lover Coronis had been unfaithful to him. In his anger, Apollo turned the crow black. Apollo and crows are closely linked in legend, for during the war waged by the giants on the gods, Apollo turned himself into a crow. The four brightest stars in Corvus, γ (gamma), β (beta), δ (delta), and ε (epsilon) Corvi, form a quadrilateral shape somewhat reminiscent of the Keystone of Hercules.

α (alpha) Corvi, 12h 08m −24°.7, (Alchiba), mag. 4.0, is a white star 49 l.y. away.

β (beta) Crv, 12h 34m −23°.4, (Kraz), mag. 2.6, is a yellow giant star 148 l.y. away.

γ (gamma) Crv, 12h 16m −17°.5, (Gienah, "wing"), mag. 2.6, the brightest star in the constellation by a few hundredths of a magnitude, is a blue-white giant 154 l.y. away.

δ (delta) Crv, 12h 30m −16°.5, (Algorab, "the raven"), is a wide double star for small telescopes. The brighter component, visible to the naked eye, is a mag.-2.9 blue-white star 86 l.y. away, which is accompanied by a mag.-8.4 star often described as purplish in color.

ε (epsilon) Crv, 12h 10m −22°.6, mag. 3.0, is an orange giant star 310 l.y. away.

Σ 1669 (Struve 1669), 12h 41m −13°.0, is a neat pair of white stars 265 l.y. away, appearing to the naked eye as a single star of mag. 5.2 but divisible by small telescopes into near-identical components of mags. 5.9 and 6.0. The secondary, also known as VV Corvi, is slightly variable.

NGC 4038–9, 12h 02m −18°.9, the Antennae, are a pair of spiral galaxies of 10th mag. that collided a few hundred million years ago. Streams of gas and stars thrown off in the collision, visible on photographs such as the one below, stretch for 250,000 l.y. or more and form the antennae that give the galaxies their popular name. The galaxies lie around 60 million l.y. away.

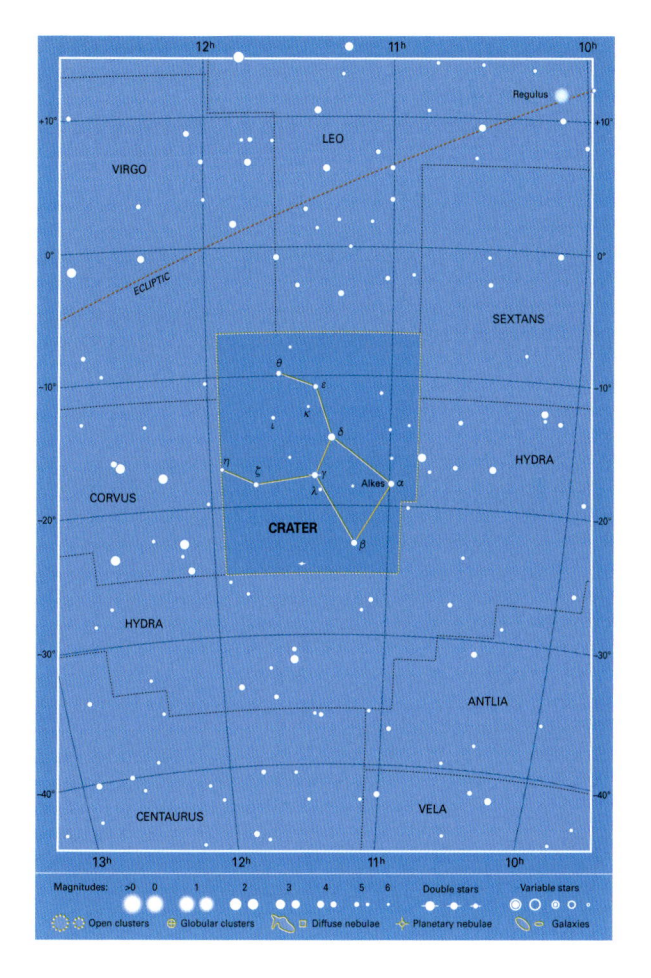

Crater, the Cup

An ancient constellation representing the chalice of Apollo, associated in Greek legend with neighboring Corvus and Hydra (see opposite and page 141). The stars of Crater are faint and the constellation contains no objects of particular interest.

pha) Crateris, 11h 00m −18°.3, (Alkes, "the cup"), mag. 4.1, is an orange giant star 160 l.y. away.

β (beta) Crt, 11h 12m −22°.8, mag. 4.4, is a blue-white star 317 l.y. away.

γ (gamma) Crt, 11h 25m −17°.7, mag. 4.1, is a white star 86 l.y. away. It has an 8th-mag. companion visible in small telescopes.

δ (delta) Crt, 11h 19m −14°.8, mag. 3.6, the constellation's brightest star, is an orange giant lying 190 l.y. away.

STARGAZING FOR BEGINNERS 127

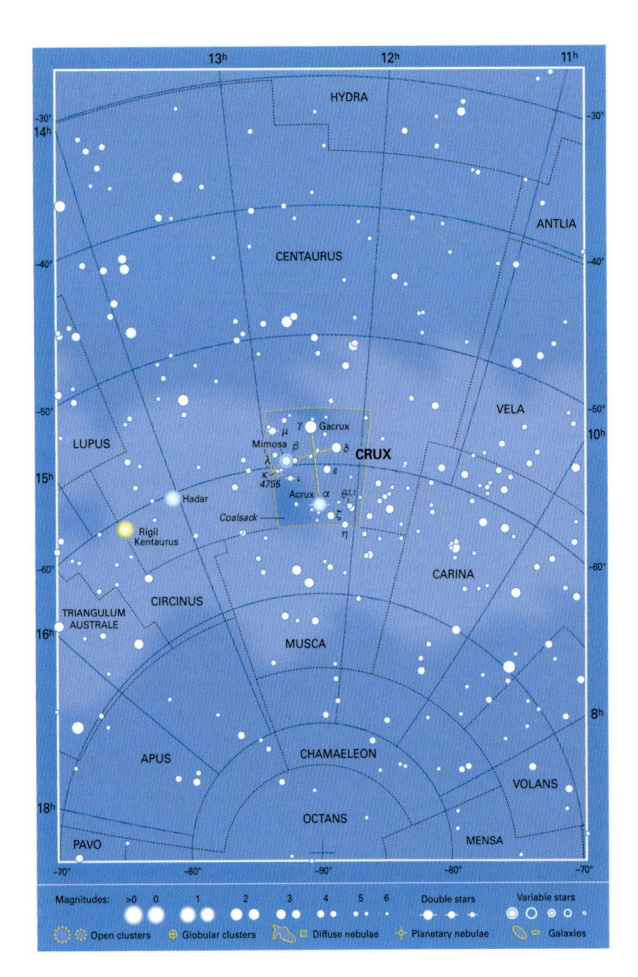

Crux, the Southern Cross

The smallest constellation in the sky, but one of the most celebrated and distinctive. Crux was visible from the Mediterranean area in ancient times, so its stars were known to Greek astronomers; the effects of precession have since carried it below the horizon from such northerly latitudes. The Greeks regarded it as part of Centaurus, but it was made a separate constellation by various seamen and astronomers in the sixteenth century. It can be useful for direction-finding at night, since its long axis, from γ (gamma) via α (alpha) Crucis, points toward the south celestial pole.

Crux lies in a dense and brilliant part of the Milky Way, and the famous dark nebula known as the Coalsack is particularly striking in silhouette against the starry background. α (alpha) Crucis is the most southerly first-magnitude star. α (alpha) and β (beta) Crucis are, by a small margin, the bluest of all first-magnitude stars.

α (alpha) Crucis, 12h 27m −63°.1, (Acrux), 320 l.y. away, appears to the naked eye as a bluish star of mag. 0.8, but a small telescope divides it into a sparkling double with components of mags. 1.3 and 1.8. There is also a wider mag.-4.8 companion that can be seen with binoculars.

β (beta) Cru, 12h 48m −59°.7, (Mimosa), mag. 1.3, is a blue giant 280 l.y. away. It is a variable star of the β Cephei type, fluctuating by less than 0.1 mag. every 5.7 hours, too small to be noticeable to the eye.

γ (gamma) Cru, 12h 31m −57°.1, (Gacrux), mag. 1.6, is a red giant star 88 l.y. away. There is a very wide 6.5-mag. unrelated companion, several times as distant, visible in binoculars.

δ (delta) Cru, 12h 15m −58°.7, mag. 2.8, the faintest of the four main stars in the Cross, is a blue-white subgiant about 450 l.y. away.

ε (epsilon) Cru, 12h 21m −60°.4, mag. 3.6, is an orange giant 230 l.y. away.

ι (iota) Cru, 12h 46m −61°.0, 125 l.y. away, is an orange giant of mag. 4.7 with a 10th-mag. companion visible in small telescopes.

μ (mu) Cru, 12h 55m −57°.2, is a wide pair of blue-white stars of mags. 4.0 and 5.2, divisible by the smallest telescopes or even good binoculars. The secondary is slightly variable. They both lie around 400 l.y away and have similar proper motions so are probably related.

NGC 4755, 12h 54m −60°.3, the Jewel Box or κ (kappa) Crucis Cluster, is one of the finest open clusters in the sky, visible to the naked eye as a hazy 4th-mag. star. Binoculars resolve the brightest individual members, which are blue supergiants of 6th and 7th mag.; three of these form a chain across the cluster like a mini Orion's belt, with an 8th-mag. red supergiant next to the middle one. The star κ Crucis itself, of mag. 5.9, is the most southerly member of this chain of supergiants. A small telescope brings at least 50 stars into view. John Herschel gave this cluster its popular name when he likened it to a piece of multicolored jewelery. The distance of the Jewel Box is about 7,000 l.y.

The Coalsack Nebula is a dark, pear-shaped dust cloud silhouetted against the Milky Way, covering nearly 7° by 5° of sky and spilling over into neighboring Centaurus and Musca. It lies an estimated 600 l.y. away. The Coalsack has no NGC or other identification number.

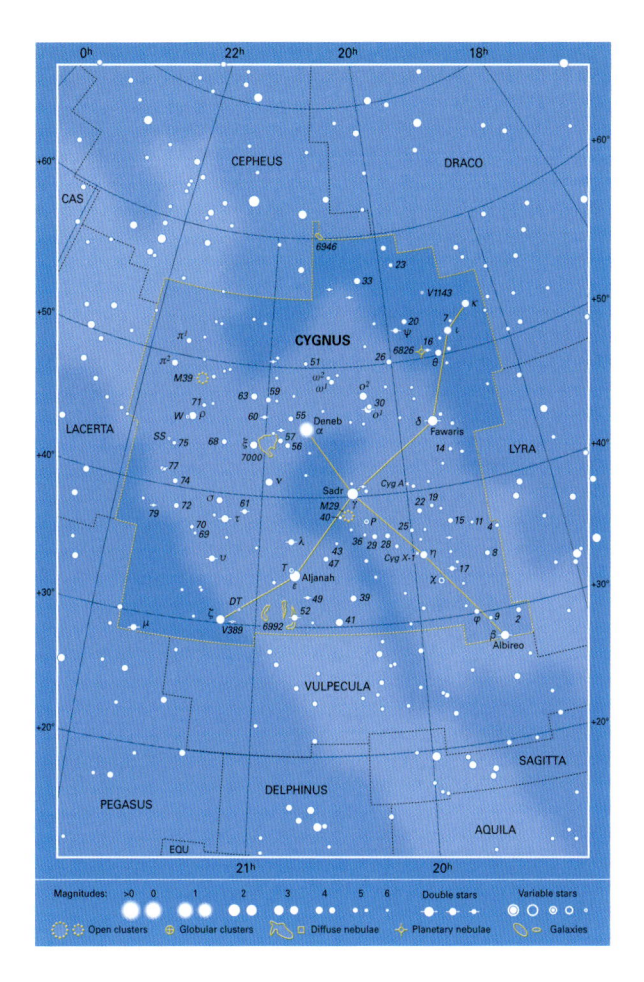

Cygnus, the Swan

Cygnus represents a swan flying down the Milky Way. In Greek mythology, the swan was the guise in which the god Zeus visited Leda, wife of King Tyndareus of Sparta; the result of their union was Pollux, one of the heavenly twins. The flying swan's tail is marked by the star Deneb (α Cygni), its beak by Albireo (β Cygni), and its wings by δ (delta) and ε (epsilon) Cygni. These stars form a distinctive cross-shape, so the constellation is also referred to as the Northern Cross; it happens to be far larger than the more famous Southern Cross. Cygnus lies in a rich part of the Milky Way, which is split here by a dark lane of dust known as the Cygnus Rift, or the Northern Coalsack. Deneb, the constellation's brightest star, forms one corner of the so-called Summer Triangle, completed by Altair and Vega.

Among the fascinating objects in Cygnus is an X-ray source called Cygnus X-1, thought to be a black hole orbiting a 9th-mag. blue supergiant star 6,000 l.y. away; it lies near η (eta) Cygni at 19h 58.4m, +35° 12'. Near γ (gamma) Cygni, at 19h 59.5m, +40° 44', is Cygnus A, a powerful radio source caused by two distant galaxies in collision.

Long-exposure photographs of the region between ε (epsilon) Cygni and the border with Vulpecula reveal beautiful swirls of gas known as the Veil Nebula, the brightest part of which, NGC 6992, is detectable in amateur instruments.

α (alpha) Cygni, 20h 41m +45°.3, (Deneb, "tail"), mag. 1.2, is a blue-white supergiant star about 1,400 l.y. away. It is slightly variable.

β (beta) Cyg, 19h 31m +28°.0, (Albireo), 400 l.y. away, is one of the sky's showpiece doubles. It consists of gloriously contrasting amber and blue-green stars, like a celestial traffic light. The brighter star, of mag. 3.1, is an orange giant, and its blue-green companion is mag. 5.1. They can be separated through good binoculars and are a beautiful sight in any amateur telescope.

γ (gamma) Cyg, 20h 22m +40°.3, (Sadr, "breast"), mag. 2.2, is a yellow-white supergiant about 1,800 l.y. away.

δ (delta) Cyg, 19h 45m +45°.1, (Fawaris), 165 l.y. away, is a blue-white giant of mag. 2.9 with a close mag.-6.3 companion, visible in small telescopes at high magnification. The stars have an orbital period of around 650 years.

ε (epsilon) Cyg, 20h 46m +34°.0, (Aljanah), mag. 2.5, is an orange giant 75 l.y. away.

μ (mu) Cyg, 21h 44m +28°.7, 72 l.y. away, is a pair of white stars of mags. 4.7 and 6.1 orbiting each other about every 790 years. They are currently closing slowly and should remain divisible in 100-mm apertures until about 2030; thereafter 150 mm will probably be necessary, particularly when they are closest in 2043–50. A wide mag.-6.9 binocular companion is an unrelated background star.

o1 (omicron1) Cyg, 20h 14m +46°.7, also known as 31 Cygni, forms perhaps the most beautiful binocular double in the heavens with 30 Cygni. The stars are orange and turquoise, mags. 3.8 and 4.8, 1,100 l.y. and 635 l.y. away, like a wider version of Albireo. A small telescope, or binoculars held steadily, shows a closer blue companion of mag. 7.0 to the brighter (orange giant) star.

χ (chi) Cyg, 19h 51m +32°.9, 520 l.y. away, is a red giant long-period variable of Mira type that varies every 400 days or so. At best it reaches mag. 3.3, brighter than any other star of its type except Mira itself, although R Hydrae can rival it. At its faintest, χ Cyg drops to 14th mag. Its diameter is about 300 Suns.

61 Cyg, 19h 56m +52°.4, 280 l.y. away, is a pair of white stars of mags. 5.0 and 7.5, divisible in small to moderate apertures.

61 Cyg, 21h 07m +38°.7, 11.4 l.y. away, is a showpiece pair of orange dwarf stars of mags. 5.2 and 6.0, orbiting each other in about 700 years, divisible in a small telescope or even binoculars. In addition to being among the closest stars to Earth, 61 Cygni was the first star to have its parallax measured, by the German astronomer Friedrich Wilhelm Bessel in 1838.

P Cyg, 20h 18m +38°.0, is an erratically variable blue supergiant that normally resides around 5th mag., although in the year 1600 it reached a peak brightness of 3rd mag. Evidently the star is so large and luminous that it is close to instability. Its average brightness has increased gradually since the eighteenth century as it evolves into a red supergiant. Its distance is uncertain, but is probably several thousand light years.

W Cyg, 21h 36m +45°.4, 630 l.y. away, is a red giant that varies semi-regularly between 5th and 7th mags. with a rough period of 130 days.

M39 (NGC 7092), 21h 32m +48°.4, is a large, loose cluster of about 30 stars of 7th mag. and fainter, arranged in a triangle. It is visible in binoculars, and is even detectable with the naked eye under ideal conditions. It lies just under 1,000 l.y. away.

NGC 6826, 19h 45m +50°.5, is an 8th-mag. planetary nebula 4,200 l.y. away, known as the Blinking Planetary because it appears to blink on and off. It is visible as a pale blue disk in 75-mm telescopes, but 150 mm is needed to show it to advantage. At its center is a 10th-mag. star. Looking alternately at this star and away again produces the "blinking" effect. NGC 6826 lies less than 1° from the wide double star 16 Cygni, which consists of two 6th-mag. creamy-white stars 69 l.y. away.

NGC 6992, 20h 56m +31°.7, is the brightest part of the Veil Nebula, the remnant of a supernova explosion about 5,000 years ago. Under ideal conditions, NGC 6992 can be seen in binoculars as a faint arc. Another part of the nebula, NGC 6960, can be located in wide-angle telescopes with low power near the 4th-mag. star 52 Cyg, but the complete Veil Nebula, also known as the Cygnus Loop, is apparent only in long-exposure photographs, spanning nearly 3° of sky. Its distance is 1,400 l.y.

NGC 7000, 20h 59m +44°.3, the North America Nebula, can be seen as a hook-shaped brightening in the Milky Way with the naked eye or binoculars in good skies. Despite its large size, 2° at its widest, the nebula can be difficult to detect because of its low surface brightness. Long-exposure photographs such as the one above clearly show its shape, resembling the continent of North America. The nebula is roughly 2,000 l.y. away, about 50 percent farther off than the star Deneb. The nebula is believed to be lit up by an extremely hot and highly obscured 13th-mag. star that lies within it.

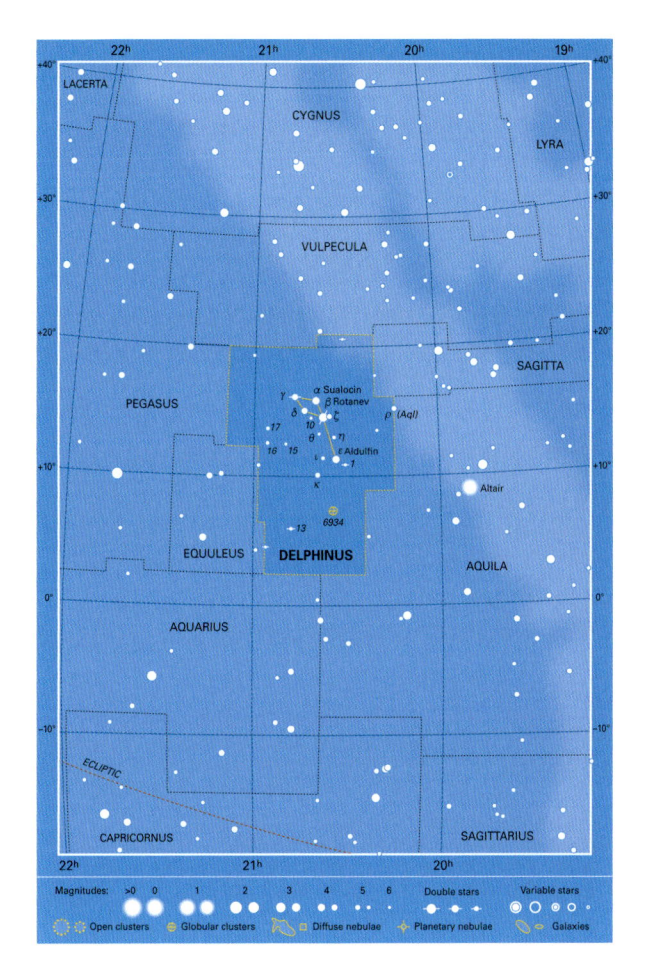

Delphinus, the Dolphin

A constellation that originated in Greek times, celebrating the long-standing relationship between humans and these intelligent aquatic mammals. In legend, dolphins were the messengers of the sea god Poseidon, who sent a dolphin to bring the sea nymph Amphitrite to his underwater palace, where he made her his wife. In another story, dolphins were credited with saving the life of Arion, a musician and poet, who was attacked on a ship. According to this tale, Arion jumped overboard to escape his attackers and was carried on the back of a dolphin to Greece, where he later identified the robbers, who were sentenced to death. In this legend, the nearby constellation Lyra represents Arion's lyre. Delphinus has a distinctive shape, its four main stars forming a diamond known as Job's Coffin. Its two brightest stars are called Sualocin and Rotanev, which backwards read Nicolaus Venator, the Latinized form of the name Niccolò Cacciatore (1780–1841), who was assistant and successor to the Italian astronomer Giuseppe Piazzi at Palermo Observatory. These names first appeared in the Palermo Catalogue of 1814, and it is usually supposed that Cacciatore himself was responsible, although it is equally possible that Piazzi could have applied the names to honor his appointed successor.

Like the small neighboring constellations Vulpecula and Sagitta, Delphinus lies in a rich area of the Milky Way and has become a favorite hunting ground for novae.

α (alpha) Delphini, 20h 40m +15°.9, (Sualocin), mag. 3.8, is a blue-white subgiant star 240 l.y. away.

β (beta) Del, 20h 38m +14°.6, (Rotanev), mag. 3.6, the constellation's brightest member, is a white subgiant star 100 l.y. away. It is a binary with an orbital period of 27 years, but the components are too close to be divided in any but the largest amateur telescopes.

γ (gamma) Del, 20h 47m +16°.1, 116 l.y. away, is a showpiece binary consisting of golden and yellow-white stars of mags. 4.3 and 5.0, neatly separated in a small telescope. Their estimated orbital period is over 2,000 years. In the same telescopic field of view, but about 5 l.y. farther from us, lies a fainter binary, Σ 2725 (Struve 2725), consisting of stars of mags. 7.5 and 8.1 divisible in small telescopes with high magnification.

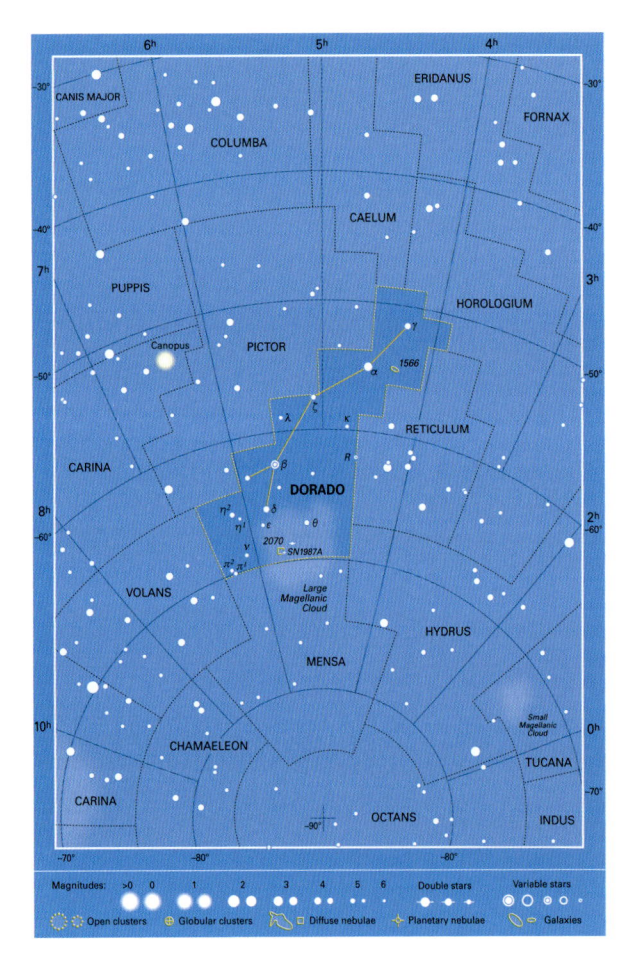

Dorado, the Goldfish

This constellation was introduced at the end of the sixteenth century by the Dutch navigators Pieter Dirkszoon Keyser and Frederick de Houtman to commemorate the dolphinfish (dorado or mahi-mahi) of tropical waters. Its most notable feature is the Large Magellanic Cloud (LMC), the larger of the two satellite galaxies that accompany our Milky Way. In 1987, the first supernova visible to the naked eye since 1604, Supernova 1987A, erupted in the LMC, at 5h 35m, −69°.3.

α (alpha) Doradus, 4h 34m −55°.0, mag. 3.3, is a blue-white star 170 l.y. away.

β (beta) Dor, 5h 34m −62°.5, a yellow-white supergiant 1,100 l.y. away, is one of the brightest Cepheid variables, ranging from mag. 3.4 to 4.1 every 9 days 20 hours.

R Dor, 4h 37m −62°.1, 178 l.y. away, is a red giant that varies semi-regularly between mags. 4.8 and 6.3 with a period of around 6 months.

Large Magellanic Cloud (LMC), 5h 24m −69°, is a mini-galaxy 170,000 l.y. away, a satellite of the Milky Way, containing perhaps 10 billion stars. To the naked eye it appears as a fuzzy elongated patch 6° in diameter, 12 times the apparent size of the Moon; its actual diameter is about 25,000 l.y. Binoculars and telescopes show individual stars, nebulae (notably the Tarantula Nebula, described right), and clusters.

NGC 2070, 5h 39m −69°.1, the Tarantula Nebula, is a looping cloud of hydrogen gas about 1,000 l.y. in diameter in the LMC. To the naked eye the nebula appears as a fuzzy star, also known as 30 Doradus. The nebula's popular name comes from its spider-like shape. At the center of the nebula is a cluster of supergiant stars, called R136, the light from which makes the nebula glow. The Tarantula is larger and brighter than any nebula in the Milky Way. If it were as close to us as the Orion Nebula, the Tarantula would fill the whole constellation of Orion and would cast shadows on Earth.

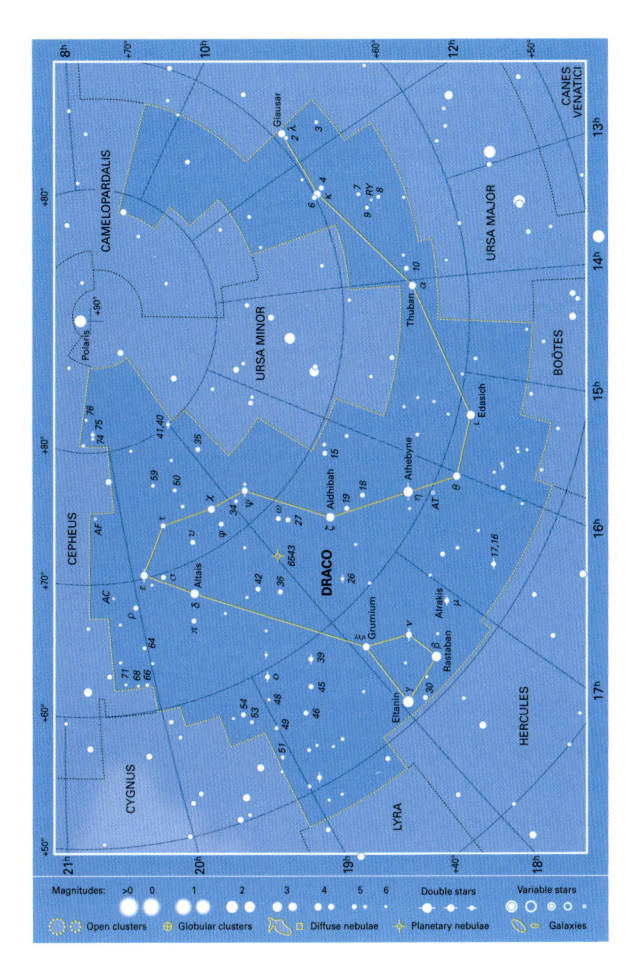

Draco, the Dragon

Dragons feature in many ancient legends, so it is not surprising to find such a monster in the sky. This one is said to be Ladon, the dragon slain by Hercules as a prelude to stealing the golden apples from the garden of the Hesperides. In the sky, one foot of Hercules rests upon the dragon's head, while the creature's body lies coiled around the north celestial pole. Although Draco is one of the largest and most ancient constellations, it is indistinct, with no stars brighter than 2nd magnitude. Draco contains the north pole of the ecliptic, i.e. one of the two points 90° from the plane of the Earth's orbit, at 18h 00m, +66.5°.

α (alpha) Draconis, 14h 04m +64°.4, (Thuban, "serpent's head"), mag. 3.7, is a blue-white giant star 260 l.y. away. It was the pole star in about 2800 BCE, but has lost that place to Polaris because of the effect of precession (see page 29).

β (beta) Dra, 17h 30m +52°.3, (Rastaban, also "serpent's head"), mag. 2.8, is a yellow giant or supergiant nearly 400 l.y. distant.

γ (gamma) Dra, 17h 57m +51°.5, (Eltanin, "the serpent"), mag. 2.2, is an orange giant 154 l.y. away, and the brightest star in the constellation. From observations of this star, the English astronomer James Bradley discovered the effect known as the aberration of starlight in 1728.

μ (mu) Dra, 17h 05m +54°.5, (Alrakis), 89 l.y. away, is a close double star with matching cream-colored components of mag. 5.7 that orbit every 400 years or so. The two stars are currently moving apart and becoming progressively easier to split in small apertures, although high magnification will be needed.

ν (nu) Dra, 17h 32m +55°.2, is a pair of identical white stars of mag. 4.9, easily visible in the smallest of telescopes and regarded as one of the finest binocular pairs. They lie 98 l.y. away.

o (omicron) Dra, 18h 51m +59°.4, 340 l.y. away, is a mag. 4.6 orange giant star with a mag. 8.3 companion visible in small telescopes.

ψ (psi) Dra, 17h 42m +72°.1, 74 l.y. away, is a mag. 4.6 yellow-white star with a yellow mag. 5.5 companion visible in small telescopes or even binoculars.

16–17 Dra, 16h 36m +52°.9, is a wide pair of blue-white stars of mags. 5.5 and 5.1, about 420 l.y. away, easily found in binoculars. Telescopes of 60-mm aperture with high magnification split the brighter star, 17 Dra, into a binary of mags. 5.4 and 6.4, making this a striking triple system.

39 Dra, 18h 24m +58°.6, 190 l.y away, is an impressive triple system. The two brightest members, at mags. 5.0 and 7.9, appear in binoculars as a wide blue-and-yellow pair. A telescope of 60-mm aperture with high magnification reveals that the brighter star has a closer companion of mag. 8.1.

40–41 Dra, 18h 00m +80°.0, is an easy pair of orange dwarf stars for small telescopes, mags. 5.9 and 5.6. They lie about 145 l.y. away.

NGC 6543, 17h 59m +66°.6, is a 9th-mag. planetary nebula around 4,000 l.y. away, one of the brightest such nebulae, showing in amateur telescopes as an irregular blue-green disk like an out-of-focus star. It is now known as the Cat's Eye Nebula from its appearance in Hubble Space Telescope photographs.

STARGAZING FOR BEGINNERS 133

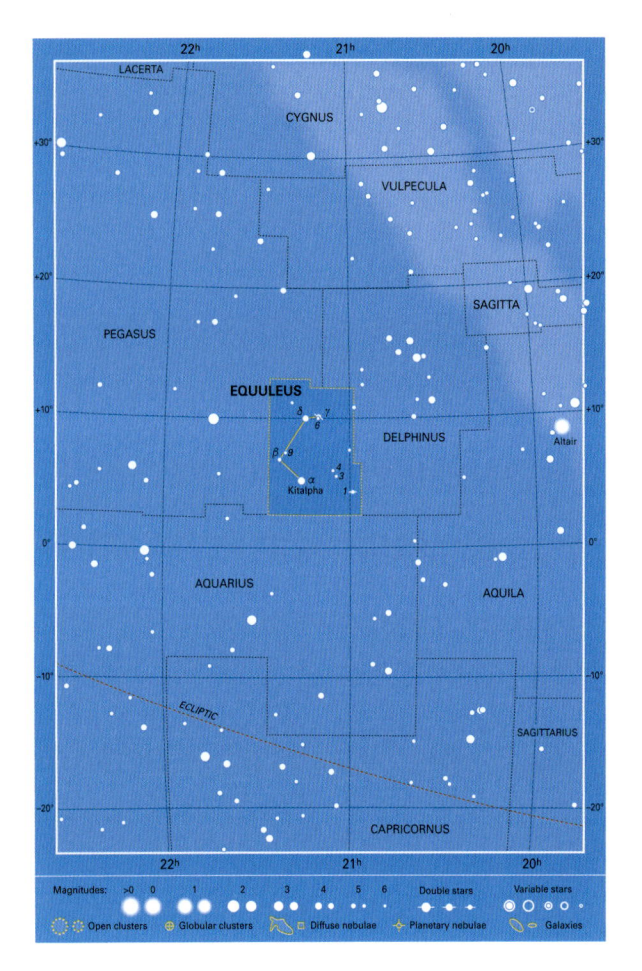

Equuleus, the Little Horse

The second-smallest constellation in the sky, Equuleus was one of the 48 constellations listed by the Greek astronomer Ptolemy in his book the Almagest in the second century CE. It is thought to have been invented either by him or by his predecessor Hipparchus. Only the head of the horse is shown, next to the much larger horse, Pegasus, and there are no legends that identify it.

α (alpha) Equulei, 21h 16m +5°.2, (Kitalpha, "the section of the horse"), mag. 3.9, is a yellow giant 190 l.y. away.

γ (gamma) Equ, 21h 10m +10°.1, is a mag. 4.7 white star 115 l.y. away. Through binoculars an apparent mag. 6.1 companion, 6 Equ, can be seen, but this is an unrelated background star over three times as far away.

ε (epsilon) Equ, 20h 59m +4°.3, also known as 1 Equ, is a triple star 176 l.y. away. In small telescopes it appears as a white-and-yellow pair of mags. 5.3 and 7.1. The brighter component is also a close binary, of mags. 6.0 and 6.3; these stars orbit each other every 104 years. They should become separable in 220-mm apertures after about 2040 as they move apart.

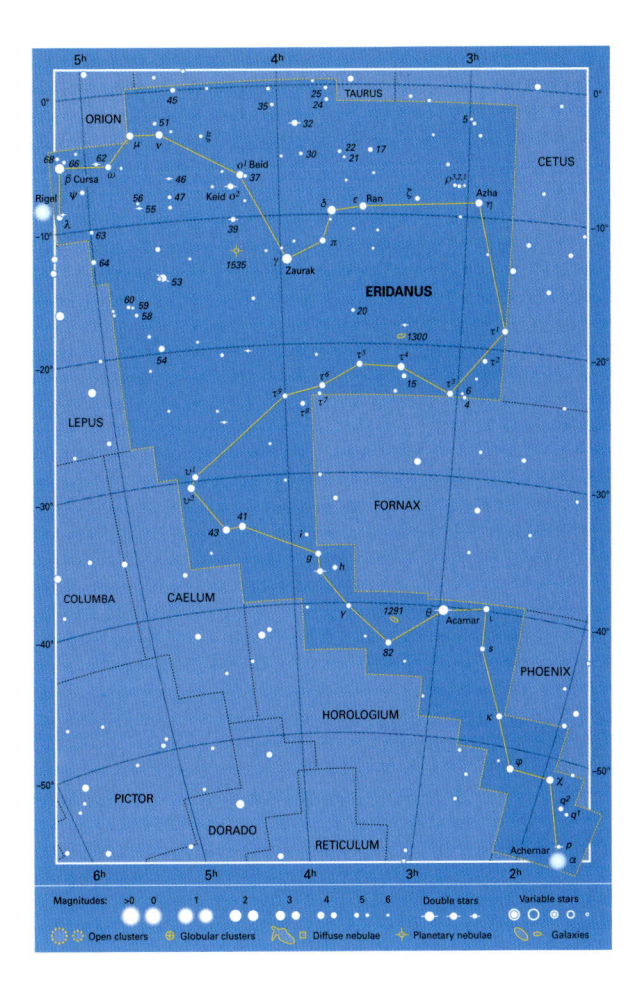

Eridanus, the River

An extensive constellation, the sixth-largest in the sky, but often overlooked because of its lack of any distinctive shape. It meanders from Taurus in the north to Hydrus in the south. In mythology, Eridanus was the river into which Phaethon fell after trying to drive the chariot of his father, Helios the sun god. But it also supposedly represents a real river. Early mythologists identified it with the Nile, but later Greek writers said it was the Po in Italy.

Originally Eridanus included the stars of what is now Fornax and it stretched only as far as θ (theta) Eridani, which was given the name Achernar, from the Arabic meaning "the river's end." In more recent times, Eridanus has been extended to nearly 60° south (below the horizon from Greece), and the title Achernar has been transferred to another star farther south. The star originally called Achernar is now known as Acamar, a name that comes from the same Arabic original as Achernar.

The constellation contains several interesting galaxies, all too distant and too faint to be picked up easily in amateur telescopes. One celebrated example is NGC 1300, located at 3h 19.7m, −19° 25'; photographs show that it is a beautiful 10th-mag. barred spiral. It lies 69 million l.y. away.

α (alpha) Eridani, 1h 38m −57°.2, (Achernar, "the river's end"), mag. 0.5, is a blue-white main-sequence star 140 l.y. away.

β (beta) Eri, 5h 08m −5°.1, (Cursa, "the footstool," referring to its position under the foot of Orion), mag. 2.8, is a blue-white star 90 l.y. away.

ε (epsilon) Eri, 3h 33m −9°.5, (Ran), mag. 3.7, an orange main-sequence star 10.5 l.y. away, is among the most Sun-like of the nearby stars. It is orbited every seven years by a planet three times the mass of Jupiter.

θ (theta) Eri, 2h 58m −40°.3, (Acamar), 165 l.y. away, is a striking pair of blue-white stars of mags. 3.2 and 4.1, divisible in small telescopes.

o2 (omicron2) Eri, 4h 15m −7°.7, 16 l.y. away, also known as 40 Eridani, is a remarkable triple star. A small telescope shows that the mag. 4.4 orange main-sequence primary has a wide mag. 9.5 white dwarf companion, the most easily seen white dwarf in the sky. Small telescopes reveal that the white dwarf has an 11th-mag. companion that is a red dwarf, thereby completing a most interesting trio. The white dwarf and red dwarf orbit each other every 230 years, and are easily split in small apertures.

32 Eri, 3h 54m −3°.0, 310 l.y. away, is a beautiful double star for small telescopes, consisting of a yellow giant of mag. 4.7 and a blue-green mag. 6.1 main-sequence companion.

39 Eri, 4h 14m −10°.3, 230 l.y. away, is an orange giant of mag. 5.0 with a mag. 8.5 companion divisible in small telescopes.

p Eri, 1h 40m −56°.2, 26 l.y. away, is a beautiful wide duo of orange dwarf stars, mags. 5.7 and 5.8, with an orbital period of nearly 500 years.

NGC 1535, 4h 14m −12°.7, is a small 9th-mag. planetary nebula about 4,500 l.y. away. Small telescopes show it, but apertures of 150 mm are needed to appreciate its blue-gray disk.

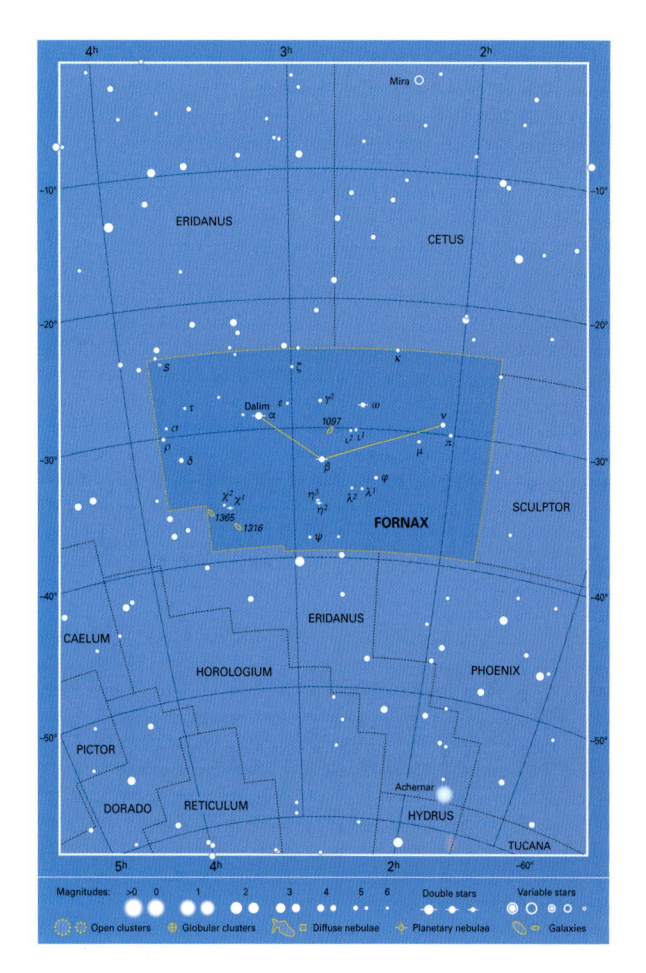

Fornax, the Furnace

A barren constellation introduced in the 1750s by Nicolas Louis de Lacaille, originally under the name of Fornax Chimiae, the Chemical Furnace. It contains a dwarf member of our Local Group of galaxies, called the Fornax Dwarf, some 450,000 l.y. from the Milky Way but too faint to see in amateur telescopes. Fornax also contains a compact cluster of galaxies about 60 million l.y. away, the brightest member of which is the 9th-mag. peculiar galaxy NGC 1316, also known as the radio source Fornax A, located at 3h 22.7m, −37° 12′. Another well-known member of the Fornax cluster is NGC 1365.

α (alpha) Fornacis, 3h 12m −29°.0, (Dalim), 46 l.y. away, is a binary consisting of a yellow-white main-sequence star of mag. 4.0 with a deeper yellow mag. 7.2 companion of suspected variability. The pair orbit each other every 270 years or so and will remain divisible in small telescopes throughout the twenty-first century.

β (beta) For, 2h 49m −32°.4, mag. 4.5, is a yellow giant star 177 l.y. away.

NGC 1097, 2h 46m −30°.6, is a 9th-mag. barred spiral galaxy with a bright nucleus, visible in medium-sized telescopes. It lies about 50 million l.y. away.

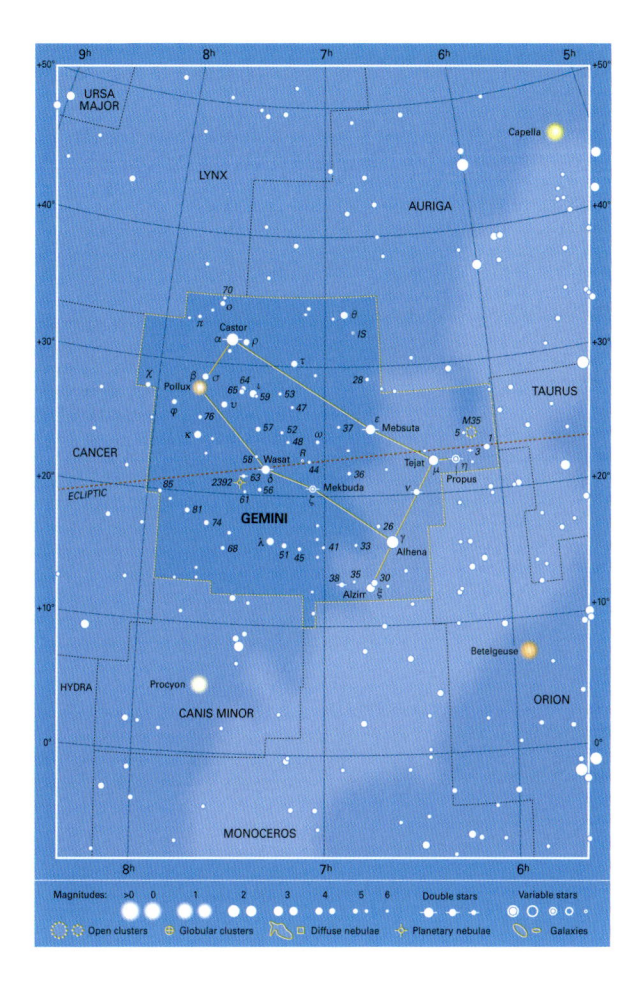

Gemini, the Twins

A constellation that dates from ancient times, representing a pair of male twins. We know them as Castor and Pollux, members of the Argonauts' crew and of mixed parentage: both were sons of Queen Leda of Sparta, but by different fathers. Castor's father was her husband King Tyndareus, while the father of Pollux was the god Zeus. The twins were the protectors of mariners, appearing in ships' rigging as the electrical phenomenon now known as St Elmo's fire. In the sky, the stars Castor and Pollux provide a useful yardstick for measuring angular distances—they are exactly 4°.5 apart.

Gemini is a member of the zodiac, and the Sun passes through the constellation from late June to late July. Each year the Geminid meteors, one of the year's richest and brightest showers, radiate from a point near Castor, reaching a maximum around December 13–14, when up to 100 meteors per hour may be seen.

α (alpha) Geminorum, 7h 35m +31°.9, (Castor), 51 l.y. away, is an astounding multiple star, consisting of six separate components. To the naked eye, Castor appears as a blue-white star of mag. 1.6, somewhat fainter than Pollux. A 60-mm telescope with high magnification splits Castor into two components of mags. 1.9 and 3.0, which orbit each other every 460 years. Separation of these stars is increasing, and they will become progressively easier to split throughout the twenty-first century. Both stars are spectroscopic binaries. Small telescopes also show a wide red dwarf companion to Castor; this is itself an eclipsing binary of Algol type, varying between mags. 8.9 and 9.6 in 19.5 hours, completing the six-star system.

β (beta) Gem, 7h 45m +28°.0, (Pollux), mag. 1.1, is the brightest star in the constellation. Some astronomers have speculated that Pollux, being labeled β Geminorum, was once fainter than Castor and has since brightened, or Castor has faded. But the truth is that Johann Bayer, who allotted the Greek letters in 1603, did not distinguish carefully which star was the brighter, and so caused unnecessary confusion. Pollux is an orange giant 34 l.y. away.

γ (gamma) Gem, 6h 38m +16°.4, (Alhena), mag. 1.9, is a blue-white subgiant star 110 l.y. away.

δ (delta) Gem, 7h 20m +22°.0, (Wasat), 61 l.y. away, is a creamy-white star of mag. 3.5 with an orange dwarf companion of mag. 8.2. The brightness contrast makes the pair difficult in telescopes below about 75-mm aperture. Their estimated orbital period is around 1,400 years.

ε (epsilon) Gem, 6h 44m +25°.1, (Mebsuta), mag. 3.0, is a yellow supergiant about 850 l.y. away. Binoculars, or a small telescope, reveal a wide companion of mag. 9.6.

ζ (zeta) Gem, 7h 04m +20°.6, (Mekbuda), about 1,060 l.y. away, is both a variable star and a binocular double. A yellow supergiant, it is a Cepheid variable, fluctuating between mags. 3.6 and 4.2 every 10.2 days. Binoculars or small telescopes reveal a wide companion of mag. 7.7, which is unrelated.

η (eta) Gem, 6h 15m +22°.5, (Propus), about 400 l.y. away, is another double-variable. A red giant star, it fluctuates in semi-regular manner between mags. 3.1 and 3.7 in about 8 years. It has a close 6th-mag. companion that orbits it every 1,000 years or so and requires a large telescope to distinguish it from the primary's glare.

κ (kappa) Gem, 7h 44m +24°.4, 140 l.y. away, is a mag. 3.6 yellow giant, with an 8th-mag. companion made difficult in small telescopes because of the extreme brightness difference.

ν (nu) Gem, 6h 29m +20°.2, mag. 4.1, is a blue giant 545 l.y. away with a wide 8th-mag. companion visible in binoculars or small telescopes.

38 Gem, 6h 55m +13°.2, 96 l.y. away, is a double star for small telescopes, with white and yellow components of mags. 4.8 and 7.8.

M35 (NGC 2168), 6h 09m +24°.3, is a large, 5th-mag. open cluster visible to the naked eye or through binoculars, noticeably elongated in shape. It consists of about 200 stars covering a similar area of sky as the Moon, and is 2,900 l.y. away. Even a small telescope at low magnification will show the members of this outstanding cluster to be arranged in curving chains. Nearby in the sky, but actually more than 10,000 l.y. beyond it, is NGC 2158, a very rich open cluster that appears as a small, faint patch of light requiring at least 100-mm aperture to distinguish.

NGC 2392, 7h 29m +20°.9, is an 8th-mag. planetary nebula known as the Eskimo or Clown Face Nebula because it looks somewhat like a face surrounded by a fringe when seen through a large telescope. A small telescope shows it as a blue-green ellipse about the same apparent size as the disk of Saturn, with a central star of 10th mag. (see page 152 for a photograph). NGC 2392 lies 6,000 l.y. away.

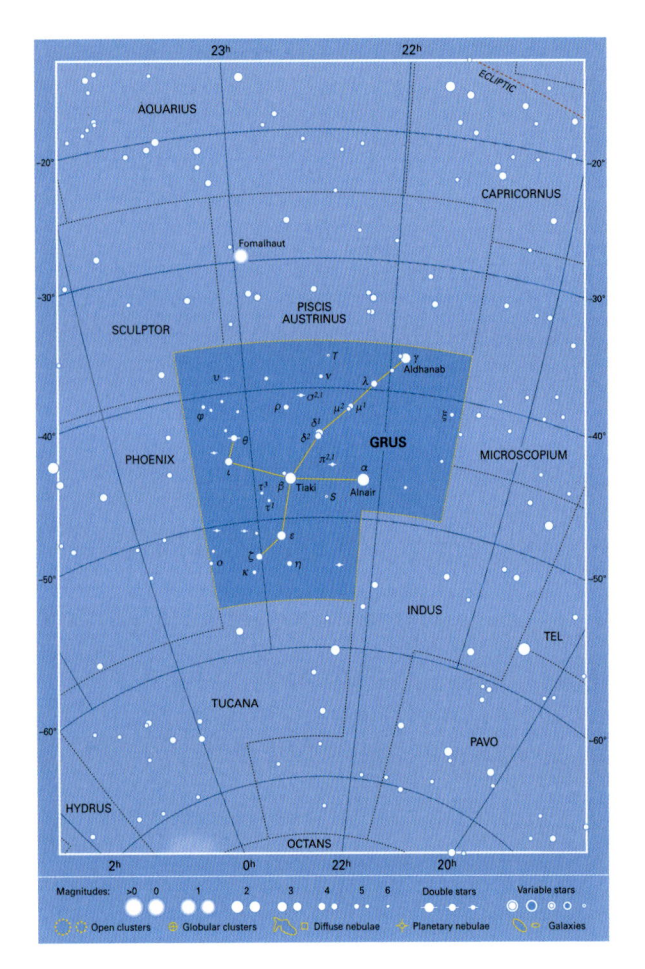

Grus, the Crane

One of the 12 constellations introduced at the end of the sixteenth century by the Dutch navigators Pieter Dirkszoon Keyser and Frederick de Houtman. It represents a water bird, the long-necked crane, possibly the sarus crane of India and southeast Asia, which is the largest species of crane. The stars δ (delta) and μ (mu) Gruis are striking naked-eye doubles.

α (alpha) Gruis, 22h 08m −47°.0, (Alnair, "the bright one"), mag. 1.7, is a blue-white main-sequence star 100 l.y. away.

β (beta) Gru, 22h 43m −46°.9, (Tiaki), is a red giant 175 l.y. away that varies between about mags. 1.9 and 2.3 every 37 days or so.

γ (gamma) Gru, 21h 54m −37°.4, (Aldhanab), mag. 3.0, is a blue-white star 180 l.y. away.

δ1 δ2 (delta1 delta2) Gru, 22h 29m −43°.5, is a naked-eye pairing of two unrelated stars: δ1, mag. 4.0, is a yellow giant 325 l.y. away; δ2 is a mag. 4.1 red giant 355 l.y. away.

μ1 μ2 (mu1 mu2) Gru, 22h 16m −41°.3, is another naked-eye double of unrelated yellow giants that appear in the same line of sight by chance: μ1 is of mag. 4.8, 350 l.y. away; μ2 is of mag. 5.1, 225 l.y. away.

π1 π2 (pi1 pi2) Gru, 22h 23m −45°.9, is a binocular duo of unrelated stars. π1 is a deep-red semi-regular variable that ranges between mags. 5.3 and 7.0 every 195 days or so; it lies about 525 l.y. away. π2 is a white main-sequence star of mag. 5.6, 130 l.y. away.

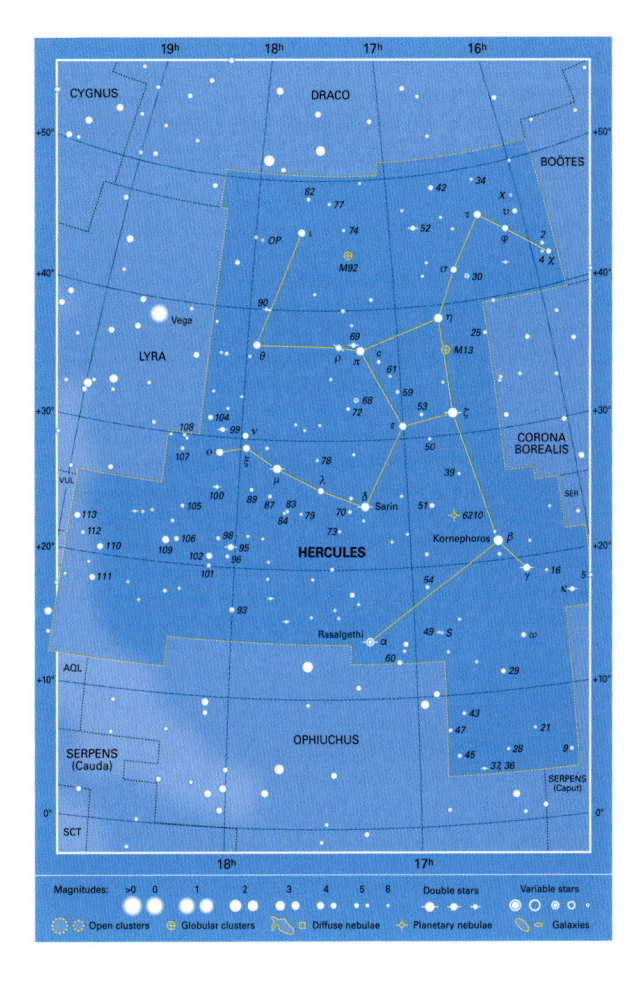

Hercules

Hercules represents the Greek mythological hero famous for his 12 labors. Originally, though, the constellation was visualized as an anonymous kneeling man, with one foot on the head of the celestial dragon, Draco, which adjoins to the north. Some legends identify the constellation with the ancient Sumerian superman Gilgamesh. Despite being the fifth-largest constellation, Hercules is not prominent. But it is stocked with an abundance of double stars for users of small telescopes, plus one of the brightest and richest globular clusters, M13, which is easily found on one side of the so-called Keystone, a quadrilateral consisting of ε, ζ, η, and π Herculis that marks the pelvis of Hercules.

α (alpha) Herculis, 17h 15m +14°.4, (Rasalgethi, "the kneeler's head"), about 350 l.y. distant, is a red giant star some 400 times the Sun's diameter, one of the largest naked-eye stars. Like most red giants it is erratically variable, in this case fluctuating from about mag. 2.7 to mag. 3.6. It is actually a double star, with a mag. 5.3 blue-green companion visible in small telescopes. The estimated orbital period of the pair is 3,600 years.

β (beta) Her, 16h 30m +21°.5, (Kornephoros, "club bearer"), mag. 2.8, the brightest star in the constellation, is a yellow giant 140 l.y. away.

γ (gamma) Her, 16h 22m +19°.2, mag. 3.8, is a white giant star 190 l.y. away, with a wide, unrelated 10th-mag. companion visible in small telescopes.

δ (delta) Her, 17h 15m +24°.8, (Sarin), mag. 3.1, is a blue-white star 75 l.y. away. Small telescopes show a nearby mag. 8.3 star, which is physically unrelated.

ζ (zeta) Her, 16h 41m +31°.6, 35 l.y. away, is a mag. 2.9 yellow-white star with a close mag. 5.3 orange companion that orbits it in 34.5 years. They should be within range of 150 mm for several years either side of 2025–6, when they are their widest, but they then become progressively more difficult to split as they close up.

κ (kappa) Her, 16h 08m +17°.0, is a mag. 5.0 yellow giant 395 l.y. away with an unrelated mag. 6.2 orange giant companion, around 415 l.y. away, easily seen in small telescopes.

ρ (rho) Her, 17h 24m +37°.1, about 375 l.y. away, is a pair of blue-white stars of mags. 4.5 and 5.4, visible in small telescopes.

30 Her, 16h 29m +41°.9, also known as g Her, is a red giant that varies semi-regularly between mags. 4.3 and 6.3 every 3 months or so. It lies about 380 l.y. away.

68 Her, 17h 17m +33°.1, also known as u Her, 950 l.y. away, is an eclipsing binary of Beta Lyrae type, varying between mag. 4.7 and 5.4 in just over 2 days.

95 Her, 18h 02m +21°.6, 420 l.y. away, is a double star for small telescopes, consisting of two giant stars of mags. 4.9 and 5.2, appearing silver and gold.

100 Her, 18h 08m +26°.1, is an easy duo for small telescopes, consisting of identical blue-white stars of mag. 5.8, like a pair of celestial cat's eyes, both 210 l.y. away but seemingly not forming a true binary.

M13 (NGC 6205), 16h 42m +36°.5, is a 6th-mag. globular cluster of 300,000 stars, the brightest of its kind in northern skies. It can be seen with the naked eye and is unmistakable in binoculars, spanning half the apparent width of the full Moon. The cluster lies 25,000 l.y. away and has a true diameter of over 100 l.y. Telescopes of 100-mm aperture resolve individual stars throughout the cluster, giving a mottled, sparkling effect.

M92 (NGC 6341), 17h 17m +43°.1, is a globular cluster only slightly inferior to its more famous neighbor, M13, which overshadows it. M92 is easily seen in binoculars as a fuzzy star. It is smaller and more condensed than M13, and needs a larger telescope to resolve its stars. It lies 29,000 l.y. away and has an estimated age of around 14 billion years, making it the oldest globular known.

NGC 6210, 16h 45m +23°.8, is a 9th-mag. planetary nebula which a telescope of 75 mm or larger shows as a blue-green ellipse. It lies nearly 7,000 l.y. away.

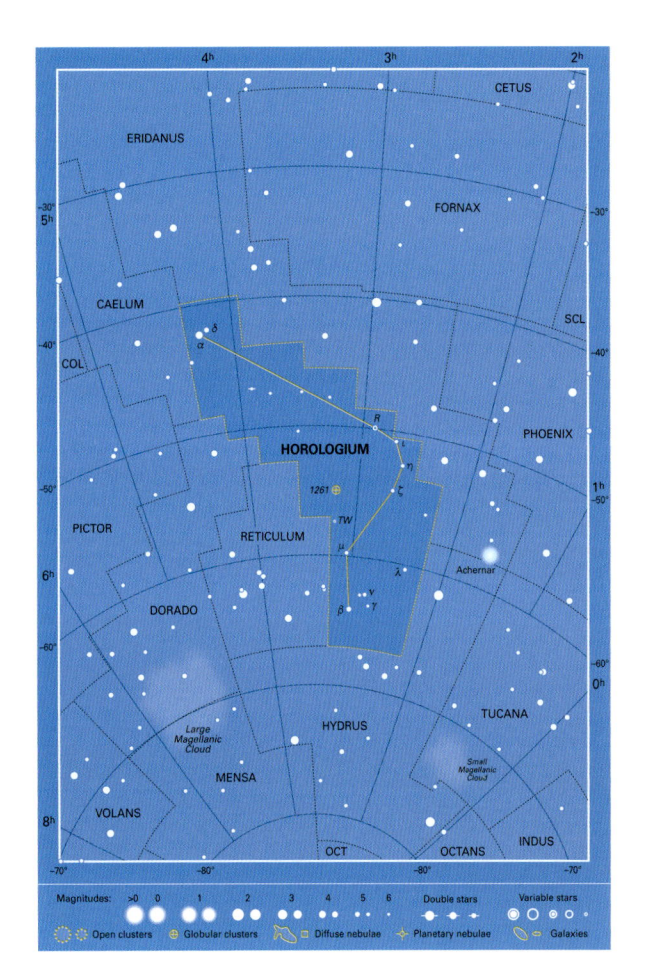

Horologium, the Pendulum Clock

One of the constellations representing mechanical instruments—in this case, the pendulum clock he used for timing observations—introduced in the 1750s by Nicolas Louis de Lacaille. As with so many of Lacaille's constellations, Horologium is faint and obscure.

α (alpha) Horologii, 4h 14m −42°.3, mag. 3.9, is an orange giant star 118 l.y. away.

β (beta) Hor, 2h 59m −64°.1, mag. 5.0, is a giant white star 320 l.y. away.

R Hor, 2h 54m −49°.9, is a red giant variable of the Mira type, ranging between extremes of mag. 4.7 and 14.3 in 13 months or so. It lies about 750 l.y. away.

TW Hor, 3h 13m −57°.3, is a deep-red pulsating variable of semi-regular type, ranging between mag. 5.5 and 6.0 with a period variously given as 5 or 9 months. It lies 1,500 l.y. away.

NGC 1261, 3h 12m −55°.2, is an 8th-mag. globular cluster around 50,000 l.y. away.

α (alpha) Hydrae, 9h 28m −8°.7, (Alphard, "the solitary one"), mag. 2.0, is an orange giant 180 l.y. away.

β (beta) Hya, 11h 53m −33°.9, mag. 4.3, is a blue-white star 310 l.y. away.

γ (gamma) Hya, 13h 19m −23°.2, mag. 3.0, is a yellow giant 130 l.y. away.

δ (delta) Hya, 8h 38m +5°.7, mag. 4.1, is a blue-white star 180 l.y. away.

ε (epsilon) Hya, 8h 47m +6°.4, 130 l.y. away, is a beautiful but difficult double star of contrasting colors, needing high power on a telescope of at least 75-mm aperture. The stars are yellow and blue, of mags. 3.5 and 6.7, and form a genuine binary with an estimated orbital period of about 370 years.

27 Hya, 9h 20m, −9°.6, mag. 4.8, is an orange giant 225 l.y. away with a wide mag. 7.0 companion visible in binoculars. The companion is 220 l.y. away but the two stars are too far apart to form a true binary. Small telescopes split this companion into components of mags. 7 and 11.

54 Hya, 14h 46m −25°.4, 100 l.y. away, also known as m Hya, is an easy double for small telescopes, consisting of yellow and purple stars of mags. 5.1 and 7.2.

R Hya, 13h 30m −23°.3, is a red giant variable star similar to Mira in Cetus that fluctuates between 4th and 10th mags. every 360 days or so. At its best it can reach mag. 3.5, making it one of the brightest Mira stars. It lies 485 l.y. away.

U Hya, 10h 38m −13°.4, 680 l.y. away, is a deep-red variable star that fluctuates semi-regularly between mags. 4.6 and 5.4 every 390 days.

M48 (NGC 2548), 8h 14m −5°.8, is a large open cluster of about 80 stars, 2,500 l.y. away, just visible to the naked eye under clear skies and a fine sight in binoculars. It is somewhat triangular in shape, wider than the apparent size of the full Moon, and is well shown by small telescopes under low power.

M68 (NGC 4590), 12h 39m −26°.7, is an 8th-mag. globular cluster visible as a fuzzy star in binoculars and just resolved with apertures of 100 mm. It lies 29,000 l.y. away.

M83 (NGC 5236), 13h 37m −29°.9, is a large, face-on spiral galaxy of 8th mag., visible in a small telescope. It has a small, bright nucleus and signs of a central bar, similar to our own Galaxy. Its spiral arms can be traced with an aperture of 150 mm M83 has been the site of more known supernovae than any other Messier object, six in all. It is one of the brightest and closest galaxies outside the Local Group, about 15 million l.y. away. It is sometimes popularly termed the Southern Pinwheel.

NGC 3242, 10h 25m −18°.6, is a 9th-mag. planetary nebula of similar apparent size to the disk of Jupiter; it is popularly termed the Ghost of Jupiter. This often-overlooked object, 4,400 l.y. away, is prominent enough to be picked up in small telescopes at low magnification as a blue-green disk, while larger instruments show a bright inner disk surrounded by a fainter halo.

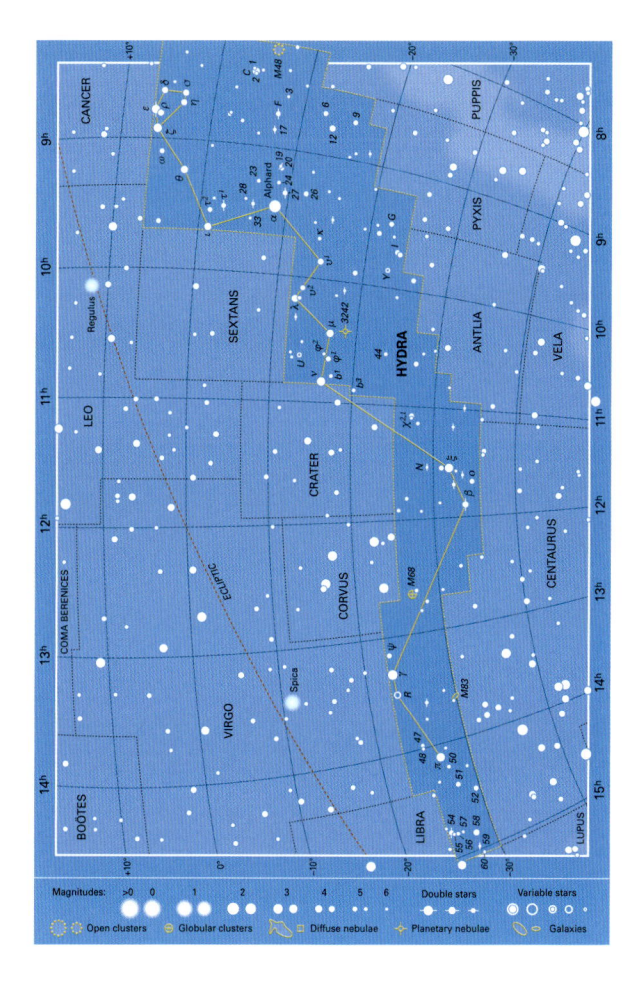

Hydra, the Water Snake

The largest constellation in the sky, but by no means easy to identify because of its faintness. Apart from its brightest star, Alphard, which marks the heart of the Water Snake, Hydra's only readily recognizable feature is its head, an attractive group of six stars. Hydra winds its way from the head in the northern celestial hemisphere, on the borders of Cancer, to the tip of its tail, south of the celestial equator adjacent to Libra and Centaurus, a total length of over 100°.

In mythology, Hydra is usually identified with the multiheaded monster slain by Hercules. Another legend links it with the adjoining constellations of Corvus the Crow and Crater the Cup, which are found on its back. In this story, the bird returned to the god Apollo with Hydra in its claws as an excuse for its aborted mission to fetch water in the cup.

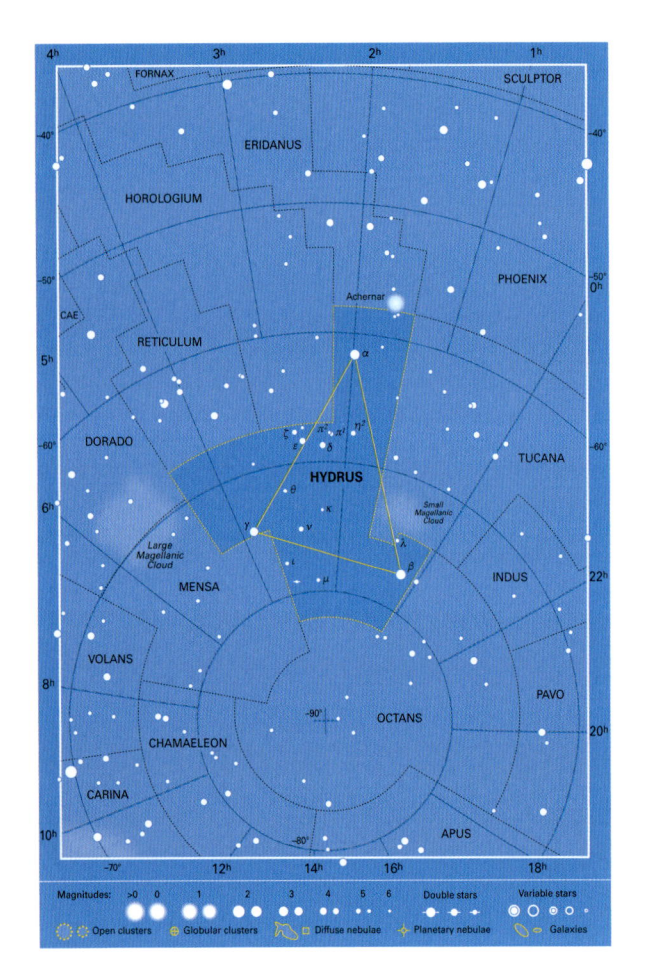

Hydrus, the Lesser Water Snake

The Dutch navigators Pieter Dirkszoon Keyser and Frederick de Houtman introduced this constellation at the end of the sixteenth century as a smaller southern counterpart to the great Water Snake, Hydra. It represents the sea snakes that the explorers would have seen on their voyages. Lying between the two Magellanic Clouds, Hydrus almost bridges the gap between Eridanus and the south celestial pole. There is little to interest the casual observer.

α (alpha) Hydri, 1h 59m −61°.6, mag. 2.8, is a white subgiant 63 l.y. away.

β (beta) Hyi, 0h 26m −77°.3, mag. 2.8, the constellation's brightest member by a few tenths of a magnitude, is a yellow-white main-sequence star 24 l.y. away.

γ (gamma) Hyi, 3h 47m −74°.2, mag. 3.3, is a red giant star 215 l.y. away.

π1 π2 (pi1 pi2) Hyi, 2h 14m −67°.8, is a binocular pair of red and orange giants, unrelated to each other: π1 is mag. 5.6 and lies 715 l.y. away; π2, mag. 5.7, is 495 l.y. distant.

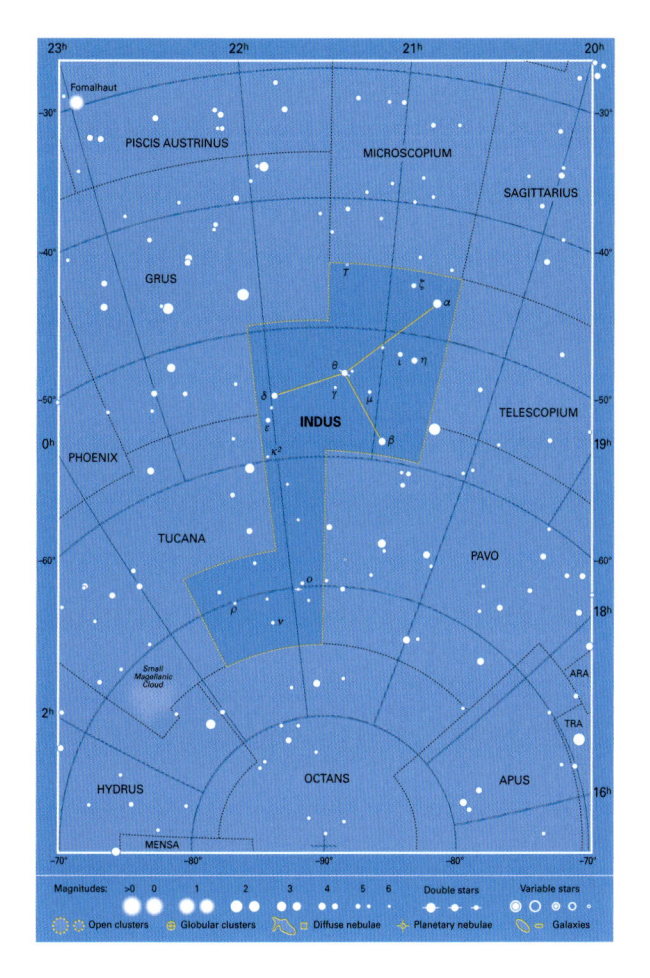

Indus, the Indian

A constellation representing a native hunter, introduced at the end of the sixteenth century by the Dutch navigators Pieter Dirkszoon Keyser and Frederick de Houtman. On old star charts he was depicted holding aloft a spear as though pursuing prey, but it is not known whether he is supposed to be a native of Africa, the East Indies, or South America. None of the constellation's stars is brighter than 3rd magnitude and none of them have names.

α (alpha) Indi, 20h 38m −47°.3, mag. 3.1, is an orange giant star 100 l.y. away.

β (beta) Ind, 20h 55m −58°.5, mag. 3.7, is an orange giant 610 l.y. away.

δ (delta) Ind, 21h 58m −55°.0, mag. 4.4, is a white star 190 l.y. away.

ε (epsilon) Ind, 22h 03m −56°.8, mag. 4.7, is an orange dwarf star somewhat smaller and cooler than the Sun. At a distance of 11.9 l.y., it is one of the Sun's closest neighbors.

θ (theta) Ind, 21h 20m −53°.4, 99 l.y. away, is a pair of white stars of mags. 4.5 and 6.9, divisible in a small telescope.

T Ind, 21h 20m −45°.0, is a deep-red variable star that ranges semi-regularly between mags. 5.8 and 6.5 every 9 months or so. It lies about 2,000 l.y. away.

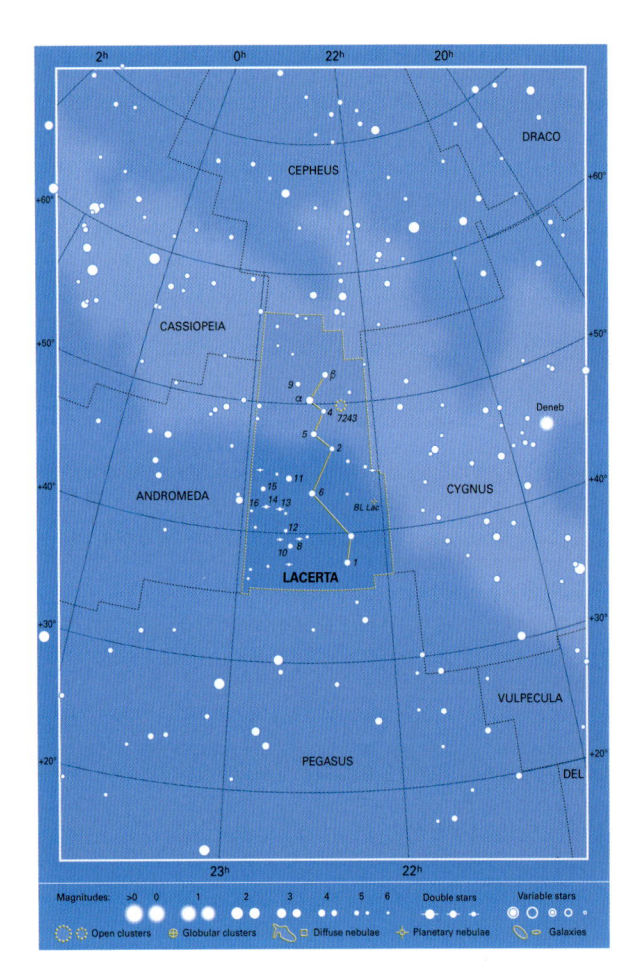

Lacerta, the Lizard

An inconspicuous constellation sandwiched between Cygnus and Andromeda like a lizard between rocks, introduced in 1687 by the Polish astronomer Johannes Hevelius. An alternative figure that previously occupied this area was Sceptrum, the Sceptre and Hand of Justice, created in 1679 by the Frenchman Augustin Royer to commemorate King Louis XIV. In 1787 the German Johann Elert Bode called this region Friedrichs Ehre (Frederick's Glory, later Latinized to Honores Friderici), in honor of King Frederick the Great of Prussia. Both these alternatives have been discarded.

The constellation's most celebrated object is BL Lacertae, location 22h 02.7m, +42° 17', originally thought to be a peculiar 14th-magnitude variable star. It is the prototype of a group of objects, the BL Lac objects or Lacertids, believed to be giant elliptical galaxies with variable centers, lying far off in the Universe and related to quasars and other galaxies with active nuclei.

Three naked-eye novae appeared within the boundaries of Lacerta during the twentieth century, in 1910, 1936, and 1950. Brightest of them was the nova of 1936, now known as CP Lacertae, which peaked at around mag. 2.

α (alpha) Lacertae, 22h 31m +50°.3, mag. 3.8, is a blue-white main-sequence star 104 l.y. away.

β (beta) Lac, 22h 24m +52°.2, mag. 4.4, is a yellow giant 169 l.y. away.

NGC 7243, 22h 15m +49°.9, is a scattered open cluster of similar apparent size to the full Moon, suitable for small telescopes, consisting of a few dozen stars of 8th mag. and fainter, 2,800 l.y. away.

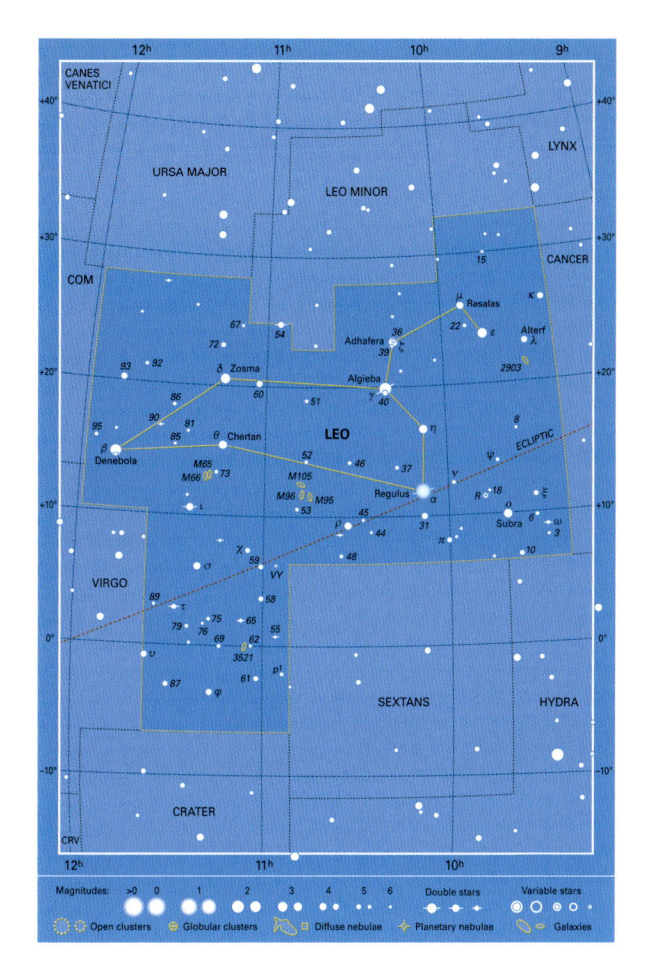

Leo, the Lion

One of the few constellations that looks like the figure it is supposed to represent—in this case, a crouching lion. The Lion's head is outlined by the so-called Sickle, consisting of six stars from ε (epsilon) to α (alpha) Leonis; the Lion's body stretches out behind, its tail marked by β (beta) Leonis. Mythologically, this is the Lion reputedly slain by the hero Hercules as the first of his 12 labors. The Sun passes through the constellation from mid-August to mid-September.

Leo contains the third-nearest star to the Sun, CN Leonis (also called Wolf 359), a red dwarf 7.9 l.y. away. It is of magnitude 13.5, although it is a flare star and can occasionally brighten by up to a magnitude; it is located at 10h 56.5m, +7° 01', near the border with Sextans. Leo contains numerous distant galaxies, plus three faint dwarf members of our Local Group, beyond the reach of amateur telescopes.

Every November, the Leonid meteors radiate from a point near γ (gamma) Leonis. Usually the numbers are low, peaking at about 10 per hour on November 17–18, but the shower's activity increases dramatically at 33-year intervals when the parent comet, Tempel–Tuttle, returns to perihelion. High Leonid rates were last seen in 1999–2002.

α (alpha) Leonis, 10h 08m +12°.0, (Regulus, "the little king"), mag. 1.4, is a blue-white main-sequence star 79 l.y. away. Binoculars or small telescopes will show a wide companion of mag. 8.2.

β (beta) Leo, 11h 49m +14°.6, (Denebola, "the lion's tail"), mag. 2.1, is a blue-white main-sequence star 36 l.y. away.

γ (gamma) Leo, 10h 20m +20°.0, (Algieba, "the forehead"), 130 l.y. away, consists of a pair of golden-yellow giant stars of mags. 2.4 and 3.6, orbiting in about 550 years. They form an exceptionally handsome double in small telescopes, one of the finest in the sky. In binoculars, an unrelated mag. 4.8 yellowish foreground star, 40 Leonis, is visible nearby.

δ (delta) Leo, 11h 14m +20°.5, (Zosma), mag. 2.5, is a blue-white star 58 l.y. away.

ε (epsilon) Leo, 9h 46m +23°.8, mag. 3.0, is a yellow giant 230 l.y. away.

ζ (zeta) Leo, 10h 17m +23°.4, (Adhafera), mag. 3.4, is a giant white star 232 l.y. away. To the north of it, binoculars show 35 Leonis, an unrelated foreground star of mag. 6.0. To the south a wider third star, 39 Leonis, mag. 5.8 and also in the foreground, can be seen in binoculars, making this an optical triple.

ι (iota) Leo, 11h 24m +10°.5, 78 l.y. away, is a challenging double star. It appears to the naked eye as a yellow-white star of mag. 4.0, but actually consists of components of mags. 4.1 and 6.7 orbiting every 184 years. Currently the stars are moving apart and becoming progressively easier to separate, coming within range of all but the very smallest apertures when at their widest around 2053 to 2065.

τ (tau) Leo, 11h 28m +2°.9, 530 l.y. away, is a yellow giant of mag. 5.0 with a mag. 7.5 companion visible in binoculars and small telescopes.

54 Leo, 10h 56m +24°.7, 305 l.y. away, is a double for small telescopes, consisting of blue-white components of mags. 4.5 and 6.3.

R Leo, 9h 48m +11°.4, is a red giant variable of Mira type, lying about 230 l.y. away. It appears strongly red when at maximum and is roughly 450 times larger than our Sun. R Leonis normally varies between 6th and 10th mags. with an average period of 310 days, but on occasion can become as bright as mag. 4.4.

M65, M66 (NGC 3623, NGC 3627), 11h 19m +13°.0, are a pair of spiral galaxies about 30 million l.y. away. At 9th mag. they can be detected in large binoculars under clear conditions, but at least 100-mm aperture at low power is required for their elongated shape and condensed centers to be clearly seen.

M95, M96 (NGC 3351, NGC 3368), 10h 44m +11°.7, 10h 47m +11°.8, are a pair of spiral galaxies of 10th and 9th mag. respectively, about 35 million l.y. away, visible as circular nebulosities in small telescopes. Larger apertures show that M95 has a central bar. About 1° away lies the smaller M105 (NGC 3379), 10h 48m +12°.6, a 9th-mag. elliptical galaxy at a similar distance.

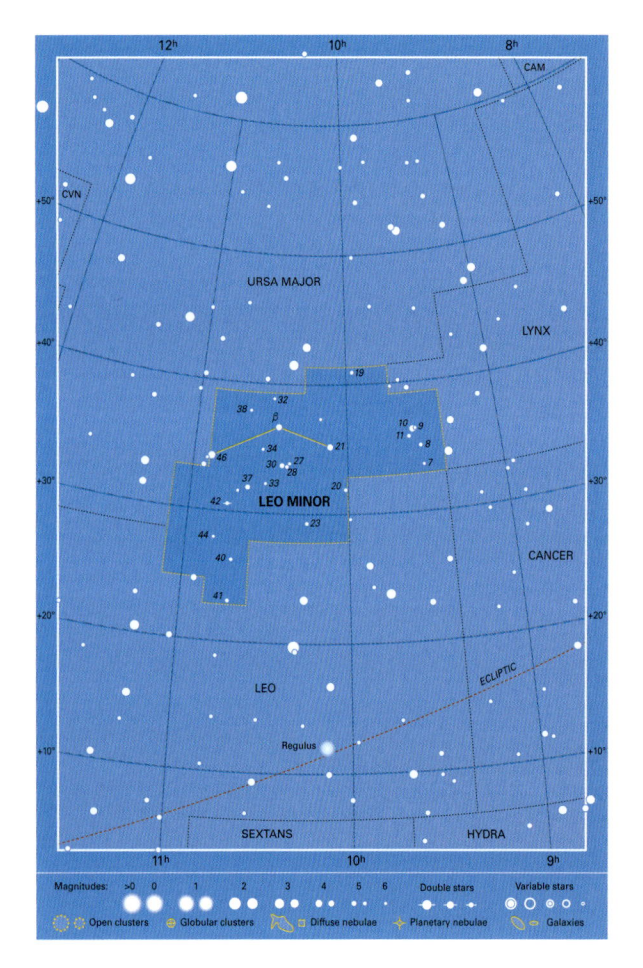

Leo Minor, the Lesser Lion

Leo Minor lies between the larger and brighter constellations of Leo to the south and Ursa Major to the north. Johannes Hevelius, the Polish astronomer, introduced Leo Minor in 1687. There is little of interest in it. Surprisingly there is no star labeled α (alpha). This is the result of an error by the English astronomer Francis Baily, who assigned the Greek letter β (beta) to the second-brightest star of Leo Minor in his British Association Catalogue of 1845, but failed to letter the brightest star, 46 LMi, through an oversight.

β (beta) Leonis Minoris, 10h 28m +36°.7, mag. 4.2, is a yellow giant star 155 l.y. away.

46 LMi, 10h 53m +34°.2, mag. 3.8, the brightest star in the constellation, is an orange giant 99 l.y. away.

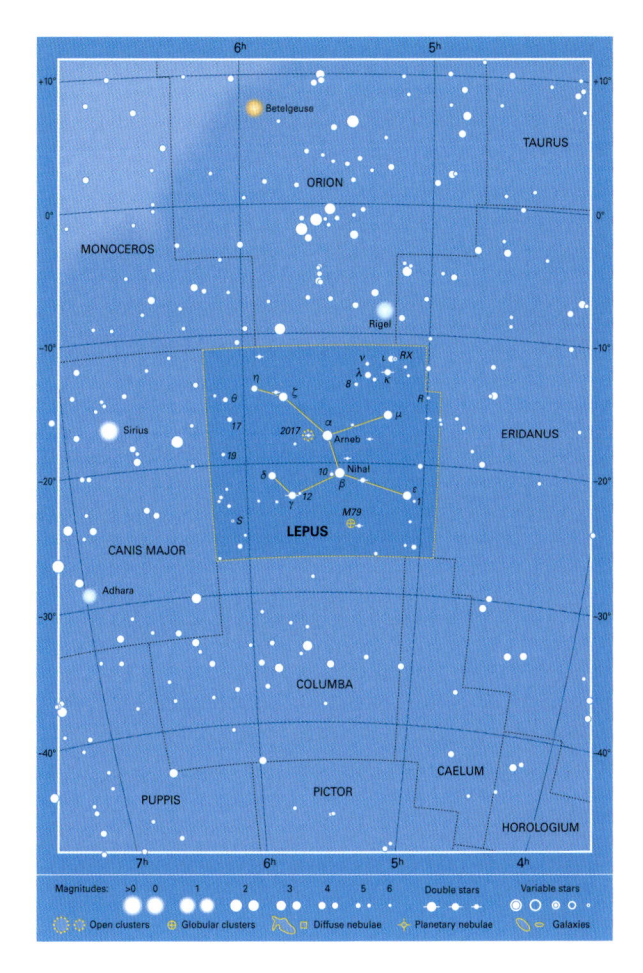

Lepus, the Hare

Lepus is a constellation known since ancient Greek times. It represents a hare, cunningly located at the feet of its hunter, Orion, and pursued endlessly across the sky by Canis Major, the hunter's dog. In one Greek legend, the inhabitants of the island of Leros began to breed hares but the population exploded and the island quickly became overrun with them. The hares destroyed the crops, reducing the inhabitants to starvation. After a concerted effort, the islanders drove the hares off the island. They put the image of a hare among the stars as a reminder that one can easily end up with too much of a good thing. The Hare is also associated in many legends with the Moon. For instance, the familiar figure of the man in the Moon is sometimes interpreted as a hare or rabbit, so perhaps Lepus is another incarnation of the lunar hare. Lepus is overshadowed by Orion's brilliance, but is not without interest for amateur observers.

α (alpha) Leporis, 5h 33m −17°.8, (Arneb, "hare"), mag. 2.6, is a white supergiant star, about 2,200 l.y. away.

β (beta) Lep, 5h 28m −20°.8, (Nihal), mag. 2.8, is a yellow giant star 156 l.y. away.

γ (gamma) Lep, 5h 44m −22°.5, 29 l.y. away, is an attractive binocular duo consisting of a yellow-white main-sequence star of mag. 3.6 with an orange companion of mag. 6.3.

δ (delta) Lep, 5h 51m −20°.9, mag. 3.8, is an orange giant 115 l.y. away.

ε (epsilon) Lep, 5h 05m −22°.4, mag. 3.2, is an orange giant 210 l.y. away.

κ (kappa) Lep, 5h 13m −12°.9, 470 l.y. away, is a mag. 4.4 blue-white main-sequence star with a close 7th-mag. companion, difficult to see in the smallest telescopes because of the magnitude contrast.

R Lep, 5h 00m −14°.8, about 1,500 l.y. away, is an intensely red star known as Hind's Crimson Star, after the English observer John Russell Hind, who described it in 1845 as "like a drop of blood on a black field." R Leporis is a Mira-type variable that ranges from mag. 5.5 at its brightest to as faint as 12th mag. in a period of around 440 days.

RX Lep, 5h 11m −11°.8, 540 l.y. away, is a red giant that varies semi-regularly between mags. 5.1 and 6.7 every 80 days or so.

M79 (NGC 1904), 5h 24m −24°.5, is a small but rich globular cluster 37,000 l.y. away, visible as a fuzzy 8th-mag. star in small telescopes. Nearby in the same low-power field is the multiple star Herschel 3752, consisting of a mag. 5.4 primary with two companions, a close one of mag. 6.6 and a wide one of mag. 9.3, all visible in small telescopes.

NGC 2017, 5h 39m −17°.8, is a small but remarkable star cluster, also known as the multiple star Herschel 3780. Modest amateur telescopes reveal a group of six well-spaced stars ranging from 6th to 11th mag. In addition, the brightest star has a mag. 7.8 companion that requires a telescope of at least 200-mm aperture to split, while an aperture of at least 100 mm shows that one of the 9th-mag. stars is a close double. There is also a 12th-mag. component that should be visible with 100 mm or upward, so this is actually a group of at least nine stars. However, it is not a true cluster: the stars lie at a range of distances and are moving in different directions.

STARGAZING FOR BEGINNERS 147

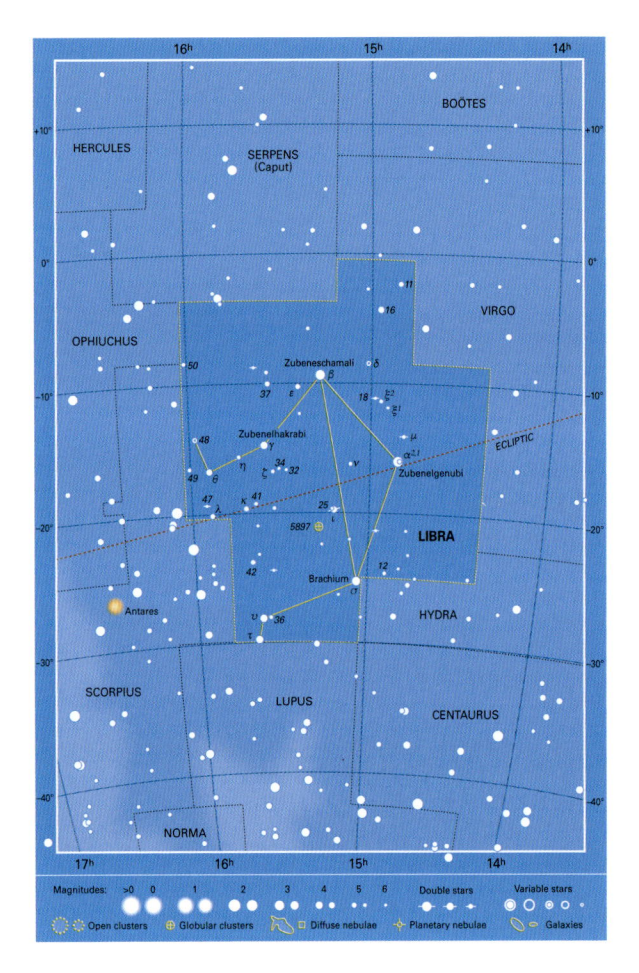

Libra, the Scales

A faint, easily overlooked constellation of the zodiac, through which the Sun passes during November. The ancient Greeks knew it as the Claws of the Scorpion, an extension of neighboring Scorpius, and this old identification lives on in the Arabic-derived names Zubenelgenubi and Zubeneschamali ("southern claw" and "northern claw," respectively), applied to the stars α (alpha) and β (beta) Librae.

The Romans made this region into a separate constellation during the time of Julius Caesar in the first century BCE. Since then the Scales have come to be regarded as the symbol of justice, held aloft by the goddess of justice, Astraeia, in the shape of the neighboring figure of Virgo.

Libra once contained the September equinox, the point at which the Sun passes south of the celestial equator each year. Because of precession, this point moved into neighboring Virgo around 730 BCE, but the September equinox is still sometimes referred to as the First Point of Libra. Despite its relative faintness, Libra contains several stars of interest.

α (alpha) Librae, 14h 50m −16°.0, (Zubenelgenubi, "the southern claw"), 76 l.y. away, is a wide binocular double consisting of a blue-white star of mag. 2.7 with a white companion of mag. 5.2.

β (beta) Lib, 15h 17m −9°.4, (Zubeneschamali, "the northern claw"), mag. 2.6, the brightest in the constellation, is celebrated as one of the few bright stars to show a distinct greenish tinge. It lies 185 l.y. away.

γ (gamma) Lib, 15h 36m −14°.8, (Zubenelhakrabi), mag. 3.9, is an orange giant 165 l.y. away.

δ (delta) Lib, 15h 01m −8°.5, 350 l.y. away, is an eclipsing variable of the Algol type. It varies between mags. 4.9 and 5.9 in 2 days 8 hours.

ι (iota) Lib, 15h 12m −19°.8, is a multiple star 380 l.y. away. Its main blue-white component, of mag. 4.5, has a wide 10th-mag. companion that is difficult to see in the smallest telescopes because of the brightness difference. Apertures of 60 mm or above with high magnification should split this fainter companion into two stars of 11th mag. The brightest component of ι Lib is itself a binary with a 23-year period, but too close for amateur telescopes to divide. Binoculars show a mag. 6.1 star nearby called 25 Librae, which is a foreground object, 290 l.y. away.

μ (mu) Lib, 14h 49m −14°.1, 246 l.y. away, is a close double star consisting of components of mags. 5.6 and 6.6, divisible in a telescope of 75-mm aperture with high magnification.

48 Lib, 15h 58m −14°.3, mag. 4.9, is a blue giant 455 l.y. away with an abnormally high speed of rotation that causes it to throw off rings of gas from its equator, as a result of which it varies irregularly by a few tenths of a magnitude in similar fashion to γ (gamma) Cassiopeiae and Pleione in Taurus. It also bears the variable-star designation FX Lib.

NGC 5897, 15h 17m −21°.0, is a large but loosely scattered 9th-mag. globular cluster about 31,000 l.y. away, unspectacular in small instruments.

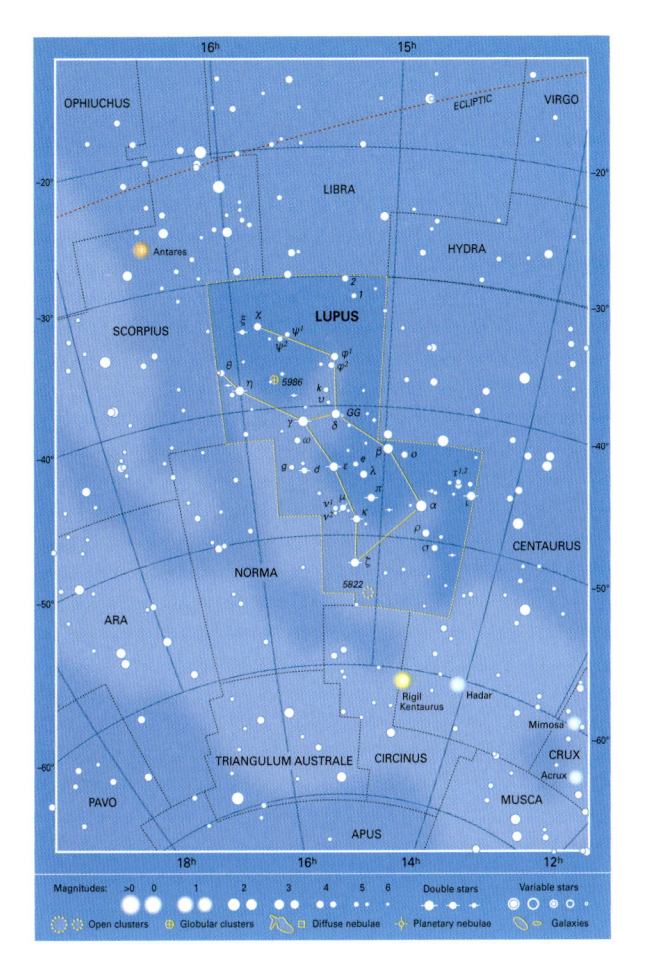

Lupus, the Wolf

Lupus is stocked with numerous interesting objects, although it is often overlooked in favor of its more spectacular neighbors Scorpius and Centaurus. The constellation was regarded by the Greeks and Romans as an unspecified wild animal, impaled on a pole by Centaurus the Centaur. Some mythologists said that Centaurus was about to sacrifice the animal on the altar, represented by nearby Ara. The constellation's identification with a wolf seems to have become common in Renaissance times. Lupus lies in the Milky Way and is rich in double stars. None of its stars is named.

α (alpha) Lupi, 14h 42m −47°.4, mag. 2.3, is a blue giant star 465 l.y. away.

β (beta) Lup, 14h 59m −43°.1, mag. 2.7, is a blue giant 380 l.y. away.

γ (gamma) Lup, 15h 35m −41°.2, mag. 2.8, is a blue-white star 420 l.y. away. It is a close binary with a 190-year orbital period, divisible only in apertures of 200 mm and above.

ε (epsilon) Lup, 15h 23m −44°.7, 510 l.y. away, is a blue-white star of mag. 3.4 with a wide mag. 9.1 companion visible in a small telescope. The primary is itself a close double, divisible only with large apertures.

η (eta) Lup, 16h 00m −38°.4, 440 l.y. away, is a double star consisting of a mag. 3.4 blue-white primary with a mag. 7.5 companion, not easy to see in a small telescope because of the magnitude contrast.

κ (kappa) Lup, 15h 12m −48°.7, 185 l.y. away, is an easy double star for small telescopes, consisting of blue-white components of mags. 3.8 and 5.5.

μ (mu) Lup, 15h 19m −47°.9, 335 l.y. away, is a multiple star. Small telescopes reveal a mag. 4.2 blue-white primary with a wide mag. 6.8 companion. The primary itself is a close binary, consisting of two near-identical stars of mags. 4.9 and 5.0 with an orbital period of about 770 years. At their closest, between the years 2050 and 2060, apertures of at least 250 mm will be needed to separate them.

ξ (xi) Lup, 15h 57m −34°.0, is a neat pair of mag. 5.1 and 5.6 blue-white main-sequence stars, both 212 l.y. away, well seen in a small telescope.

π (pi) Lup, 15h 05m −47°.1, about 475 l.y. away, appears to the naked eye as mag. 3.9, but telescopes of 75-mm aperture and above show that it consists of two close and near-identical blue-white stars each of mag. 4.6.

GG Lup, 15h 19m −40°.8, 490 l.y. away, is an eclipsing binary of Algol type, ranging between mags. 5.6 and 6.1 with a period of 1.85 days.

NGC 5822, 15h 05m −54°.3, is a large, loose open cluster of about 150 faint stars, 2,750 l.y. away, visible in binoculars or small telescopes.

NGC 5986, 15h 46m −37°.8, is a 7th-mag. globular cluster, 30,000 l.y. distant, visible as a rounded patch in a small telescope.

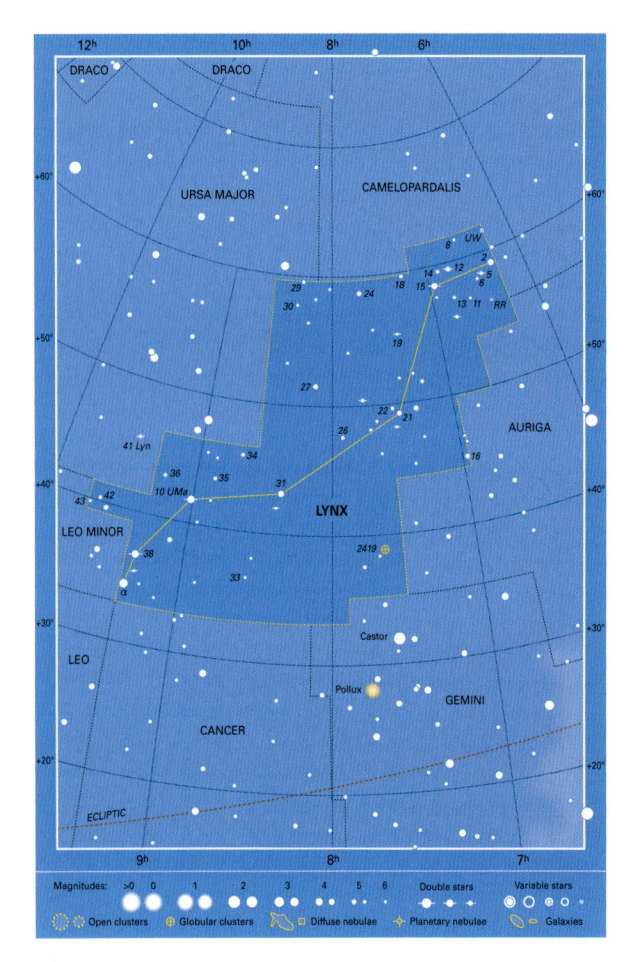

Lynx, the Lynx

A decidedly obscure constellation, despite its considerable size (larger than Gemini, for example), introduced in 1687 by the Polish astronomer Johannes Hevelius to fill the gap between the traditional figures of Ursa Major and Auriga. Hevelius named it Lynx because, he said, one would need the eyes of a lynx to see it—a reference to the fact that his own eyesight was exceptionally keen. Indeed, Hevelius continued to measure star positions with naked-eye instruments throughout his life, even though others had by then moved on to telescopes. Despite the constellation's faintness, owners of small telescopes will find many exquisite double stars within it.

α (alpha) Lyncis, 9h 21m +34°.4, mag. 3.1, is an orange giant star 220 l.y. away.

5 Lyn, 6h 27m +58°.4, 625 l.y. away, is an orange giant of mag. 5.2 with a wide, unrelated mag. 7.8 companion, visible in a small telescope.

12 Lyn, 6h 46m +59°.4, 240 l.y. away, is a fascinating triple star. A small telescope will show a mag. 4.9 blue-white star with a fainter mag. 7.1 companion. Telescopes of 75-mm aperture and above reveal that the brighter component is itself a binary, of mags. 5.4 and 6.0, which orbit each other every 700 years or so.

15 Lyn, 6h 57m +58°.4, 178 l.y. away, is a close double star for telescopes of 150-mm aperture and above. The components are of mags. 4.5 and 5.5, the brighter star appearing a deep yellow color. They form a genuine binary with an orbital period of 260 years.

19 Lyn, 7h 23m +55°.3, is an attractive triple star for small telescopes, with blue-white components of mags. 5.8 and 6.7. The very wide third star is of mag. 7.6. The trio lies about 650 l.y. away.

38 Lyn, 9h 19m +36°.8, 125 l.y. away, is a pair of mag. 3.9 and 6.1 blue-white stars, difficult in the smallest telescopes because of their closeness. They form a genuine binary with a period of thousands of years.

41 Lyn, 9h 29m +45°.6, 275 l.y. away, is actually over the border in Ursa Major (as a result of which the designation "41 Lyncis" has now fallen out of use, but we retain it here for identification purposes). It is a yellow giant of mag. 5.4, and small telescopes reveal a wide companion of mag. 7.8. An 11th-mag. star nearby forms a triangle, making this an apparent triple.

NGC 2419, 7h 38m +38°.9, is an unusually remote globular cluster, of the type known as an intergalactic tramp. It is of 10th mag. and lies some 300,000 l.y. from the center of our Galaxy, farther than either of the Magellanic Clouds.

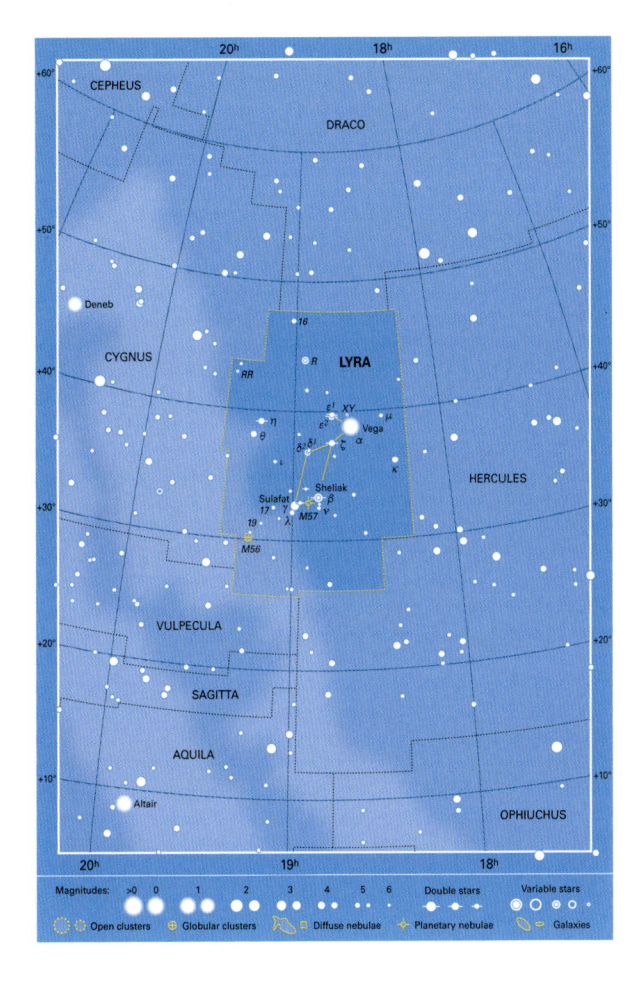

Lyra, the Lyre

A constellation dating from ancient times, representing the stringed instrument invented by Hermes and subsequently given by his half-brother Apollo to the great musician Orpheus. This constellation has also been visualized as an eagle or vulture. Although small, Lyra is bright and prominent. It contains the fifth-brightest star in the sky, Vega, which forms one corner of the northern Summer Triangle (Deneb in Cygnus and Altair in Aquila mark the other two corners). Because of precession, Vega will be the pole star between 13,000 and 14,000 CE, although it will come no closer than 5°.7 to the pole. In a separate effect, our Sun's motion around the Galaxy is carrying us in the general direction of Vega at a velocity of 20 km/s relative to nearby stars. The Lyrid meteor shower emanates from this constellation each year, reaching a peak of about 10 per hour on April 21-2.

α (alpha) Lyrae, 18h 37m +38°.8, (Vega, "the swooping eagle"), mag. 0.03, is a brilliant blue-white main-sequence star 25 l.y. away. It is the fifth-brightest star in the sky, and is surrounded by a disk of dust from which planets may be forming.

β (beta) Lyr, 18h 50m +33°.4, (Sheliak, "the harp"), 960 l.y. away, is a remarkable multiple star. Small telescopes easily resolve it as a double star of cream and blue components. The fainter, blue star is of mag. 6.7, while the brighter star is an eclipsing binary that varies between mags. 3.3 and 4.4 in 12.9 days. β Lyrae is the prototype of a class of eclipsing variables in which the stars are so close together that gravity distorts them into egg-shapes, and hot gas spirals off them into space.

γ (gamma) Lyr, 18h 59m +32°.7, (Sulafat, "the tortoise"), mag. 3.2, is a blue-white giant about 600 l.y. away.

δ1 δ2 (delta1 delta2) Lyr, 18h 54m +37°.0, is a wide naked-eye or binocular double consisting of two unrelated stars: δ1, a blue-white main-sequence star of mag. 5.6, 1,160 l.y. away; and δ2, a red giant 740 l.y. away, which varies erratically between mags. 4.1 and 4.4.

ε1 ε2 (epsilon1 epsilon2) Lyr, 18h 44m +39°.7, 160 l.y. away, is a celebrated quadruple star known popularly as the Double Double. Binoculars or even keen eyesight separate it into two stars, ε1 and ε2, of mags. 4.7 and 4.6 respectively But a telescope of 60- to 75-mm aperture and high magnification reveals that each star is itself double, the two pairs being oriented almost at right angles to each other. The ε1 pair has mags. of 5.0 and 6.1 and an orbital period of about 2,800 years; the ε2 pair, slightly closer together, has mags. of 5.2 and 5.4, and a period of about 720 years. Quadruple stars are rare, and this is the finest of them.

ζ (zeta) Lyr, 18h 45m +37°.6, 158 l.y. away, is a double star, easily split in small telescopes or binoculars into components of mags. 4.4 and 5.6.

η (eta) Lyr, 19h 14m +39°.1, 1,400 l.y. away, is a blue-white star of mag. 4.4 with a wide mag. 8.6 companion visible in a small telescope.

R Lyr, 18h 55m +43°.9, 310 l.y. away, is a red giant that varies semi-regularly between mags. 3.8 and 4.4 every 6 or 7 weeks.

RR Lyr, 19h 25m +42°.8, 820 l.y. away, is the prototype of an important class of variable stars used as "standard candles" for indicating distances in space. RR Lyrae variables are often found in globular clusters and are thus known as cluster-type variables. Related to Cepheid variables, they are giant stars that pulsate in size, varying by about one magnitude usually in less than a day. RR Lyrae itself varies from mag. 7.2 to 8.1 in 13.6 hours.

M57 (NGC 6720), 18h 54m +33°.0, the Ring Nebula, is a famous 9th-mag. planetary nebula 2,500 l.y. away, conveniently placed between β (beta) and γ (gamma) Lyrae. On photographs taken through large telescopes it looks like a celestial smoke ring, but in fact it is a cylinder seen end-on. A small telescope shows it as a noticeably elliptical misty disk, but a larger aperture is needed to see the central hole. It is one of the brightest planetary nebulae and appears larger in the sky than the planet Jupiter. The nebula's true diameter is about one light year.

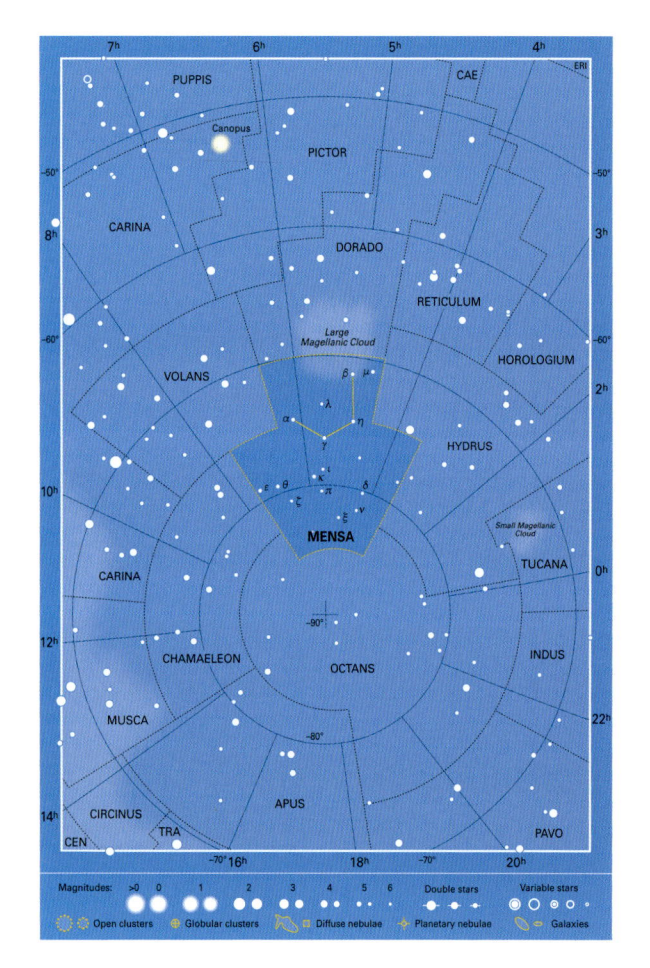

Mensa, the Table Mountain

The faintest of all the constellations, with no star brighter than magnitude 5.0. Mensa was introduced by the Frenchman Nicolas Louis de Lacaille to commemorate Table Mountain at the Cape of Good Hope, from where he surveyed the southern skies in 1751–2. Part of the Large Magellanic Cloud strays from neighboring Dorado over the border into Mensa, reminding Lacaille of the white cloud known as the tablecloth that frequently caps the real Table Mountain. Unfortunately, there is little else of interest here.

α (alpha) Mensae, 6h 10m −74°.8, mag. 5.1, is a yellow star similar to the Sun, 33 l.y. away.

β (beta) Men, 5h 03m −71°.3, mag. 5.3, is a yellow giant 680 l.y. away.

γ (gamma) Men, 5h 32m −76°.3, mag. 5.2, is an orange giant 106 l.y. away.

η (eta) Men, 4h 55m −74°.9, mag. 5.4, is an orange giant 660 l.y. away.

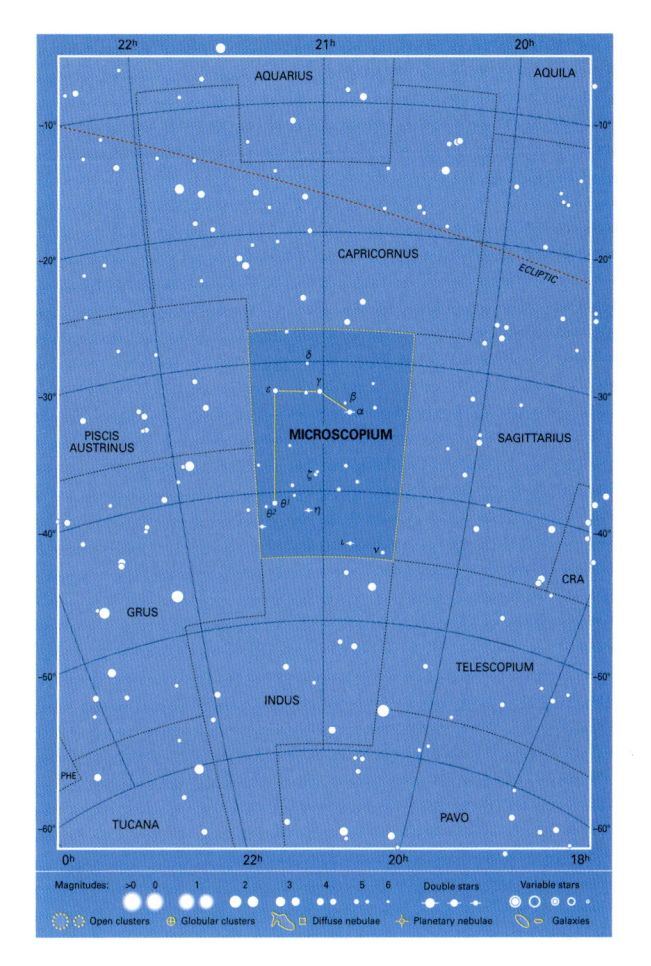

Microscopium, the Microscope

Another of the southern-hemisphere constellations representing scientific instruments that were introduced in the 1750s by the Frenchman Nicolas Louis de Lacaille. The instrument in this case was an early form of compound microscope (i.e. one that uses more than one lens). As with so many of Lacaille's constellations, Microscopium is little more than a filler, encompassing a few faint stars between better-known figures.

α (alpha) Microscopii, 20h 50m −33°.8, mag. 4.9, is a yellow giant star 395 l.y. away. It has a 10th-mag. companion, visible in small telescopes.

γ (gamma) Mic, 21h 01m −32°.3, mag. 4.7, is a yellow giant 255 l.y. away.

ε (epsilon) Mic, 21h 18m −32°.2, mag. 4.7, is a blue-white main-sequence star 170 l.y. away.

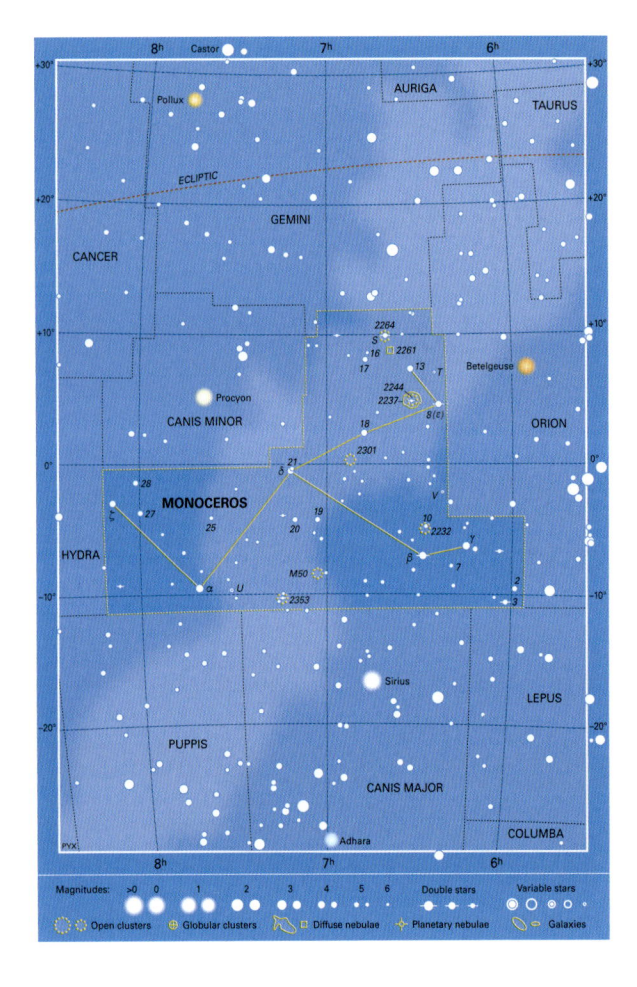

Monoceros, the Unicorn

A faint but fascinating constellation between Orion and Canis Minor, introduced by the Dutch theologian and astronomer Petrus Plancius in 1612, possibly because there are several references to a unicorn in the Old Testament of the Bible, although these are now regarded as mistranslations. Its location in the Milky Way ensures that it is well stocked with nebulae and clusters. Among the constellation's most celebrated features is Plaskett's Star, a mag. 6.1 spectroscopic binary named after the Canadian astronomer John Stanley Plaskett (1865–1941), who found in 1922 that it was the most massive pair of stars then known; according to current data, it consists of two blue giants or supergiants of masses 54 and 56 times that of the Sun, orbiting each other every 14.4 days. Plaskett's Star lies at 6h 37.4m, +6° 08', near the open cluster NGC 2244, of which it may be an outlying member.

α (alpha) Monocerotis, 7h 41m −9°.6, mag. 3.9, is an orange giant star 145 l.y. away.

β (beta) Mon, 6h 29m −7°.0, 680 l.y. away, is rated as perhaps the finest triple star in the heavens. The smallest of telescopes should separate the three components, of mags. 4.6, 5.0, and 5.3, on a steady night. They form a curving line of blue-white stars, the faintest two being the closest together.

δ (delta) Mon, 7h 12m −0°.5, mag. 4.2, is a blue-white star 384 l.y. away. It has a wide naked-eye companion, 21 Mon, of mag. 5.4 and some 200 light years closer.

8 Mon, 6h 24m +4°.6, also known as ε (epsilon) Mon, 130 l.y. away, is an easy double star for small telescopes. It consists of blue-white and yellow components of mags. 4.4 and 6.6 in an attractive low-power field.

S Mon, 6h 41m +9°.9, also known as 15 Mon, is an intensely luminous blue-white star of mag. 4.7, situated 2,400 l.y. away in the star cluster NGC 2264 (see page 188). It is a double star, with a close companion of mag. 7.8 visible in a small telescope. S Mon is slightly variable, fluctuating erratically by about 0.1 mag.

M50 (NGC 2323), 7h 03m −8°.3, is an open cluster of about 80 stars, half the apparent size of the full Moon, visible in binoculars and small telescopes. Apertures of 100 mm or so resolve it into a ragged patch of stars of 8th mag. and fainter, including an orange giant near its southern edge. M50 is 3,250 l.y. away.

NGC 2232, 6h 27m −4°.7, is a scattered cluster of 20 stars for binoculars, containing the mag. 5.0 blue-white star 10 Mon. The cluster, which covers the same area of sky as the full Moon, lies 1,060 l.y. away.

NGC 2237, NGC 2244, 6h 32m +4°.9, is a complex combination of a faint diffuse nebula, known as the Rosette Nebula, and a cluster of stars, all about 5,300 l.y. away. Long-exposure photographs show the nebula as a pink loop, twice the apparent diameter of the full Moon. Visual observations with large amateur telescopes reveal only the brightest parts of the nebula. The associated cluster, NGC 2244, consists of stars that have been born from the Rosette Nebula's gas; it is just visible to the naked eye and is an easy binocular object. The six most prominent stars of the cluster form a rectangular shape, although the brightest of them, 12 Mon, mag. 5.8, is not a true member but an unrelated foreground star. The cluster is likely to be the only part of this celebrated object visible in a small telescope, but the pale outline of the nebula can just be made out in binoculars under clear, dark skies.

NGC 2261, 6h 39m +8°.7, Hubble's Variable Nebula, is a small, faint, fan-shaped nebula containing the remarkable variable star R Mon. Its erratic brightness fluctuations, from mag. 9.5 down to about 12th mag., may be caused by the pangs of its birth from the surrounding nebula. This star and its nebula are open to study only by larger amateur telescopes; the star's distance is uncertain but it may be associated with the nearby NGC 2264 complex (see below), which would place it at about 2,400 l.y.

NGC 2264, 6h 41m +9°.9, is another combination of star cluster and nebula. The cluster, visible in binoculars, has about 40 members, including the 5th-mag. S Mon. The associated nebula, known as the Cone Nebula because of its tapered shape, shows up well only in photographs such as the one at the right and is beyond the reach of visual observation with amateur telescopes. NGC 2264 is 2,400 l.y. away.

NGC 2301, 6h 52m +0°.5, is a binocular cluster of 80 or so stars of 8th mag. and fainter, the brightest of them arranged in a vertical chain. It lies 2,800 l.y. away.

NGC 2353, 7h 15m −10°.3, is an open cluster for small telescopes, consisting of about 30 stars of 9th mag. and fainter seemingly arranged in a spiral pattern. It lies 4,000 l.y. away.

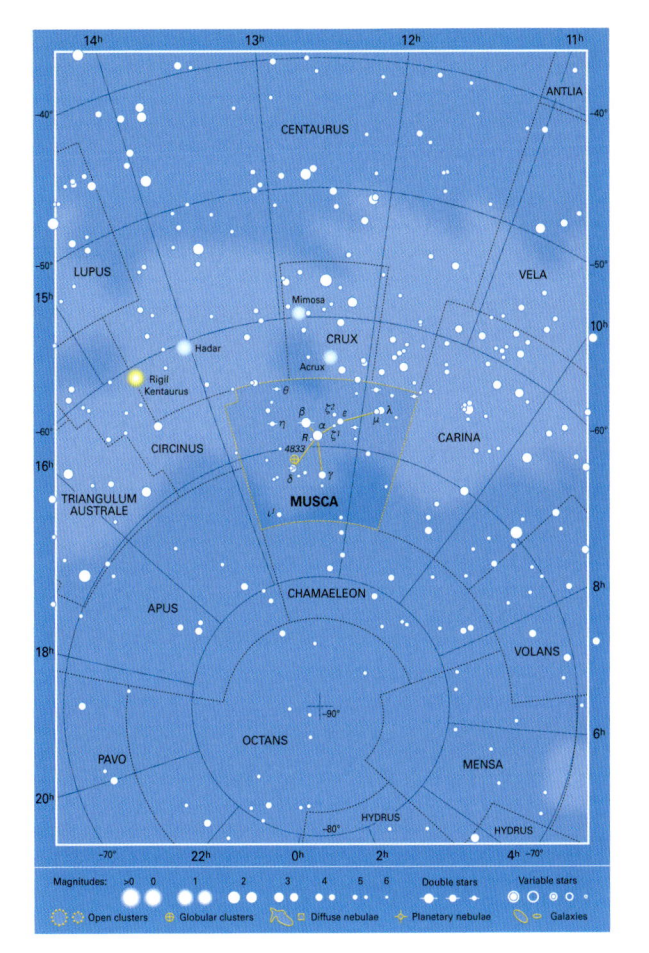

Musca, the Fly

A small southern constellation lying at the foot of the Southern Cross. It is one of the 12 constellations introduced at the end of the sixteenth century by the Dutch navigators Pieter Dirkszoon Keyser and Frederick de Houtman. It was also widely known under the alternative name of Apis, the Bee, which is how it was shown in several old star atlases. There is little of note in Musca other than part of the dark Coalsack Nebula, which spills over the border from Crux.

α (alpha) Muscae, 12h 37m −69°.1, mag. 2.6, is a blue-white star 315 l.y. away.

β (beta) Mus, 12h 46m −68°.1, 315 l.y. away, is a close pair of stars of mags. 3.6 and 4.0, requiring 100-mm aperture and high magnification to split. The orbital period of the pair is about 190 years.

δ (delta) Mus, 13h 02m −71°.5, mag. 3.6, is an orange giant 87 l.y. away.

θ (theta) Mus, 13h 08m −65°.3, is a double star of mags. 5.7 and 7.6 for small telescopes. The brighter star is a blue supergiant, while its companion is a Wolf–Rayet star, a rare type of very hot star; it is the second-brightest such star in the sky, γ (gamma) Velorum being the brightest of all. It lies around 7,000 l.y. away.

NGC 4833, 13h 00m −70°.9, is a fairly large, 7th-mag. globular cluster 20,000 l.y. away, visible in binoculars and small telescopes and resolvable into stars with an aperture of 100 mm.

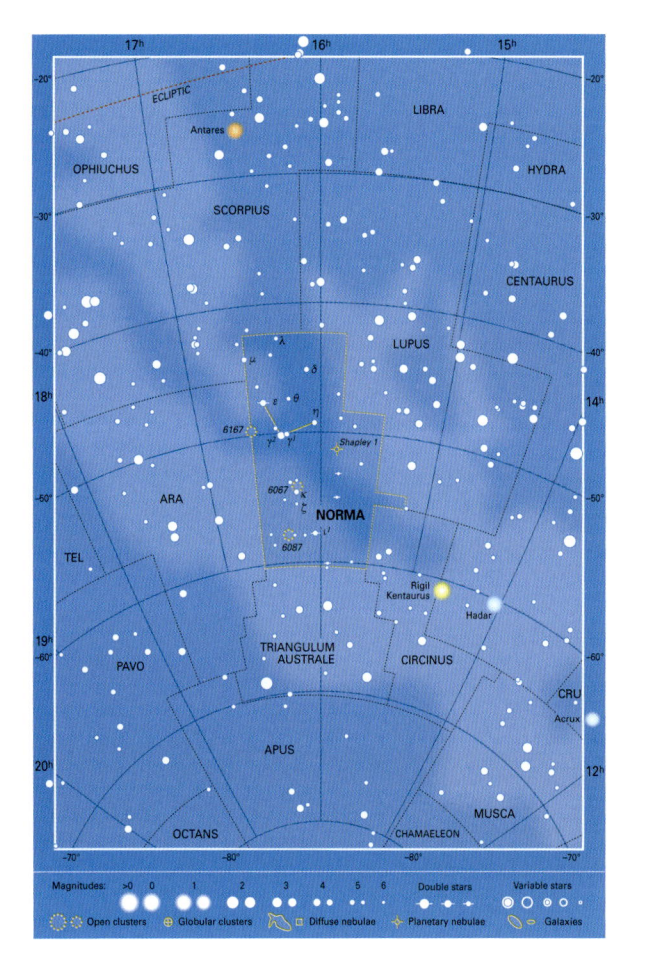

Norma, the Set Square

A superfluous constellation invented in the 1750s by Nicolas Louis de Lacaille, originally under the name Norma et Regula, the Set Square and Ruler. Lacaille placed it next to the Compasses (Circinus) and the Southern Triangle (Triangulum Australe), an earlier invention of Keyser and de Houtman, which Lacaille visualized as a builder's level. The stars of Norma were previously part of Ara, Lupus, and Scorpius, but since Lacaille's time the boundaries of Norma have been altered, so that the stars that he named α (alpha) and β (beta) Normae have been reabsorbed into Scorpius; in the chart on page 174 these two stars are labeled N and H Scorpii. Norma lies in a rich region of the Milky Way.

γ2 (gamma2) Normae, 16h 20m −50°.2, mag. 4.0, is a yellow giant 138 l.y. away, the brightest star in the constellation. Next to it lies the far more distant yellow-white supergiant γ1 (gamma1) Normae, mag. 5.1, around 1,500 l.y. away.

δ (delta) Nor, 16h 06m −45°.2, mag. 4.7, is a white star 105 l.y. away.

ε (epsilon) Nor, 16h 27m −47°.6, 530 l.y. away, is a double star with components of mags. 4.5 and 6.1 visible in small telescopes.

ι1 (iota1) Nor, 16h 04m −57°.8, 128 l.y. away, appears in small telescopes as a double star of mags. 4.7 and 8.0. In addition, the brighter star is itself a very close binary, divisible only in very large telescopes, with a 27-year orbital period.

NGC 6087, 16h 19m −57°.9, is a loose, large binocular cluster of about 40 stars, 3,200 l.y. away, with chains of stars extending from it like a spider's legs. At the center lies its brightest star, the Cepheid variable S Nor, which ranges from mag. 6.1 to 6.8 in 9.8 days.

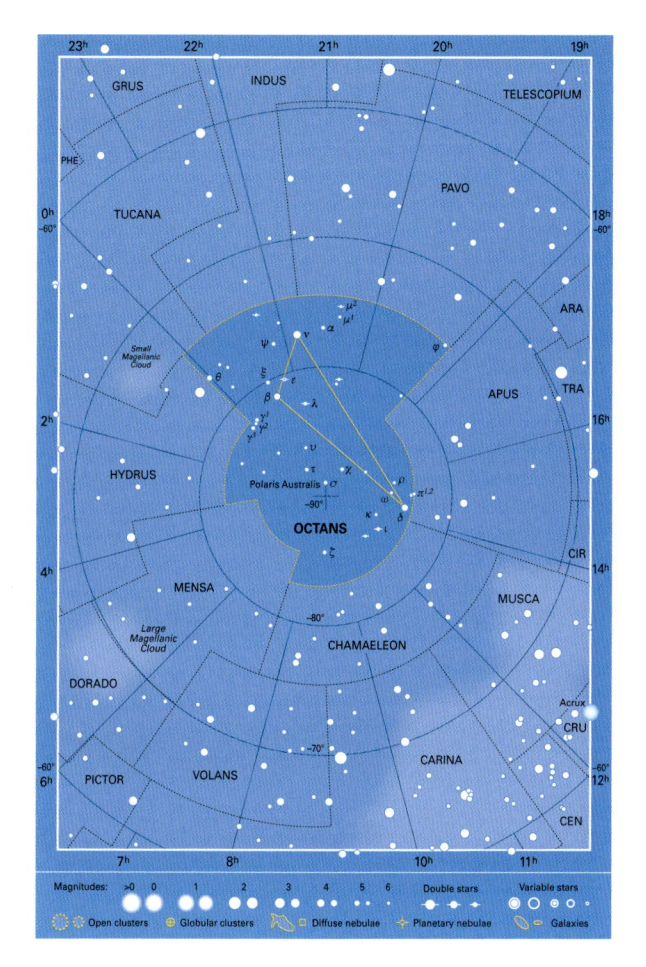

Octans, the Octant

The constellation that contains the south pole of the sky. Despite this privileged position, Octans is faint and unremarkable. The south celestial pole forms an almost exact equilateral triangle with 5th-mag. τ (tau) and χ (chi) Octantis. There is no southern equivalent of Polaris, the north pole star; the nearest naked-eye star to the southern pole is 5th-mag. σ (sigma) Octantis. Currently this star lies about 1° from the pole, but the distance is increasing due to precession (see chart left); σ Octantis was closest to the pole, under ¾°, around 1860.

Octans commemorates the navigational instrument known as the octant, a forerunner of the sextant, invented in 1730 by the English instrument maker John Hadley (1682–1744). The octant consisted of an arc of 45°, i.e. an eighth of a circle, from which came its name. The constellation itself was introduced in the 1750s by Nicolas Louis de Lacaille, and its dullness is a memorial to his lack of imagination.

α (alpha) Octantis, 21h 05m −77°.0, mag. 5.1, is a spectroscopic binary consisting of white and yellow giants 147 l.y. away.

β (beta) Oct, 22h 46m −81°.3, mag. 4.1, is a white star 149 l.y. away.

δ (delta) Oct, 14h 27m −83°.7, mag. 4.3, is an orange giant 300 l.y. away.

ε (epsilon) Oct, 22h 20m −80°.4, also known as BQ Oct, is a red giant that varies semi-regularly between mags. 4.6 and 5.3 with an approximate period of 8 weeks. It lies 290 l.y. away.

λ (lambda) Oct, 21h 51m −82°.7, 407 l.y. away, is a double star with yellow and white components of mags. 5.5 and 7.2, individually visible in small telescopes.

ν (nu) Oct, 21h 41m −77°.4, mag. 3.7, the brightest star in the constellation, is an orange giant 63 l.y. away.

σ (sigma) Oct, 21h 09m −89°.0, (Polaris Australis), mag. 5.4, the nearest naked-eye star to the south celestial pole, is a white giant or subgiant 294 l.y. away.

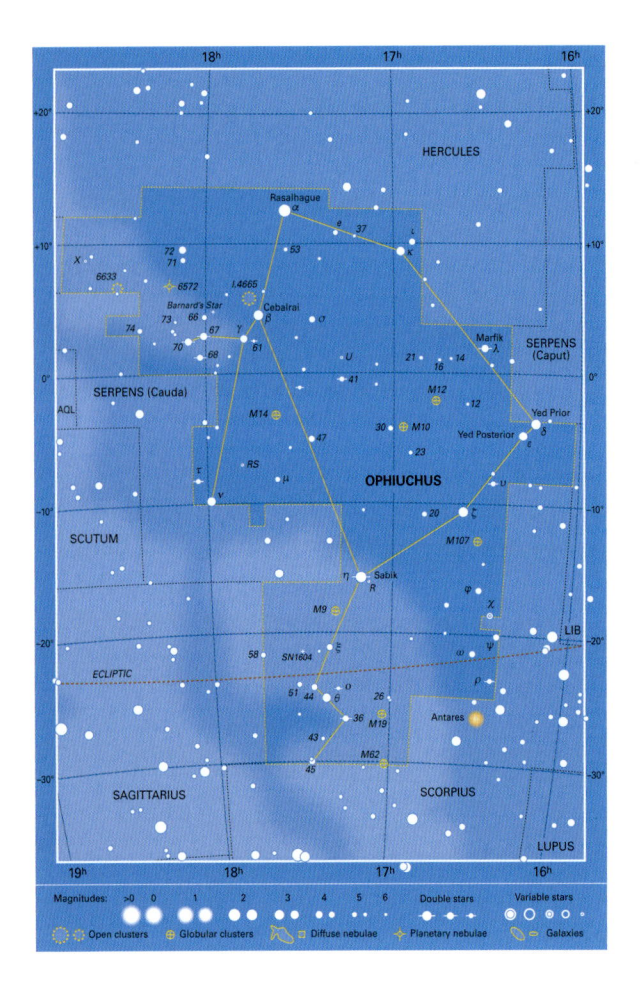

Ophiuchus, the Serpent Holder

An ancient constellation, representing a man encoiled by a serpent (the constellation Serpens). Ophiuchus is usually identified as Aesculapius, a mythical healer who was a forerunner of Hippocrates; his reputed powers included the ability to raise the dead. The serpent he holds is a symbol of this power, since snakes are seemingly reborn every year when they shed their skin. Perhaps the most celebrated star in Ophiuchus is Barnard's Star, a mag. 9.5 red dwarf 5.96 l.y. away, the second-closest star to the Sun. It lies at 17h 57.8m, +4° 42', and is named after the American astronomer E. E. Barnard (1857–1923), who found in 1916 that it has the greatest proper motion of any star, covering the apparent diameter of the Moon every 180 years.

The southernmost regions of Ophiuchus extend into rich starfields of the Milky Way, in the direction of the center of the Galaxy; consequently the constellation is replete with star clusters. Ophiuchus was the site of the last supernova seen to erupt in our Galaxy, popularly called Kepler's Star, which appeared in 1604 at 17h 30.6m, −21° 29', reaching mag. −3 at its brightest.

α (alpha) Ophiuchi, 17h 35m +12°.6, (Rasalhague, "head of the serpent collector"), mag. 2.1, is a white giant star 49 l.y. away.

β (beta) Oph, 17h 43m +4°.6, (Cebalrai, "the shepherd's dog"), mag. 2.8, is an orange giant 83 l.y. away.

γ (gamma) Oph, 17h 48m +2°.7, mag. 3.7, is a blue-white main-sequence star 97 l.y. away.

δ (delta) Oph, 16h 14m −3°.7, (Yed Prior, "the preceding star of the hand"), mag. 2.7, is a red giant 171 l.y. away.

ε (epsilon) Oph, 16h 18m −4°.7, (Yed Posterior, "the following star of the hand"), mag. 3.2, is a yellow giant 107 l.y. away.

ζ (zeta) Oph, 16h 37m −10°.6, mag. 2.6, is a blue main-sequence star 365 l.y. away.

η (eta) Oph, 17h 10m −15°.7, (Sabik), mag. 2.4, is a blue-white main-sequence star 88 l.y. away.

ρ (rho) Oph, 16h 26m −23°.4, 450 l.y. away, is a striking multiple star for small instruments. The brightest component, of mag. 5.1, has a close companion of mag. 5.7 visible in a small telescope at high magnification; either side of this pair are wide binocular companions of mags. 6.8 and 7.3.

τ (tau) Oph, 18h 03m −8°.2, 180 l.y. away, is a close pair of cream-colored stars, mags. 5.3 and 5.9, orbiting every 260 years and gradually closing; at least 100-mm aperture will be required to separate them by the year 2040, and 150 mm by 2050.

36 Oph, 17h 15m −26°.6, 19 l.y. away, is a pair of near-identical mag. 5.1 orange dwarf stars divisible with small apertures. Their calculated orbital period is 470 years.

70 Oph, 18h 05m +2°.5, 17 l.y. away, is a celebrated double star, consisting of yellow and orange components of mags. 4.2 and 6.2 that orbit each other in a period of 88 years. They are easily divisible in the smallest apertures and will remain so throughout the first half of the 21st century, being at their widest around 2025. For a diagram showing how the positions and separations of the pair change throughout the twenty-first century, see page 282.

RS Oph, 17h 50m −6°.7, is a recurrent nova seen to have erupted seven times, one short of the record held by the fainter U Scorpii. Normally around 12th mag., RS Oph flared up to naked-eye brightness in 1898, 1933, 1958, 1967, 1985, 2006, and 2021. Its distance is uncertain but is probably over 8,000 l.y.

M10 (NGC 6254), 16h 57m −4°.1, is a 7th-mag. globular cluster visible in binoculars or a small telescope. It is 16,000 l.y. away, similar to its neighbor M12 (see below). Individual stars in the cluster can be resolved with telescopes of 75-mm aperture.

M12 (NGC 6218), 16h 47m −1°.9, is a 7th-mag. globular cluster 16,000 l.y. away, visible in binoculars or a small telescope. In small apertures it appears slightly larger and less easy to resolve than its neighbor M10, and its stars are more loosely scattered. There are several other globular clusters in Ophiuchus worthy of attention, but M10 and M12 are the finest.

NGC 6572, 18h 12m +6°.8, is a planetary nebula of 9th mag., visible in apertures of 75 mm or more as a tiny blue-green ellipse. It lies 6,000 l.y. away.

NGC 6633, 18h 28m +6°.6, is a scattered binocular cluster of about 30 stars of 8th mag. and fainter, covering an area similar to that of the full Moon. It is 1,300 l.y. away.

IC 4665, 17h 46m +5°.7, is a loose and irregular open cluster of two dozen or so stars of 7th mag. and fainter, 1,150 l.y. away, larger than the apparent size of the Moon and best seen in binoculars.

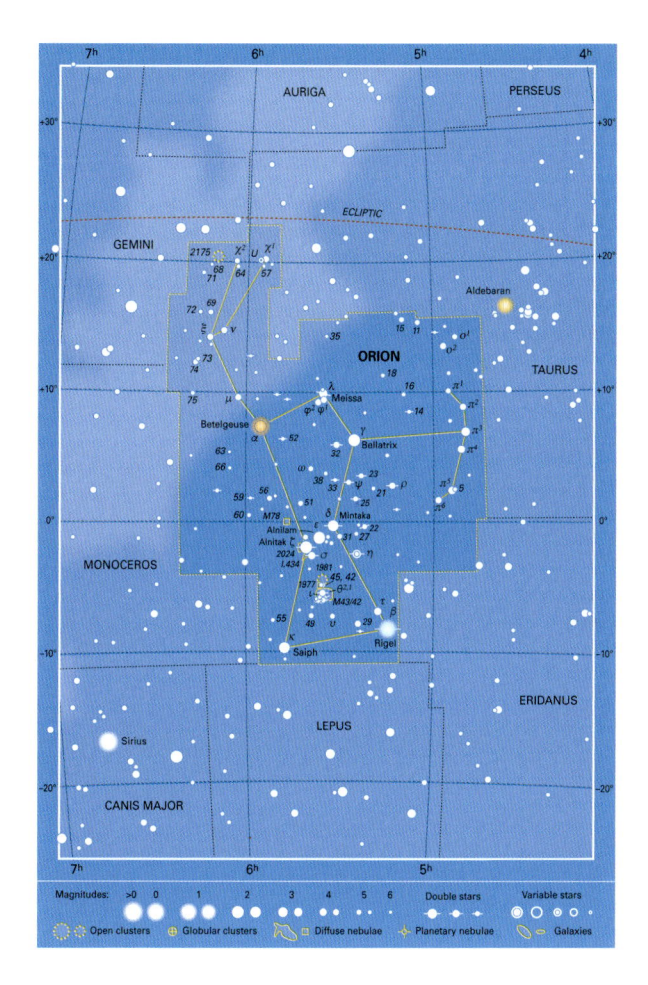

Orion, the Hunter

Without doubt the brightest and grandest constellation of all, crammed with objects of interest for all sizes of instrument. Orion's impressiveness stems in large measure from the fact that it contains an area of star formation in a nearby arm of our Galaxy, centered on the famous Orion Nebula. The Greek poet Homer described Orion as a giant hunter, armed with an unbreakable club of solid bronze. Legend tells that the boastful Orion was stung to death by a scorpion, and was placed in the sky so that he sets at the same time as his slayer, represented by the constellation Scorpius, rises.

In the sky, Orion is depicted brandishing his club and shield at the snorting Taurus, the Bull, while the hunter's dogs (the constellations Canis Major and Canis Minor) follow at his heels, in pursuit of the hare (Lepus). The Orion Nebula, M42, marks the hunter's sword, hanging from his belt. The belt itself is formed by a distinctive line of three bright stars, while the outline of his body is picked out by α (alpha), β (beta), γ (gamma), and κ (kappa) Orionis. Orion is one of the constellations in which the star labeled α is not the brightest; that honor goes to β Orionis, better known as Rigel. Each year the Orionid meteors, caused by dust from Halley's Comet, radiate from a point near the border with Gemini. As many as 25 Orionid meteors per hour may be seen around October 21.

α (alpha) Orionis, 5h 55m +7°.4, (Betelgeuse, corrupted from an Arabic name referring to a hand), 500 l.y. away, is a red supergiant star about 500 times the size of the Sun, and the most obviously variable of all first-magnitude stars. It fluctuates erratically in size, changing in brightness as it does so from about mag. 0.0 to 1.3; a major dimming in 2019–20 to mag. 1.6 is attributed to the ejection of a dust cloud. Its average brightness is around mag. 0.4 or 0.5, fainter than Rigel.

β (beta) Ori, 5h 15m −8°.2, (Rigel, "foot"), at mag. 0.1 the brightest star in Orion, is a blue-white supergiant 860 l.y. away; note its color contrast with the ruddy hue of Betelgeuse. Rigel has a mag. 6.8 companion, difficult to see in small telescopes, particularly in poor seeing, because of glare from Rigel itself.

γ (gamma) Ori, 5h 25m +6°.3, (Bellatrix, "the female warrior"), mag. 1.6, is a blue giant star 250 l.y. away.

δ (delta) Ori, 5h 32m −0°.3, (Mintaka, "belt"), about 700 l.y. away, is a complex multiple star. It is a blue giant that appears of mag. 2.4 to the naked eye. Binoculars or small telescopes reveal a wide companion of mag. 6.8. The brighter star is also an eclipsing binary that varies by about 0.1 mag. every 5.7 days.

ε (epsilon) Ori, 5h 36m −1°.2, (Alnilam, "the string of pearls"), mag. 1.7, is a blue supergiant about 2,000 l.y. away.

ζ (zeta) Ori, 5h 41m −1°.9, (Alnitak, "the girdle"), about 750 l.y. away, is a blue supergiant that appears to the naked eye as mag. 1.8. But telescopes of 75-mm aperture and above reveal a close companion of mag. 3.7 that is estimated to orbit it every 1,500 years. There is also a much wider 10th-mag. star.

η (eta) Ori, 5h 24m −2°.4, about 1,500 l.y. away, is a complex multiple–variable. A telescope of at least 75-mm aperture at high magnification is needed to show that it consists of two close stars, of mags. 3.6 and 4.9. The brighter star is also an eclipsing binary, varying by 0.3 mag. every 8 days.

θ1 (theta1) Ori, 5h 35m −5°.4, about 1,300 l.y. away, is a multiple star at the heart of the Orion Nebula, from which it has recently formed and which it now illuminates. This star is popularly known as the Trapezium, because a small telescope shows four stars here; but 100-mm aperture also reveals two others, of 11th mag. The four main stars of the Trapezium are of mags. 5.1, 6.7, 6.7, and 8.0. Nearby lies θ2 (theta2) Orionis, a binocular double of mags. 5.1 and 6.4.

ι (iota) Ori, 5h 35m −5°.9, (Hatysa), 2,300 l.y. away, is a double star on the southern edge of the Orion Nebula, divisible in a small telescope. Its components are of mags. 2.8 and 7.7. Also visible in the same field of view is a wider double of blue-white stars, Σ 747 (Struve 747), of mags. 4.7 and 5.5.

κ (kappa) Ori, 5h 48m −9°.7, (Saiph, "sword"), mag. 2.1, is a blue supergiant around 650 l.y. away.

λ (lambda) Ori, 5h 35m +9°.9, (Meissa), is a blue giant of mag. 3.5 with a mag. 5.5 companion visible in small telescopes under high magnification. The stars lie about 1,100 l.y. away.

σ (sigma) Ori, 5h 39m −2°.6, about 1,000 l.y. away, is perhaps the most impressive of all Orion's stellar treasures. To the naked eye it appears as a blue-white star of mag. 3.8, but small telescopes reveal much more. On one side of the star are two blue-white companions each of mag. 6.6, the wider of which can be glimpsed in binoculars; it is an eclipsing binary with a range of about 0.1 mag. On the opposite side is a closer mag. 8.8 companion, which is more difficult to see because of glare from the primary. The effect is like a planet with moons. To complete the picture, in the same telescopic field of view is a faint triple star called Σ 761 (Struve 761), consisting of a narrow triangle of 8th- and 9th-mag. stars. This is an extraordinarily rich and unexpected sight, to be returned to again and again.

U Ori, 5h 56m +20°.2, 930 l.y. away, is a huge red giant variable of Mira type, many hundreds of times larger than the Sun, whose brightness ranges between 5th and 13th mags. in about 1 year.

M42, M43 (NGC 1976, NGC 1982), 5h 35m −5°.4, the Orion Nebula, is one of the greatest deep-sky wonders—a gigantic cloud of gas and dust, 1,500 l.y. away and 20 l.y. in diameter, from which a star cluster is being born. Behind the visible part of the nebula, illuminated by the stars of the Trapezium (θ1 Ori), radio and infrared astronomers have detected an even larger dark cloud in which more stars are forming. M42 covers an area greater than 1° × 1°, and is indisputably the finest diffuse nebula in the sky, clearly visible to the naked eye as a hazy cloud. Binoculars and small telescopes reveal some of the more prominent wreaths and swirls of gas, which become more complex and breathtaking with increasing aperture. Although color photographs such as the one on page 196 depict the nebula as red and blue, to the eye it appears distinctly greenish; this is because the human eye has a different color sensitivity to that of photographic film. A dark lane of dust separates M42 from M43, a smaller and rounder patch to the north that is centered on a 7th-mag. star. Although M43 has a separate catalog number from M42, it is really part of the same huge gas cloud.

M78 (NGC 2068), 5h 47m +0°.0, is a small, elongated reflection nebula lying almost exactly on the celestial equator. Through small to moderate apertures it looks like a short-tailed comet, with a 10th-mag. double star at its head.

NGC 1977, 5h 36m −4°.9, 1,300 l.y. away, is an elongated nebulosity just north of the Orion Nebula. The mag. 4.6 blue star 42 Orionis, also known as c Orionis, lies in front of it. This nebula would be more celebrated were it not so overshadowed by M42.

NGC 1981, 5h 35m −4°.4, is a scattered cluster of about 20 stars of 6th mag. and fainter, 1,300 l.y. away, to the north of the nebulosity NGC 1977. Included in this cluster is the double star Σ 750 (Struve 750), a neat pair of mag. 6.4 and 8.4 components.

NGC 2024, 5h 41m −2°.4, is a mushroom-shaped cloud of gas about 0.5° wide, surrounding the star ζ (zeta) Orionis. Running south from ζ Ori is a strip of nebulosity, IC 434, into which is indented the celebrated Horsehead Nebula, a dark cloud of obscuring dust shaped like a horse's head. Although long-exposure photographs show NGC 2024 and the Horsehead Nebula well, they are notoriously difficult to detect with amateur telescopes.

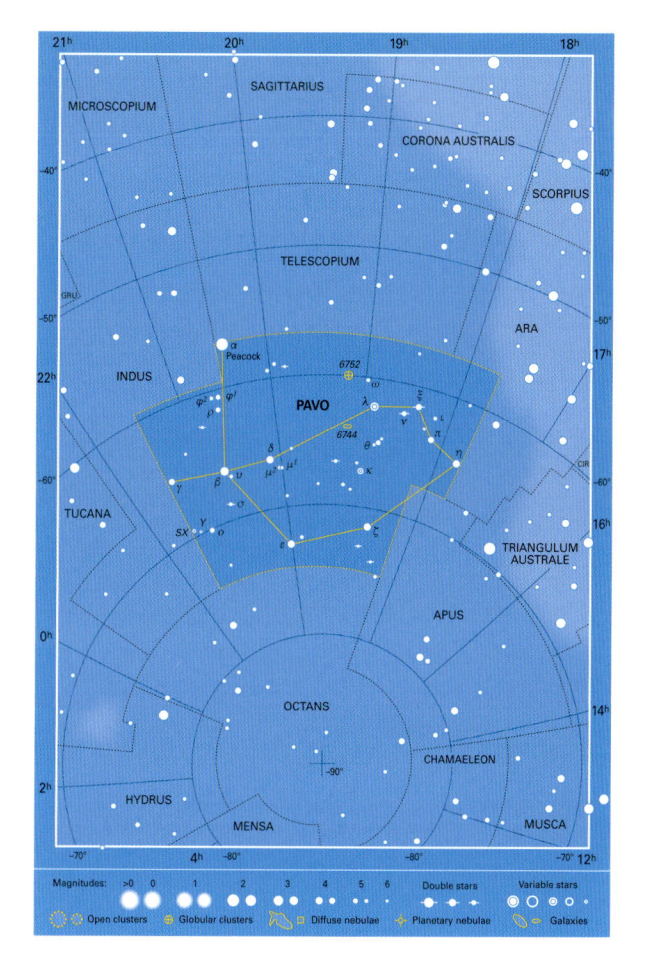

Pavo, the Peacock

One of the 12 new southern constellations introduced at the end of the sixteenth century by the Dutch navigators Pieter Dirkszoon Keyser and Frederick de Houtman. It is one of several exotic birds in this region of sky, the others being Apus, Grus, Phoenix, and Tucana. Pavo is probably meant to represent the Java green peacock, which Keyser and de Houtman would have encountered on their voyages of exploration in the East Indies, rather than the common blue, or Indian, peacock seen in parks. Pavo and the other new southern constellations were first depicted in 1598 on a globe by the Dutch astronomer and cartographer Petrus Plancius.

α (alpha) Pavonis, 20h 26m −56°.7, (Peacock), mag. 1.9, is a blue-white star 180 l.y. away.

β (beta) Pav, 20h 45m −66°.2, mag. 3.4, is a white star 137 l.y. away.

δ (delta) Pav, 20h 09m −66°.2, mag. 3.6, is a yellow subgiant 20 l.y. away.

η (eta) Pav, 17h 46m −64°.7, mag. 3.6, is an orange giant 370 l.y. away.

κ (kappa) Pav, 18h 57m −67°.2, 620 l.y. away, is one of the brightest Cepheid variables. It is a yellow-white supergiant, varying between mags. 3.9 and 4.8 in 9.1 days.

ξ (xi) Pav, 18h 23m −61°.5, 740 l.y. away, is a red giant of mag. 4.4 with a close companion of mag. 8.1, lost in the primary's glare in small telescopes.

SX Pav, 21h 29m −69°.5, 415 l.y. away, is a red giant that varies semi-regularly between mags. 5.2 and 6.0 every 7 weeks or so.

NGC 6744, 19h 10m −63°.9, is a 9th-mag. spiral galaxy for small telescopes, with a short central bar and widespread arms. It is one of the largest barred spirals, but its outer regions are faint and in modest apertures only the brightest central portion will be seen. It is about 30 million l.y. away.

NGC 6752, 19h 11m −60°.0, is a large 5th-mag. globular cluster, visible in binoculars and resolvable in 75-mm telescopes, covering half the apparent diameter of the Moon. A double star of 8th and 9th mags. at one edge is a foreground object. The cluster lies 13,000 l.y. away.

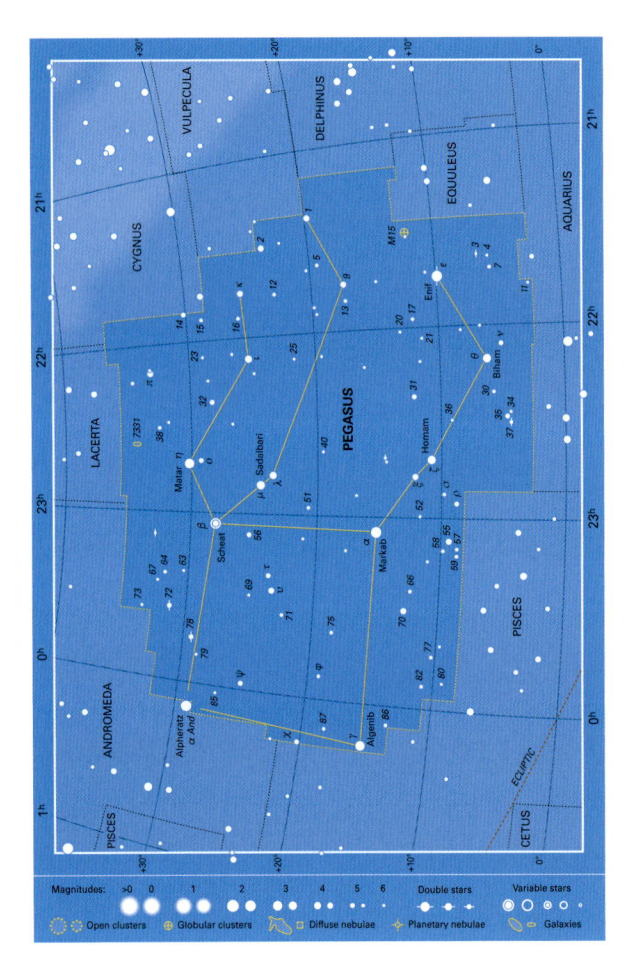

Pegasus, the Winged Horse

An ancient constellation representing the winged horse of Greek mythology, born from the blood of Medusa after she was slain by the hero Perseus, who lies nearby in the sky. Only the front half of the horse is shown. Its body is represented by the constellation's most famous feature, the Great Square, the corners of which are marked by four stars. One of these stars, once known as δ (delta) Pegasi, is now assigned to Andromeda and is known as α (alpha) Andromedae. The Great Square of Pegasus is over 15° wide and 13° high, yet contains surprisingly few naked-eye stars for such a large area. The brightest of them are: υ (upsilon), mag. 4.4; τ (tau), mag. 4.6; ψ (psi), mag. 4.7; 56, mag. 4.7; φ (phi), mag. 5.1; 71, mag. 5.4; and 75, mag. 5.5.

α (alpha) Pegasi, 23h 05m +15°.2, (Markab, "shoulder"), mag. 2.5, is a blue-white giant star 133 l.y. away.

β (beta) Peg, 23h 04m +28°.1, (Scheat, "shin"), 196 l.y. away, is a red giant that varies from mag. 2.3 to 2.7 every six weeks or so. At maximum it is the brightest star in the constellation, just outshining Epsilon Pegasi.

γ (gamma) Peg, 0h 13m +15°.2, (Algenib, "the side"), mag. 2.8, is a blue-white subgiant 390 l.y. away. It is a pulsating variable of the β Cephei type, but its fluctuations every 3 hours 40 minutes are only about 0.1 mag., too slight to be discernible with the naked eye.

ε (epsilon) Peg, 21h 44m +9°.9, (Enif, "nose"), 690 l.y. away, is an orange supergiant of mag. 2.4. A small telescope, or even good binoculars, reveals a wide mag. 8.7 companion star. ε Peg is slightly variable, by about 0.1 mag., but has also undergone two unexplained fluctuations: one night in November 1847 it was seen about a magnitude fainter than normal, while on a night in September 1972 it was seen at about magnitude 0.7 for 4 minutes before fading back to normal.

ζ (zeta) Peg, 22h 41m +10°.8, (Homam), mag. 3.4, is a blue-white star 204 l.y. away.

η (eta) Peg, 22h 43m +30°.2, (Matar), mag. 2.9, is a yellow giant 195 l.y. away.

π (pi) Peg, 22h 10m +33°.2, is a very wide binocular duo of white and yellow stars, both giants. They are of mags. 4.3 and 5.6, and both lie 290 l.y. away.

1 Peg, 21h 22m +19°.8, mag. 4.1, is an orange giant 156 l.y. away, with a wide 8th-mag. companion visible in small telescopes.

51 Peg, 22h 57m +20°.8, mag. 5.5, is a yellow subgiant or main-sequence star 51 l.y. away. In 1995 it became the first star other than the Sun known to have a planet orbiting it; the planet's estimated mass is about half that of Jupiter.

M15 (NGC 7078), 21h 30m +12°.2, is an outstanding 6th-mag. globular cluster, 34,000 l.y. distant, at the limit of naked-eye visibility but easily seen in binoculars; a 6th-mag. star nearby is a sure guide to its location. A small telescope shows it as a glorious misty sight in an attractive field. Apertures of 150 mm or so resolve its outer regions into a mottled ground of sparkling stars, and larger telescopes show stars all the way to the bright and condensed core.

NGC 7331, 22h 37m +34°.4, is a 10th-mag. spiral galaxy seen nearly edge-on, visible under good conditions in apertures of 100 mm or so as an elongated smudge. It lies about 50 million l.y. away.

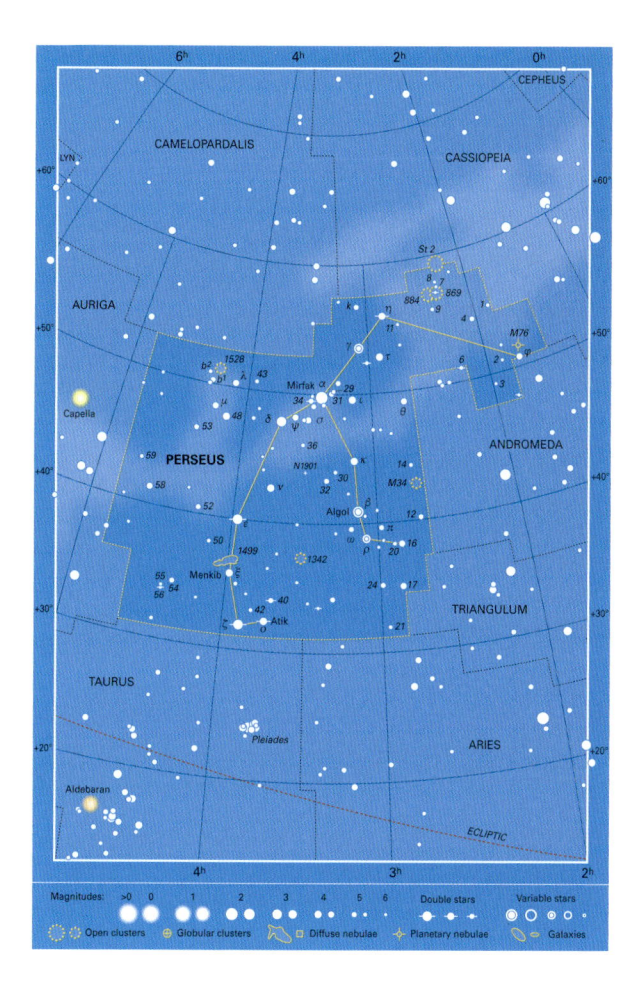

Perseus

Perseus was the hero of Greek mythology who rescued the chained maiden Andromeda from the clutches of the sea monster Cetus. Prior to that, Perseus had slain Medusa the Gorgon, whose head he is pictured holding in one hand. The Gorgon's head is marked by the winking star Algol, or β (beta) Persei, sometimes imagined as Medusa's evil eye. Perseus lies in a rich part of the Milky Way and is well worth sweeping with binoculars; note, in particular, the Double Cluster, a sparkling pair of open clusters near the border with Cassiopeia. In 1901 Nova Persei flared up to mag. 0.2 at 3h 31.2m, +43° 54', throwing off a shell of gas that is now visible in large telescopes. NGC 1499, the California Nebula, so named because its shape resembles California, spans the apparent width of five full Moons north of 4th-mag. ξ (xi) Persei, an exceedingly hot blue giant or supergiant that illuminates it. Despite its considerable size, the California Nebula is elusive visually, but shows up well in long-exposure photographs.

At 3h 19.8m, +41° 31', lies the radio source Perseus A, associated with the 12th-mag. peculiar galaxy NGC 1275, which is at the center of the Perseus cluster of galaxies, 250 million l.y. away. Near γ (gamma) Persei lies the radiant of the Perseid meteors, the most glorious meteor shower of the year: around 12–13 August, as many as 75 bright meteors can be seen flashing from Perseus each hour.

α (alpha) Persei, 3h 24m +49°.9, (Mirfak, "elbow"), mag. 1.8, is a yellow-white supergiant 505 l.y. away. Binoculars reveal a brilliant scattering of stars covering 3° in this region, forming a loose cluster, Melotte 20. α Per itself seems to form the head of a snaking chain of stars within this cluster.

β (beta) Per, 3h 08m +41°.0, (Algol, "the demon"), 90 l.y. away, is one of the most celebrated variable stars in the sky. It is the prototype of the eclipsing binary class of variables, in which two close stars periodically eclipse each other as they orbit their common center of gravity. Algol's eclipses occur every 2.87 days, when the star's apparent brightness sinks from mag. 2.1 to 3.4 before returning to normal about 10 hours later.

γ (gamma) Per, 3h 05m +53°.5, mag. 2.9, is a yellow giant 245 l.y. away. It is an eclipsing binary with the unusually long period of 14.6 years, dimming by 0.3 mag. for 10 days. This star's variability was first detected in 1990.

δ (delta) Per, 3h 43m +47°.8, mag. 3.0, is a blue giant 520 l.y. away.

ε (epsilon) Per, 3h 58m +40°.0, 640 l.y. away, is a blue star of mag. 2.9 with an unrelated 9th-mag. companion that is difficult to see through the smallest telescopes because of the brightness contrast.

ζ (zeta) Per, 3h 54m +31°.9, 750 l.y. away, is a blue supergiant of mag. 2.9 with a mag. 9.2 companion visible in a small telescope.

η (eta) Per, 2h 51m +55°.9, 880 l.y. away, is an orange giant or supergiant of mag. 3.8 with a mag. 8.5 blue companion that forms an attractive double for small telescopes. The field of view contains a sprinkling of background stars.

ρ (rho) Per, 3h 05m +38°.8, 308 l.y. away, is a red giant that varies between mags. 3.3 and 4.0 in semi-regular fashion every 7 weeks or so.

M34 (NGC 1039), 2h 42m +42°.8, is a bright open cluster at the limit of naked-eye visibility, containing about 60 stars splashed over an area larger than the apparent size of the full Moon. Binoculars resolve it into stars, and it is well seen in a small telescope, many of the stars seeming to form pairs. M34 lies 1,700 l.y. away.

M76 (NGC 650–1), 1h 42m +51°.6, the Little Dumbbell, is a planetary nebula, the faintest object in Charles Messier's famous list of deep-sky objects. At about magnitude 10 it is difficult to see, although it can be picked up in 100-mm apertures on a dark night. It is relatively large for a planetary nebula, similar in size to the Ring Nebula in Lyra, but smaller than its namesake the Dumbbell Nebula in Vulpecula, the elongated shape of which it resembles. Each end of the Little Dumbbell has a separate NGC number. It lies about 3,500 l.y. away.

NGC 869, NGC 884, 2h 19m +57°.2, 2h 22m +57°.1, the famous Double Cluster in Perseus, also known as h and χ (chi) Persei. These are two open star clusters visible to the naked eye and superb in binoculars, each covering an area equal to the full Moon. NGC 869 is the brighter and richer of the pair, containing an estimated 200 stars, while NGC 884 appears more scattered. They both lie just over 8,000 l.y. away and are relatively young—only a few million years old. Small telescopes have an advantage when observing these objects, for at low powers both clusters fit into the same field of view, which is not always the case with larger and more powerful telescopes. Most of the stars in the clusters are blue-white, but there are several red giants to be spotted in and around NGC 884. In binoculars, note a curving chain of stars leading northward toward a large starry patch known as the cluster Stock 2 (abbreviated St 2). This whole area is a breathtaking sight in all apertures.

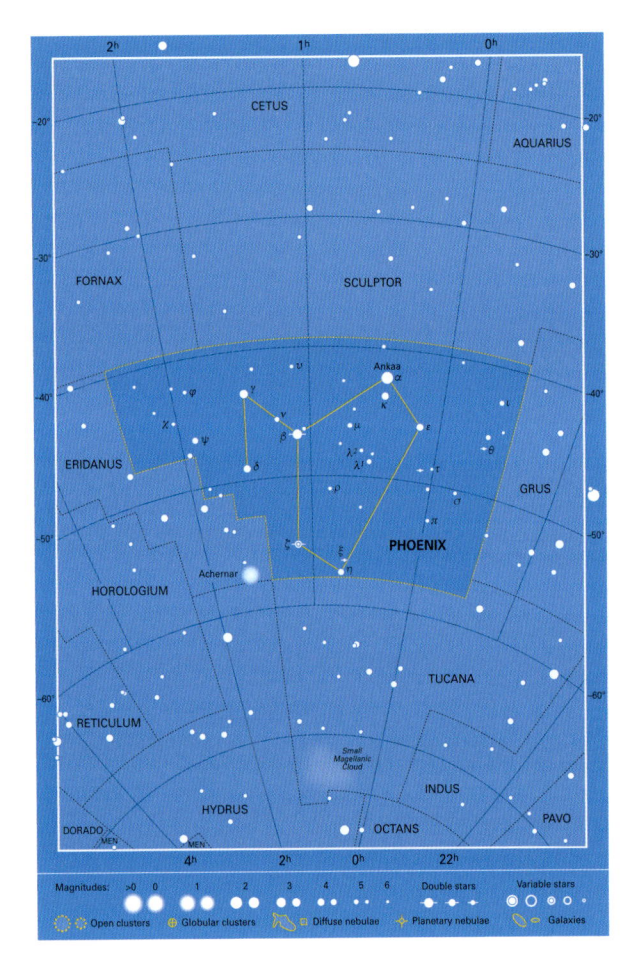

Phoenix, the Phoenix

An inconspicuous constellation near the southern end of Eridanus, representing the mythical bird that was regularly reborn from its own ashes. It was introduced at the end of the sixteenth century by the Dutch navigators Pieter Dirkszoon Keyser and Frederick de Houtman and was first depicted in 1598 on a celestial globe by Petrus Plancius.

α (alpha) Phoenicis, 0h 26m −42°.3, (Ankaa), mag. 2.4, is an orange giant star 85 l.y. away.

β (beta) Phe, 1h 06m −46°.7, appears to the naked eye as a yellow star of mag. 3.3. In fact, it is a binary of orbital period 170 years with well-matched components of mags. 4.0 and 4.1, currently moving apart. They become divisible in 150-mm aperture by 2030 and in 100 mm after 2045. The distance is uncertain.

γ (gamma) Phe, 1h 28m −43°.6, mag. 3.4, is a red giant 245 l.y. away.

ζ (zeta) Phe, 1h 08m −55°.2, 299 l.y. away, is a complex variable and multiple star. The main star is a blue-white eclipsing binary of the same type as Algol that rises and falls between mags. 3.9 and 4.4 every 1.67 days. It has a companion of mag. 8.2 visible in a small telescope. There is also a much closer mag. 6.8 companion visible only in large apertures.

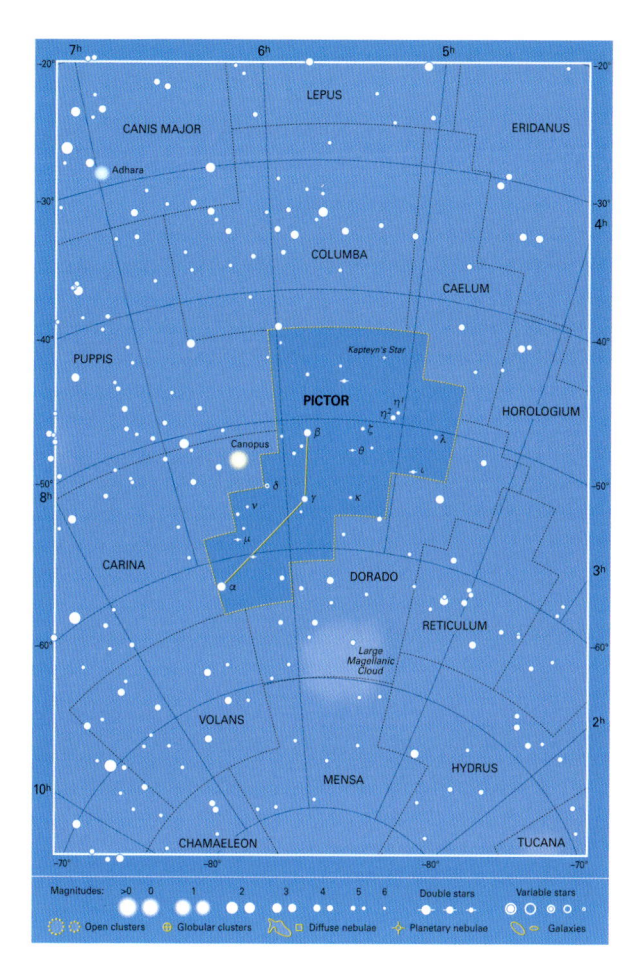

Pictor, the Painter's Easel

A faint constellation overshadowed by the neighboring brilliant star Canopus in Carina on one side and the Large Magellanic Cloud in Dorado on the other. The constellation was invented in the 1750s by Nicolas Louis de Lacaille, who originally called it Equuleus Pictorius, which has since been shortened. At 5h 11.7m, −45° 01' lies the mag. 8.9 red dwarf known as Kapteyn's Star, 12.8 l.y. away, named after the Dutch astronomer who discovered in 1897 that it has the second-largest proper motion of any known star (the record is held by Barnard's Star in Ophiuchus). Kapteyn's Star moves 1° in 415 years.

α (alpha) Pictoris, 6h 48m −61°.9, mag. 3.2, is a white main-sequence star 97 l.y. away.

β (beta) Pic, 5h 47m −51°.1, mag. 3.9, is a blue-white main-sequence star 63 l.y. away. This star became famous in 1984 when astronomers photographed a disk of dust and gas around it, thought to be a planetary system in the process of formation.

γ (gamma) Pic, 5h 50m −56°.2, mag. 4.5, is an orange giant 185 l.y. away.

δ (delta) Pic, 6h 10m −55°.0, is a blue giant or subgiant 1,500 l.y. away. It is an eclipsing binary of the β (beta) Lyrae type, varying from mag. 4.7 to 4.9 every 1.67 days.

ι (iota) Pic, 4h 51m −53°.5, is an easy double of mags. 5.6 and 6.2 for small telescopes, distances 120 and 145 l.y.

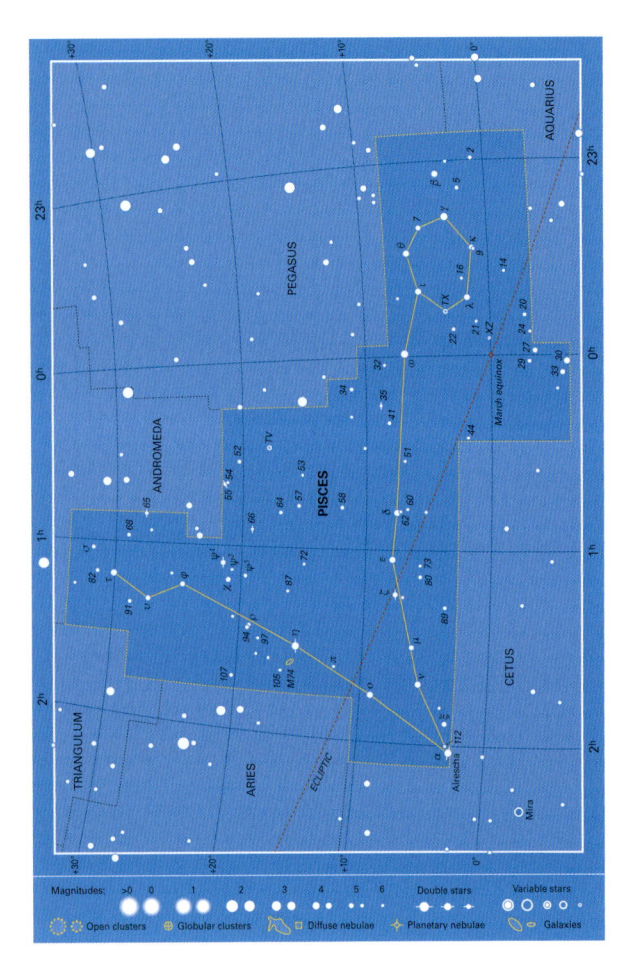

Pisces, the Fishes

An ancient constellation representing two fishes with their tails tied together by cords; the knot where the two cords join is marked by the star α (alpha) Piscium. One legend identifies the constellation with Aphrodite and her son Eros, who swam away from the attack of the monster Typhon in the guise of fishes.

The Sun is in Pisces from mid-March to late April, so the constellation contains the March (or vernal) equinox —the point at which the Sun crosses the celestial equator into the northern celestial hemisphere each year. This point originally lay in neighboring Aries, but the effect of precession has now carried it into Pisces; eventually, in 2597 CE, it will pass on into Aquarius.

The constellation's most distinctive feature is a ring of seven stars called the Circlet, marking the body of one of the fish; clockwise from north the seven stars are θ (theta), 7, γ (gamma), κ (kappa), λ (lambda), TX (or 19), and ι (iota) Piscium.

α (alpha) Piscium, 2h 02m +2°.8, (Alrescha, "the cord"), 160 l.y. away, appears to the naked eye as a star of mag. 3.8, but in fact is a challenging double with an orbital period of over 3,000 years. Its components, of mags. 4.1 and 5.1, are gradually closing but will remain within range of 100-mm aperture telescopes for at least the first half of the twenty-first century. The brighter star is a spectroscopic binary, and the fainter one may be also. Their color is bluish white, although some observers see the brighter star as greenish.

β (beta) Psc, 23h 04m +3°.6, mag. 4.5, is a blue-white main-sequence star 410 l.y. away.

γ (gamma) Psc, 23h 17m +3°.3, mag. 3.7, is a yellow giant 135 l.y. away.

ζ (zeta) Psc, 1h 14m +7°.6, 133 l.y. away, is a wide double of mags. 5.2 and 6.1, divisible in the smallest telescopes.

η (eta) Psc, 1h 31m +15°.3, mag. 3.6, is the brightest star in the constellation. It is a yellow giant, 375 l.y. away.

κ (kappa) Psc, 23h 27m +1°.3, mag. 4.9, is a blue-white star 160 l.y. away. It forms a wide binocular double with the yellow giant 9 Piscium, mag. 6.3, which lies about 250 l.y. farther off.

ρ (rho) Psc, 1h 26m +19°.2, mag. 5.3, is a white main-sequence star 84 l.y. away that forms an easy binocular duo with the unrelated orange giant 94 Piscium, mag. 5.5, 300 l.y. away.

ψ1 (psi1) Psc, 1h 06m +21°.5, is a wide pair of blue-white main-sequence stars of mags. 5.3 and 5.5 about 290 l.y. away, visible in small telescopes or even good binoculars.

TV Psc, 0h 28m +17°.9, 540 l.y. away, is a red giant that varies semi-regularly between mags. 4.7 and 5.4 every 7 weeks.

TX Psc, 23h 46m +3°.5, also known as 19 Psc, about 900 l.y. away, is a deep-red irregular variable star in the Circlet of Pisces, visible with the naked eye or binoculars. It fluctuates between mags. 4.8 and 5.2 with no set period.

M74 (NGC 628), 1h 37m +15°.8, is a spiral galaxy of 9th mag., presented face-on to us. In dark conditions it can be glimpsed through a small telescope as a pale disk with a star-like nucleus and faint foreground stars dotted across it, but it needs an aperture of at least 150 mm to be well seen. M74 lies some 30 million l.y. away.

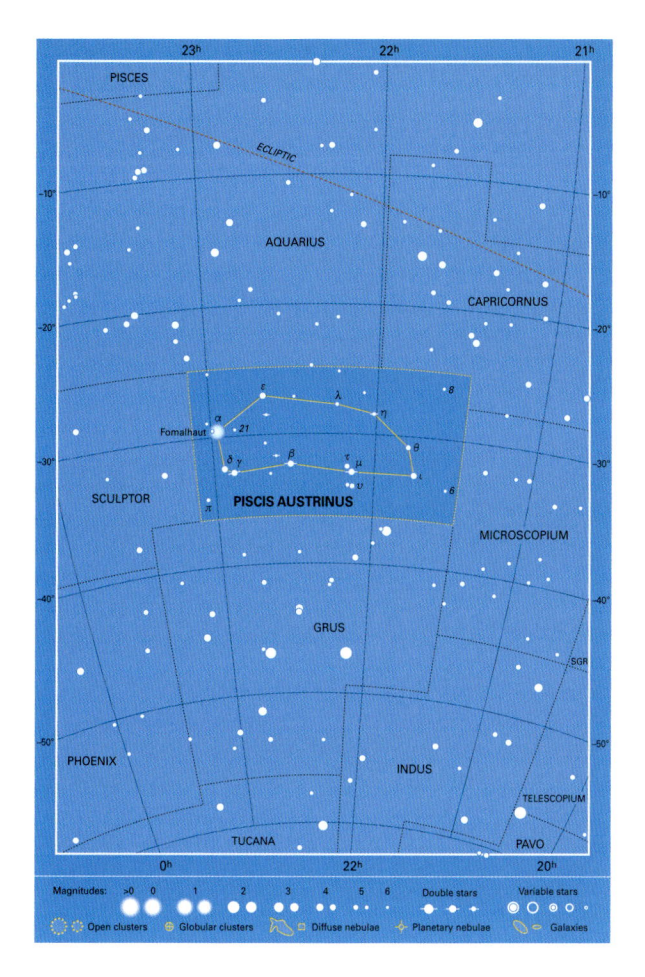

Piscis Austrinus, the Southern Fish

This constellation has been known since ancient times and is visualized as a fish lying on its back, drinking the stream of water flowing southward from the urn of neighboring Aquarius. It mouth is marked by the bright star Fomalhaut. This fish was said to be the parent of the two smaller zodiacal fishes, represented by Pisces.

α (alpha) Piscis Austrini, 22h 58m −29°.6, (Fomalhaut, "the fish's mouth"), mag. 1.2, is a blue-white main-sequence star 25 l.y. away. It is surrounded by a disk of cool dust from which a planetary system may be forming.

β (beta) PsA, 22h 32m −32°.3, 150 l.y. away, is a wide double star consisting of a mag. 4.3 blue-white primary and a mag. 7.1 companion visible in small telescopes.

γ (gamma) PsA, 22h 53m −32°.9, 225 l.y. away, is a double star of mags. 4.5 and 8.2, made difficult to split in small telescopes by the magnitude contrast.

η (eta) PsA, 22h 01m −28°.5, is a close pair of blue-white stars of mags. 5.7 and 6.8, requiring an aperture of at least 100 mm and high magnification to divide. Their distances are 860 and 720 l.y.

STARGAZING FOR BEGINNERS 167

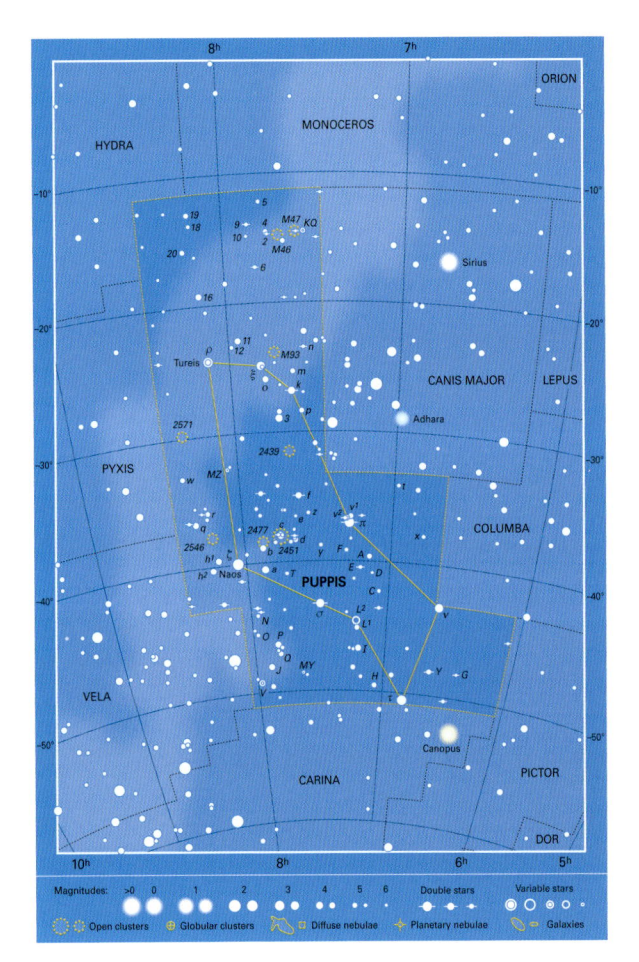

Puppis, the Stern

This is the largest of the three sections into which the ancient constellation of Argo Navis, the ship of the Argonauts, was dismembered in 1763 by Nicolas Louis de Lacaille; the other sections are Carina and Vela. Lacaille used one set of Greek letters to label the stars in all three parts, and it so happens that the labeling of the stars he assigned to Puppis begins with ζ (zeta). Puppis lies in the Milky Way and contains rich starfields for sweeping with binoculars.

ζ (zeta) Puppis, 8h 04m −40°.0, (Naos, "ship"), mag. 2.2, is a brilliant blue supergiant 1,100 l.y. away, the hottest (and hence bluest) of all naked-eye stars, with a surface temperature of about 40,000°C (72,000°F).

ξ (xi) Pup, 7h 49m −24°.9, mag. 3.3, is a yellow supergiant about 1,200 l.y. away. Binoculars reveal a wide, unrelated mag. 5.3 yellow companion, 200 l.y. away.

π (pi) Pup, 7h 17m −37°.1, mag. 2.7, is an orange supergiant about 800 l.y. away.

ρ (rho) Pup, 8h 08m −24°.3, (Tureis), mag. 2.8, is a yellow-white giant 64 l.y. away. It is a variable of the δ (delta) Scuti type, fluctuating by 0.2 mag. with a period of 3 hours 23 minutes.

k Pup, 7h 39m −26°.8, is a striking double star with blue-white components of mags. 4.4 and 4.5, distances 390 and and 360 l.y., easily divisible in small telescopes.

L1 L2 Pup, 7h 13m −45°.2, is an optical double consisting of two unrelated and contrasting stars. L1 is mag. 4.9, a blue-white main-sequence star 190 l.y. away. L2 is a red giant semi-regular variable, 185 l.y. away, that fluctuates between 3rd and 8th mags. every 140 days or so. Despite their similar distances the stars are moving in different directions and are unrelated.

V Pup, 7h 58m −49°.2, is an eclipsing binary of the β (beta) Lyrae type, 1,200 l.y. away. It varies from mag. 4.4 to 4.9 with a period of 35 hours.

M46 (NGC 2437), 7h 42m −14°.8, is a 6th-mag. open cluster of about 100 faint stars of remarkably uniform brightness, most of them around 10th mag. With its neighbor M47 it is visible to the naked eye as a brighter knot in the Milky Way. In binoculars it appears as a smudgy patch two-thirds the size of the full Moon, while a small telescope shows it as a sprinkling of stardust. M46 lies 5,400 l.y. away. On its northern edge lies the 10th-mag. planetary nebula NGC 2438. This is not a member of the cluster, but is a foreground object about 2,400 l.y. from us.

M47 (NGC 2422), 7h 37m −14°.5, is a scattered naked-eye cluster of similar apparent size to the full Moon. It contains three dozen or so stars, the brightest being of mag. 5.7. M47 lies 1,600 l.y. away, less than a third the distance of its richer neighbor M46.

M93 (NGC 2447), 7h 45m −23°.9, is a 6th-mag. binocular cluster, 3,400 l.y. away, consisting of 80 stars of 8th mag. and fainter, arranged in a wedge shape.

NGC 2451, 7h 45m −38°.0, is a large and bright open cluster of 40 stars 630 l.y. away, well seen in binoculars, centered on the mag. 3.6 orange giant c Puppis, which may actually be a foreground star.

NGC 2477, 7h 52m −38°.5, is a large 6th-mag. open cluster consisting of a swarm of faint stars, looking in binoculars like a loose globular, seemingly with arms. This excellent object would undoubtedly have featured on Messier's list had he lived farther south. It lies 4,900 l.y. away.

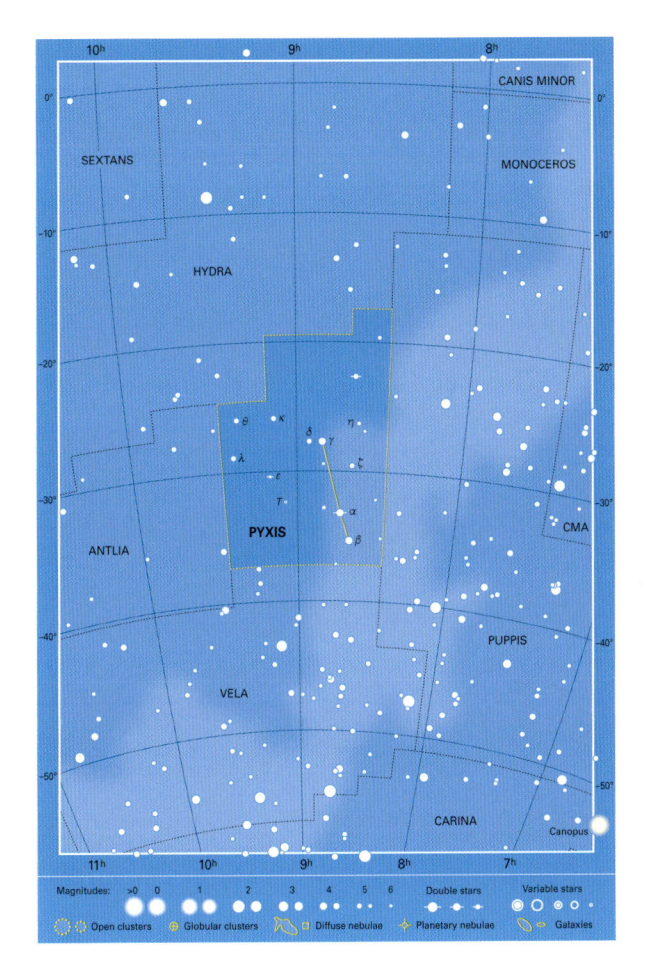

Pyxis, the Compass

A small, faint constellation invented by Nicolas Louis de Lacaille in the 1750s, representing a magnetic compass as used by sailors. It lies near Puppis, the stern of the Argonauts' ship, and was formed from stars that Ptolemy had cataloged as being part of the mast of Argo; but that ship, of course, would not have had a magnetic compass, so Pyxis cannot really be considered a genuine part of the old Argo. This area has also been known as Malus, the mast of Argo, a suggestion made in 1844 by the English astronomer John Herschel, although it was not widely adopted. The German astronomer Johann Bode introduced an additional constellation here called Lochium Funis, the log and line, representing a device used for measuring distance traveled at sea, which he visualized as curving around the compass.

Pyxis contains no objects of particular interest to users of small telescopes, despite the fact that it lies in the Milky Way, and its brightest stars are only of 4th magnitude.

α (alpha) Pyxidis, 8h 44m −33°.2, mag. 3.7, is a blue giant star 880 l.y. away.

β (beta) Pyx, 8h 40m −35°.3, mag. 4.0, is a yellow giant 416 l.y. away.

γ (gamma) Pyx, 8h 51m −27°.7, mag. 4.0, is an orange giant 210 l.y. away.

T Pyx, 9h 05m −32°.4, is a recurrent nova that has undergone six recorded eruptions, in 1890, 1902, 1920, 1944, 1966, and 2011. Normally it is of 15th mag., but brightens to 6th or 7th magnitude. Further outbursts may be expected.

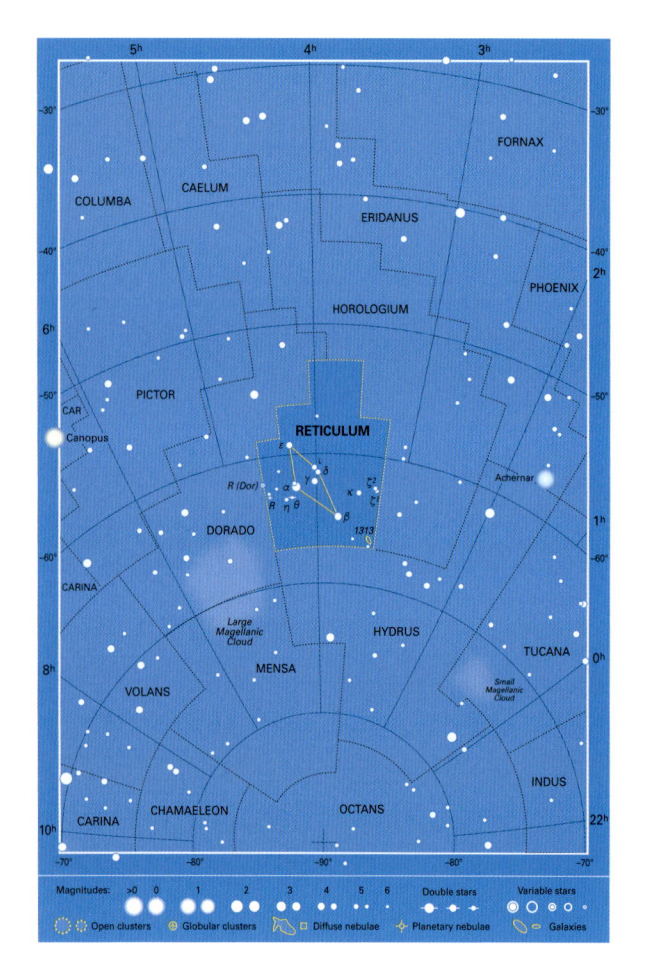

Reticulum, the Net

A constellation introduced in the 1750s by the French astronomer Nicolas Louis de Lacaille to commemorate a device in his telescope's eyepiece known as a reticle, which assisted him in measuring star positions during his surveys of the southern sky. The reticle consisted of a diamond-shaped aperture that helped him measure the position of stars as they passed through the field of view. In the sky it lies conveniently next to Lacaille's observatory clock, Horologium, which he used for timing the passage of stars through the reticle. Reticulum lies near the Large Magellanic Cloud, but is not prominent.

α (alpha) Reticuli, 4h 14m −62°.5, mag. 3.4, is a yellow giant star 160 l.y. away.

β (beta) Ret, 3h 44m −64°.8, mag. 3.8, is an orange giant 107 l.y. away.

ζ1 ζ2 (zeta1 zeta2) Ret, 3h 18m −62°.5, 39 l.y. away, is a wide naked-eye or binocular double of near-identical yellow main-sequence stars similar to the Sun, of mags. 5.5 and 5.2 respectively.

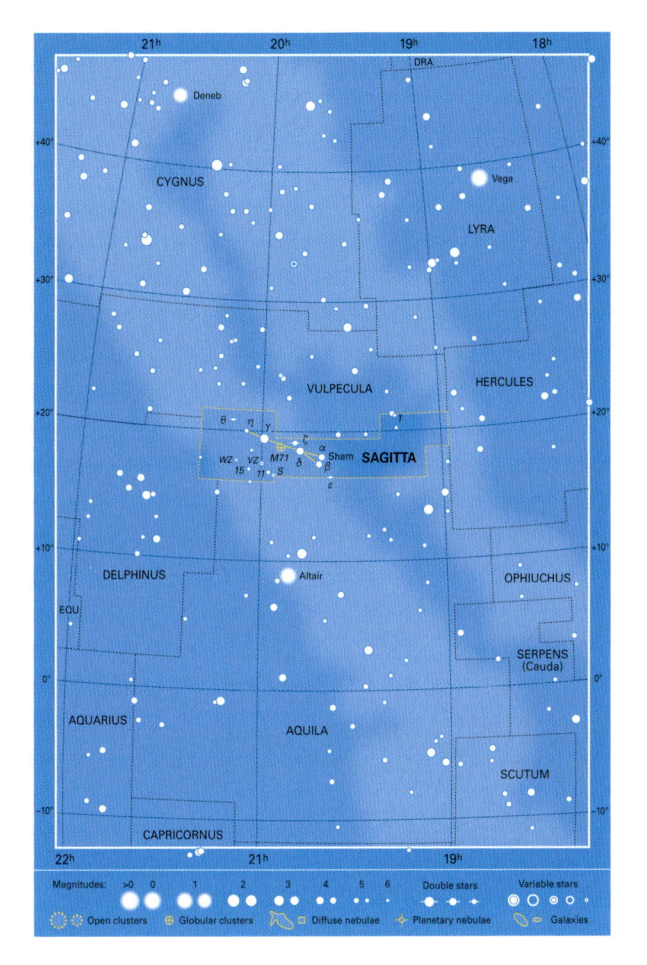

Sagitta, the Arrow

Despite its diminutive size—it is the third-smallest constellation—this arrow-shaped group is distinctive enough to have been recognized by the ancient Greeks. In the sky, the Arrow seems to be flying between Cygnus the Swan and Aquila the Eagle; in one legend, the arrow was shot by Hercules, although it has also been seen as the arrow of Apollo or Eros, the god of love. Like its northerly neighbor Vulpecula, Sagitta lies in a rich part of the Milky Way.

α (alpha) Sagittae, 19h 40m +18°.0, (Sham, "arrow"), mag. 4.4, is a yellow giant 425 l.y. away.

β (beta) Sge, 19h 41m +17°.5, mag. 4.4, is a yellow giant 440 l.y. away.

γ (gamma) Sge, 19h 59m +19°.5, mag. 3.5, the brightest star in the constellation, is a red giant 288 l.y. away.

δ (delta) Sge, 19h 47m +18°.5, mag. 3.8, is a red giant about 550 l.y. away.

ζ (zeta) Sge, 19h 49m +19°.1, 255 l.y. away, is a blue-white main-sequence star of mag. 5.0 with a mag. 9.0 companion, divisible in a small telescope.

S Sge, 19h 56m +16°.6, is a yellow supergiant Cepheid variable around 2,000 l.y. away, ranging between mag. 5.2 and 6.0 in 8.4 days.

VZ Sge, 20h 00m +17°.5, about 950 l.y. away, is a pulsating red giant that varies irregularly between mags. 5.3 and 5.6.

WZ Sge, 20h 08m +17°.7, is a so-called dwarf nova that flared up from 15th mag. to 7th or 8th mag. in 1913, 1946, 1978, and 2001; its location is worth checking in case of another outburst.

M71 (NGC 6838), 19h 54m +18°.6, is an 8th-mag. globular cluster 13,000 l.y. away, visible as a small, somewhat elongated misty patch in binoculars or a small telescope and resolvable with 100-mm aperture. Its stars are scattered and it lacks a central condensation, so it looks more like a dense open cluster than a typical globular.

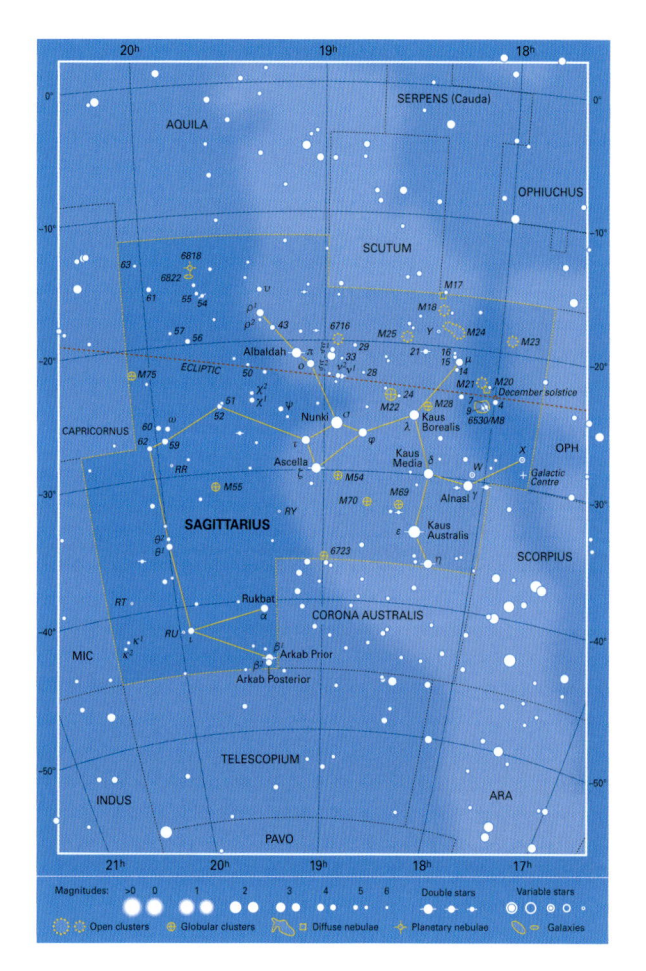

Sagittarius, the Archer

An ancient constellation depicting a centaur, a creature half man, half beast, with a raised bow and arrow. It is an older constellation than the other celestial centaur, Centaurus, and is different in character. Whereas Centaurus is identified as a scholarly, beneficent creature, Sagittarius is depicted with a threatening look, aiming his arrow at the heart of Scorpius, the Scorpion. The bow is marked by the stars λ (lambda), δ (delta), and ε (epsilon) Sagittarii, while γ (gamma) is the tip of the arrow. The shape outlined by the eight main stars of Sagittarius is popularly referred to as the Teapot, while λ (lambda), φ (phi), σ (sigma), τ (tau), and ζ (zeta) Sagittarii form a ladle shape known as the Milk Dipper—a suitable implement to dip into this rich region of the Milky Way.

The center of our Galaxy lies in Sagittarius, so the Milky Way starfields are particularly rich here, as well as in neighboring Scutum and Scorpius. The actual center of the Galaxy is marked by a radio and infrared source known as Sagittarius A, at 17h 46.1m, −28° 51'. The main attraction of Sagittarius is its clusters and nebulae. Charles Messier cataloged a total of 15 objects in Sagittarius, more than in any other constellation; only a selection of them can be mentioned here. The Sun passes through the constellation from mid-December to mid-January, and thus lies in Sagittarius at the December solstice, its farthest point south of the equator. Please see opposite page (page 173) for full details.

α (alpha) Sagittarii, 19h 24m −40°.6, (Rukbat, "knee"), mag. 3.9, is one of several instances in which the star labeled α in a constellation is not the brightest. It is a blue-white main-sequence star 180 l.y. away.

β1 β2 (beta1 beta2) Sgr, 19h 23m −44°.5, (Arkab Prior and Arkab Posterior), is a pair of unrelated naked-eye stars. β1 Sgr, a blue-white star of mag. 4.0, 390 l.y. away, has a mag. 7.2 companion visible in a small telescope. β2 Sgr is a white star of mag. 4.3, 134 l.y. away. All three stars appear in the same line of sight by chance.

γ (gamma) Sgr, 18h 06m −30°.4, (Alnasl, "the point," i.e. of the arrow), mag. 3.0, is an orange giant 97 l.y. away.

δ (delta) Sgr, 18h 21m −29°.8, (Kaus Media, "middle of the bow"), mag. 2.7, is an orange giant 415 l.y. away.

ε (epsilon) Sgr, 18h 24m −34°.4, (Kaus Australis, "southern part of the bow"), mag. 1.8, is the brightest star in Sagittarius. It is a blue-white giant 145 l.y. away.

λ (lambda) Sgr, 18h 28m −25°.4, (Kaus Borealis, "northern part of the bow"), mag. 2.8, is an orange giant or subgiant 78 l.y. away.

σ (sigma) Sgr, 18h 55m −26°.3, (Nunki), mag. 2.1, is a blue-white star 228 l.y. away.

W Sgr, 18h 05m −29°.6, is a yellow supergiant Cepheid variable ranging between mags. 4.3 and 5.1 with a period of 7.6 days. It lies about 1,400 l.y. away.

X Sgr, 17h 48m −27°.8, is a yellow-white giant Cepheid variable ranging between mags. 4.2 and 4.9 in 7 days. It lies about 1,200 l.y. away.

Y Sgr, 18h 21m −18°.9, is a yellow-white giant Cepheid that varies between mags. 5.3 and 6.2 in 5.8 days. It lies 1,650 l.y. away.

RR Sgr, 19h 56m −29°.2, is a red giant variable of Mira type, with a range from mag. 5.4 to 14.0 and a period of about 11 months. It lies about 1,200 l.y. away.

RY Sgr, 19h 17m −33°.5, is the southern equivalent of R Coronae Borealis—a star like a reverse nova that normally shines at around 6th mag., but which can suddenly and unpredictably drop to 14th mag. It lies over 5,000 l.y. away.

M8 (NGC 6523), 18h 04m −24°.4, the Lagoon Nebula, is a famous gaseous nebula, visible to the naked eye, elongated in shape and encompassing the star cluster NGC 6530. M8 is a fine object for binoculars or telescopes, covering the area of three full Moons, with a dark rift down its center. In the eastern half of the nebula is NGC 6530, a cluster of about 25 stars of 7th mag. and fainter, formed recently from the surrounding gas. The opposite (western) side of the nebula is dominated by two main stars, the brighter of which is the hot, blue star 9 Sgr, mag. 6.0. Photographs show the nebula as an intense red, but visually it appears milky-white. M8 is about 4,000 l.y. away.

M17 (NGC 6618), 18h 21m −16°.2, the Omega, Horseshoe, or Swan Nebula, is another gaseous nebula. Binoculars show it as a wedge-shaped object of similar apparent width to the full Moon, while in larger instruments it appears arch-shaped, variously likened to a Greek capital omega (Ω), a horseshoe, or a swan, which accounts for its range of popular names. M17 lies 5,400 l.y. away. About 1° south of it is M18 (NGC 6613), 5,100 l.y. distant, a small, loose cluster of 20 stars of 9th mag. and fainter, unimpressive in binoculars.

M20 (NGC 6514), 18h 03m −23°.0, the Trifid Nebula, is a cloud of glowing gas far less impressive visually than photographically. Moderate-sized telescopes show it as only a diffuse patch of light centered on the double star HN 40, of 8th and 9th mags., which was evidently born from it and now illuminates it. The Trifid Nebula gets its name from three dark lanes of dust that trisect it, well shown in photographs but elusive in small apertures. It lies about 4,100 l.y. away.

M21 (NGC 6531), 18h 05m −22°.5, is a spider-like open cluster in the same low-power field of view as M20, containing about 70 stars of 7th mag. and fainter, 4,100 l.y. away.

M22 (NGC 6656), 18h 36m −23°.9, is a large, rich 5th-mag. globular cluster, one of the finest in the entire heavens and ranked third only to ω (omega) Centauri and 47 Tucanae. Visible to the naked eye, M22 is an excellent binocular object and a fine sight in small telescopes, which reveal its noticeably elliptical outline. A telescope of 75-mm aperture will begin to resolve its outer regions, and larger apertures show its brightest stars to be reddish. Its nucleus is not as condensed as that of many other globulars. M22 lies 10,600 l.y. away. At the edge of the same binocular field is the smaller 7th-mag. globular M28 (NGC 6626) about 5,000 l.y. more distant.

M23 (NGC 6494), 17h 57m −19°.0, is a widely spread open cluster of fairly uniform appearance, just at the limit of resolution in binoculars. It is elongated in shape, consisting of a field of stars of 9th- to 11th-mag., some arranged in arcs. It lies 2,400 l.y. away.

M24, 18h 18m −18°.5, is a rich and extensive Milky Way starfield south of M17 and M18, grainy and shimmering in binoculars. Some observers restrict the name M24 to a small cluster of faint stars in its northern half known also as NGC 6603, but this is not what Messier meant. The whole Milky Way star cloud that makes up M24 measures about 2° × 1° and is one of the most prominent parts of the Milky Way to the naked eye.

M25 (IC 4725), 18h 32m −19°.2, 2,200 l.y. away, is a scattered cluster of about 30 stars, well seen in binoculars. The most prominent members form two bars across the cluster's center. Its brightest star is U Sgr, a yellow supergiant Cepheid that varies from mag. 6.3 to 7.1 with a period of 6 days 18 hours.

M55 (NGC 6809), 19h 40m −31°.0, is a 6th-mag. globular cluster, nebulous-looking in binoculars with little central condensation. Small telescopes resolve individual stars and show a dark notch on one side. M55 lies 16,000 l.y. away.

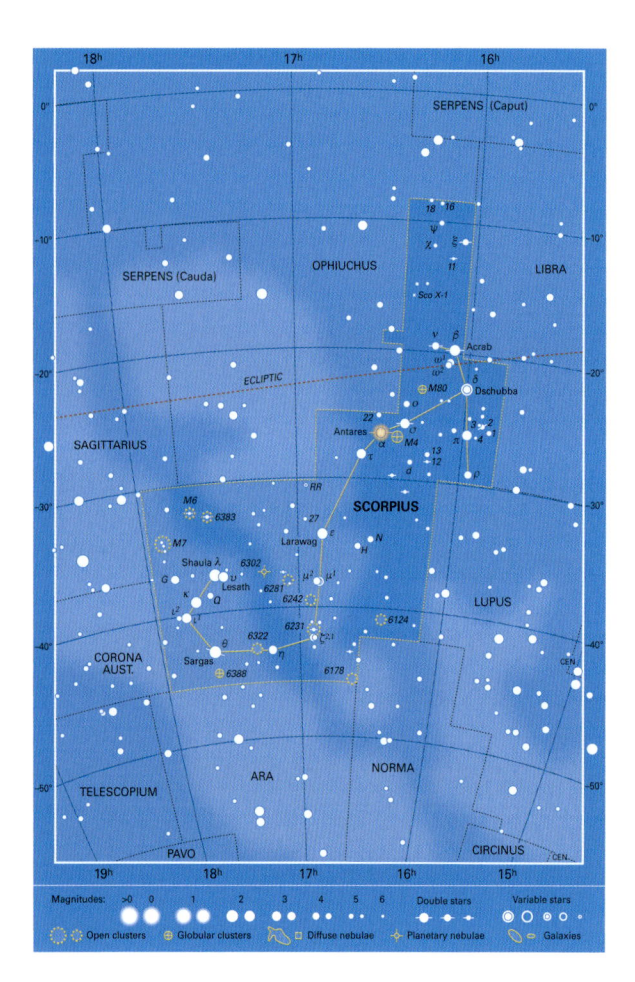

Scorpius, the Scorpion

A resplendent constellation, lying in a rich area of the Milky Way and packed with exciting objects for users of small telescopes. In mythology, Scorpius was the scorpion whose sting killed Orion. In the sky, Orion still flees from the scorpion, for Orion sets below the horizon as Scorpius rises.

Originally, in ancient Greek times and earlier, Scorpius was a much larger constellation, but in the first century BCE the stars that comprised its claws were made into the separate constellation of Libra by the Romans. Despite the loss of its claws, Scorpius still bears a clear resemblance to the creature after which it is named, with a distinctive curve of stars forming its stinging tail. Its heart is marked by the bright star Antares, a name that can be translated either as "rival of Mars" or "like Mars," in reference to its strong red color. North of β (beta) Scorpii, at 16h 19.9m, −15° 38′, lies the strongest X-ray source in the sky, Scorpius X-1. This has been identified with an 11th-magnitude spectroscopic binary some 7,500 l.y. away, consisting of a neutron star accreting matter from a companion.

A scattering of several hundred bright stars centered between about 400 and 500 l.y. from us extends from Scorpius via Lupus into Centaurus and Crux; this is known as the Scorpius–Centaurus (or Sco–Cen) association. Antares is a prominent member. The Sun passes briefly through Scorpius during the last week of November. Please see opposite page (page 175) for full details.

α (alpha) Scorpii, 16h 29m −26°.4, (Antares), about 550 l.y. away, is a red supergiant 400 times the diameter of the Sun. It is a semi-regular variable, fluctuating between mags. 0.8 and 1.2 with an approximate period of 6 years. Antares has a mag. 5.4 blue companion that requires at least 75-mm aperture and the steadiest atmospheric conditions to be visible against the primary's glare. The orbital period of the companion around Antares is estimated to be around 2,600 years.

β (beta) Sco, 16h 05m −19°.8, (Acrab, "scorpion"), is a striking double star divisible in the smallest telescopes, consisting of two blue-white main-sequence stars of mags. 2.6 and 4.9, both about 400 l.y. away and with similar proper motions.

δ (delta) Sco, 16h 00m −22°.6, (Dschubba, "forehead"), is a blue-white subgiant about 500 l.y. away. Normally of mag. 2.3, in 2000 it began to brighten, apparently as a result of throwing off a shell of gas, reaching a maximum of around mag. 1.6 in 2002. It subsequently subsided to near-normal brightness but may surge again in future.

ε (epsilon) Sco, 16h 50m −34°.3, (Larawag), mag. 2.3, is an orange giant 64 l.y. away.

ζ1 ζ2 (zeta1 zeta2) Sco, 16h 54m −42°.4, is a naked-eye double of unrelated stars, ζ2 being a mag. 3.6 orange giant 135 l.y. away, and ζ1 a blue supergiant that varies erratically between mags. 4.7 and 4.9; ζ1 is probably an outlying member of the open cluster NGC 6231 (see page 229).

θ (theta) Sco, 17h 37m −43°.0, (Sargas), mag. 1.9, is a white giant 300 l.y. away with a mag. 5.3 companion visible in small telescopes.

λ (lambda) Sco, 17h 34m −37°.1, (Shaula, "sting"), mag. 1.6, is a blue subgiant about 570 l.y. away.

μ1 μ2 (mu1 mu2) Sco, 16h 52m −38°.0, is a naked-eye double of unrelated stars. μ1, about 1,700 l.y. away, is an eclipsing binary that varies from mag. 2.9 to 3.2 in 34 hours 43 minutes; μ2 is a blue subgiant of mag. 3.5, 475 l.y. away.

ν (nu) Sco, 16h 12m −19°.5, 450 l.y. away, is a quadruple star similar to the famous Double Double in Lyra. A small telescope, or even powerful binoculars, shows ν Sco as a wide double, with blue-white components of mags. 4.0 and 6.3. Telescopes of 60 mm and above should reveal at high magnification that the fainter star is itself a close double of mags. 6.6 and 7.2. The brighter star is an even closer double of mags. 4.4 and 5.3, requiring an aperture of 100 mm to divide.

ξ (xi) Sco, 16h 04m −11°.4, 91 l.y. away, is a celebrated multiple star. A small telescope shows it as a white star of mag. 4.2 with a mag. 7.3 yellow companion that form a binary with an estimated period of around 1,500 years. Also visible in the same field is a fainter and wider pair, Σ 1999 (Struve 1999), composed of mag. 7.4 and 8.0 stars that are gravitationally connected to ξ Sco. Therefore, at first sight, ξ Sco looks like another Double Double. But the brightest star is itself a close pair, consisting of yellow-white stars of mags. 4.9 and 5.2 orbiting each other with a period of 46 years. Apertures of 150 mm should be sufficient to divide them until 2035, and 220 mm until about 2040, but at their closest around 2042 only very large apertures will separate them.

ω1 ω2 (omega1 omega2) Sco, 16h 07m −20°.7, is a pair of unrelated stars distinguishable with the naked eye: ω1, a blue main-sequence star of mag. 4.0, 470 l.y. away, and ω2, a yellow giant of mag. 4.3, 290 l.y. away.

RR Sco, 16h 57m −30°.6, is a red giant variable of the Mira type, ranging between 5th and 12th mag. every 9 months or so. It lies about 920 l.y. away.

M4 (NGC 6121), 16h 24m −26°.5, is a 6th-mag. globular cluster appearing almost as large as the full Moon. It looks like a woolly ball in binoculars, but is not as easy to spot as its magnitude would suggest because its light is spread over a large area. In 100-mm telescopes individual stars are resolved and there is a noticeable bar of stars running north–south across its center. M4 is more loosely scattered than many globulars and does not have a strong central condensation. It is the closest globular to us, just under 6,000 l.y. away.

M6 (NGC 6405), 17h 40m −32°.2, is an impressive 4th-mag. open cluster of about 80 stars arranged in radiating chains, popularly called the Butterfly Cluster. Binoculars and small telescopes resolve the main stars that form the butterfly shape. The brightest of them, BM Sco, on one of the "wings," is an orange supergiant semi-regular variable that ranges between 5th and 7th mags. every 27 months or so. M6 lies 1,500 l.y. away.

M7 (NGC 6475), 17h 54m −34°.8, the most southerly object in the Messier catalog, is a huge, scattered 3rd-mag. open cluster of 80 or so stars individually of 6th mag. and fainter, visible to the naked eye as a brighter knot in the Milky Way. The apparent diameter is over twice that of the full Moon, and it is easily resolved in binoculars. With M6 at the edge of the same binocular field, this is an exceptionally rich sight. The central group of stars is arranged in an X-shape, with scattered outliers that form triangular surroundings like a Christmas tree, set against the backdrop of a dense star cloud. An outstanding cluster and a classic for small apertures. M7 lies 900 l.y. away, just over half the distance of M6; the two clusters are not related.

M80 (NGC 6093), 16h 17m −23°.0, is a small, 7th-mag. globular cluster visible in binoculars or a small telescope, appearing like the fuzzy head of a comet. It lies 32,000 l.y. away.

NGC 6231, 16h 54m −41°.8, is a naked-eye open cluster of over 100 stars in a rich area of the Milky Way that is well worth sweeping with binoculars. The brightest stars of the group are of 6th mag. and give the impression of a mini-Pleiades in binoculars and small telescopes. The 5th-mag. blue supergiant ζ1 (zeta1) Sco is probably an outlying member of this cluster, which lies 5,500 l.y. away. NGC 6231 is connected to a larger, scattered cluster of fainter stars visible in binoculars, known both as Trumpler 24 and Harvard 12, which lies 1° to the north. The chain of stars linking the two clusters delineates one of the spiral arms of our Galaxy.

NGC 6242, 16h 56m −39°.5, is a wedge-shaped open cluster for small telescopes containing about two dozen stars of 7th mag. and fainter. It lies 4,300 l.y. away.

NGC 6281, 17h 05m −37°.9, is an open cluster visible in small telescopes, consisting of over 50 stars of 8th mag. and fainter arranged in a rectangular shape, 1,700 l.y. away.

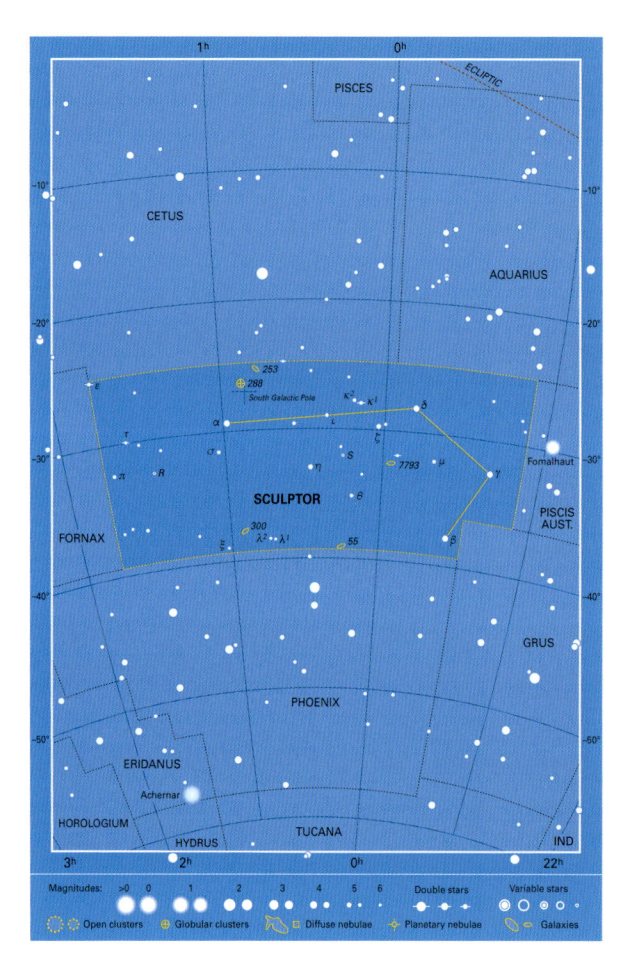

Sculptor, the Sculptor

One of the faint and half-forgotten constellations introduced in the 1750s by Nicolas Louis de Lacaille to fill in the southern skies. It represents a sculptor's studio. Sculptor contains the south pole of our Galaxy, 90° from the plane of the Milky Way. In this direction we can look out into deep space, unobscured by stars or dust, and see many faint galaxies. Among these is a very faint member of our Local Group, the Sculptor Dwarf, detectable only in long-exposure photographs taken through large telescopes.

α (alpha) Sculptoris, 0h 59m −29°.4, mag. 4.3, is a blue-white giant 780 l.y. away.

β (beta) Scl, 23h 33m −37°.8, mag. 4.4, is a blue-white giant or subgiant 183 l.y. away.

γ (gamma) Scl, 23h 19m −32°.5, mag. 4.4, is an orange giant 190 l.y. away.

δ (delta) Scl, 23h 49m −28°.1, mag. 4.6, is a blue-white main-sequence star 143 l.y. away.

ε (epsilon) Scl, 1h 46m −25°.1, 91 l.y. away, is a binary of mags. 5.3 and 8.5, divisible in small telescopes. The estimated orbital period is 2,200 years.

κ1 (kappa1) Scl, 0h 09m −28°.0, 223 l.y. away, is a tight pair of white stars, of mags. 6.1 and 6.2, at the limit of resolution of a 75-mm telescope.

R Scl, 1h 27m −32°.5, is a deep-red semi-regular variable star that ranges from 5th to 9th mags. with a period of roughly a year. It lies 1,280 l.y. away.

S Scl, 0h 15m −32°.0, is a red giant variable of Mira type, ranging from mag. 5.5 to 13.6 in around a year. It lies about 1,400 l.y. away.

NGC 55, 0h 15m −39°.2, is an 8th-mag. spiral galaxy seen nearly edge-on, so it appears elongated. One half is more prominent than the other. It is similar in size and shape to NGC 253 (see below), although not quite as bright. Its distance is 6 million l.y.

NGC 253, 0h 48m −25°.3, is a 7th-mag. spiral galaxy seen nearly edge-on, which gives it a cigar-shaped appearance, as seen in the photograph below. Nearly 0.5° long, it can be picked up in binoculars under dark skies but at least 100-mm aperture is required to distinguish the mottling caused by dust clouds in its spiral arms. NGC 253 lies 13 million l.y. away.

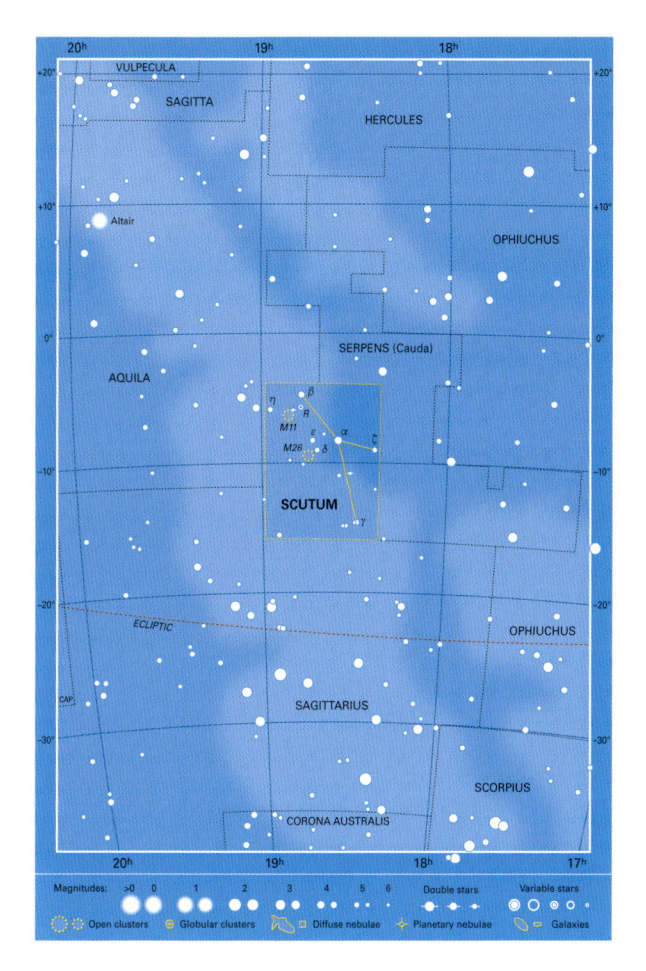

Scutum, the Shield

A faint constellation between Aquila and Serpens, introduced in 1684 by the Polish astronomer Johannes Hevelius under the title Scutum Sobiescianum, Sobieski's Shield, in honor of his patron, King John III Sobieski of Poland. Rich Milky Way starfields are the main attraction here, notably the Scutum star cloud in the northern half of the constellation, about 6° across and generally considered to be the brightest part of the Milky Way outside Sagittarius. The prominent open cluster M11 lies near a notch of dark nebulosity at the northern edge of the Scutum star cloud.

α (alpha) Scuti, 18h 35m −8°.2, mag. 3.8, is an orange giant 184 l.y. away.

δ (delta) Sct, 18h 42m −9°.1, 201 l.y. distant, is the prototype of a rare class of variable stars that pulsate in size every few hours, producing small-amplitude brightness changes. δ Scuti itself is a white giant that varies from mag. 4.6 to 4.8 with a period of 4 hours 39 minutes.

R Sct, 18h 48m −5°.7, is a pulsating orange supergiant about 1,300 l.y. away that varies between mags. 4.2 and 8.6 every 5 months or so.

M11 (NGC 6705), 18h 51m −6°.3, the Wild Duck Cluster, is a showpiece open cluster of about 200 stars, half the apparent width of the full Moon. At 6th mag. it is at the limit of naked-eye visibility, but binoculars show it as a misty patch. In a telescope with a magnification of around ×100 it breaks up into a sparkling field of faint stardust. The cluster gets its popular name from the fact that its brightest members form a distinct fan-shape, resembling a flight of ducks, an effect that is noticeable visually through a telescope but which becomes lost in long-exposure photographs. An 8th-mag. star, slightly brighter than the rest, lies at the fan's apex, with a faint double star nearby. M11 is 7,600 l.y. away.

M26 (NGC 6694), 18h 45m −9°.4, is an open cluster for small telescopes, of similar size to M11 but containing only about two dozen stars and hence much fainter. It lies 6,200 l.y. away.

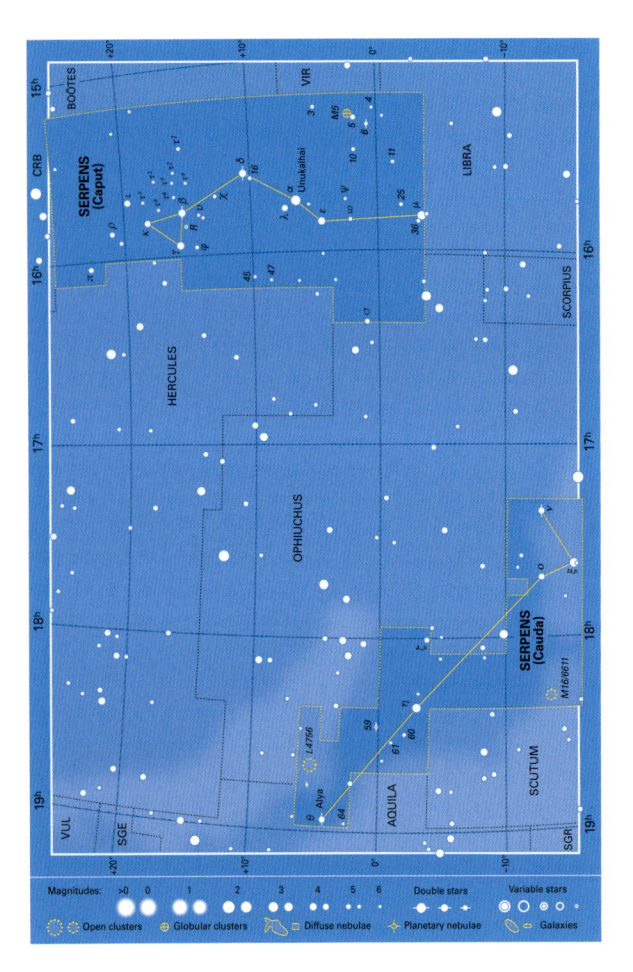

Serpens, the Serpent

An ancient constellation, representing a snake wound around the body of Ophiuchus. Serpens is actually split into two halves, one either side of Ophiuchus: Serpens Caput, the head, which is the larger and more prominent half; and Serpens Cauda, the tail. It is the only constellation to be split in two, but both halves count as one constellation.

α (alpha) Serpentis, 15h 44m +6°.4, (Unukalhai, "the serpent's neck"), mag. 2.6, is an orange giant star 74 l.y. away.

β (beta) Ser, 15h 46m +15°.4, 150 l.y. distant, is a blue-white subgiant star in the serpent's head of mag. 3.7, with a 10th-mag. companion visible in small telescopes. An unrelated mag. 6.7 background star, 29 Ser, about 500 l.y. farther off, is visible just north of it in binoculars.

γ (gamma) Ser, 15h 56m +15°.7, mag. 3.8, is a white main-sequence star 37 l.y. away.

δ (delta) Ser, 15h 35m +10°.5, 300 l.y. away, is a white subgiant or giant of mag. 4.1 with a close mag. 5.1 companion visible in a small telescope at high magnification.

η (eta) Ser, 18h 21m −2°.9, mag. 3.3, is an orange giant or subgiant star 60 l.y. away.

θ (theta) Ser, 18h 56m +4°.2, (Alya), 133 l.y. away, is an elegant pair of white stars of mags. 4.6 and 5.0, easily split in the smallest telescopes.

ν (nu) Ser, 17h 21m −12°.8, 204 l.y. away, is a blue-white star of mag. 4.3 with a wide 9th-mag. companion visible in good binoculars or small telescopes.

τ1 (tau1) Ser, 15h 26m +15°.4, 830 l.y. away, a red giant of mag. 5.2, is the brightest member of a loose scattering of eight stars of 6th mag. near β Ser, all visible in binoculars.

R Ser, 15h 51m +15°.1, is a red giant variable of Mira type, ranging between mags. 5.2 and 14.4 in approximately a year. It lies about 2,200 l.y. away.

M5 (NGC 5904), 15h 19m +2°.1, is a 6th-mag. globular cluster 23,000 l.y. away, easily visible in binoculars or small telescopes. It is rated as one of the finest globulars in the northern sky, second only to the famous M13 in Hercules. A telescope of 100-mm aperture reveals its brilliant, condensed center and mottled outer regions with curving chains of stars. Close to M5 (but actually in the foreground) lies 5 Ser, a mag. 5.1 yellow-white subgiant with a mag. 10 companion.

M16 (NGC 6611), 18h 19m −13°.8, is a hazy-looking open star cluster 5,800 l.y. distant, of similar apparent size to the full Moon, embedded in the larger Eagle Nebula. The cluster is a grouping of about 60 stars of 8th mag. and fainter at the limit of resolution in binoculars. Small telescopes show that most of its members congregate in a V-shape in the northern half. The surrounding Eagle Nebula adds a touch of haziness to the cluster when seen in binoculars. The nebula is too faint to be seen well in amateur telescopes, but shows up beautifully in long-exposure photographs. It was the subject of a famous Hubble Space Telescope image called the "pillars of creation," taken in 1995 and re-photographed in 2014.

IC 4756, 18h 39m +5°.4, is a scattered open cluster of 8th-mag. stars and fainter, about two Moon diameters wide, visible in binoculars. It lies 1,600 l.y. away.

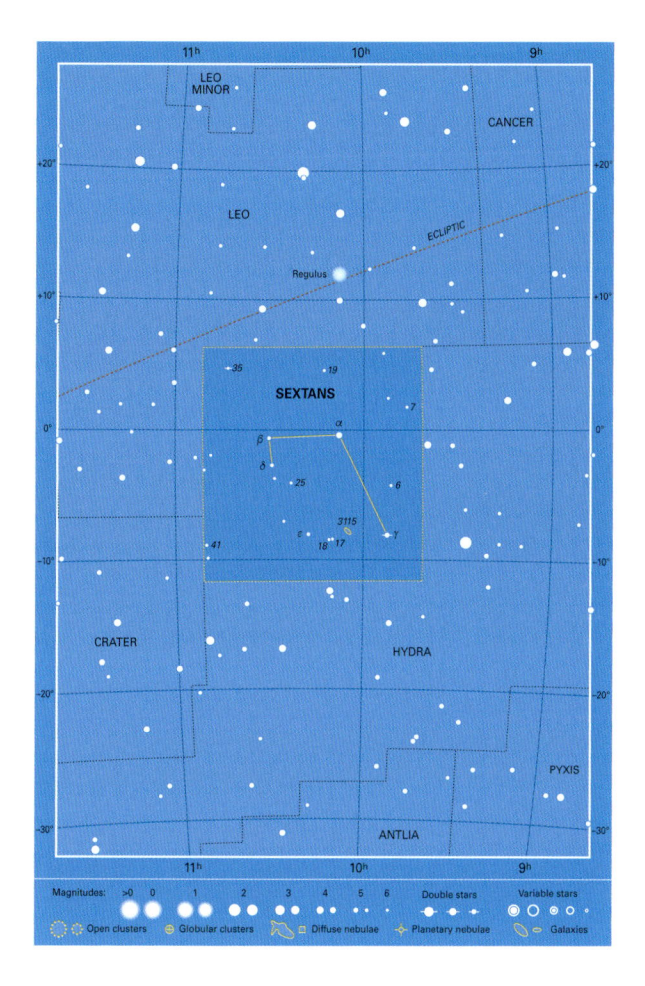

Sextans, the Sextant

A faint and insignificant constellation south of Leo, introduced in 1687 by the Polish astronomer Johannes Hevelius. It commemorates the instrument he used for measuring star positions. Hevelius continued to make naked-eye sightings of star positions with his sextant long after telescopes were available, and it was perhaps to demonstrate how good his eyesight was that he formed Sextans out of stars no brighter than mag. 4.5.

α (alpha) Sextantis, 10h 08m −0°.4, mag. 4.5, is a blue-white giant star 425 l.y. away.

β (beta) Sex, 10h 30m −0°.6, mag. 5.1, is a blue-white main-sequence star 370 l.y. away.

γ (gamma) Sex, 9h 53m −8°.1, mag. 5.1, is a blue-white main-sequence star 280 l.y. away.

δ (delta) Sex, 10h 30m −2°.7, mag. 5.2, is a blue-white main-sequence star 345 l.y. away.

17, 18 Sex, 10h 10m −8°.4, is a neat pairing of unrelated stars, mags. 5.9 and 5.6, distances 600 and 560 l.y., easily divisible in binoculars.

NGC 3115, 10h 05m −7°.7, is a 9th-mag. galaxy known as the Spindle Galaxy, some 30 million l.y. away. It is of the type known as a lenticular galaxy, with a disk of stars surrounding a central bulge but no spiral arms. Moderate-sized amateur telescopes show its elongated outline and brighter center.

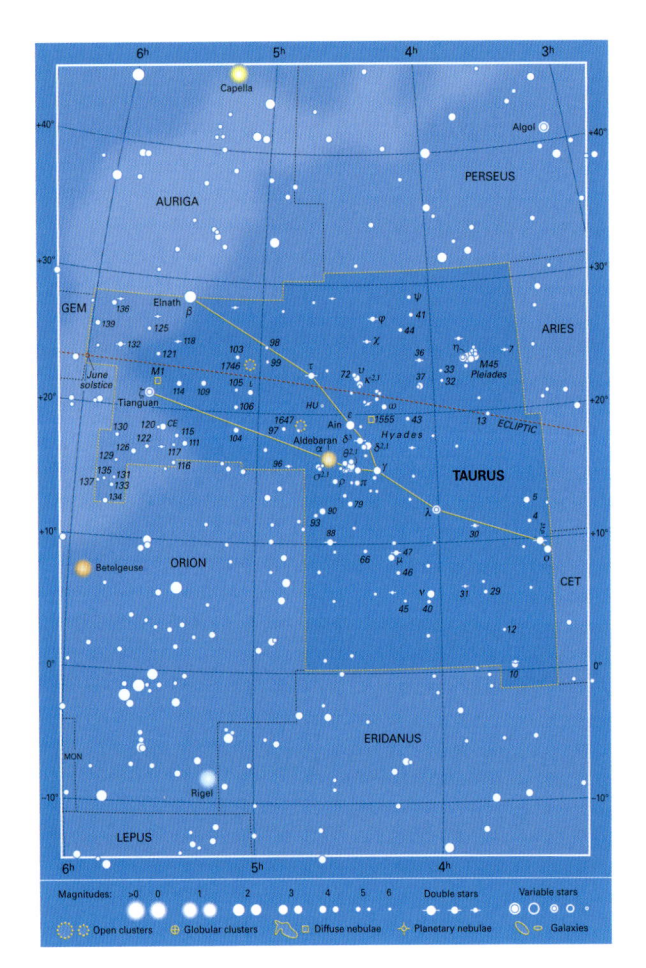

Taurus, the Bull

One of the most ancient constellations, recognized since the dawn of civilization. In Greek mythology, Taurus represents the animal disguise adopted by Zeus to carry off Princess Europa to Crete. Only the front half of the bull is depicted in the sky, its face being formed by the V-shaped cluster of stars known as the Hyades. Its glinting red eye is marked by the star Aldebaran, and its long horns are tipped by the stars β (beta) and ζ (zeta) Tauri. As well as the Hyades, Taurus contains another celebrated star cluster, the Pleiades or Seven Sisters. Taurus was the site of a supernova seen from Earth in 1054 CE that gave rise to the Crab Nebula, M1. At 4h 22.0m, +19° 32', lies the faint Hind's Variable Nebula, NGC 1554–5, discovered in the nineteenth century by the English astronomer John Russell Hind; within this nebula lies the 10th-mag. star T Tauri, 473 l.y. away, prototype of a class of irregular variables that are stars still in the process of formation.

Each year, the Taurid meteors radiate from south of the Pleiades, reaching a maximum of about 10 per hour around November 12. The Sun passes through the constellation from mid-May to late June, and is in Taurus at the June solstice; precession carried the position of the June solstice into Taurus from Gemini at the end of 1989. Please see opposite page (page 181) for full details.

α (alpha) Tauri, 4h 36m +16°.5, (Aldebaran, "the follower," i.e. of the Pleiades), is an orange giant irregular variable that fluctuates between about mags. 0.75 and 0.95. Although it appears to be part of the Hyades cluster, it is in fact an unrelated foreground star, 67 l.y. away.

β (beta) Tau, 5h 26m +28°.6, (Elnath, "the butting one"), mag. 1.7, is a blue-white giant star 134 l.y. away.

ζ (zeta) Tau, 5h 38m +21°.1, (Tianguan), 450 l.y. away, is a blue giant that is slightly variable, ranging erratically between about mags. 2.8 and 3.2.

θ1 θ2 (theta1 theta2) Tau, 4h 29m +15°.9, is a naked-eye or binocular double in the Hyades cluster, consisting of yellow and white giants of mags. 3.8 and 3.4 respectively, distances 152 and 150 l.y. θ2 is the brightest member of the Hyades.

κ1 κ2 (kappa1 kappa2) Tau, 4h 25m +22°.3, are a pair of white stars of mags. 4.2 and 5.3 that form a naked-eye or binocular duo, 156 and 148 l.y. away respectively. Both are outlying members of the Hyades.

λ (lambda) Tau, 4h 01m +12°.5, 485 l.y. away, is an eclipsing binary of the Algol type, varying between mags. 3.4 and 3.9 every 4 days.

σ1 σ2 (sigma1 sigma2) Tau, 4h 39m +15°.8, is a wide binocular double of blue-white stars on the outskirts of the Hyades, mags. 5.1 and 4.7 and distances 190 and 180 l.y. respectively.

φ (phi) Tau, 4h 20m +27°.4, is an optical double divisible in small telescopes, consisting of a mag. 5.0 orange giant 285 l.y. away, and a mag. 7.4 white star 346 l.y. away.

χ (chi) Tau, 4h 23m +25°.6, 291 l.y. away, is a double star for small telescopes, with blue and gold components of mags. 5.4 and 8.4.

The Hyades, 4h 27m +16°, is a large and bright open cluster of about 200 stars covering over 5° of sky. The brightest members form a distinctive V-shape, easily visible to the naked eye. In Greek mythology the Hyades were the daughters of Atlas and Aethra, and half-sisters of the Pleiades. Because of its considerable size, the cluster is best studied with binoculars rather than a telescope. The bright star Aldebaran is not a member of the Hyades, but is superimposed on it by chance; the brightest true member is actually θ2 (theta2) Tauri (see separate entry above). The center of the cluster lies 155 l.y. away; the distance of the Hyades is important, for it marks the first step in our distance scale of the Galaxy.

M1 (NGC 1952), 5h 35m +22°.0, is the Crab Nebula, the remains of a star that exploded as a supernova. It can be glimpsed through binoculars on clear, dark nights. Despite its fame, the Crab Nebula is a disappointing object for small telescopes, appearing as an elliptical 8th-mag. wisp of nebulosity several times the apparent diameter of the disk of Jupiter—in fact the Crab Nebula can be missed because it is larger than expected. At the heart of the nebula, beyond the reach of amateur telescopes, is a 16th-mag. object, the remains of the star that exploded. This faint object is now known to be a pulsar that spins 30 times a second, emitting flashes at radio, optical, and other wavelengths as it does so. The Crab Nebula and pulsar lie about 6,500 l.y. away.

M45, 3h 47m +24°, the Pleiades, is the brightest and most famous star cluster in the sky; it is popularly termed the Seven Sisters, after a group of mythological nymphs, the daughters of Atlas and Pleione. Approximately seven stars are visible to the naked eye, arranged into a mini-dipper shape covering three full-Moon widths of sky; binoculars bring dozens more into view. About 100 stars belong to the cluster, which is centered about 440 l.y. away. Unlike the stars of the Hyades, which are older and more evolved, the Pleiades formed within the last 50 million years and include many young blue giants. The brightest member is η (eta) Tauri (Alcyone), mag. 2.9. Other prominent members are 16 Tau (Celaeno), mag. 5.5; 17 Tau (Electra), mag. 3.7; 19 Tau (Taygeta), mag. 4.3; 20 Tau (Maia), mag. 3.9; 21 Tau (Asterope), mag. 5.8; 23 Tau (Merope), mag. 4.2; 27 Tau (Atlas), mag. 3.6; and BU Tau (Pleione), a so-called shell star that throws off rings of gas at irregular intervals, causing it to fluctuate unpredictably between mags. 4.8 and 5.5. The whole of the Pleiades cluster is embedded in a faint nebulosity, which reflects the light of the hot blue stars. This nebula is noticeable in long-exposure photographs and under very clear conditions its brightest part, around Merope, may be glimpsed in binoculars or small telescopes. This nebulosity was long thought to be the remains of the cloud from which the cluster formed, but now it seems that it is an entirely separate cloud into which the stars have since drifted by chance.

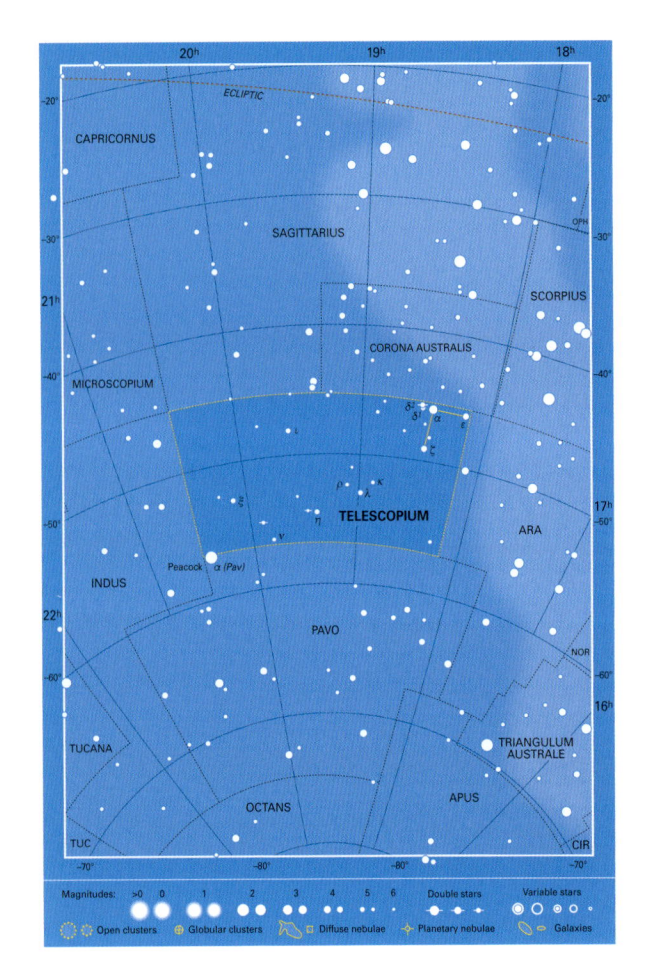

Telescopium, the Telescope

A constellation invented in the 1750s by the Frenchman Nicolas Louis de Lacaille to honor the most important of astronomical instruments, the telescope. Lacaille had in mind a specific instrument, the great refractor used by J. D. Cassini at Paris Observatory which, in common with other large refractors of the day, had an exceptionally long tube to combat chromatic aberration (false color) of the image and was suspended from a mast. Lacaille's version of Telescopium originally extended northward, including stars from Corona Australis, Sagittarius, Scorpius, and Ophiuchus, but modern astronomers have cut off the top of the telescope's tube and supporting mast. As with so many of Lacaille's constellations, it is faint and contrived, containing little to interest owners of small telescopes.

α (alpha) Telescopii, 18h 27m −46°.0, mag. 3.5, is a blue-white subgiant star 280 l.y. away.

δ1 δ2 (delta1 delta2) Tel, 18h 32m −45°.9, is a pair of blue-white stars of mags. 4.9 and 5.1, visible separately in binoculars. They are unrelated to each other, being 710 and 1,100 l.y. from us respectively.

ε (epsilon) Tel, 18h 11m −46°.0, mag. 4.5, is an orange giant 400 l.y. away.

ζ (zeta) Tel, 18h 29m −49°.1, mag. 4.1, is a yellow giant 122 l.y. away.

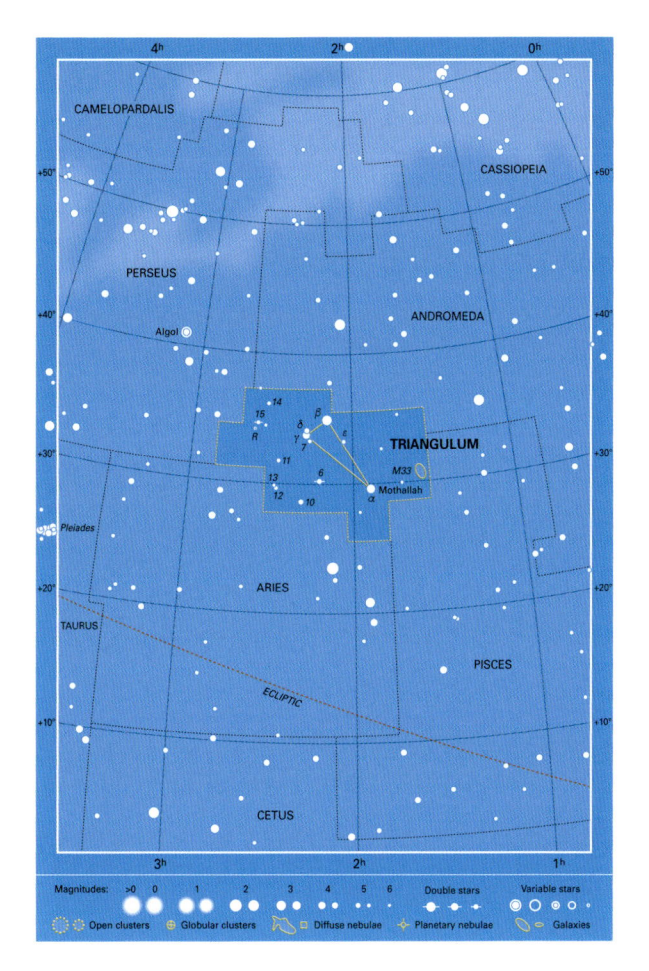

Triangulum, the Triangle

A small but distinctive constellation lying between Andromeda and Aries, consisting of three main stars that form a shape like a thin Greek capital delta; as a result, the Greeks referred to it as Deltoton. It has also been visualized as the Nile river delta and the island of Sicily. Its most important feature is the spiral galaxy M33, the third-largest member of our Local Group of galaxies, after the Andromeda Galaxy and our own Milky Way.

α (alpha) Trianguli, 1h 53m +29°.6, (Mothallah, "triangle"), mag. 3.4, is a yellow-white star 63 l.y. away.

β (beta) Tri, 2h 10m +35°.0, mag. 3.0, the constellation's brightest star, is a white giant 127 l.y. away.

γ (gamma) Tri, 2h 17m +33°.8, mag. 4.0, is a blue-white main-sequence star 112 l.y. away.

6 Tri, 2h 12m +30°.3, about 280 l.y. away, is a mag. 5.2 yellow giant with a close mag. 6.6 companion visible in a small telescope.

R Tri, 2h 37m +34°.3, is a red giant variable of Mira type, ranging from mag. 5.4 to 12.6 every 9 months or so. It lies about 1,340 l.y. away.

M33 (NGC 598), 1h 34m +30°.7, is a spiral galaxy 2.7 million l.y. away in our Local Group, sometimes popularly termed the Pinwheel Galaxy. Presented almost face-on, it covers a larger area of sky than the full Moon. Despite its size and proximity, M33 is not prominent visually because its light is spread over such a large area. M33 is best picked up on a dark night in binoculars or a small telescope using low power to enhance the contrast. Unlike most galaxies, it does not have a noticeably stellar nucleus. Quite large amateur telescopes are needed to trace the spiral arms.

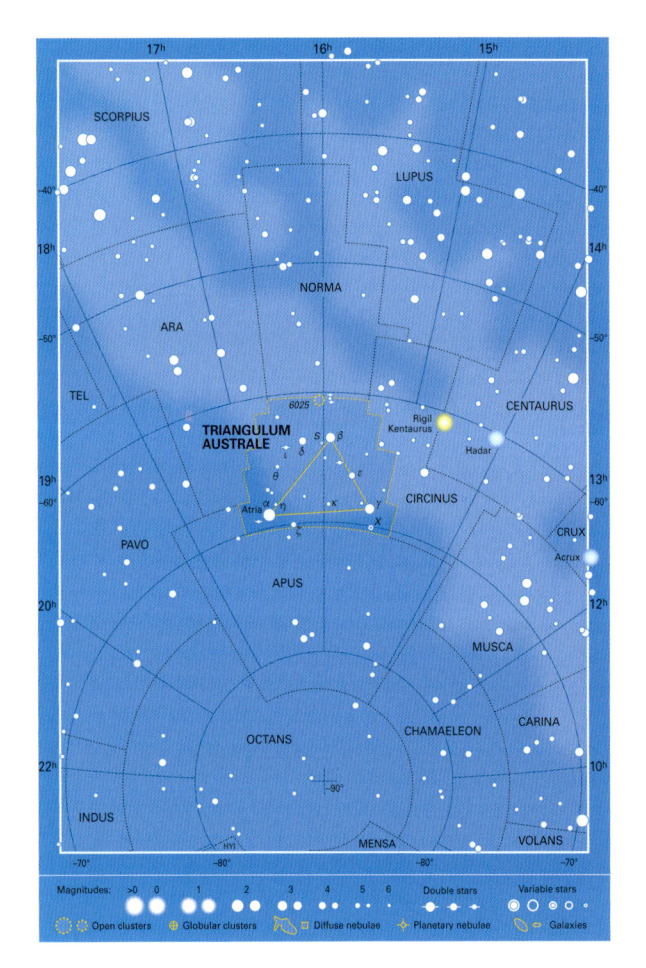

Triangulum Australe, the Southern Triangle

A small but readily distinguishable constellation near α (alpha) Centauri, introduced at the end of the sixteenth century by the Dutch navigators Pieter Dirkszoon Keyser and Frederick de Houtman. The French astronomer Nicolas Louis de Lacaille visualized it as a surveyor's level with an attached plumbline and placed two of his own related inventions next to it: Circinus (the Compasses) and Norma (the Set Square and Ruler). Its three main stars are brighter than those of its northern equivalent, Triangulum, although the constellation itself is smaller.

α (alpha) Trianguli Australis, 16h 49m −69°.0, (Atria), mag. 1.9, is an orange giant star 390 l.y. away.

β (beta) TrA, 15h 55m −63°.4, mag. 2.8, is a white star 40 l.y. away.

γ (gamma) TrA, 15h 19m −68°.7, mag. 2.9, is a blue-white star 190 l.y. away.

NGC 6025, 16h 04m −60°.5, is a binocular cluster, elongated in shape, containing about 60 stars of 7th mag. and fainter, 2,600 l.y. away.

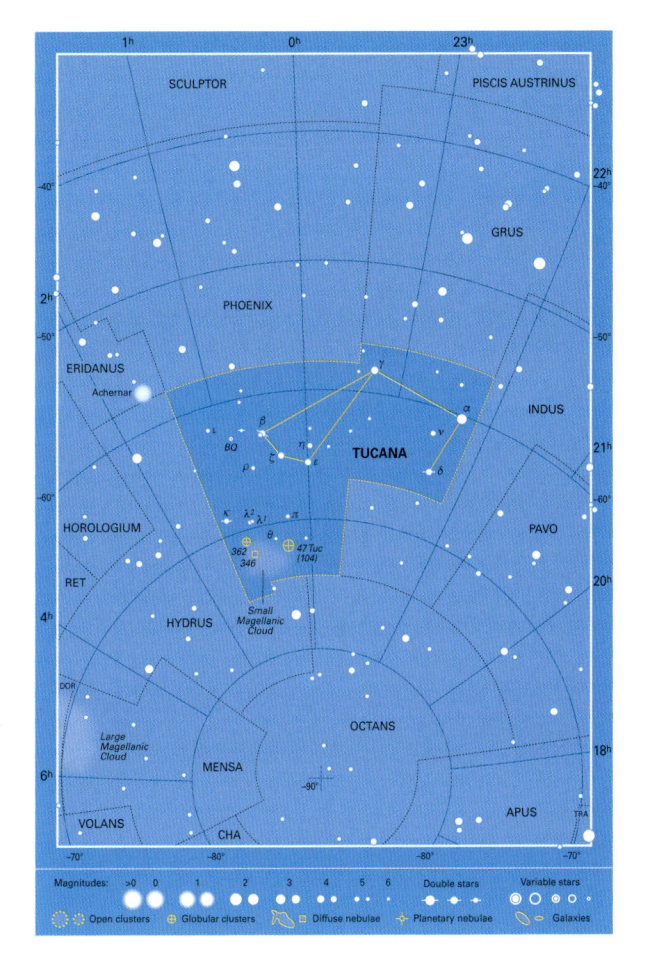

Tucana, the Toucan

A constellation near the south pole of the sky, introduced in the late sixteenth century by the Dutch navigators Pieter Dirkszoon Keyser and Frederick de Houtman. It represents a toucan, the South American bird with the large beak. Its most notable features are the Small Magellanic Cloud, actually a neighboring mini-galaxy, and the globular cluster known as 47 Tucanae.

α (alpha) Tucanae, 22h 19m −60°.3, mag. 2.8, is an orange giant star 184 l.y. away.

β (beta) Tuc, 0h 32m −63°.0, is a complex multiple star. Binoculars or small telescopes show that it consists of two almost identical blue-white stars, β1 and β2, mags. 4.3 and 4.5; β2 is itself a close binary with a period of 45 years, requiring a large aperture to divide even when at their widest, around 2039–40. Nearby lies a mag. 5.1 white star that on older charts and catalogs was labeled β3. All three stars share the same proper motion through space, but their distances of 143, 165, and 150 l.y. suggest they are not truly connected.

γ (gamma) Tuc, 23h 17m −58°.2, mag. 4.0, is a white main-sequence star 74 l.y. away.

δ (delta) Tuc, 22h 27m −65°.0, mag. 4.5, is a blue-white star 280 l.y. away with a 9th-mag. companion visible in small telescopes.

κ (kappa) Tuc, 1h 16m −68°.9, is a double star consisting of white and orange components of mags. 4.9 and 7.6 about 70 l.y. away visible in small telescopes. This pair moves through space with a wide mag. 7.3 star, itself a close binary with an orbital period of 85 years, just divisible in 100-mm aperture.

47 Tuc (NGC 104), 0h 24m −72°.1, is a prominent globular cluster the same apparent size as the full Moon, visible to the naked eye as a fuzzy 4th-mag. star; on early charts, it was actually cataloged as a star and given a stellar designation. Among globular clusters it is second only to ω (omega) Centauri in size and brightness. Telescopes of 100-mm aperture begin to resolve 47 Tuc, and even binoculars show the brilliant blaze of its star-packed core. Its diameter is about 200 light years and it is among the closest globulars to us, 14,000 l.y. away.

NGC 362, 1h 03m −70°.8, is a 6th-mag. globular cluster visible in binoculars near the northern edge of the Small Magellanic Cloud, but not associated with it. NGC 362 actually lies 29,000 l.y. away, in our own Galaxy.

Small Magellanic Cloud (SMC), 0h 53m −73°, is a satellite galaxy of the Milky Way, as is its larger sibling, the Large Magellanic Cloud in Dorado. The Small Magellanic Cloud appears to the naked eye as a nebulous, tadpole-shaped patch 3.5° across. Binoculars and small telescopes resolve star clusters and glowing gas clouds within it, although they are smaller and less impressive than those in the Large Magellanic Cloud. It lies about 200,000 l.y. away but is elongated along our line of sight.

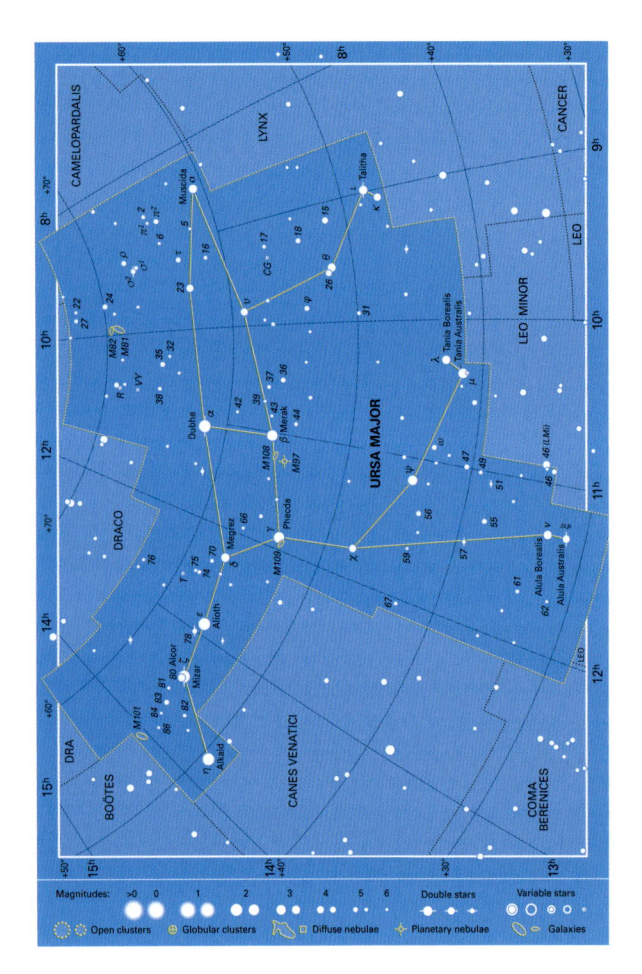

Ursa Major, the Great Bear

The third-largest constellation in the sky. Its central feature is the seven stars that make up the familiar shape variously called the Plough or the Big Dipper, the best known of all star patterns, although why so many people, including the Indigenous North Americans, visualized this group as a bear remains a mystery. In Europe the pattern was seen as a wagon or chariot. Others, notably the Arabs, viewed the dipper shape not as a bear, but as a bier or coffin. In Greek mythology the bear represented Callisto, who was turned into a bear in punishment for her illicit love affair with Zeus.

The two end stars in the dipper's bowl, Merak and Dubhe, are known as the Pointers, since they indicate the direction of Polaris, the north pole star in neighboring Ursa Minor. The curving handle of the Big Dipper points to the bright star Arcturus in Boötes. At 11h 03.3m, +35° 58', lies the mag. 7.5 red dwarf Lalande 21185, which is the Sun's fourth-closest stellar neighbor, 8.3 l.y. away. Its designation comes from its number in a catalog drawn up by the eighteenth-century French astronomer Joseph Lalande.

With the exception of Alkaid and Dubhe, the stars of the Big Dipper are traveling together through space, along with a number of other stars in the region; together, these stars make up the so-called Ursa Major moving cluster. Ursa Major contains numerous galaxies, but only a few of them are easily visible in amateur telescopes. Please see opposite page (page 187) for full details.

α (alpha) Ursae Majoris, 11h 04m +61°.8, (Dubhe, "the bear"), mag. 1.8, is a yellow-orange giant 123 l.y. away. It has a close mag. 4.9 companion that orbits it in 44 years. Large apertures are needed to divide them, particularly when at their closest separation, around 2047.

β (beta) UMa, 11h 02m +56°.4, (Merak, "flank"), mag. 2.4, is a blue-white star 84 l.y. away.

γ (gamma) UMa, 11h 54m +53°.7, (Phecda, "thigh"), mag. 2.4, is a blue-white star 83 l.y. away.

δ (delta) UMa, 12h 15m +57°.0, (Megrez, "root of the tail"), mag. 3.3, is a blue-white star 81 l.y. away.

ε (epsilon) UMa, 12h 54m +56°.0, (Alioth), mag. 1.8, is a blue-white star with a peculiar spectrum, 83 l.y. away.

ζ (zeta) UMa, 13h 24m +54°.9, (Mizar), mag. 2.2, is a celebrated multiple star. Good eyesight, or binoculars, reveals a mag. 4.0 companion, Alcor (80 Ursae Majoris). Both are about 81 l.y. away but are too far apart to be a genuine binary. However, a small telescope reveals that Mizar has another companion of mag. 3.9 closer to it, which definitely is related. This companion was first seen by the Italian astronomer Giovanni Riccioli in 1650, making Mizar the first double star to be discovered telescopically. Mizar itself was also the first star discovered to be a spectroscopic binary, by the American astronomer Edward C. Pickering in 1889. The companion of Mizar is another spectroscopic binary, as is Alcor, making this a highly complex group.

η (eta) UMa, 13h 48m +49°.3, (Alkaid, from the Arabic for "leader of the mourners"), mag. 1.9, is a blue-white main-sequence star 104 l.y. away.

ξ (xi) UMa, 11h 18m +31°.5, (Alula Australis), 29 l.y. away, was the first binary star to have its orbit computed. Its two yellow components (both also spectroscopic binaries), of mags. 4.3 and 4.8, orbit each other with a period of 60 years. They are within range of the smallest apertures for most of their orbit, but 150 mm will be needed for a few years either side of 2052 when they are at their closest.

M81 (NGC 3031), 9h 56m +69°.1, is a beautiful 7th-mag. spiral galaxy, one of the brightest in the sky, visible in binoculars. A small telescope shows it as a roundish, softly glowing patch noticeably brighter toward the center. Being tilted at an angle to us it appears somewhat elliptical in outline, covering over half a Moon's width at its longest. In the same telescopic field of view 0.5° to the north is M82 (see separate entry below); the two galaxies are about 12 million l.y. from us.

M82 (NGC 3034), 9h 56m +69°.7, is a neighbor galaxy of M81, a quarter as bright and less than half its size but still visible in binoculars. In a small telescope it appears as an elongated blur and can actually seem more prominent than M81 because of its higher surface brightness. Detailed studies by professional astronomers have shown that M82 is an edge-on spiral galaxy mottled by dust clouds and experiencing a burst of star formation as a result of a recent interaction with M81.

M97 (NGC 3587), 11h 15m +55°.0, is an elusive 11th-mag. planetary nebula known as the Owl Nebula because of two dark patches like eyes that give it the appearance of an owl's face when seen through a large telescope. In moderate apertures it appears as only a pale disk over three times the size of Jupiter, and will probably need an aperture of at least 75 mm to be seen at all. The Owl lies about 2,600 l.y. away.

M101 (NGC 5457), 14h 03m +54°.3, is a spiral galaxy visible in binoculars as a pale, rounded smudge about half the apparent size of the full Moon; because of its large size it is less prominent than its quoted brightness of 8th mag. would suggest. Long-exposure photographs show it as a face-on galaxy with widely spread spiral arms, but these are not apparent in small telescopes, which tend to show only the elliptical central region. M101 lies 22 million l.y. away.

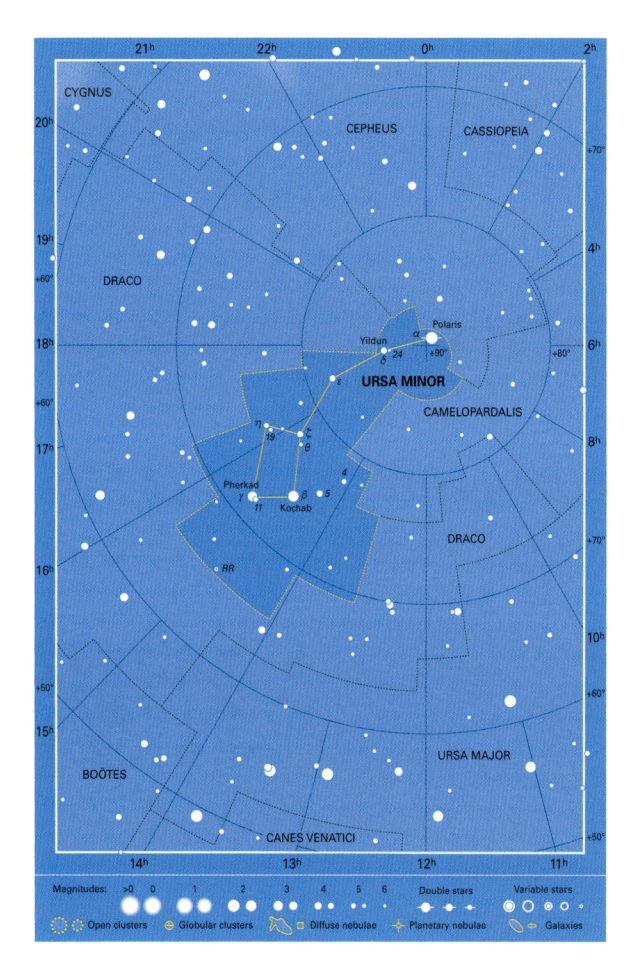

Ursa Minor, the Little Bear

A constellation said to have been introduced in about 600 BCE by the Greek astronomer Thales. Ursa Minor currently contains the north celestial pole, which lies within 1° of the conveniently placed 2nd-magnitude star popularly known as Polaris, α (alpha) Ursae Minoris. Precession will carry the pole closest to Polaris around 2100 CE, when the two will be separated by just under 0.5°, after which it will start to move away again toward Cepheus, crossing the border in 2234 CE (see chart left).

Ursa Minor is also termed the Little Dipper because its seven brightest stars outline a shape like a smaller version of the Big Dipper in Ursa Major. The stars β (beta) and γ (gamma) Ursae Minoris (Kochab and Pherkad) in the Little Dipper's bowl are called the Guardians of the Pole.

α (alpha) Ursae Minoris, 2h 32m +89°.3, (Polaris), mag. 2.0, is a yellow-white supergiant 430 l.y. away. It is a Cepheid variable—in fact, the nearest Cepheid to us. Its pulsations diminished during the twentieth century, stabilizing by the 1990s at a few hundredths of a magnitude. Polaris is also a double star, with a mag. 8.2 companion visible in a small telescope. Binoculars and small telescopes show a ring of 8th- to 11th-mag. stars some ¾° across, on which Polaris appears to be strung like some lustrous pearl on a necklace.

β (beta) UMi, 14h 51m +74°.1, (Kochab), mag. 2.1, is an orange giant 130 l.y. away.

γ (gamma) UMi, 15h 21m +71°.8, (Pherkad), mag. 3.0, is a blue-white giant star 485 l.y. away. The mag. 5.0 orange giant 11 UMi that appears near it, as seen by the naked eye or in binoculars, is unrelated, lying 410 l.y. away.

ε (epsilon) UMi, 16h 46m +82°.0, mag. 4.2, 330 l.y. away, is a yellow giant eclipsing binary that varies in a period of 39.5 days by under 0.1 mag., not discernible to the naked eye.

η (eta) UMi, 16h 18m +75°.8, mag. 5.0, is a white main-sequence star 97 l.y. away. A wide mag. 5.5 companion, 19 UMi, is an unrelated background star some 500 l.y. farther off.

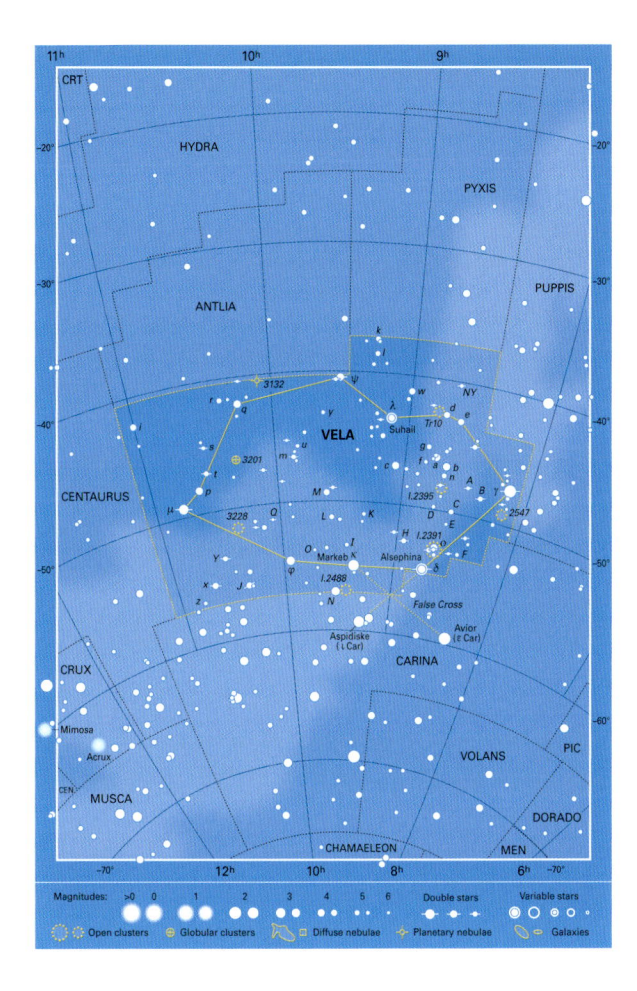

Vela, the Sails

Formerly part of the ancient constellation Argo Navis, the ship of Jason and the Argonauts, Vela was made into a separate constellation (along with Carina and Puppis) by the French astronomer Nicolas Louis de Lacaille in 1763. Vela represents the ship's sails, while Carina is the keel or hull, and Puppis the stern. Lacaille allocated Greek letters to the stars in these three sections as though they were still part of the whole ship; α (alpha) and β (beta) were given to the two brightest stars in Carina, so the lettering in Vela begins with γ (gamma). After he had run out of Greek letters, Lacaille labeled the remaining main stars of each constellation with Roman letters.

The stars κ (kappa) and δ (delta) Velorum, in conjunction with ι (iota) and ε (epsilon) Carinae, form a shape known as the False Cross, sometimes mistaken for the real Southern Cross. Vela lies in a part of the Milky Way rich in faint nebulosity visible in long-exposure photographs. This nebulosity—named the Gum Nebula after the Australian astronomer Colin S. Gum (1924–60), who drew attention to it in 1952—is believed to be the remains of one or more supernovae that occurred long ago. Another remnant of a supernova in the constellation is the Vela Pulsar, which flashes 11 times per second, one of the few pulsars that can be seen flashing optically as well as at radio wavelengths.

γ (gamma) Velorum, 8h 10m −47°.3, is an interesting multiple star. Binoculars or small telescopes divide it into two blue-white components of mags. 1.8 and 4.2, both about 1,200 l.y. away. The brighter of these is the brightest known example of a Wolf–Rayet type, a rare class of stars with very hot surfaces that seem to be ejecting gas. There are also two wider companions of 7th and 9th mags. for small telescopes.

δ (delta) Vel, 8h 45m −54°.7, (Alsephina), mag. 2.0, is a blue-white main-sequence star 80 l.y. away with a close mag. 5.6 binary companion with an orbital period of 147 years. They are currently widening and should come within range of 150-mm apertures by 2035. The brighter star is an eclipsing binary that dips to mag. 2.4 for a few hours every 45 days.

κ (kappa) Vel, 9h 22m −55°.0, mag. 2.5, is a blue-white star 570 l.y. away.

λ (lambda) Vel, 9h 08m −43°.4, (Suhail), mag. 2.2, is an orange supergiant, irregularly variable (by less than 0.2 mag.), 545 l.y. away.

H Vel, 8h 56m −52°.7, 393 l.y. away, is a neat double of mags. 4.7 and 7.7, difficult in the smallest telescopes because of the magnitude contrast.

NGC 2547, 8h 11m −49°.3, 1,300 l.y. away, is an open cluster of about 80 stars of mag. 6.5 and fainter, just visible to the naked eye and best seen in binoculars.

NGC 3132, 10h 08m −40°.5, is a relatively large and bright planetary nebula of 8th mag., known as the Eight-Burst Nebula. Through a small telescope it appears larger than Jupiter, with a central star of 10th mag. It lies 2,500 l.y. away.

NGC 3228, 10h 22m −51°.7, is an open cluster of about 15 faint stars for binoculars and small telescopes, 1,600 l.y. away.

IC 2391, 8h 40m −53°.1, is a large naked-eye cluster of 50 stars 495 l.y. away, scattered around the mag. 3.6 blue-white star ο (omicron) Vel, a small-amplitude variable of β Cephei type. About 1° from it is the binocular cluster NGC 2669, over 3,000 l.y. farther away.

IC 2395, 8h 41m −48°.2, is a binocular cluster of 40 stars, 2,400 l.y. away. A mag. 5.5 star that seems to be the brightest member, HX Velorum, is possibly a background star. Also visible 0.5° to the south is the 8th-mag. open cluster NGC 2670, 5,000 l.y. away.

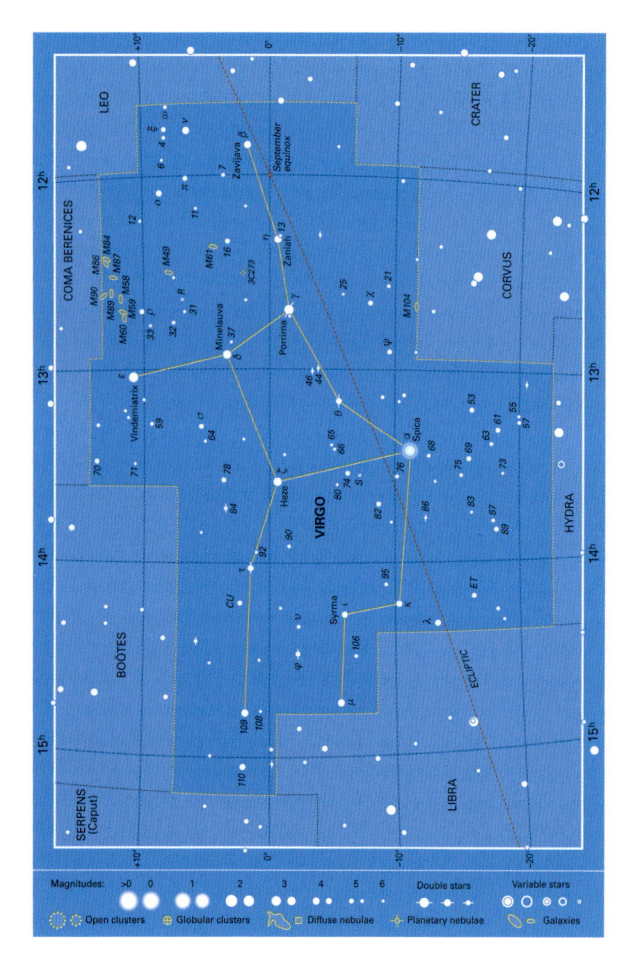

Virgo, the Maiden

The second-largest constellation in the sky (only Hydra is bigger), and the largest in the zodiac. Virgo is usually identified as Dike, the goddess of justice, the scales of justice being represented by neighboring Libra. But another legend sees her as Demeter, the corn goddess, and in the sky she is pictured holding an ear of wheat (the star Spica). The Sun passes through the constellation from mid-September to early November, and thus is within Virgo's boundaries at the time of the September equinox each year, when the Sun moves south of the celestial equator.

Virgo contains the nearest major cluster of galaxies to us, which spills over into neighboring Coma Berenices; the area is sometimes known as "the realm of the galaxies." The Virgo Cluster lies 55 million l.y. away and contains about 3,000 members, several dozen of which are visible in apertures of 150 mm or so, although they appear as little more than hazy patches; some of the brightest members are mentioned below. Another famous feature in Virgo, unrelated to the Virgo Cluster of galaxies, is the optically brightest quasar, 3C 273, located at 12h 29.1m, +2° 03'. It appears as a 13th-mag. blue star and is estimated to lie about 3 billion l.y. away. Please see opposite page (page 191) for full details.

α (alpha) Virginis, 13h 25m −11°.2, (Spica, "ear of wheat"), mag. 1.0, is a blue-white giant or subgiant star 250 l.y. away. It is a spectroscopic binary, tidally distorted by its companion so that it varies slightly with a period of 4 days as it rotates, although by no more than 0.1 mag.

β (beta) Vir, 11h 51m +1°.8, (Zavijava), mag. 3.6, is a yellow-white main-sequence star 36 l.y. away.

γ (gamma) Vir, 12h 42m −1°.4, (Porrima), 39 l.y. away, is a celebrated double star. Together, the stars shine as mag. 2.7. But a small telescope reveals γ Vir to consist of a matching pair of white stars, both of mag. 3.5. They orbit each other every 169 years and were last closest in 2005, when a 250-mm telescope was needed to split them. Now moving apart rapidly, the two stars are divisible in the smallest apertures and will remain so for the rest of the twenty-first century.

δ (delta) Vir, 12h 56m +3°.4, (Minelauva), mag. 3.4, is a red giant 198 l.y. away.

ε (epsilon) Vir, 13h 02m +11°.0, (Vindemiatrix,"grape gatherer"), mag. 2.8, is a yellow giant 108 l.y. away.

θ (theta) Vir, 13h 10m −5°.5, mag. 4.4, is a blue-white star 290 l.y. away with an unrelated 9th-mag. companion visible in a small telescope.

τ (tau) Vir, 14h 02m +1°.5, mag. 4.2, a blue-white star 227 l.y. away, forms a wide optical double for small telescopes with a mag. 9.4 companion.

φ (phi) Vir, 14h 28m −2°.2, 121 l.y. away, is a yellow subgiant of mag. 4.8 with a 10th-mag. companion, difficult to see in the smallest telescopes because of the brightness contrast.

M49 (NGC 4472), 12h 30m +8°.0, is an 8th-mag. elliptical galaxy visible as a rounded glow in a 75-mm telescope under low power. It is one of the largest and brightest members of the Virgo Cluster of galaxies.

M58 (NGC 4579), 12h 38m +11°.8, is a 10th-mag. barred spiral galaxy with a noticeably brighter core.

M59 (NGC 4621), 12h 42m +11°.6, is a 10th-mag. elliptical galaxy with a star-like nucleus, lying about a quarter of the way from M60 to M58.

M60 (NGC 4649), 12h 44m +11°.5, is a 9th-mag. elliptical galaxy, one of the most prominent members of the Virgo Cluster, detectable with 75-mm aperture.

M84 (NGC 4374), 12h 25m +12°.9, and M86 (NGC 4406), 12h 26m +12°.9, are a pair of 9th-mag. elliptical galaxies appearing in the same telescopic field as fuzzy patches with noticeably brighter cores. M86 is slightly the larger of the two and noticeably elongated, whereas M84 appears round.

M87 (NGC 4486), 12h 31m +12°.4, is a celebrated giant elliptical galaxy. It is also a strong radio source, known as Virgo A, and an X-ray source. Photographs taken through large telescopes show a jet of matter emerging from M87. In amateur telescopes, M87 appears as a rounded, 9th-mag. glow with a noticeable nucleus.

M90 (NGC 4569), 12h 37m +13°.2, is a large 9th-mag. spiral galaxy, tilted at an angle to us so that it appears elongated.

M104 (NGC 4594), 12h 40m −11°.6, is an 8th-mag. spiral galaxy seen edge-on so that it appears elongated. It is popularly known as the Sombrero because its appearance in long-exposure photographs has been likened to that of a Mexican hat. However, with its bulging nucleus ringed by tightly coiled spiral arms, its appearance is more reminiscent of Saturn. Apertures of 150 mm reveal a dark lane of dust along its rim, silhouetted against the brighter arms and nucleus. The Sombrero is not in the Virgo Cluster but somewhat closer, about 30 million l.y. away.

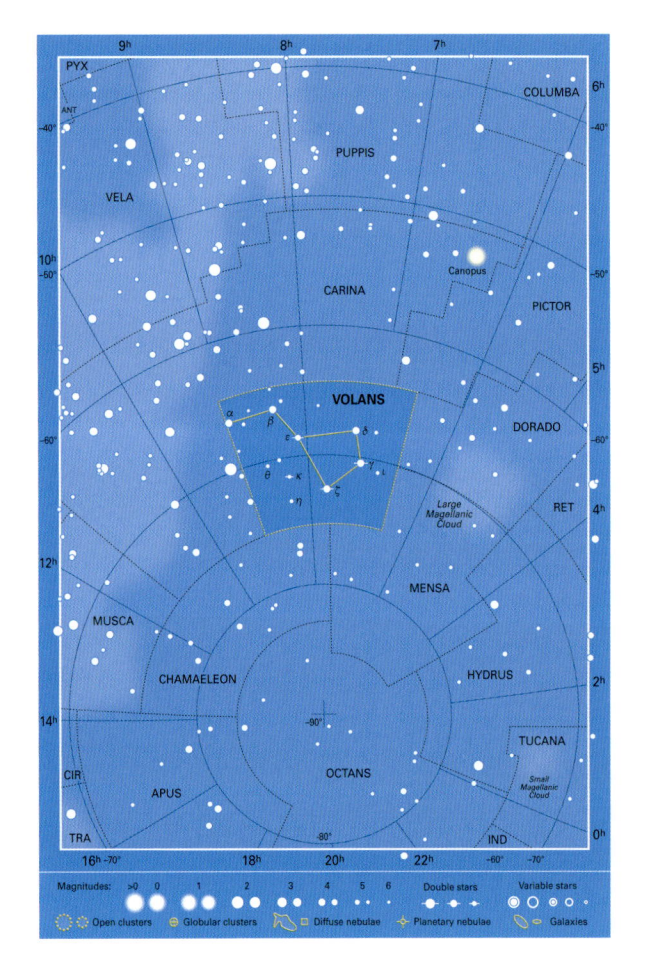

Volans, the Flying Fish

This figure was invented at the end of the sixteenth century by the Dutch navigators Pieter Dirkszoon Keyser and Frederick de Houtman. It represents the type of fish found in tropical waters that can leap out of the water and glide through the air on wings. In the sky it is depicted being stalked by the predatory dolphinfish or mahi-mahi, represented by the constellation Dorado. None of its stars is particularly bright, but there are two fine double stars for small telescopes.

α (alpha) Volantis, 9h 02m −66°.4, mag. 4.0, is a blue-white star 126 l.y. away.

β (beta) Vol, 8h 26m −66°.1, mag. 3.8, is an orange giant 108 l.y. away.

γ (gamma) Vol, 7h 09m −70°.5, 135 l.y. away, is a pair of gold and cream stars of mags. 3.7 and 5.6, beautiful in a small telescope.

δ (delta) Vol, 7h 17m −68°.0, mag. 4.0, is a yellow-white supergiant 740 l.y. away.

ε (epsilon) Vol, 8h 08m −68°.6, 645 l.y. away, is a mag. 4.4 blue-white giant or subgiant with a mag. 7.3 companion visible in a small telescope.

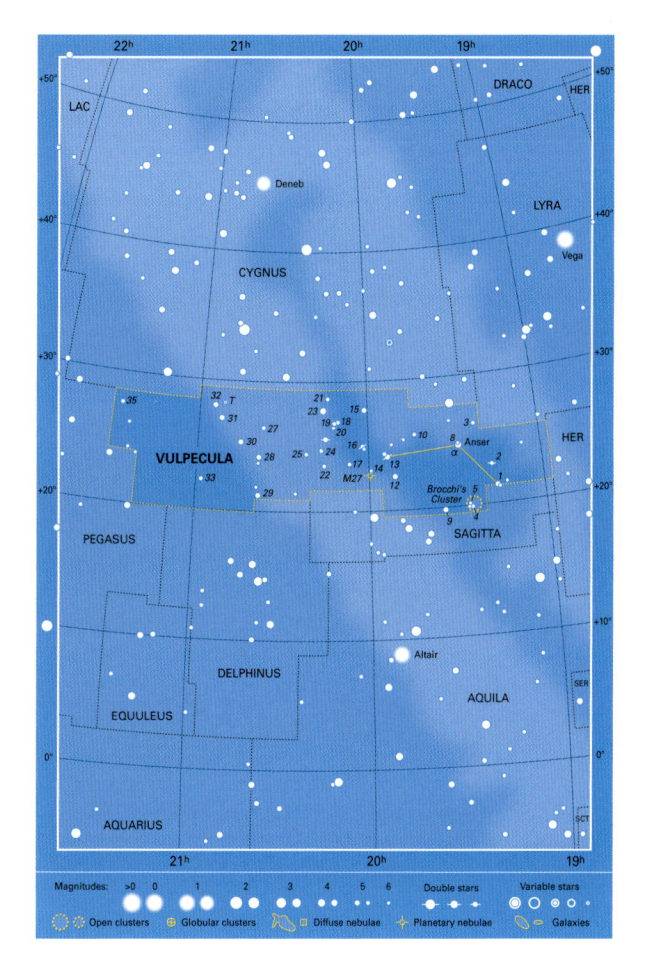

Vulpecula, the Fox

A faint constellation at the head of Cygnus. It was originated in 1687 by the Polish astronomer Johannes Hevelius, who called it Vulpecula cum Ansere, the Fox and Goose. Since then the goose has fled, leaving the fox on its own. The first pulsar, or flashing radio source, was discovered in Vulpecula in 1967 by radio astronomers at Cambridge, England. It lies about 1.5° north of the notable grouping called Brocchi's Cluster.

α (alpha) Vulpeculae, 19h 29m +24°.7, (Anser), mag. 4.4, is a red giant 290 l.y. away. An unrelated mag. 5.8 orange giant, 8 Vul, 460 l.y. away, is seen nearby in binoculars.

T Vul, 20h 51m +28°.3, is a yellow-white supergiant Cepheid variable that ranges between mags. 5.4 and 6.1 every 4.4 days. It lies about 1,900 l.y. away.

M27 (NGC 6853), 20h 00m +22°.7, the Dumbbell Nebula, is a large and bright planetary nebula, reputedly the most conspicuous of its kind, visible in binoculars as an elliptical misty glow. Its dumbbell shape is better seen through telescopes, which also reveal its greenish tinge. M27 is of 8th mag. and spans about one-quarter the apparent diameter of the full Moon. M27 is about 1,300 l.y. away.

Brocchi's Cluster (Collinder 399), 19h 25m +20°.2, is a striking binocular group of stars on the border with Sagitta, popularly known as the Coathanger because of its distinctive shape. An almost straight line of six stars extends for three Moon diameters; from the center of this line emerges a curve of four more stars, forming the Coathanger's hook. The brightest member of the group, lying in the hook, is 4 Vul, an orange giant of mag. 5.2. Distances of the individual stars range from 235 l.y. to about 1,700 l.y., and all have different motions through space, so this is not a true cluster but a chance alignment.

STARS, CLUSTERS, AND NEBULAE

Stars are balls of gas made incandescent by energy from nuclear reactions deep in their interiors. They come in a wide range of sizes and brightnesses, from faint dwarfs a hundredth of the Sun's diameter to dazzling supergiants hundreds of times larger than the Sun. They range in temperature from intensely hot blue-white specimens with surface temperatures of more than 36,000°F (20,000°C) (to cool red stars with surfaces at around 5,400°F (3,000°C). The Sun, which is a medium-temperature yellow star (surface temperature 9,900°F [5,500°C]), turns out to be pretty average in all respects.

Stars are born from massive clouds of gas and dust within our Galaxy. An interstellar gas cloud is termed a *nebula* (plural: *nebulae*), from the Latin for "cloud." A nebula is not uniformly distributed in space, but contains denser knots—the seeds of future stars. If the knot is dense enough it begins to contract under the inward pull of its own gravity. As it gets smaller and denser, it heats up, until the temperature and pressure at the center of the shrinking blob become so great that nuclear reactions begin. The gas blob has switched on to become a true star, generating its own heat and light for millions or, more usually, billions of years.

Several star-spawning clouds are well within reach of observation by amateurs. Most famous is the Orion Nebula, marking the sword in the constellation of Orion the Hunter. This nebula is visible as a hazy green glow to the naked eye; binoculars show it more clearly, while in photographs it appears complex and colorful (see page 196). Embedded within the Orion Nebula is a star called θ1 (theta1) Orionis, recently born from the surrounding gas. Small telescopes divide this star into four components that form a group known as the Trapezium. Energy emitted by the brightest of these four stars makes the surrounding nebula shine. Behind the bright, visible part of the cloud is an even larger, still-dark area where stars are being born at this moment. The Orion Nebula is estimated to contain enough matter to produce hundreds of stars: it's a star cluster in the making.

The Orion Nebula is by no means unique in our own Galaxy, and there are plenty of similar nebulae in other galaxies, too. One such example is the Tarantula Nebula in the southern constellation Dorado, which dwarfs the Orion Nebula. The Tarantula is part of a small nearby galaxy called the Large Magellanic Cloud.

Opposite This enhanced NASA image of the Lagoon Nebula displays the dense areas of seeds of future stars. It is a famous gaseous nebula, visible to the naked eye, and encompasses the star cluster NGC 6530.
Above right At the heart of Orion Nebula is the multiple star Orionis, also known as Trapezium.

Star clusters and associations

One celebrated group of stars that has come into being recently is the Pleiades cluster, popularly known as the Seven Sisters, in the constellation Taurus, the Bull. At least five members of the Pleiades can be distinguished by normal eyesight; binoculars and small telescopes bring dozens more members into view. The whole cluster is estimated to contain about 100 stars. The brightest and youngest of these formed no more than 2 million years ago, making them extremely youthful by astronomical standards.

The Pleiades is an example of a type of cluster referred to as an *open cluster* or *galactic cluster*. About 1,000 such clusters are known to astronomers, and the most prominent of them are listed in this book. Not far from the Pleiades in Taurus is a larger and older open cluster, the Hyades, which is estimated to be about 500 million years old. Being older than the Pleiades, the stars have had more time to drift apart. Eventually, most open clusters disperse completely. The Sun was probably a member of such a cluster when it was born 4,6 billion years ago.

Much larger than open clusters are *stellar associations*, vast scatterings of young stars hundreds of light years across. It's no coincidence that most of the bright stars in Orion lie at similar distances from us, for they are members of such an association, centered on the Orion Nebula, about 1,500 light years away. Three times closer to us is the extensive Scorpius–Centaurus association, which extends across 60° or more of sky from Scorpius via Lupus into Centaurus and Crux. Antares is the brightest member; other prominent members are β (beta) Centauri, α (alpha), and β (beta) Crucis, and the open cluster IC 2602 in Carina. Associations arise from particularly large clouds of gas and dust in the spiral arms of our Galaxy.

An altogether different type of cluster, containing much older stars, is a *globular cluster*; these are described on page 209.

Above left Orion Nebula, M42, is a cradle of star formation.
Above right Veil Nebula is made up of the tenuous remains of a star that shattered itself in a supernova explosion around 5,000 years ago.

Star sizes and lifetimes

Nebulae are composed of a 10:1 mixture of hydrogen and helium, the primary constituents of the Universe, so, naturally enough, stars have this same composition. Stars get their energy from nuclear reactions that transform hydrogen into helium. In the reactions, four hydrogen atoms are crushed together to make one atom of helium; an uncontrolled version of the same reaction occurs in a hydrogen bomb.

There are certain limits on the size of a star. A gas blob with less than about 8 percent of the Sun's mass (equivalent to about 80 times the mass of Jupiter, the largest planet in the Solar System) cannot become a true star, because conditions in its interior will never reach the extremes necessary for nuclear reactions to begin.

Below this threshold, the gas blob would become a *brown dwarf*, a not-quite star that shines feebly from energy released during its contraction. Smaller still, below about 1 percent of the Sun's mass, the gas blob would be regarded as a planet. In other words, if gaseous Jupiter had possessed at least 10 times its current mass it would have been not a planet but a brown dwarf, and with a mass over 80 times its current value it would have been a small star, in which case our Sun would have had a faint companion.

At the other end of the scale, the largest stars have masses of about 100 times that of the Sun. It was once thought that stars more massive than this would produce so much energy that they would literally disintegrate, but this may not be true in all cases. A few stars are known that seem to have masses greater than 100 Suns, one example being η (eta) Carinae in the southern constellation Carina; this star has varied erratically in brightness in the past, as though on the verge of instability.

A star's most vital statistic is its mass, for this factor affects everything else about it: its temperature, its brightness, and its lifetime. The stars with the least mass are, not surprisingly, the coolest; they are known as *red dwarfs*. A typical red dwarf, such as Barnard's Star, the second-closest star to the Sun, has a mass about a tenth that of the Sun and glows a dull red with a surface temperature of about 5,400°F (3,000°C). Even though Barnard's Star is only 6 light years away, it's too faint to be seen with the naked eye.

Surprisingly enough, stars with the lowest mass live the longest. Their nuclear fires burn so slowly that they can survive for as much as a million million years —a hundred times as long as the Sun. The Sun itself, which by definition is of 1 solar mass, has a surface temperature of 9,900°F (5,500°C), and is expected to live for about 1 billion years. It's currently in the prime of its life.

Moving up the weight scale, a star such as Sirius, which is twice the Sun's mass, can live for only about 1 billion years, a tenth of the Sun's age. The surface temperature of Sirius is a blue-white 19,800°F (11,000°C). Larger and hotter still, the star Spica in the constellation Virgo has a mass of about 11 Suns and a surface temperature of around 43,200°F (24,000°C). The lifetime of this intensely hot, highly luminous star is less than 1 percent of the lifetime of the Sun.

Star temperatures and colors

A star's color is a direct indicator of its temperature. The most precise way to measure a star's temperature is to study the spectrum of its light, which is done by splitting up the light in a device called a spectroscope. Stars are classified into a sequence of so-called *spectral types* according to their temperature, as shown in the table on page 201. The bluest and hottest stars are categorized as spectral types O and B. Stars of type O are rare and none is close enough to us to appear really bright. However, B-type stars are more common: prominent examples are α (alpha) and β (beta) Crucis, β (beta) Centauri, and Spica; these are in fact the bluest of the 1st-magnitude stars.

Then come the cooler blue-white A-type stars, which include Sirius, followed by stars of F type, which appear white or yellow-white; Procyon is an example. G-type stars are yellow; they include the Sun, α (alpha) Centauri, and τ (tau) Ceti. Cooler still are K-type stars, such as Aldebaran, which have orange hues. Coolest of all are stars with an M-type spectrum, examples being Antares and Betelgeuse, which are the reddest 1st-magnitude stars. Red dwarfs also have M-type spectra but none of them are bright enough to be visible to the naked eye.

Each spectral type is subdivided into ten steps from 0 to 9; on this more precise scale, the Sun ranks as a G2 star. The seemingly haphazard lettering sequence used for spectral types is the result of a previous classification scheme that was rearranged and shortened to produce the present system. The sequence of stellar spectral types can be remembered by the mnemonic "Oh Be A Fine Girl/Guy, Kiss Me."

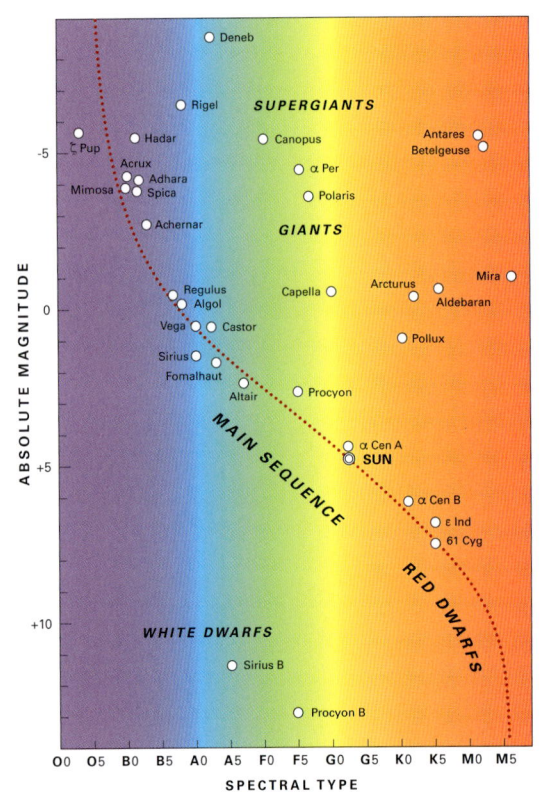

HERTZSPRUNG–RUSSELL DIAGRAM

Above The Hertzsprung-Russell Diagram, shown here, displays the true brightness of stars (their absolute magnitude), plotted against their temperature (spectral type). Stars of different types fall in different areas of the diagram.

Star colors are necessarily subjective, given the individual characteristics of different people's eyes and the varying conditions under which stars are viewed. For example, astronomers define Vega, of spectral type A0, as pure white, yet to most eyes it has a distinctly bluish cast, as do the components of Castor, also of type A. At the other end of the scale, few so-called red giant or supergiant stars appear truly red, but more a deep orange or tan color. The 1st-magnitude stars on the charts in this book are colored according to their spectral types rather than how they actually appear to the eye.

An apparent paradox is that our Sun, whose light is usually thought of as white, is classified as a yellow star. In fact, what makes the daytime Sun appear white is its overpowering brilliance. If we could see it from a distance, so it was much fainter, it would indeed appear yellowish. It turns out that a star of truly white appearance to the eye has a spectral type of around F0, similar to that of Canopus, 3,600°F (2,000°C) hotter than the Sun.

The star colors described in the constellation notes in this book are an indication of how stars appear to an observer, although the real colors are subtle and your own visual impressions may well disagree. Indeed, you'll find that the apparent intensity of the coloration varies from night to night with changing atmospheric conditions. For faint stars, you'll be unlikely to see any color at all without optical aid.

When the spectral type of stars is plotted against their actual luminosity (absolute magnitude), all stars that are in stable, hydrogen-burning middle age are found to lie in a well-defined band known as the *main sequence*. Such a plot of stellar brightness against spectral type is known as a Hertzsprung–Russell diagram, after the Danish astronomer Ejnar Hertzsprung and the American Henry Norris Russell who both devised it in 1911–13. An example is shown on left.

A star's position along the main sequence is fixed by its mass, with the least massive stars at the bottom and the most massive stars at the top. The Sun, as befits its middle-of-the-road nature, lies about halfway along the main sequence. Although most stars lie on the main sequence, a number of particularly bright stars lie above and to the right of it, while a few faint stars lie below and to the left of it. These stars are all in late stages of evolution. We can best understand what's happening to them by following the future evolution predicted for the Sun.

Evolving stars

As mentioned earlier, the Sun formed about 4.6 billion years ago and is now about halfway through its expected life span. In a few thousand million years, though, it will start to run out of hydrogen at its core. In search of more hydrogen to use as fuel, the nuclear reactions inside the Sun will start to migrate outwards, releasing more energy. Eventually, when they're surrounded by a shell of burning hydrogen, even the helium atoms in the Sun's core will enter into nuclear reactions of their own, fusing together to form carbon atoms. With all this extra energy being given off, the Sun will become much brighter than it is today and will start to swell alarmingly in size. But as its outer layers expand they'll also cool, becoming redder in color, and the Sun will turn into a *red giant*, similar to the bright stars Aldebaran and Arcturus. At its largest, the red giant Sun will grow to at least 100 times its present diameter, engulfing Mercury, Venus, and perhaps even the Earth within its outer layers. Needless to say, all life on our planet will long since have become extinct.

On the Hertzsprung–Russell diagram, the Sun's increase in brightness will move it upward, off the main sequence, and its change in spectral type will in addition move it to the right. Stars at the top end of the main sequence, which are far more massive than the Sun, become so big and bright at this stage of their evolution that they're referred to not as mere giants but as *supergiants*. Prominent examples of red supergiants are Betelgeuse and Antares, both of which are hundreds of times larger than the Sun. Other stars that have not yet evolved enough to become red in color but which are nevertheless firmly in the supergiant bracket are Rigel, Deneb, and Polaris.

To distinguish whether a star is, say, a giant or supergiant, or lies on the main sequence, astronomers assign stars a luminosity class (see the table on page 201) in addition to the spectral type. For example, a G-type main-sequence star such as the Sun is classified GV, whereas a G-type giant such as Capella is GIII, and the G-type supergiant Sadalmelik (Alpha Aquarii) is GIb.

As an aside, it should be noted that main-sequence stars are commonly referred to as dwarfs, even though the most massive of them may be several times larger in diameter than the Sun. Hence, although the Sun is in most ways an average star, in astronomical terminology it's classified as a dwarf.

Taken together, the spectral type and luminosity class define the main properties of a star as it exists at present. But those properties change as the star ages. Stars spend only a small percentage of their total lifetime in the red giant phase, which in the case of stars like the Sun will amount to no more than a few hundred million years. A red giant is a star that has grown old and is about to die.

Star death

Once a red giant has swollen beyond a certain size its distended outer layers drift off into space, forming a stellar smoke ring known rather confusingly as a *planetary nebula*, even though it has nothing to do with planets. The name was first used in 1785 by William Herschel, because they looked like the small, rounded disks of planets as seen through his telescope. Probably the best-known of all planetary nebulae is the Ring Nebula in Lyra (M57), although it's not the easiest to see. Much larger is the Dumbbell Nebula, M27, in Vulpecula, which can be picked up in binoculars on a clear, dark night. Two small but bright planetary nebulae for amateur telescopes are NGC 6826 in Cygnus and NGC 7662 in Andromeda.

At the center of a planetary nebula, the core of the former red giant is exposed as a small, intensely hot star. Once the surrounding gases of the planetary nebula have dispersed, usually after a few thousand years, the central star remains as a so-called *white dwarf*.

A white dwarf is only about the diameter of the Earth, but contains most of the matter of the original star; only about 10 percent of the star's mass is lost in the planetary nebula stage. White dwarfs are therefore exceptionally dense bodies. A teaspoonful of white dwarf material would have a mass of thousands of pounds. Over thousands of millions of years, white dwarfs slowly cool off and fade into oblivion.

Being so small, white dwarfs are very faint. Not one is visible to the naked eye. Both the nearby bright stars Sirius and Procyon have white dwarf companions, but Procyon's companion is too close to its parent to be distinguishable in amateur telescopes, and the companion of Sirius can be glimpsed only under the most favorable conditions. The easiest white dwarf to see is a companion of the star o2 (omicron2) Eridani (also known as 40 Eridani); a small telescope will show it. Of added interest is a fainter third member of this system, a red dwarf, which is also visible in amateur telescopes.

Our Sun, it seems, is destined to go through the stage of being a planetary nebula before fading away as a white dwarf. But stars with several times the Sun's mass, toward the top end of the main sequence, suffer a far more spectacular end. As we've seen, they first become dazzling supergiants rather than mere giants. They don't get a chance to reach the planetary nebula stage. So massive are they that the nuclear reactions at their centers continue in runaway fashion until the star becomes unstable and explodes. Such an explosion is known as a *supernova*.

Supernovae and their remnants

In a supernova eruption a star's brightness increases millions of times, so that for a few days the star can rival the brilliance of an entire galaxy. The shattered outer layers of the star are thrown off into space at speeds of around 3,100 miles (5,000 kilometers) per second. In 1054 CE, astronomers on Earth saw a star erupt as a supernova in the constellation Taurus. The star became brighter than Venus and was visible in daylight for three weeks. It finally faded from naked-eye view more than a year after it had first appeared.

At the site of that explosion lies one of the most famous objects in the heavens: the Crab Nebula, the shattered remains of the star that erupted. The Crab Nebula is visible as a smudgy patch in amateur telescopes, but is best seen in long-exposure photographs taken with large instruments. Over the next 50,000 years or so the gases of the Crab Nebula will disperse into space, forming delicate traceries like those of the Veil Nebula in Cygnus (see page 130), itself the remains of a former supernova.

The last supernova observed in our Galaxy was in 1604. This star, in the constellation Ophiuchus, reached a maximum magnitude of around −3, brighter than Jupiter. It was studied by the German astronomer Johannes Kepler, and is often known as Kepler's Star.

Hundreds of supernovae in other galaxies have been seen telescopically since then, but only one has become bright enough to be visible to the naked eye. That was Supernova 1987A, which erupted in our neighboring galaxy, the Large Magellanic Cloud. It was first spotted in on February 24, 1987 and rose to a maximum magnitude of 2.9 in late May, eventually fading from naked-eye view by the end of the year. Another supernova in our Galaxy is long overdue. When it comes, it should be a spectacular sight. Many astronomers dream of seeing it outshining the other stars in the sky, dazzling the eye and casting shadows.

A star might not be entirely destroyed in a supernova explosion, though. Sometimes the central core of the exploded star is left as an object even smaller and denser than a white dwarf, termed a *neutron star*. In such an object, the protons and electrons of the star's atoms have been crushed by the tremendous forces of the supernova so that they combine to form the particles known as neutrons. A typical neutron star is a mere 12 miles (20 kilometers) in diameter, but contains as much mass as one or two Suns. Being so small and dense, neutron stars can spin very rapidly without flying apart. Each time they spin we receive a flash of radiation like a lighthouse beam. Astronomers have detected radio pulses from well over 3,000 such sources, which they term *pulsars*; one lies at the center of the Crab Nebula. The Crab Pulsar flashes 30 times per second; others pulse hundreds of times a second, while the slowest take over 10 seconds. Most neutron stars are too faint to be seen optically, but the pulsar in the Crab Nebula has been seen flashing in step with the radio pulses.

Black holes

If the core of the exploded star has a mass of more than three Suns, then even a neutron star is not the end for it. Instead, it becomes something still more bizarre: a *black hole*. No force can shore up a dead star weighing more than 3 solar masses against the inward pull of its own gravity. It continues to shrink, becoming ever smaller and denser until its gravity becomes so great that nothing can escape from it, not even its own light. It has dug its own grave— a black hole.

Since a black hole is by definition invisible, it is of only academic interest to amateur observers. However, professional astronomers have detected X-ray emissions from various sources that they believe are caused by hot gas plunging into the bottomless pits of black holes. The best-known candidate for a black hole, Cygnus X-1, orbits a 9th-magnitude star in the neck of Cygnus, the Swan.

Stellar spectral types

Type	Assigned color	Temperature range (°C)	Examples
O	Blue	40,000–25,000	ζ Puppis (supergiant)
B	Blue	25,000–11,000	Spica (giant) Regulus (main sequence) Rigel (supergiant)
A	Blue-white	11,000–7,500	Vega (main sequence) Sirius (main sequence) Deneb (supergiant)
F	White	7,500–6,000	α Persei (supergiant) Procyon (subgiant) Polaris (supergiant)
G	Yellow	6,000–5,000	Sun (main sequence) α Centauri (main sequence) τ Ceti (main sequence) Capella (giant)
K	Orange	5,000–3,500	ε Eridani (main sequence) Arcturus (giant) Aldebaran (giant)
M	Red	3,500–3,000	Barnard's Star (main sequence) Antares (supergiant) Betelgeuse (supergiant)

Stellar luminosity classes

Ia0	Exceptionally bright supergiant
Ia	Bright supergiant
Iab	Less bright supergiant
Ib	Supergiant
II	Bright giant
III	Giant
IV	Subgiant
V	Main sequence
VI or sd	Subdwarf

DOUBLE AND MULTIPLE STARS

To the naked eye, stars appear as solitary, isolated objects. But the great majority of them actually have one or more companion stars, too faint or too close to be seen separately with the naked eye but which can be distinguished telescopically. Many attractive double and multiple stars are within the range of binoculars and small telescopes, and the best are described in the constellation notes in this book.

There are two sorts of double star. In one type, the two stars are not truly associated but happen to lie in the same line of sight by chance; in this arrangement, termed an *optical double*, one star may be many times farther from us than the other. Such doubles are comparatively rare. In most cases, the two stars are physically linked by gravity, forming a genuine *binary* system. The two stars in a binary orbit their mutual center of gravity, which can take decades, centuries, or even millennia, depending how far apart they are. There are also triple, quadruple, and even larger families of stars, and in such cases the orbits can become very complicated.

The closest star to the Sun, α (alpha) Centauri, is a triple system—a bright binary accompanied by a much fainter red dwarf known as Proxima. Another celebrated multiple star is the so-called Double Double, ε (epsilon) Lyrae. Binoculars show it to be a wide double, but modest-sized amateur telescopes reveal that each of these stars is itself double, making a quadruple system.

Even more remarkable is Castor, a system of six stars all linked by gravity. Amateur telescopes show that Castor consists of two bright blue-white stars close together, with a fainter third star some way off. Professional observations have revealed that each of these three stars is itself a *spectroscopic binary*. In a spectroscopic binary, the two stars are too close together to be seen individually in any telescope, but analysis of the light from the star reveals the companion's presence. The two brightest stars of Castor take around 460 years to orbit each other, whereas their spectroscopic companions whirl around them in only a few days. Sometimes the members of a binary system eclipse each other, as seen from Earth. Such *eclipsing binaries* are dealt with in the section on variable stars (pages 204–7).

Observing double stars

For amateurs, the attraction of observing double stars lies in separating (or "splitting") close doubles and in comparing their brightnesses and colors. Some pairs display a beautiful color contrast, as with the yellow and blue-green components of Albireo, β (beta) Cygni. A similar pair with contrasting colors but smaller separation is γ (gamma) Andromedae. For a pair with almost identical colors, look at 61 Cygni. Star colors are subtle, but can become more noticeable if you defocus the telescope slightly, or tap it gently to make the image vibrate.

The closer together two stars are, the larger the aperture of telescope needed to split them. A table giving the resolutions of various apertures can be found on page 206, but this not the whole story since the relative brightness of the two components also affects their observability. If one star is much fainter than the other, it will be swamped by the primary star's light, and so will be more difficult to see than a companion that's of similar brightness to the main star.

Another factor to take into account is that the separation of binary pairs with relatively short orbital periods can change markedly over a few years as the stars orbit each other. The changing positions of four fast-moving binaries is shown in the diagram on the opposite page.

The notes in this book on objects of interest give guidance as to the likely aperture needed to split various double stars. But there can be no hard-and-fast rules. So much depends on the observer's eyesight, the quality of telescope, and the observing conditions. The only way to find out for sure is to look yourself.

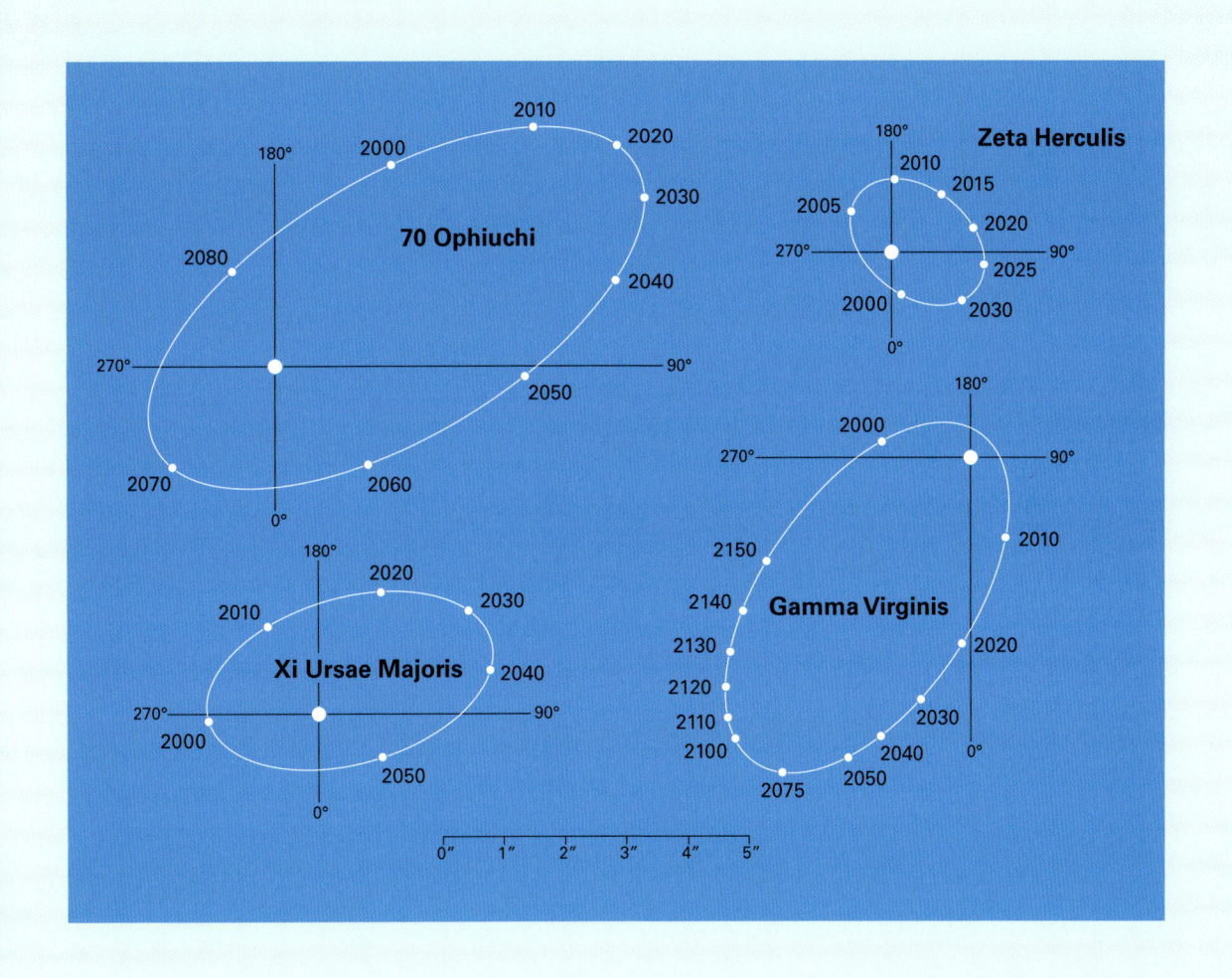

Above The diagram above shows the orbits of four fast-moving binaries: 70 Ophiuchi (period 88 years), Zeta Herculis (34.5 years), Ursae Majoris (60 years), and Gamma Virginis (169 years).

VARIABLE STARS

Certain stars change in brightness over time and are known as *variable stars*. Amateur astronomers can make valuable observations of such stars. The observer estimates the brightness of the variable by comparing it with nearby stars of known constant magnitude, and the observations are plotted on a graph to form what's known as a *light curve*, illustrating how the star's brightness varies with time. Such a graph can reveal much about the nature of the star under study. Light curves of three famous variables are shown on pages 205 and 206.

As might be expected, the usual reason that a star appears to vary is because of actual changes in its light output, but that's not the only possible explanation. In some cases the star is a member of a binary system in which one star periodically eclipses the other. For this to happen, the stars' orbits must be oriented virtually edge-on to us. The first such *eclipsing binary* to be noticed, and still the most famous, is Algol in the constellation Perseus; its variability was discovered in the 1660s by Geminiano Montanari of Italy.

Eclipsing variables

Algol consists of a blue-white main-sequence star, from which most of the light comes, orbited by a fainter orange subgiant. Every 2.87 days Algol's brightness drops from magnitude 2.1 to 3.4 in five hours as the fainter star obscures the brighter one, returning to normal after another five hours as the star re-emerges. Compare it at maximum with α (alpha) Persei, magnitude 1.8, and at minimum with δ (delta) Persei, magnitude 3.0 (see the chart on page 206). There's also a secondary minimum half an orbit later when the brighter star passes in front of the fainter one, but the drop in light is too small for the eye to notice. Most variable-star observers start their careers by following the changes in Algol's brightness; predictions of its eclipses are issued by astronomical societies and in astronomy magazines.

An extreme form of eclipsing binary is β (beta) Lyrae, in which the stars are so close together that they've been distorted into elongated shapes by each other's gravity. Beta Lyrae varies between magnitudes 3.3 and 4.4 every 12.9 days. Its brightness continually changes even outside of eclipses as the elongated stars move around their orbits. Compare it with nearby γ (gamma) Lyrae, constant at magnitude 3.2, and κ (kappa) Lyrae, magnitude 4.3.

Pulsating variables

Of the stars that vary intrinsically in brightness, most do so because of physical changes in their size. These are known as *pulsating variables* (not to be confused with pulsars). Of particular importance to astronomers are the so-called *Cepheid variables*, named after their prototype δ (delta) Cephei.

Cepheid variables are yellow supergiant stars that go through a cycle of pulsation in periods ranging from about 2 to 40 days, varying in brightness by up to one magnitude as they do so. δ Cephei itself ranges between magnitudes 3.5 and 4.4 every 5.4 days, and is thus an easy object for amateur observation. Good comparison stars are ε (epsilon) Cephei, magnitude 4.2, and ζ (zeta) Cephei, magnitude 3.4 (see the chart on page 206). Other bright Cepheids for easy amateur observation are η (eta) Aquilae (magnitude range 3.5 to 4.3, period 7.2 days); ζ (zeta) Geminorum (3.6 to 4.2, 10.2 days); and, in the southern hemisphere, l Carinae (3.3 to 4.1, 35.6 days).

The importance of Cepheid variables is that their period of pulsation is directly related to their absolute magnitude: the brighter the Cepheid, the longer it takes to complete one cycle of variation. Astronomers therefore have an easy way of finding these stars' absolute magnitude, simply by observing their period of fluctuation. When the absolute magnitude is compared with the apparent magnitude, the star's distance can easily be computed. Cepheid variables are thus important distance indicators within our own Galaxy and to other galaxies.

Opposite top Comparison chart for Algol—(beta) Persei, the prototype eclipsing variable that varies between magnitudes 2.1 and 3.4 every 2.87 days. **Opposite bottom** Comparison chart for (delta) Cephei, the prototype Cepheid variable that varies between magnitudes 3.5 and 4.4 every 5.4 days. The magnitudes of comparison stars are given without decimal points to prevent confusion between the points and faint stars.

STARGAZING FOR BEGINNERS 205

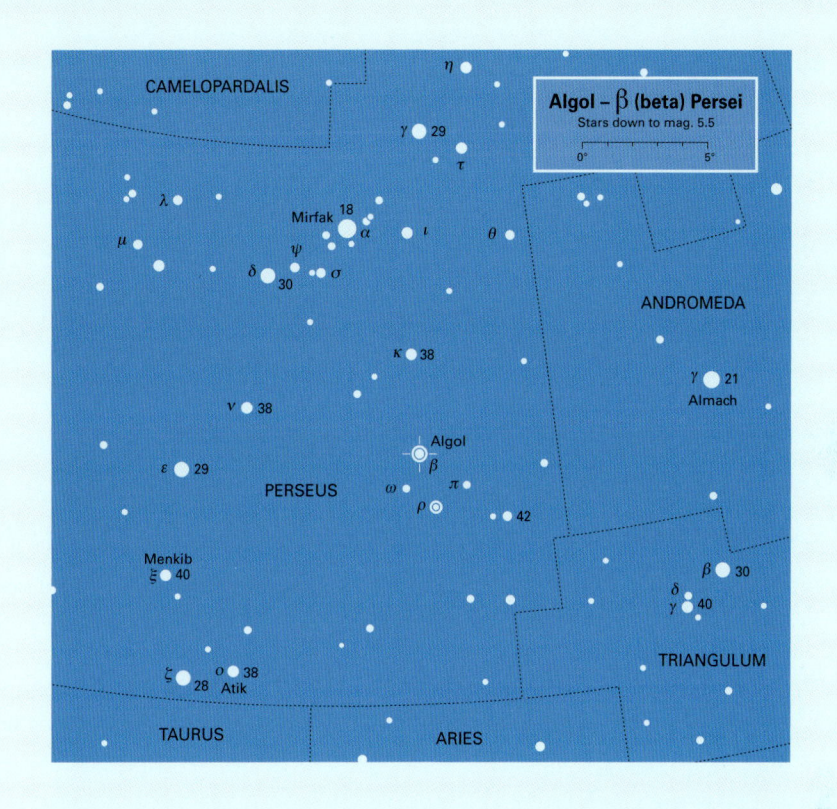

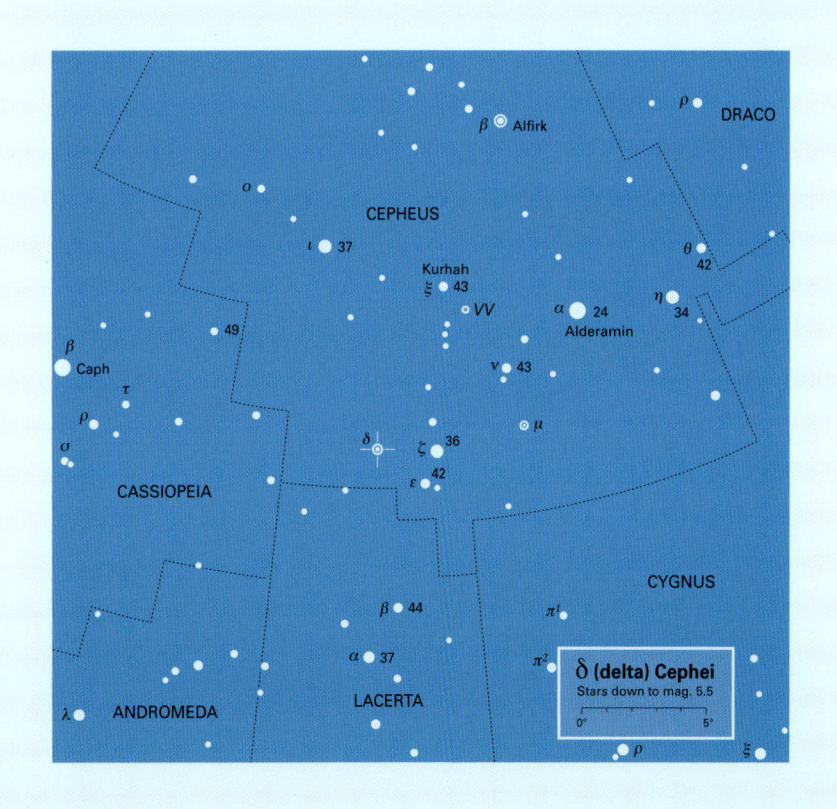

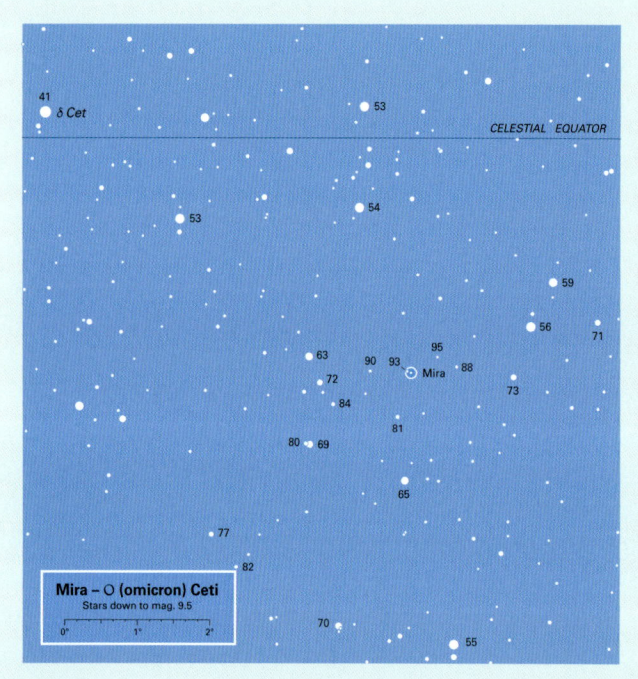

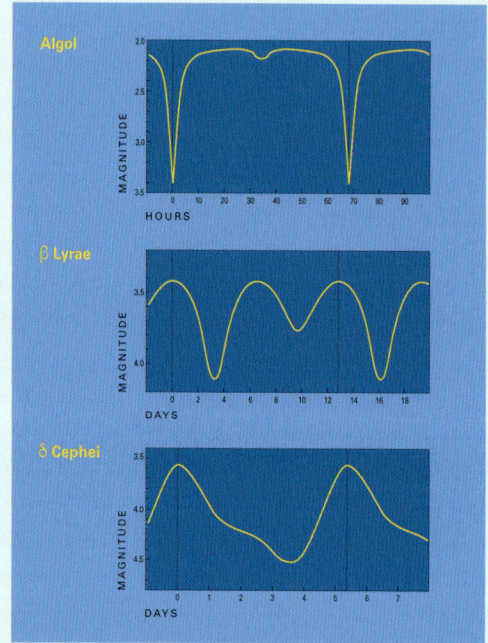

Above left This is a finder chart for the variable star Mira, also known as (omicron) Ceti. The numbers against the surrounding stars are their magnitudes with the decimal points omitted to avoid confusion. These stars can be used to estimate the magnitude of Mira.

Above right This diagram shows the light curves of three well-known variable stars: Algol, (beta) Lyrae, and (delta) Cephei. Their brightness rises and falls in a predictable fashion over the course of a few days. Note that the rise to maximum of Cephei is quicker than its subsequent decline. The small secondary minimum Algol midway between the main fades is undetectable to the naked eye.

Telescopic limiting magnitudes

		Limiting magnitudes
50 mm	(2 inches)	11.2
60 mm	(2.4 inches)	11.6
75 mm	(3 inches)	12.1
100 mm	(4 inches)	12.7
150	(6 inches)	13.6
220	(8.5 inches)	14.4

Another important class of pulsating stars are the *RR Lyrae variables*. These are old blue stars frequently found in globular clusters, and they vary by about 0.5 to 1.5 magnitudes in less than a day. Their prototype, RR Lyrae itself, varies from magnitude 7.1 to 8.1 and back in 0.57 days. Stars of this type all have similar absolute magnitudes, which makes them valuable distance indicators.

Minor types of pulsating variables include β (beta) Cephei stars and δ (delta) Scuti stars, both of which have short periods of only a few hours and ranges of variation too small to be noticeable to the naked eye.

Each type of variable falls in a specific area on the Hertzsprung–Russell diagram, and represents stars of different mass at various stages of their evolution. Variability of one kind or another seems to be an inevitable consequence of the ageing process of all stars—including, one day, our own Sun.

Long-period variables

Red giants and red supergiants are old stars that frequently prove to be variable. They pulsate, but not with anything like the same regularity as the types of stars mentioned above. The most abundant type of variables known are the Mira stars, also termed *long-period variables*, which have periods ranging from about three months to two years, and amplitudes of several magnitudes. This is the type of variable our Sun is expected to evolve into, billions of years from now.

Their prototype is Mira, ο (omicron) Ceti, in the constellation Cetus, the Whale, a red giant that ranges between about 3rd and 9th magnitude every 11 months or so, although the exact period and amplitude differ slightly from cycle to cycle. A finder chart is given on page 206. Another prominent Mira-type variable is χ (chi) Cygni, in the neck of Cygnus, the Swan.

More erratic are the *semi-regular variables*, typically with periods of about 100 days and amplitudes of 1 or 2 magnitudes. Finally, the *irregular variables* have little or no discernible pattern at all to their fluctuations. All these stars are red giants and red supergiants that have reached a stage of instability, oscillating in size and brightness. Prominent examples of semi-regular and irregular variables are Antares, Betelgeuse, α (alpha) Herculis, and μ (mu) Cephei.

Novae

Most spectacular of all variable stars are the *novae*, which suddenly and unexpectedly erupt by perhaps 10 magnitudes (10,000 times) or more, sometimes becoming visible to the naked eye where no star appeared before. Their name comes from the Latin meaning "new," for they were once thought to be genuinely new stars. Now we know that they're merely old, faint stars undergoing a temporary outburst. Despite their name, they're not related to supernovae, which are massive stars that explode for different reasons, as explained on page 200.

According to current theory, novae are close binary stars, one member being a white dwarf. Gas spilling from the companion star onto the white dwarf is thrown off in an eruption. The star does not disrupt itself in a nova outburst. In fact, some novae have undergone more than one recorded outburst—notably RS Ophiuchi and T Pyxidis—and perhaps all novae recur, given time.

A nova rises to maximum brightness in a few days. Amateur astronomers are often the first to spot such eruptions and to notify professional observatories. After a few days or weeks at maximum comes a slow decline over several months, as the nova slowly sinks back to its previous obscurity, sometimes punctuated by minor additional outbursts. Following the progress of a nova is one of the most fascinating aspects of variable-star observation. Prominent naked-eye novae occur only about once a decade, but many fainter ones are visible through binoculars.

THE MILKY WAY AND OTHER GALAXIES

Our Sun and all the stars visible in the night sky are members of a vast aggregation of stars known as the Galaxy (given a capital G to distinguish it from any other galaxy). Our Galaxy is spiral in shape, with arms composed of stars and nebulae winding outward from a central bulge of stars. It's about 100,000 light years in diameter; the Sun lies in a spiral arm some 30,000 light years from the Galaxy's center. Astronomers estimate that the Galaxy contains at least 250 billion stars.

Most of the stars in the Galaxy lie in a disk about 2,000 light years thick. Seen from our position within the Galaxy, this disk of stars appears as a faint, hazy band crossing the sky on clear, dark nights. We call this band the Milky Way, and the name Milky Way is often used for our entire Galaxy. The starfields of the Milky Way are particularly dense in the region of Sagittarius, which is the direction of the Galaxy's center. Note that the plane of the Milky Way is tilted at 63° with respect to the celestial equator. This results from a combination of the tilt of the Earth's axis, and the fact that the plane of the Earth's orbit around the Sun is tilted with respect to the plane of the Galaxy.

Globular clusters

Dotted around our Galaxy in a spherical halo are more than 100 ball-shaped conglomerations of stars known as *globular clusters*. These each contain from a hundred thousand to several million stars, all bound together by gravity. Several globular clusters are visible with the naked eye or binoculars. The brightest are ω (omega) Centauri and 47 Tucanae, both in the southern hemisphere; in the northern hemisphere the best example is M13 in Hercules.

To the naked eye and in binoculars, these objects appear as softly glowing patches of light. Moderate-sized telescopes start to resolve some of the individual red giant stars within them, giving the clusters a speckled appearance. Globular clusters formed early in the history of the Galaxy and contain some of the most ancient stars known, 10 billion years or more old—over twice the age of the Sun.

The Local Group

Our Galaxy has two small and irregularly shaped companion galaxies called the Magellanic Clouds. To the naked eye they appear like detached portions of the Milky Way, in the southern constellations Dorado and Tucana. The Large Magellanic Cloud contains about one-tenth the number of stars in our Galaxy and lies about 170,000 light years away. The Small Magellanic Cloud has only about a fifth as many stars as the Large Magellanic Cloud, and lies somewhat farther off, about 200,000 light years. Both Clouds contain numerous star clusters and bright nebulae, and are rich territories for sweeping with instruments of all sizes. One can only imagine the magnificent view of our Galaxy that any astronomers living in the Magellanic Clouds would have.

Countless galaxies are dotted like islands throughout the Universe as far as the largest telescopes can see. Most galaxies are members of clusters containing anything from a dozen to many thousands of galaxies. Our own Milky Way is the second-largest member of a small cluster of over 80 galaxies, most quite small and faint, known as the Local Group.

The largest galaxy in the Local Group is visible to the naked eye as a fuzzy, elongated patch in the constellation Andromeda. The Andromeda Galaxy, also commonly known as M31, is estimated to contain about twice as many stars as our own Galaxy, and to be about 25 percent greater in diameter. It lies about 2.5 million light years away.

Long-exposure photographs reveal that the Andromeda Galaxy is a spiral, but tilted so that we see it almost edge-on. Amateur telescopes show that the Andromeda Galaxy has two small companion galaxies —M32 and M110, its equivalent of the Magellanic Clouds—although several additional fainter galaxies are visible through large telescopes.

The only other member of the Local Group within easy reach of amateur instruments is M33 in Triangulum, another spiral galaxy, somewhat farther from us than the Andromeda Galaxy and containing considerably fewer stars. M33 can be picked up in binoculars under clear, dark skies.

The nearest rich cluster of galaxies to the Local Group lies some 50 million light years away in the constellation of Virgo, part of it spilling over into neighboring Coma Berenices. Of the Virgo Cluster's 3,000 or so known members, dozens are within the reach of amateur telescopes.

Types of galaxy

Astronomers classify galaxies into three main types: elliptical, spiral, and barred spiral (see the diagram on page 210). *Elliptical galaxies* range in shape from virtually spherical, which are designated E0, to flattened lens shapes, designated E7. They include both the largest and the smallest galaxies in the Universe. Supergiant ellipticals composed of up to 10 million million stars are the most luminous galaxies known. An example is M87 in the Virgo Cluster. At the other end of the scale, dwarf ellipticals resemble

Opposite Astronomers estimate that our Galaxy is about 100,000 light years in diameter, containing an unfathomable 250 billion stars.

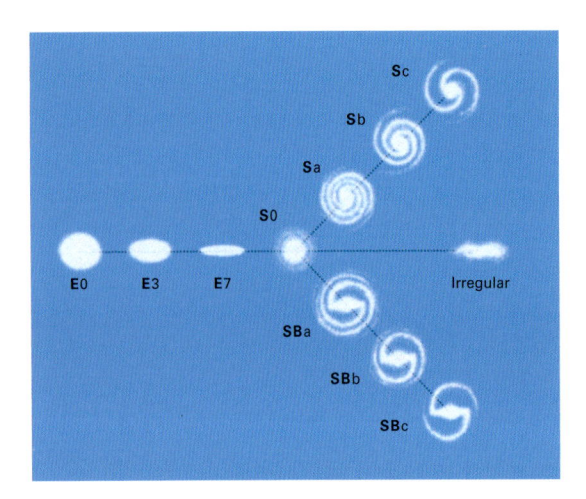

large globular clusters. Dwarf ellipticals may be the most abundant type of galaxy in the Universe, but their faintness makes them difficult to see.

Spiral galaxies (type S), such as the great Andromeda Galaxy, M31, have arms winding out from a central bulge. There are usually two arms, but sometimes more. In *barred spirals* (type SB), such as NGC 1300 in Eridanus, the arms emerge from the ends of a bar of stars and gas that runs across the galaxy's center. Most of the galaxies we see in the Universe are large, bright spirals. Spiral and barred spiral galaxies are subdivided according to how tightly their arms are wrapped around the galaxy's core: types Sa and SBa have the tightest-wound arms, while types Sc and SBc have the loosest-wound arms. The Andromeda Galaxy is of type Sb. Until recently, our own Galaxy was thought to fall midway between Sb and Sc, but it's now generally classified as a barred spiral, although the bar is short.

In addition to these three main types there are certain galaxies classified as *irregular*; the two Magellanic Clouds are usually reckoned to fall into this class, although there's some semblance of spiral structure in the Large Magellanic Cloud.

Being faint, fuzzy objects, galaxies are best seen where skies are clear and dark, away from city haze and stray light. Low magnification is best, to increase the contrast of the galaxy against the sky background. Through a telescope you should see the nucleus of a galaxy as a starlike point surrounded by the misty halo of the rest of the galaxy. Don't expect to see prominent spiral arms, as show up in long-exposure photographs, except through larger apertures at dark sites (see page 324). Galaxies present themselves to us at all angles, so even spirals can seem to be elliptical in shape when viewed edge-on.

Peculiar galaxies

Strange things are going on within some galaxies. For example, certain galaxies give off vast amounts of energy as radio waves; these *radio galaxies* include the supergiant ellipticals M87 in Virgo and NGC 5128 (pictured [above]) in Centaurus. NGC 5128, in particular, looks like an elliptical galaxy that's been caught merging with a spiral. Some spiral galaxies have unusually bright nuclei. These are termed *Seyfert galaxies*, after the American astronomer Carl Seyfert, who was the first to draw attention to them, in 1943. M77 in Cetus (see the photograph on page 211) is the brightest Seyfert galaxy.

Most peculiar of all are the objects known as *quasars*, which emit as much energy as hundreds of normal galaxies from an area less than a light year in diameter. Despite their exceptional nature, quasars are not at all exciting visually, which is why they were overlooked until 1963. The brightest quasar, 3C 273, appears as an unimposing 13th-magnitude star in Virgo.

Observations with the Hubble Space Telescope have shown that quasars are actually the highly luminous centers of galaxies far off in the Universe. Radio galaxies, Seyferts, and quasars are all now known to be related and are collectively termed *active galaxies*. Their central powerhouse is thought to be a massive black hole that gobbles up stars and gas from the surrounding galaxy, and its activity can be boosted by infalling gas when two galaxies collide. Most galaxies may have experienced such activity at their centers early in their evolution.

Above Galaxies are classified according to their shape: elliptical, type E; spiral, type S; barred spiral, type SB; and irregular. This image shows the range of galaxy shapes. It was once thought that galaxies evolved from one type into another as they aged, but this is not so.

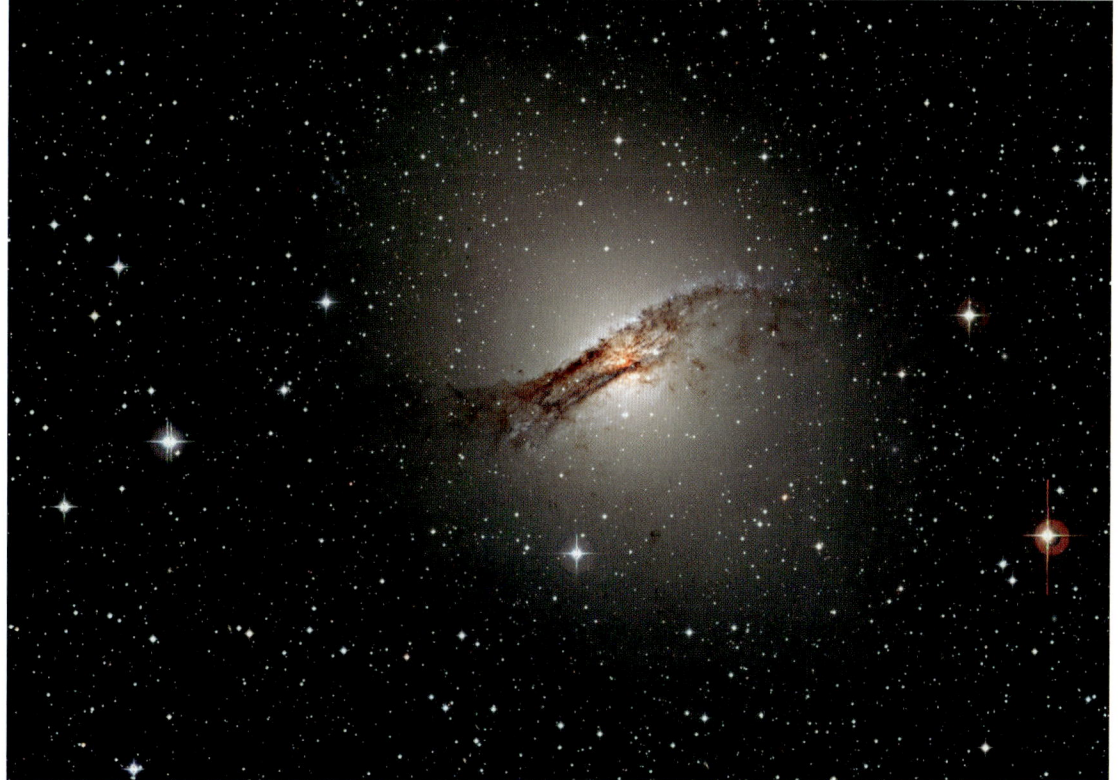

Above M77 in Cetus is the most prominent example of a Seyfert galaxy, a type of spiral with an unusually bright and active nucleus. **Below** NGC 5128, also known as Centaurus A, is a supergiant elliptical galaxy with a prominent dust band. Elliptical galaxies do not normally contain dust, so NCG 5128 is thought to result from the merger of an elliptical galaxy with a spiral.

THE SUN

Our Sun is a glowing sphere of hydrogen and helium gas 0.9 million miles (1.4 million kilometers) in diameter, 109 times the diameter of the Earth, and 745 times as massive as all the planets put together. The Sun is vital to all life on Earth because it provides the heat and light that makes our planet habitable. To astronomers, the Sun is important because it's the only star that we can observe in close-up. Studying the Sun tells us a lot that we could not otherwise know about stars in general.

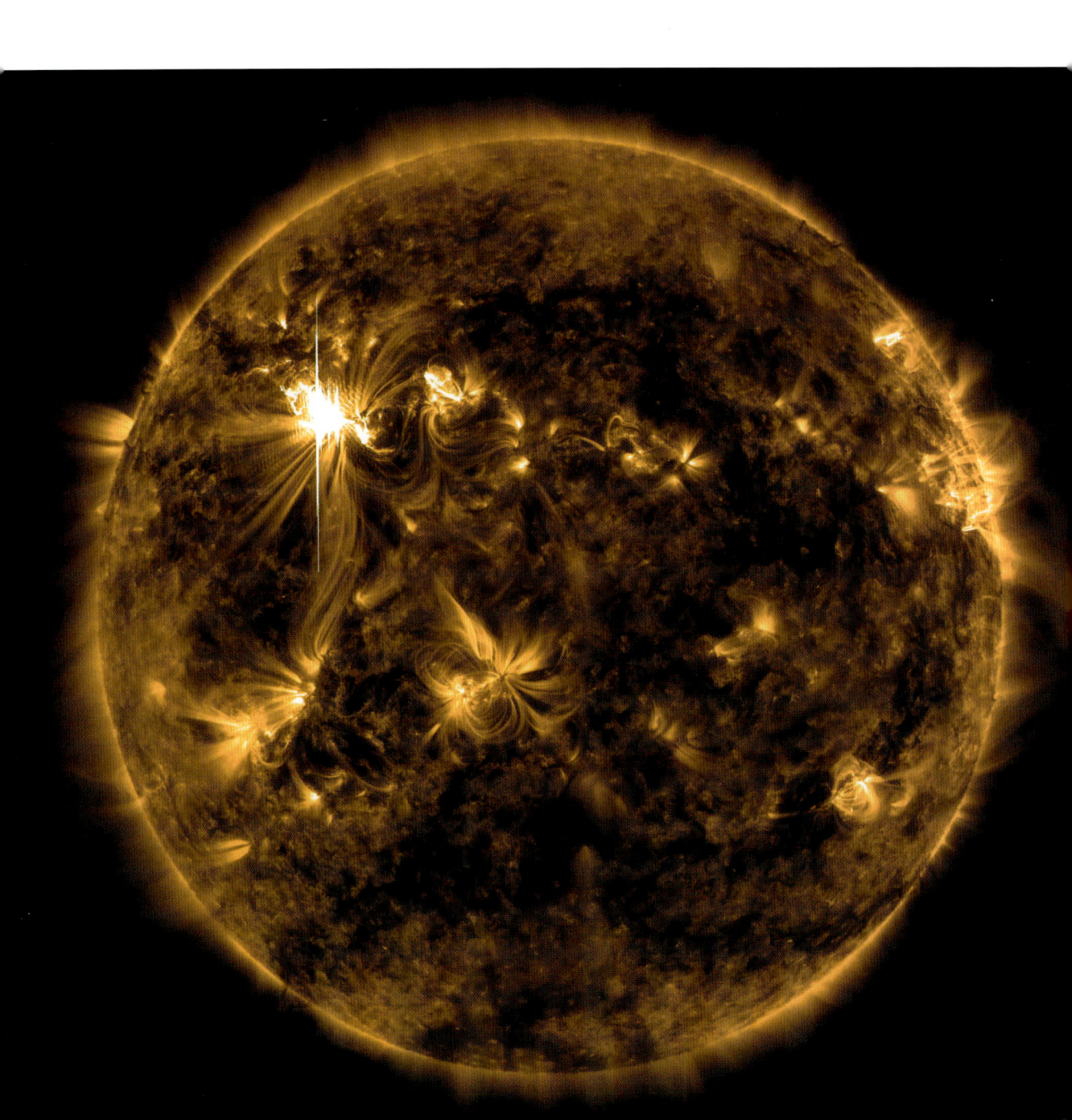

Whereas most celestial objects pose problems for observers because they're faint, the Sun presents entirely the opposite dilemma: it's so bright that it's dangerous to look at. It cannot be emphasized too strongly that anyone glancing for an instant at the Sun through any form of optical instrument, be it binoculars or telescope, risks instant blindness. Even staring with the naked eye at the Sun for a few seconds can permanently damage your sight. There's a safe and easy way to observe the Sun, and that is by projecting its image onto a white surface such as a screen or sheet of card (see above).

Some telescopes come equipped with dark Sun filters to place behind the eyepiece, but these should never be used because the focused light and heat of the Sun can crack them, with inevitably disastrous results. Safe filters for telescopic solar observation do exist, made either of glass or plastic film with a metallic coating. These filters fit over the front of the telescope tube, reducing the amount of incoming light and heat to safe levels. Filters made from a thin plastic film called Mylar give a bluish image, whereas other types give yellow or orange images.

A telescopic view of the Sun, seen either by projection or through a safe filter, displays the brilliant solar surface, the *photosphere* ("sphere of light"), consisting of seething gases at a temperature of 9,900°F (5,500°C). Although intensely hot by terrestrial standards, this is cool by comparison with the Sun's core, where the energy-generating nuclear reactions take place; there, the temperature is calculated to be about 27 million °F (15 million °C).

Seen at high magnification, the photosphere exhibits a mottled effect termed *granulation*, caused by cells of hot gas bubbling up in the photosphere like water boiling in a pan. Granules range from about 190–930 miles (300–1,500 kilometers) in diameter.

Look carefully at a projected image of the Sun, and you'll notice that the edges appear fainter than the center of the disk, an effect known as *limb darkening*. This is caused by the fact that the gases of the photosphere are somewhat transparent, so that at the center of the disk we're looking more deeply into the Sun's hotter and brighter interior than at the limbs. Brighter patches on the photosphere known as *faculae* may be visible against the limb darkening. Faculae are areas of higher-temperature gas. More prominent are dark markings known as *sunspots*, patches of cooler gas that come and go on the photosphere.

Opposite This incredible NASA-captured image is from March 2012, when the Sun erupted with one of the largest solar X-class flares of this solar cycle. This image was shared from the NASA Goddard Space Flight Center, which plays an important role by contributing scientific knowledge to advance the Agency's mission and to support scientific endeavor. **Above** To observe the sun safely, project its image onto a white surface. Never look directly at the Sun through any form of optical instrument without a specialized solar filter.

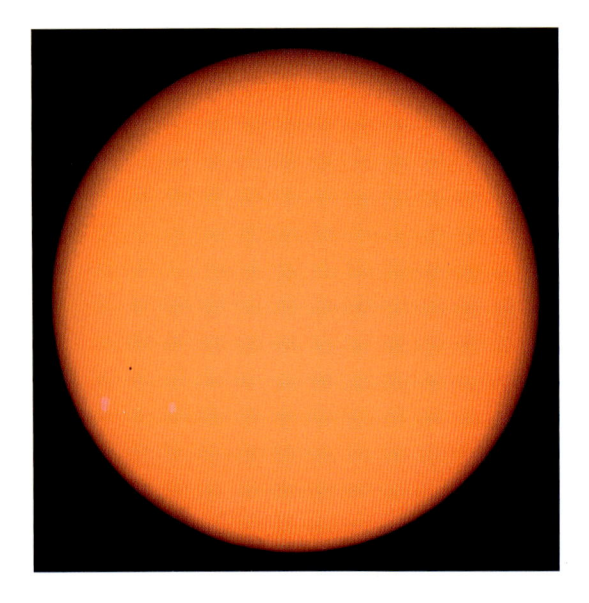

Sunspots and solar rotation

Sunspots arise where magnetic lines of force burst through the photosphere from within the Sun. The presence of a strong magnetic field blocks the outward flow of heat from inside the Sun, producing the cooler spot. Sunspots have a dark center known as the *umbra*, of temperature about 7,200°F (4,000°C), surrounded by a brighter *penumbra* at about 9,000°F (5,000°C). The temperature of a sunspot's umbra is similar to that of the surface of a red giant star such as Aldebaran, so a spot would glow quite brightly if it could be seen in isolation; in practice, sunspots appear dark by contrast with the brilliance of the surrounding photosphere.

Sunspots range in size from small pores no bigger than a large granule to enormous, complex patches many times larger than the Earth. Spots of that size are visible to the naked eye when the Sun is dimmed by the atmosphere, shortly after sunrise or before sunset, or through special solar filters.

A major sunspot takes about a week to develop to full size, then slowly dies away again over the next two weeks or so. The largest sunspots tend to form in groups long enough to span the best part of the distance from the Earth to the Moon. Spot groups can persist for one or two solar rotations (a month or two). As the Sun rotates, new spots are brought into view at one limb while others disappear around the opposite limb. By observing the progress of spots as they are carried across the Sun's face, we can measure the rotation period of the Sun.

Being gaseous and not solid, the Sun does not rotate at the same rate at all latitudes. It spins most quickly at its equator, in 25 days; this slows to about 28 days at latitude 45°; and near the poles it's slowest of all, 34 days. The average figure usually quoted, 25.38 days, refers to the rotation rate at latitude 17°. These rotation periods are relative to a fixed external point. Since the Earth is not fixed but is orbiting the Sun, a spot takes about two days longer than this to rotate once with respect to the Earth.

Usually, a spot group contains two main components, aligned east–west. The spot that leads across the disk as the Sun rotates is termed the p (preceding) spot, and is usually larger than its follower, the f spot. The p and f spots have opposite magnetic polarities, like the ends of a horseshoe magnet. The horseshoe is completed by invisible lines of force looping between the spots.

Above left Silhouette of the tiny planet Mercury transiting the disk of the Sun, photographed through a 70 mm refractor fitted with a solar filter. Planets in the solar system orbit in a nearly flat plane, and occasionally the inner planets Mercury and Venus perfectly line up with the Sun during inferior conjunction and can be seen silhouetted against the solar disk. Solar transits by Mercury occur about 14 times per century. **Above right** A close-up of Mercury silhouetted against the solar disk. The perfectly spherical shape and solid black color differ in appearance from sunspots, which are usually irregular and have a distinct penumbra around the darker, central umbra region. Mercury takes about five hours to move across the entire solar disk, whereas a sunspot takes about 14 days to cross the surface, moving with the Sun as it rotates.

Flares, CMEs, and aurorae

Sometimes the intense magnetic field lines in a complex sunspot group become entangled, releasing a sudden flash of energy known as a *flare* that may last from a few minutes to an hour. X-rays and ultraviolet radiation emitted by the flare cause effects on Earth such as radio blackouts and interference with satellites.

Some flares are accompanied by a so-called coronal mass ejection (CME), a huge bubble of hot gas thrown off from the Sun, although CMEs can also occur without flares. Atomic particles from CMEs take two to three days to reach the Earth, where they interact with the magnetosphere, giving rise to the ethereal displays known as *aurorae,* popularly termed the northern or southern lights. In an aurora, the sky glows with patches of green and red light that can take the form of arches, folded drapery, or rays, shimmering and changing shape for hours on end.

Aurorae occur in rings around the Earth's north and south geomagnetic poles called the *auroral ovals,* where atomic particles are accelerated down the Earth's lines of magnetic force onto the upper atmosphere, causing it to glow. These ovals are more or less permanent, but are normally not visible outside far northerly or southerly latitudes. However, they extend to lower latitudes at times of extreme activity, such as when the Earth is hit by CMEs.

Another common cause of aurorae, unconnected with sunspots or flares, are lower-density areas in the Sun's outer layers called *coronal holes*. The solar wind streams outward through these holes at about twice its normal speed and when these gusts hit us they cause bright and extensive auroral displays. Coronal holes are most common in the few years after solar maximum. They can last for a full solar rotation or more, so keeping an eye on coronal holes as photographed by spacecraft is one way of getting advance warning of auroral displays. A good source of information on all forms of solar activity is the website www.spaceweather.com.

Although aurorae appear bright green on photographs, to the eye they appear more grayish in color; this is because the camera is better than the eye at picking up faint colors. Aurorae can be easily photographed with modern cameras. Results will depend on the brightness of the aurora, but typical settings are ISO 1600, maximum aperture, and an exposure time of a few seconds. Remarkable results can also be obtained with the cameras on cellphones simply by pointing and clicking, as with the example shown in the photograph below.

The solar cycle

The number of sunspots on view, along with other forms of solar activity, waxes and wanes in a cycle lasting 11 years on average, although the length of individual cycles has been as short as 8 years and as long as 16 years. At times of minimum activity, the Sun may be spotless for days on end, whereas at solar maximum over 100 spots may be visible at any one time. However, the level of activity varies considerably from cycle to cycle; the average number of sunspots visible at maximum has ranged from 40 to 180, and even when the Sun is supposedly quiet, some large spots and flares can break out. Solar activity is notoriously unpredictable, which adds to its fascination for observers.

A few general rules can be deduced, however. The first spots of each new cycle appear at latitudes of about 30 to 35° north and south of the Sun's equator. As the cycle progresses, spots tend to form closer to the equator. Sunspot numbers build up to a peak, then start to decay again. As minimum approaches, the last spots of the cycle are found at latitudes between 5° and 10° north and south of the equator. At the same time—that is, around solar minimum—the first spots of the next cycle are starting to appear at higher latitudes. Sunspots are seldom found on the Sun's equator, and spots at solar latitudes greater than 40° are exceedingly rare.

Left This photograph of the Northern Lights, or aurorae, shows the spectacular colors and shapes that can be on display when atomic particles interact with the Earth's magnetosphere.

The Sun's outer layers

Above the photosphere is a tenuous layer of gas known as the *chromosphere*, about 6,200 miles (10,000 kilometers) deep. So faint is the chromosphere that it's normally invisible except with special instruments. It can, however, be seen for a few seconds at a total eclipse, when it appears as a pinkish crescent just before and after the Moon completely covers the face of the Sun. This pinkish color is caused by light from hydrogen gas, and gives rise to the layer's name, which means "color sphere."

Also visible around the edge of the Sun at total eclipses are huge tufts, or loops, of gas known as *prominences,* which extend from the chromosphere into space. They have the same characteristic rosy-pink color as the chromosphere, caused by emission from hydrogen. Like so many features of the Sun, prominences are controlled by magnetic fields. So-called *quiescent prominences* extend for 60,000 miles (100,000 kilometers) or more across the Sun, often forming graceful arches tens of thousands of miles high. When seen silhouetted against the brighter background of the Sun's disk they are termed *filaments*. Quiescent prominences can last for months.

At the other end of the scale are *eruptive prominences*, with lifetimes of only a few hours. They are actually flares seen at the limb of the Sun, ejecting material into space at speeds of up to 60 miles (1,000 kilometers) per second. All forms of solar activity—sunspots, flares, and prominences—follow the 11-year solar cycle.

The corona and solar wind

The Sun's crowning glory is its *corona*, a faint halo of gas that comes into view only when the brilliant photosphere is blotted out at a total solar eclipse. The corona is composed of exceptionally rarefied gas at a temperature of 1.8–3.6 million °F (1–2 million °C). Petal-like streamers of coronal gas extend from the Sun's equatorial zone, while shorter, more delicate plumes fan out from the polar regions. The shape of the corona changes during the solar cycle: at solar maximum, when there are more active areas on the Sun, the corona appears more rounded than at solar minimum.

Gas from the corona is continually flowing away from the Sun out into the Solar System, forming what is known as the *solar wind*. Atomic particles of the solar wind are detected streaming past the Earth at a speed of about 250 miles (400 kilometers) per second. The most obvious effect of the solar wind is to push comet tails away from the Sun. The solar wind extends outward beyond the orbit of the most distant planet, finally merging with the thin gas between the stars. In a sense, therefore, all the planets of the Solar System can be said to lie within the outer reaches of the Sun's corona.

Sun data

Diameter	865000 miles (1,392,000 km)
Mass	1.99 × 2270 lbs (1030 kg)
Mean density (water = 1)	1.41
Volume (Earth = 1)	1,304,000
Apparent magnitude	−26.7
Absolute magnitude	+4.82
Axial rotation period (average)	25.38 days
Axial inclination	7°.25
Mean distance from Earth	930,570,130 miles (149,600,000 km)

Observing the Sun

Projecting the Sun's image is the recommended method of observation for all beginners. A small refractor is ideal for this purpose. If you have a reflector larger than 4 inches (100 millimeters), stop down the aperture or heat may damage the optics. Take great care when aiming the telescope. Cover the main lens or mirror and the finder to ensure that no light accidentally enters your eyes. Squint along the tube or line up the telescope on its own shadow (do *not* look through the telescope).

Regular observers use a lightweight solar observing box attached to the eyepiece, into which the Sun's image is projected. Counting individual sunspots and sunspot groups to build up a record of changing solar activity is one of the main purposes of solar observation.

You may also wish to make a sketch of the Sun's appearance. A circle of 6 inches (150 millimeters) diameter is the standard size for solar disk drawings. Divide the circle into smaller squares 0.25 inch (1 centimeter) or so across so that the positions of sunspots can be accurately copied onto a chart.

In a projected image, north is at the top, and west at the left, for an observer in the northern hemisphere, but the other way round in the southern hemisphere. To find the exact orientation at any time, watch sunspots drift from east to west as the Earth turns.

Above This enhanced NASA image of the Sun shows the halo of the solar corona, extending in streamers and plumes of coronal gas. The shape of the corona alters during the solar cycle.

THE MOON

The Moon, the Earth's natural satellite and nearest celestial neighbor, is an object of perennial fascination for observers. Despite its small size—2,160 miles (3,475 kilometers) in diameter, roughly a quarter that of the Earth—it's so close to us, on average 238,850 miles (384,400 kilometers), that even the unaided eye can pick out the major markings that create the familiar "man in the Moon" pattern, while ordinary binoculars reveal a wealth of detail on its cratered surface. Some of the most interesting objects to look out for with small instruments are described in the notes on pages 319–327.

The Moon undergoes the most obvious changes in appearance of any celestial body, termed its *phases*. In the course of a month we see it change shape from a thin crescent to fully illuminated and back again—a result of its orbital motion around us, which affects how much of its sunlit side we can see.

When the Moon is in line with the Sun, all its sunlit side is turned away from us and the Moon is said to be *new*. A day or so later the Moon emerges as a thin crescent low in the evening sky. At this stage its night side can be seen faintly illuminated by light reflected from the Earth, a phenomenon known as *earthshine* or, more popularly, "the old Moon in the young Moon's arms."

After a week we see the Moon half-lit, a phase termed *first quarter* because the Moon is a quarter of the way around its orbit. The Moon then fills out, passing through the gibbous phase to become full, approximately 15 days after new Moon. At full the Moon lies on the opposite side of the sky from the Sun and rises around sunset. The phases then repeat in reverse order, reaching last quarter a week or so after full and then passing through a crescent back to new Moon. As the Moon's phase increases toward full it is said to be *waxing*, and when the phases are decreasing it is said to be *waning*.

Motions of the Moon

One complete cycle of lunar phases takes 29.5 days, an interval called a *synodic month* or a *lunation*. However, there's a second type of lunar month, called a *sidereal month*, that lasts 27.3 days; this is the time the Moon takes to complete one orbit of the Earth relative to a fixed point such as a distant star. The two differ because the Earth is constantly on the move around the Sun, so the Moon must complete rather more than one orbit to return to the same phase as seen from Earth.

Since the Moon's orbit is somewhat elliptical, its distance from Earth changes during the month. When at its nearest to Earth, a point called *perigee*, it appears as much as 12 percent larger than when at its farthest, *apogee*.

The Moon spins on its axis in 27.3 days, the same as its orbital period. Such an arrangement is known as

Opposite The Moon is the most obviously changeable of all the astronomical bodies, and its phases change over the course of its 28-day cycle. Even with the naked eye, stargazers can pick out many details on the lunar surface.

a *captured* or *synchronous rotation* and comes about because of the effect of the Earth's gravity. With a captured rotation, the Moon keeps the same face turned toward us at all times—so whenever we look at the Moon we always see the same features. In practice, though, we can see slightly more than half the Moon's surface because of an effect termed *libration*. The Moon's equator is tilted at about 6.5° to the plane of its orbit, so that at times we can see as much as 6.5° over the Moon's north or south pole; this is known as *libration in latitude*.

In addition, the Moon's speed of motion along its elliptical orbit changes rhythmically as it approaches and recedes from the Earth, while its axial rotation remains uniform. The Moon therefore seems to rock slightly to west and east as it orbits the Earth, so that we can peer up to 7.75° or so around each limb; this is known as *libration in longitude*. The net effect of these librations is that we can see 59 percent of the Moon's surface at one time or another.

The lunar surface

The line dividing the lit and unlit portions of the Moon is known as the *terminator*. An astronaut standing on the terminator would see the Sun either rising or setting. Objects near the terminator are thrown into sharp relief by the low angle of illumination; with no atmosphere to soften the shadows, craters and mountains appear particularly rugged. As the Sun rises higher over the moonscape, details become more washed out, but at the same time the difference in contrast between the bright highlands and the dark lowland plains becomes more noticeable.

Near full Moon, many individual formations are difficult to pick out. An exception is those craters that have bright ray systems, apparently made of pulverized rock thrown out from the impact during the crater's formation; the rays become more prominent under high illumination. Notable examples of rayed craters are Copernicus, 60 miles (96 kilometers) in diameter, and Aristarchus, 25 miles (40 kilometers) across, which is the brightest area on the Moon. Most magnificent of all is Tycho, 53 miles (85 kilometers) across, with rays that stretch for over 620 miles (1,000 kilometers).

With the exception of a few small areas near the poles, each spot on the Moon is subjected to two weeks of unbroken daylight, during which surface temperatures reach more than 210°F (100°C)—higher than the boiling point of water—followed by a two-week night when temperatures plummet to –270°F (–170°C) or less. At the poles, though, some crater floors remain permanently shaded from the Sun so temperatures never rise above freezing. Here, ice deposited by cometary impacts can exist beneath the surface.

After looking at the brilliant full Moon it comes as a surprise to realize that the lunar surface rocks are actually dark gray in color; on average, the Moon's surface reflects only 12 percent of the light that hits it. If the Moon were, for instance, covered in clouds like those of Venus, it would be over five times brighter.

Seas, mountains, and craters

Formations on the Moon bear a variety of curious names. The dark lowland plains are termed *maria* (singular: *mare*), Latin for "seas," because the first observers imagined them to be stretches of water; the name persists, even though it has been clear for centuries that the Moon possesses neither air nor liquid water.

Thus we have, for instance, Oceanus Procellarum (Ocean of Storms), Mare Imbrium (Sea of Rains), and Mare Tranquillitatis (Sea of Tranquillity). Less prominent lowlands are termed bays (Sinus), marshes (Palus), or lakes (Lacus). The mare areas were formed by giant impacts and were subsequently flooded, not by water but by lava—so perhaps the old names are not so inappropriate after all.

Mountains on the Moon are named after terrestrial ranges; hence we have the lunar Alps (Montes Alpes) and the lunar Apennines (Montes Apenninus). The craters have been named after philosophers and scientists of the past, although it's true to say that the selection has been somewhat arbitrary.

The term "crater" is rather misleading as it conjures up the vision of a deep, circular bowl. A better term for the largest craters is walled plains. For example, the 90-mile (150-kilometer)-diameter Ptolemaeus, near the Moon's center, is noticeably hexagonal in outline, as is the 71-mile (115-kilometer) Purbach farther south. These distorted shapes presumably arose from stresses in the lunar crust.

Many large craters have terraced walls, caused by slumping, and central peaks, caused by rebound of the crater floor after the impact of the meteorite that formed them. Outstanding examples of craters that show both terracing and central peaks are Theophilus, Arzachel, and Copernicus.

There is a near-continuous range of sizes from the largest craters to the smallest maria. Most crater-like in appearance of the maria is Mare Crisium, a basin some 300 miles (500 kilometers) across, twice the width of the largest craters. Some craters have been flooded with dark lava, which has left them with mare-like floors; a notable example is Plato, 60 miles (100 kilometers) wide, on the northern shore of Mare Imbrium.

Near Plato is a huge half-crater called Sinus Iridum (Bay of Rainbows), 155 miles (250 kilometers) across, which bridges the gap between craters and maria. It opens into Mare Imbrium and only a few low ridges, visible under oblique sunlight, remain where its seaward wall was overrun by the advancing lava. The wall of Sinus Iridum that still stands forms an arcuate range called the Montes Jura (Jura Mountains), 2.8 miles (4.5 kilometers) high. Sunrise on the Montes Jura, 10 days after new Moon, is one of the most impressive sights for lunar observers.

Controversy over the origin of the lunar features started almost immediately after Galileo turned his first telescope toward the Moon in 1609. One school of thought held that they were caused by volcanic action; the opposition believed the craters and maria to have been blasted out by impacts of meteorites and asteroids. It was not until the late 1960s, when space probes and astronauts first reached the Moon, that the controversy was finally resolved in favor of the impact theory, although it's accepted that there has also been a certain amount of subsequent volcanic action.

Space-probe exploration

In 1964 and 1965 three American probes called Ranger 7, 8, and 9 zoomed toward the Moon, sending back a stream of photographs showing features down to 3 feet (1 meter) across, a hundred times smaller than Earth-based telescopes could see. They revealed that even the apparently smoothest parts of the Moon's surface were pitted with small craters caused by aeons of meteorite bombardment. Landing sites for the manned missions that were then being planned would have to be chosen with particular care, to prevent the landing craft from toppling into a crater or hitting a boulder.

This initial cursory examination by the Rangers was followed by a two-pronged attack: the Surveyors, a series of craft that soft-landed automatically (the Rangers had simply crashed), and the Lunar Orbiters which, as their name implies, photographed the Moon from close orbit.

Between 1966 and 1968 these two series of probes revolutionized our knowledge of the Moon, paving the way for the manned Apollo landings. The Surveyors showed that the Moon's surface rocks had been eroded by micrometeorites into a compacted topsoil, known as the *regolith*, which fortunately was firm enough to bear the weight of astronauts and their spacecraft. Lunar Orbiter photographs allowed astronomers to compile their most detailed maps of the near and far sides of the Moon.

The Moon's far side, permanently turned away from the Earth, had first been glimpsed by Luna 3, a Soviet probe, in October 1959. Although poor by modern standards, its photographs did at least reveal the main difference between the two hemispheres of the Moon: there are scarcely any mare areas on the Moon's far side. Instead, heavily cratered bright uplands dominate the scene. This asymmetry arises because the Moon's crust is some 9 miles (15 kilometers) thicker on the far side.

Large lowland basins like Mare Imbrium do exist on the far side of the Moon, but they have not been flooded by dark lava. Volcanic lavas from the Moon's interior found it easier to leak out through the thinner crust of the Earth-facing hemisphere. The most prominent dark area on the Moon's far side is not a true mare at all, but a deep crater called Tsiolkovskiy, 115 miles (185 kilometers) in diameter, three-quarters the size of Sinus Iridum on the visible hemisphere.

The Apollo missions

Once the Lunar Orbiter probes had spied out potential landing sites, the stage was set for astronauts to follow. On July 20, 1969 the Apollo 11 lunar module called *Eagle* carried Neil Armstrong and Edwin "Buzz" Aldrin to the first manned lunar landing, in the southwestern Sea of Tranquillity. The two astronauts spent two hours exploring the Moon's surface, setting up experiments and collecting samples for geologists to study. For the first time, humans had touched another world in space.

By the time Apollo 17 concluded the series of manned landings in December 1972, astronauts had brought back to Earth over 840 pounds (380 kilograms) of Moon samples, most of which remains stored at NASA's Johnson Space Center in Houston, Texas. When divided into the overall cost of Apollo, 2 pounds (1 kilogram) of Moon rocks is worth $100 million. In addition to the Apollo samples, three Soviet automatic lunar probes have returned with a few hundred grams of Moon soil.

What have we learned from these precious specimens? The most astounding fact about the Moon rocks is their immense age. The Apollo 11 samples, for instance, proved to have an age of 3,700 million (3.7 billion) years, older than virtually any rocks on Earth—yet the site from which they came, Mare Tranquillitatis, is one of the youngest areas on the Moon. The youngest of all the rocks brought back came from the Apollo 12 site in Oceanus Procellarum; they are 3,200 million (3.2 billion) years old.

As expected, the lunar maria turned out to be covered with lava flows similar in composition to volcanic basalt on Earth. They do not resemble a rugged lava field on Earth because, over the billions of years since the lavas were laid down, the sandblasting effect of micrometeorites has eroded the surface rocks to form a layer of soil, the regolith, several feet deep.

By contrast, the highlands, sampled by later Apollo missions, consist of a paler rock called anorthosite, rare on Earth. Rocks from the highlands proved older than those from the maria, mostly dating back 4 billion years or more. Their jumbled, fragmented nature bears witness to the violent bombardment from meteorites that the Moon suffered early in its history.

Origin and evolution of the Moon

At the time of Apollo there were three main theories of the Moon's origin: that it split from our planet shortly after its formation, that it was once a separate body that was captured by the Earth's gravity, or that the Earth and Moon formed side by side, much as they are now. But all had drawbacks.

Following the Apollo landings, a fourth proposal emerged, usually termed the Giant Impact theory, which combined aspects of the others. According to this now widely accepted view, a stray body the size of Mars struck the young Earth a glancing blow, spraying debris into orbit around the Earth where it coalesced into the Moon.

Precise dating of lunar samples suggests that this collision happened about 4.5 billion years ago, some 100 million years after the Earth formed. By then, iron had sunk to the Earth's center, producing a core, so that the material thrown off the outer layers in the collision was predominantly rocky.

While many details of the Moon's origin remain obscure, we now have a much clearer picture of its subsequent history. Heat released by its rapid accumulation in orbit around the Earth (a process termed *accretion*) melted its outer layers. A scum of less-dense rock formed a primitive crust that was then battered for several hundred million years by the infall of other debris left over from the Solar System's formation. Similar scars from this mopping-up operation can also be seen on other rocky bodies in the Solar System, notably lunar-like Mercury.

This heavy bombardment created the Moon's jumbled highlands and carved out the mare basins, in the process removing much of the crust on one hemisphere—the side that ended up facing the Earth when the Moon's rotation became tidally locked.

About 4 billion years ago the storm of meteoric debris abated. Then, slowly, molten lava began to seep out from inside the Moon, solidifying to form the dark lowland maria. So-called *wrinkle ridges* of solidified lava, hundreds of miles long but only a few hundred feet high, are visible through binoculars and small telescopes when the maria are under low illumination. In particular, look out for the Serpentine Ridge in eastern Mare Serenitatis when the Moon is five to six days old, i.e. five to six days past full. Mare Tranquillitatis and Mare Imbrium also have prominent wrinkle ridges. In fact, one "crater" on Mare Tranquillitatis, called Lamont, consists of nothing more than low ridges of solidified lava.

In places, particularly in western Oceanus Procellarum, a number of blister-like domes can be found near the crater Marius, produced by the upwelling of molten lava. The largest lunar dome, or rather a complex of domes, is Rümker, 44 miles (70 kilometers) wide, farther north on Oceanus Procellarum. Volcanic cones like Vesuvius are absent on the Moon, evidently because the lunar lavas were too runny to build up into mountains.

Further evidence of geological activity is provided by valleys of various kinds. One obvious example, Vallis Alpes (the Alpine Valley), slices through the lunar Alps near Plato. The Alpine Valley seems to be a result of crustal faults, as do some narrower grooves known as *rilles*. The crater Hyginus near the Moon's center lies in the middle of a long, rimless cleft along which smaller craters have been formed by subsidence. Some lunar faults have produced shallow cliffs, as in the case of the Rupes Recta (popularly termed the Straight Wall), which stretches for 75 miles (120 kilometers) down the eastern edge of Mare Nubium.

Another type of rille, called a *sinuous rille*, snakes over mare surfaces like a meandering river. Apollo 15 landed at the edge of one of these sinuous rilles, called Hadley Rille. Such winding rilles were not formed by running water but are probably collapsed tunnels through which lava streams once flowed.

By 1 billion years ago the volcanic outpourings had ceased, leaving the Moon cold and dead. Since then it has remained virtually unchanged, save for the arrival of an occasional meteorite to punch a new crater in its surface. For instance, the rayed crater Copernicus was formed roughly 1 billion years ago, while Tycho was blasted out about 100 million years ago.

Opposite The lunar cycle goes through a series of phases: from New Moon to Waxing Crescent, First Quarter, Waxing Gibbous, Full Moon, Waning Gibbous, Last Quarter, and Waning Crescent. This image, provided from the Lick Observatory, shows the Moon in its Waxing Gibbous phase (as seen from the Northern Hemisphere), 10 days after new. Dark lowland plains (the maria), relatively devoid of craters, contrast with bright, rugged highlands. The large mare toward the top is Mare Imbrium, the horseshoe-shaped bay Sinus Iridum on its shore. South of Mare Imbrium is the prominent rayed crater Copernicus. Most of the features in this photograph can be seen with a small telescope or mounted binoculars.

Moon map 1 (Northwest section)

Aristarchus Brilliant young crater 25 miles (40 kilometers) in diameter with multiple terracing on its inner walls. The brightest area on the Moon, and the center of a major ray system. Dark bands are visible on its inner walls under high illumination. Aristarchus has been the location of numerous red glows known as transient lunar phenomena (TLPs), possibly caused by outgassing from the surface. Mountains to the north exhibit fantastic sculpturing.

Copernicus Diameter 60 miles (96 kilometers). One of the most magnificent craters on the Moon. Center of a major ray system. Terraced walls and numerous hummocky central peaks. Rays from Copernicus are splashed for more than 370 miles (600 kilometers) across the Mare Imbrium and Oceanus Procellarum.

Encke Low-walled crater 17 miles (28 kilometers) in diameter covered by rays from nearby Kepler. Brilliant under high illumination.

Euler Small, sharp crater, 16 miles (26 kilometers) across, in southwestern Mare Imbrium. Center of a minor ray system.

Fauth and Fauth A Keyhole-shaped double crater south of Copernicus, 11 miles (18 kilometers) long and nearly 1.2 miles (2 kilometers) deep. Striking under low illumination.

Harpalus Crater 25 miles (40 kilometers) in diameter on Mare Frigoris, bright under high illumination. The smaller crater Foucault, 15 miles (25 kilometers) in diameter, lies between it and Sinus Iridum.

Herodotus Companion to Aristarchus, similar in size (diameter 22 miles [36 kilometers]) but different in structure—Herodotus has a dark, lava-flooded floor and it is not a ray center. The W-shaped Vallis Schröteri (Schröter's Valley), 115 miles (185 kilometers) long, starts at a craterlet outside the northern wall of Herodotus.

Hevelius Large, bright crater 71 miles (114 kilometers) in diameter on western shore of Oceanus Procellarum, with clefts crossing the floor. The smaller, sharper crater Cavalerius (diameter 37 miles [59 kilometers]) adjoins to the north.

Kepler Major ray center in Oceanus Procellarum. Brilliant crater, 19 miles (30 kilometers) in diameter, with central peak and heavily terraced walls.

Mairan Prominent crater in highlands west of Sinus Iridum. Diameter 25 miles (40 kilometers).

Marius Dark, flat-floored ring 25 miles (40 kilometers) in diameter on Oceanus Procellarum, notable because it lies on a wrinkle ridge. There are many dome-like structures in this area where lava has bubbled up through the surface.

Oceanus Procellarum Vast dark plain with no clear-cut borders extending from Mare Imbrium southwards to Mare Humorum. Oceanus Procellarum has a maximum width of about 1,550 miles (2,500 kilometers) and occupies an area of more than 0.8 square miles (2 million square kilometers). It is dotted with numerous craters and bright rays.

Prinz U-shaped formation 29 miles (46 kilometers) in diameter, remains of a half-destroyed crater flooded by lava from Oceanus Procellarum. Notable because of sinuous rilles to the north.

Pythagoras Magnificent large crater 90 miles (145 kilometers) in diameter with terraced walls rising nearly 3 miles (5 kilometers) and prominent central peak, near the northwestern limb of the Moon.

Reiner Sharp crater 19 miles (30 kilometers) in diameter west of Kepler in Oceanus Procellarum. Particularly notable is a tadpole-shaped splash of brighter material on the dark plain to the west and north.

Reinhold Prominent crater 27 miles (43 kilometers) in diameter southwest of Copernicus, with terraced walls. A smaller, lower ring to the northeast is Reinhold B.

Rümker Remarkable formation on northwestern Oceanus Procellarum, visible only under low illumination. Rümker is an irregular, lumpy dome 45 miles (73 kilometers) wide.

Sinus Iridum Beautiful large (diameter 155 miles [250 kilometers]) bay on the Mare Imbrium. Its seaward wall has been broken down by invading lava and is reduced to a few low wrinkle ridges. Its remaining walls, known as Montes Jura (the Jura Mountains), are brilliantly illuminated by the morning Sun. The deep, 24-mile (38-kilometer)-diameter crater on the northern rim is Bianchini.

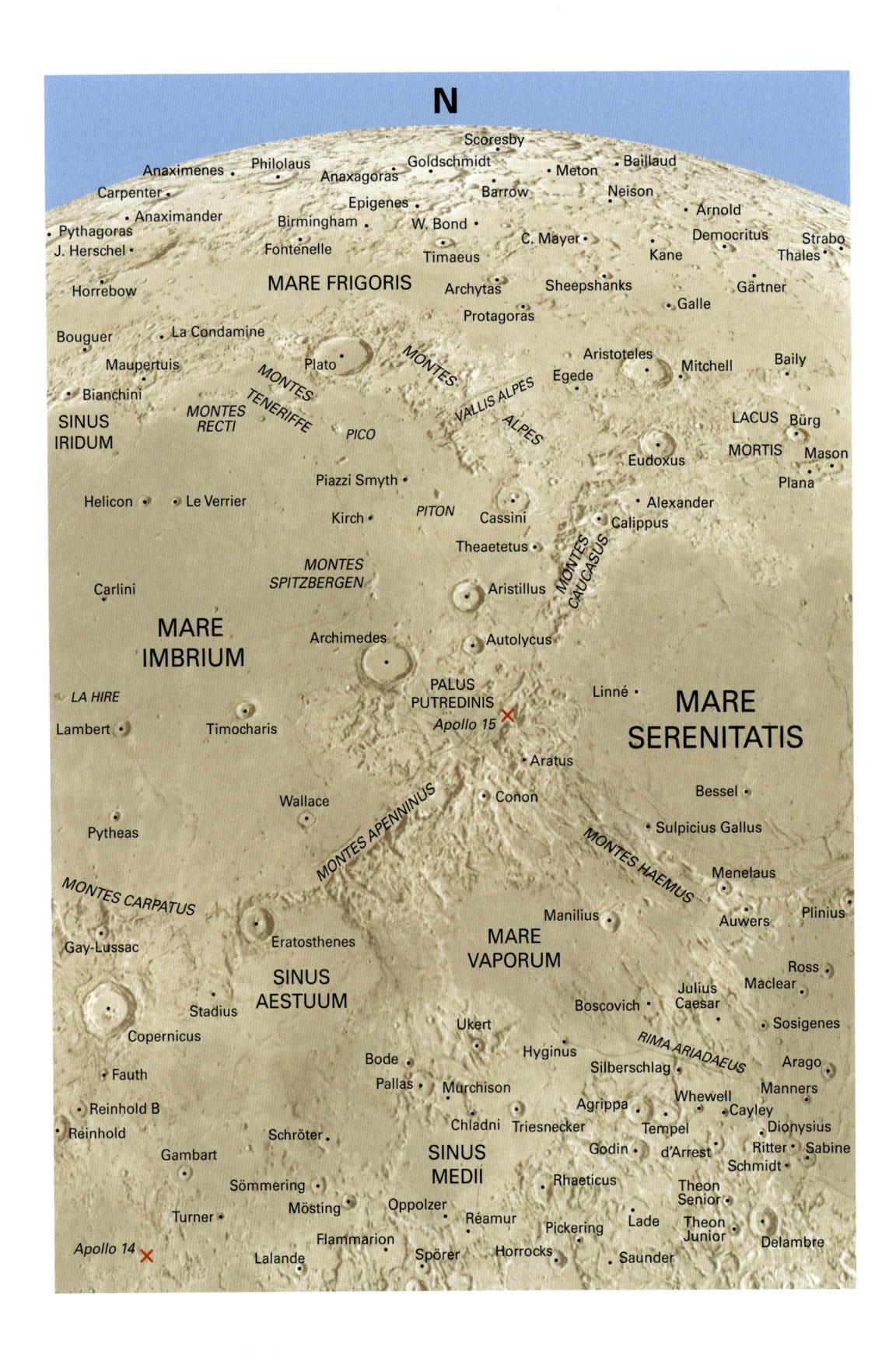

Moon map 2 (North Central section)

Agrippa Oval-shaped crater 27 miles (44 kilometers) in diameter with central peak, forming a neat pair with Godin.

Anaxagoras Crater 32 miles (52 kilometers) in diameter near the lunar north pole, center of an extensive ray system.

Archimedes Distinctive, flooded ring 50 miles (81 kilometers) in diameter in eastern Mare Imbrium, notable for its almost perfectly flat floor. Apollo 15 landed southeast of here, at the foot of the Montes Apenninus, in 1971.

Aristillus Prominent crater 34 miles (54 kilometers) in diameter in eastern Mare Imbrium with terraced walls, numerous surrounding ridges, and a complex central peak 2,950 feet (900 meters) high. Under high illumination it's seen to be surrounded by a faint ray system. Forms a pair with Autolycus to the south.

Aristoteles Magnificent, partially lava-flooded crater 55 miles (88 kilometers) in diameter, touching the smaller crater Mitchell to the east. Numerous ridges radiate from its outer walls. Makes a pair with Eudoxus to the south.

Autolycus Prominent crater 24 miles (39 kilometers) in diameter, south of Aristillus. It is the center of a faint ray system, seen under high illumination.

Cassini Flooded ring of unusual appearance with partially destroyed walls, 35 miles (57 kilometers) in diameter. Contains a bowl-shaped crater, Cassini A, 11 miles (17 kilometers) wide.

Eratosthenes Prominent, deep crater on the edge of Sinus Aestuum at the southern end of Montes Apenninus (the Apennine Mountains). Terraced walls and craterlet on central peak. Its diameter is 37 miles (59 kilometers) .

Eudoxus Rugged crater 43 miles (70 kilometers) in diameter with small central peak and terraced walls. South of Aristoteles.

Godin Smaller but deeper companion to Agrippa; 21 miles (34 kilometers) in diameter with central peak, bright walls, and faint ray system.

Hyginus Rimless crater 5.5 miles (9 kilometers) in diameter, center of a cleft or rille 124 miles (200 kilometers) long, visible in small telescopes, evidently formed by collapse of the surface. To the east is another cleft, Rima Ariadaeus.

Lambert Crater 19 miles (30 kilometers) in diameter in Mare Imbrium, with central craterlet. Situated on a wrinkle ridge. Under low illumination a larger "ghost" ring, Lambert R, is seen to the south.

Linné Bright spot on Mare Serenitatis, best seen under high illumination. At its center is a small, young crater 1.4 miles (2.2 kilometers) in diameter.

Manilius Bright crater 24 miles (38 kilometers) in diameter in Mare Vaporum, with terraced walls and central peak. Develops a ray system as illumination increases, as does its neighbor Menelaus (diameter 17 miles [27 kilometers]) on the rim of Mare Serenitatis.

Mare Imbrium Enormous circular plain 715 miles (1,150 kilometers) in diameter, dominating this section. Bounded by the Montes Alpes, Caucasus, Apenninus, and Carpatus, but open on the southwest to Oceanus Procellarum. Mare Imbrium has a double structure: traces of a smaller inner ring are visible, marked by a few isolated mountains and wrinkle ridges. Note the different shades of lava on its surface. Isolated mountains protruding from its dark floor include Pico and Piton, plus the ranges known as Montes Recti, Spitzbergen, and Teneriffe.

Mare Serenitatis A major rounded lunar sea 416 × 373 miles (670 × 600 kilometers), bounded on the northwest by Montes Caucasus and on the southwest by Montes Haemus. A bright ray from Tycho crosses the dark lava plain, passing through the crater Bessel, 10 miles (16 kilometers) in diameter. Under high illumination, Mare Serenitatis is seen to be rimmed with darker lava. Note also a major wrinkle ridge (the Serpentine Ridge) on the east side. Apollo 17 landed at the southeastern edge of Mare Serenitatis in 1972.

Plato Unmistakable large, dark-floored crater in the highlands north of Mare Imbrium, 63 miles (101 kilometers) in diameter; prominent under all conditions of illumination. Tiny craterlets pockmark its flat floor. Temporary obscurations of the surface, presumed to be caused by outgassing, have been observed in this area. Landslips appear to have detached part of the inner western wall.

Pytheas Small (diameter 12 miles [19 kilometers]) but deep and prominent crater on Mare Imbrium, rhomboidal in outline. Becomes brilliant under high illumination.

Stadius "Ghost" ring to the east of Copernicus, outlined only by a few ridges and craterlets and visible only under low illumination. It is 42 miles (68 kilometers) in diameter.

Timocharis Bright crater 21 miles (34 kilometers) in diameter on Mare Imbrium, with terraced walls and distinctive central crater. Faint ray center.

Triesnecker Crater 15 miles (25 kilometers) in diameter with a surrounding system of clefts.

Vallis Alpes (Alpine Valley) Flat-floored valley 96 miles (155 kilometers) long through the lunar Alps, connecting Mare Imbrium with Mare Frigoris.

Moon Map (Eastern Hemisphere)

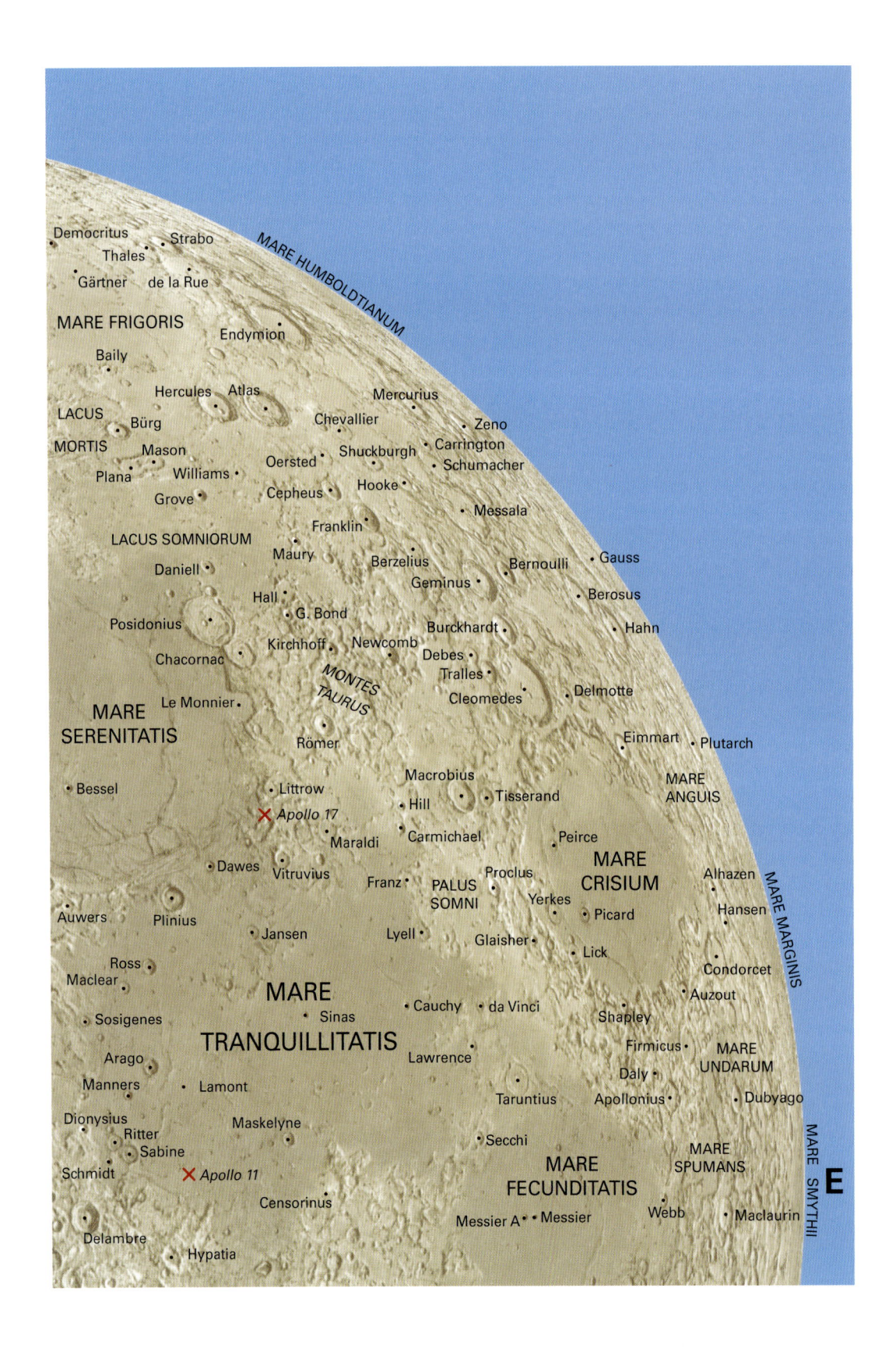

- Democritus
- Strabo
- Thales
- Gärtner
- de la Rue

MARE HUMBOLDTIANUM

MARE FRIGORIS
- Endymion
- Baily
- Hercules
- Atlas
- Mercurius
- Zeno

LACUS MORTIS
- Bürg
- Chevallier
- Mason
- Shuckburgh
- Carrington
- Plana
- Oersted
- Schumacher
- Williams
- Cepheus
- Hooke
- Grove
- Messala

LACUS SOMNIORUM
- Franklin
- Maury
- Berzelius
- Bernoulli
- Gauss
- Daniell
- Geminus
- Berosus
- Hall
- G. Bond
- Burckhardt
- Hahn
- Posidonius
- Kirchhoff
- Newcomb
- Chacornac
- Debes
- Tralles
- Delmotte
- Le Monnier
- Cleomedes

MONTES TAURUS

MARE SERENITATIS
- Römer
- Eimmart
- Plutarch
- Bessel
- Macrobius
- Littrow
- Hill
- Tisserand
- **MARE ANGUIS**
- × Apollo 17
- Maraldi
- Carmichael
- Peirce
- Dawes
- Vitruvius
- Franz
- Proclus
- **MARE CRISIUM**
- Alhazen
- Auwers
- Plinius
- **PALUS SOMNI**
- Yerkes
- Hansen
- Jansen
- Lyell
- Glaisher
- Picard
- Ross
- Lick
- Maclear
- Condorcet
- Sosigenes
- Cauchy
- da Vinci
- Auzout
- Arago
- **MARE TRANQUILLITATIS**
- Sinas
- Shapley
- Manners
- Lawrence
- Firmicus
- **MARE UNDARUM**
- Lamont
- Daly
- Dionysius
- Taruntius
- Apollonius
- Dubyago
- Ritter
- Maskelyne
- Sabine
- Schmidt
- × Apollo 11
- Secchi
- **MARE SPUMANS**
- **MARE FECUNDITATIS**
- **MARE SMYTHII**
- Delambre
- Censorinus
- Messier A • Messier
- Webb
- Maclaurin
- Hypatia

MARE MARGINIS

E

Moon map 3 (Northeast section)

Atlas Large crater, 55 miles (88 kilometers) in diameter, with terraced walls and complex floor. A ruined ring, Atlas E, abuts to the northwest. Atlas and Hercules form one of the many crater pairs in this region.

Burckhardt Complex crater 33 miles (54 kilometers) across, overlapping older formations (Burckhardt E and F) on either side.

Bürg Prominent crater despite its moderate size (diameter 25 miles [41 kilometers]). Central peak. Lies in the center of Lacus Mortis. Note nearby rilles.

Cleomedes Large, irregular-shaped crater 81 miles (131 kilometers) in diameter with partially flooded floor, to north of Mare Crisium. West wall is interrupted by 27-mile (44-kilometers)-diameter crater Tralles.

Endymion Large, dark-floored enclosure 76 miles (122 kilometers) in diameter, with walls up to 16,000 feet (4,900 meters) high.

Franklin Crater 35 miles (56 kilometers) in diameter. Forms a pair with the 24-mile (39-kilometer)-diameter Cepheus to the northwest.

Geminus Prominent crater 51 miles (82 kilometers) in diameter with central peak. Bright rays emanate from two smaller craters nearby, Messala B and Geminus C.

Hercules Flat-floored enclosure 42 miles (68 kilometers) in diameter containing the sharp, bright bowl crater Hercules G, 9 miles (14 kilometers) wide.

Le Monnier Old and flooded crater with its western wall washed away by lava from Mare Serenitatis. Diameter 42 miles (68 kilometers).

Mare Crisium Unmistakable dark lowland plain ringed by high mountains like an oversized crater, 260 × 342 miles (420 × 550 kilometers). Main feature on its flat, lava-flooded floor is the crater Picard, 14 miles (22 kilometers) across, with smaller crater Peirce to its north.

Mare Tranquillitatis An irregularly shaped lowland, 336 × 544 miles (540 × 875 kilometers). Wrinkle ridges attest to numerous lava flows over the plain. The main crater on the mare is the distorted Arago, on the western side, 16 miles (26 kilometers) in diameter. Apollo 11 made the first manned lunar landing in southwest Mare Tranquillitatis in 1969.

Plinius Distinctive crater 25 miles (41 kilometers) in diameter with complex central peak combined with a central craterlet. Stands between Mare Serenitatis and Mare Tranquillitatis, along with Dawes to the northeast, 11 miles (18 kilometers) in diameter.

Posidonius Large (diameter 59 miles [95 kilometers]) partially flooded and ruined crater on the northeast shore of Mare Serenitatis. Its floor contains a curving ridge, several rilles, and a small bowl crater. The ruined structure Chacornac, 31 miles (50 kilometers) in diameter, abuts to the southeast.

Proclus Small (17 miles [27 kilometers] wide) but brilliant crater on the western edge of Mare Crisium. High illumination shows it to be the center of a fan-shaped ray system.

Taruntius Low-walled crater in northwestern Mare Fecunditatis, 35 miles (57 kilometers) in diameter, with concentric inner ring. Crater Cameron, 7 miles (11 kilometers) across, interrupts the northwestern wall. Center of a faint ray system.

Thales Bright ray crater 19 miles (31 kilometers) in diameter, northeast of Mare Frigoris.

Moon Map — Southwest Quadrant

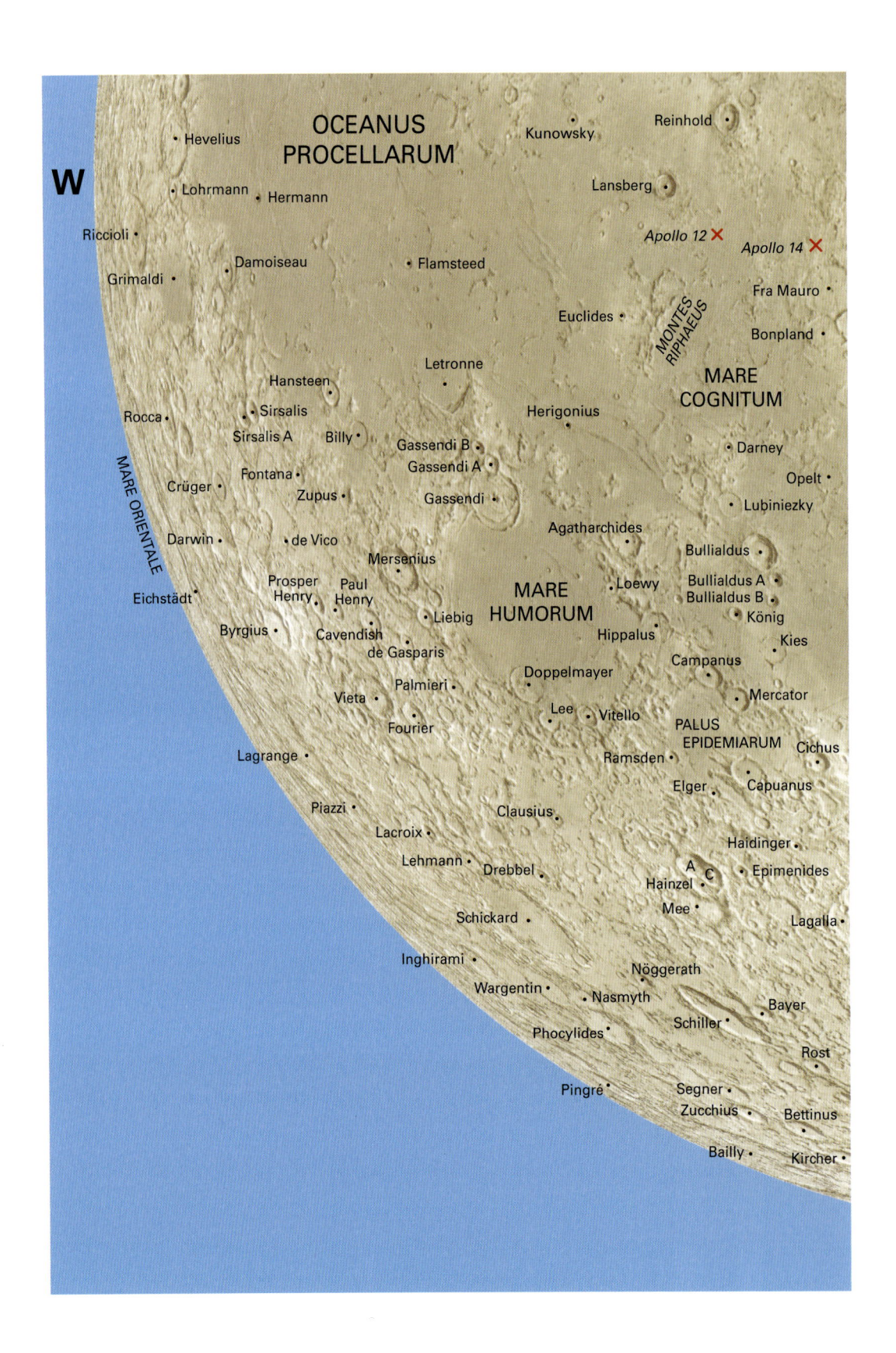

W

OCEANUS PROCELLARUM

- Hevelius
- Kunowsky
- Reinhold
- Lohrmann
- Hermann
- Lansberg
- Riccioli
- Apollo 12 ✕
- Apollo 14 ✕
- Grimaldi
- Damoiseau
- Flamsteed
- Fra Mauro
- Bonpland
- Euclides
- MONTES RIPHAEUS
- **MARE COGNITUM**
- Letronne
- Hansteen
- Rocca
- Sirsalis
- Herigonius
- Sirsalis A
- Billy
- Gassendi B
- Gassendi A
- Darney
- Opelt
- Crüger
- Fontana
- Zupus
- Gassendi
- Lubiniezky
- **MARE ORIENTALE**
- Darwin
- de Vico
- Agatharchides
- Bullialdus
- Mersenius
- Loewy
- Bullialdus A
- Bullialdus B
- Eichstädt
- Prosper Henry
- Paul Henry
- Liebig
- **MARE HUMORUM**
- König
- Byrgius
- Cavendish
- de Gasparis
- Hippalus
- Kies
- Vieta
- Palmieri
- Doppelmayer
- Campanus
- Mercator
- Lee
- Vitello
- **PALUS EPIDEMIARUM**
- Fourier
- Ramsden
- Cichus
- Lagrange
- Elger
- Capuanus
- Piazzi
- Clausius
- Haidinger
- Lacroix
- A C
- Epimenides
- Lehmann
- Drebbel
- Hainzel
- Mee
- Lagalla
- Schickard
- Inghirami
- Nöggerath
- Wargentin
- Nasmyth
- Bayer
- Phocylides
- Schiller
- Rost
- Pingré
- Segner
- Zucchius
- Bettinus
- Bailly
- Kircher

230 THE BACKYARD STARGAZER'S BIBLE

Moon map 4 (Southwest section)

Bullialdus Handsome crater in Mare Nubium with terraced walls and complex central peak. Diameter 38 miles (61 kilometers). Two smaller craters, Bullialdus A (15 miles [25 kilometers]) and Bullialdus B (13 miles [21 kilometers]), form a chain extending south.

Flamsteed Modest (diameter 12 miles [19 kilometers]) crater on Oceanus Procellarum with a much larger ring of eroded hills to its north, Flamsteed P, which appears bright against the mare background at full Moon.

Gassendi Large, 69-mile (111-kilometer)-diameter, partially flooded ring on northern border of Mare Humorum. Complex internal pattern of clefts, ridges, and hillocks. Deeper crater Gassendi A (20 miles [32 kilometers] wide) interrupts the north wall, with the smaller Gassendi B (15 miles [25 kilometers]) lying farther north.

Grimaldi Vast, 146-mile (235-kilometer)-diameter, dark-floored enclosure at the western limb of the Moon with broad, crater-strewn walls. Nearer the limb is a smaller dark patch, marking the floor of Riccioli, 97 miles (156 kilometers) in diameter.

Hainzel Curious keyhole-shaped formation, composed of three craters fused together; the two smaller ones are Hainzel A and C.

Hippalus Lava-flooded bay 35 miles (57 kilometers) in diameter on the shores of Mare Humorum, in an area with many rilles.

Lansberg Prominent crater 24 miles (39 kilometers) in diameter with massive walls and a central peak, on Oceanus Procellarum. Apollo 12 landed southeast of Lansberg in 1969.

Letronne Large bay, 73 miles (118 kilometers) wide, on the south side of Oceanus Procellarum. Its seaward-facing wall has evidently been washed away by invading dark lava.

Mare Humorum Rounded lowland plain 260 miles (420 kilometers) across with Gassendi on its northern rim. It is ringed by clefts and wrinkle ridges. On the south it invades the rings Doppelmayer and Lee, although Vitello escapes destruction. On the east is the ringed bay Hippalus, associated with much surface faulting.

Schickard Major dark-floored enclosure 132 miles (212 kilometers) across. South of Schickard are the overlapping craters Nasmyth (diameter 48 miles [78 kilometers]) and Phocylides (71 miles [115 kilometers]). Adjoining it to the southwest is the extraordinary plateau Wargentin, 53 miles (85 kilometers) wide, evidently a crater filled to the brim with solidified lava.

Schiller Curious footprint-shaped enclosure, 111 miles (179 kilometers) long by 40 miles (65 kilometers) wide.

Sirsalis and Sirsalis A Twin craters 27 and 30 miles (44 and 48 kilometers) wide near a 249-mile (400-kilometer)-long cleft, Rima Sirsalis, which stretches south towards 76-mile (122-kilometer) Darwin.

Moon map (southern portion) with labeled craters and features:

- **SINUS MEDII** region (top): Reinhold, Gambart, Schröter, Chladni, Triesnecker, Agrippa, Tempel, Cayley, Whewell, Dionysius, Sömmering, Godin, d'Arrest, Ritter, Schmidt, Sabine, Turner, Mösting, Oppolzer, Rhaeticus, Théon Senior, Réaumur, Pickering, Lade, Théon Junior, Delambre, Apollo 14 ✗, Flammarion, Spörer, Horrocks, Saunder, Taylor, Fra Mauro, Lalande, Herschel, Gyldén, Hipparchus, Alfraganus, Bonpland, Parry, Müller, Hind, Apollo 16 ✗, Zöllner, Palisa, Ptolemaeus, Halley, Anděl, Dollond, Kant
- **MARE COGNITUM** / **MARE NUBIUM**: Guericke, Davy, Albategnius, Ritchey, Descartes, Cyrillus, Alphonsus, Klein, Burnham, Abulfeda, Darney, Lassell, Parrot, Vogel, Tacitus, Opelt, Alpetragius, Airy, Argelander, Almanon, Catharina, Lubiniezky, Arzachel, Geber, Bulliardus, Donati, Bulliardus A, Thebit, Faye, Abenezra, Fermat, Bulliardus B, Nicollet, Birt, Delaunay, Azophi, Sacrobosco, König, La Caille, Playfair, Kies, Blanchinus, Pons, Purbach, Apianus, Pontanus, Mercator, Hesiodus, Regiomontanus, Werner, Wilkins, Pitatus, Poisson, Goodacre, Zagut, Lindenau, Weiss, Aliacensis, Gemma Frisius, Celsius, Rabbi Levi, Cichus, Hell, Walther, Capuanus, Gauricus, Deslandres, Nonius, Kaiser, Riccius, Wurzelbauer, Lexell, Fernelius, Buch, Büsching, Ball, Miller, Stöfler, Haidinger, Heinsius, Sasserides, Orontius, Huggins, Nasireddin, Faraday, Maurolycus, Nicolai, Epimenides, Wilhelm, Tycho, Pictet, Saussure, Barocius, Spallanzani, Lagalla, Brown, Proctor, Licetus, Clairaut, Breislak, Dove, Montanari, Street, Heraclitus, Ideler, Pitiscus, Longomontanus, Maginus, Cuvier, Baco, Hommel, Vlacq, Bayer, Porter, Lilius, Asclepi, Schiller, Deluc, Jacobi, Tannerus, Rosenberger, Rost, Clavius, Zach, Kinau, Nearch, Hagecius, Segner, Bettinus, Scheiner, Rutherfurd, Cysatus, Pentland, Mutus, Helmholtz, Zucchius, Kircher, Blancanus, Grüemberger, Curtius, Manzinus, Boussingault, Bailly, Wilson, Klaproth, Moretus, Simpelius, Boguslawsky, Casatus, Newton, Short, Schomberger
- **RUPES RECTA**, **RUPES ALTAI**

S

232 THE BACKYARD STARGAZER'S BIBLE

Moon map 5 (South Central section)

Abulfeda Prominent smooth-floored crater with sculptured inner walls, 39 miles (62 kilometers) in diameter. Apollo 16 landed in the highlands north of here in 1972.

Albategnius Large (diameter 81 miles [131 kilometers]) walled enclosure with central peak. The prominent crater Klein, 27 miles (43 kilometers) in diameter, breaks its southwest rim.

Aliacensis Prominent crater with irregular outline, 50 miles (80 kilometers) in diameter. Forms a pair with the slightly smaller crater Werner.

Alpetragius Bowl-shaped crater 12,795 feet (3,900 meters) deep on outer slopes of Alphonsus, with large central dome. Diameter 25 miles (40 kilometers).

Alphonsus Large enclosure 69 miles (111 kilometers) in diameter with complex walls and a ridge running through its center. Numerous craterlets and clefts cover the floor; several dark patches are visible under high illumination. Alphonsus has been the site of reported obscurations, possibly caused by the release of gas from the surface.

Arzachel Magnificent crater 60 miles (97 kilometers) in diameter with terraced walls and prominent central peak. To its east is a noticeable crater with a central peak, like a smaller version of Alpetragius, called Parrot C, 19 miles (30 kilometers) in diameter.

Barocius Large formation 52 miles (83 kilometers) in diameter southeast of Maurolycus. Its northeast wall is broken by Barocius B, diameter 23 miles (37 kilometers). To the southwest is Clairaut, 48 miles (77 kilometers) in diameter, between Barocius and Cuvier.

Birt Sharp, bright crater 10 miles (16 kilometers) in diameter on eastern Mare Nubium. Telescopes reveal a smaller crater (diameter 4 miles [7 kilometer]), Birt A, on the east wall, and under low illumination a rille to the west.

Blancanus Crater 66 miles (106 kilometers) in diameter, south of Clavius.

Clavius Magnificent walled plain 144 miles (231 kilometers) in diameter. Note the distinctive arc of smaller craters across its convex floor. Its south wall is interrupted by the 31-mile (50-kilometer) crater Rutherfurd, and its northeast wall by 32-mile (51-kilometer) Porter.

Delambre Prominent crater 32 miles (51 kilometers) in diameter with irregular interior, southwest of Mare Tranquillitatis.

Deslandres Huge, low and eroded formation 141 miles (227 kilometers) in diameter, southeast of Mare Nubium. The partially ruined ring Lexell, diameter 40 miles (64 kilometers), opens onto its southern side, and the prominent crater Hell (diameter 21 miles [33 kilometers]) lies on its western floor.

Fra Mauro Largest member, 60 miles (97 kilometers) wide, of an old, eroded crater group north of Mare Nubium, also including Bonpland (diameter 37 miles [59 kilometers]), Parry (29 miles [47 kilometers]), and Guericke (38 miles [61 kilometers]). Apollo 14 landed just north of Fra Mauro in 1971.

Heraclitus Curious elongated formation 53 miles (86 kilometers) long with central ridge, south of Stöfler. Its southern end is rounded off by the crater Heraclitus D. Between Heraclitus and Stöfler is the crater Licetus, diameter 47 miles (75 kilometers). Touching Heraclitus to the east is Cuvier, 48 miles (77 kilometers) in diameter.

Herschel Deep (12,795 feet [3,900 meters]) crater with elongated central peak, north of Ptolemaeus; diameter 24 miles (39 kilometers). Farther north is the slightly smaller Spörer, 16 miles (26 kilometers), a partially filled ring.

Hipparchus Large, eroded enclosure 89 miles (144 kilometers) in diameter north of Albategnius. Its central "peak" is actually a small ruined crater. On its northeastern floor is Horrocks, 19 miles (30 kilometers) in diameter and 9,185 feet (2,800 meters) deep. Between Hipparchus and Albategnius lies the crater Halley (diameter 22 miles [35 kilometers]), east of which is Hind, 18 miles (29 kilometers) wide and 9,185 feet (2,800 meters) deep.

Longomontanus Large walled plain in the rugged southern uplands of the Moon, diameter 90 miles (145 kilometers). A ridge to the east forms a crescent-shaped enclosure known as Longomontanus Z.

Maginus Major walled plain 97 miles (156 kilometers) in diameter north of Clavius. Convex floor with small peaks in center. Southwest wall is interrupted by smaller crater Maginus C, 29 miles (47 kilometers) wide.

Mare Nubium Irregular dark lowland plain, covered with numerous wrinkle ridges and ghost craters. On its southern shore is the flooded crater Pitatus, and in the southwest are the dark-floored pair Campanus and Mercator (both 29 miles [46 kilometers] in diameter). Its most celebrated feature is a 72-mile (116-kilometers)-long fault on the eastern side called the Rupes Recta, or Straight Wall, between the craters Birt and Thebit. The Rupes Recta appears to run north–south through the remains of an old flooded crater of which only the eastern half remains. At the southern end of the Rupes Recta are the Stag's Horn Mountains, apparently the remains of a flooded crater.

Maurolycus Distinctive crater, 71 miles (115 kilometers) wide, with twin central peak. Its walls rise to 16,400 feet (5,000 meters). It partially obliterates a smaller, unnamed formation to the north.

Moretus Crater with a central peak in the jumbled uplands southeast of Clavius. Diameter 71 miles (114 kilometers). Closer to the south pole are the craters Short (diameter 42 miles [68 kilometers]) and Newton (diameter 52 miles [84 kilometers]), heavily foreshortened.

Pitatus Large, dark-floored ring 63 miles (101 kilometers) across on the southern shore of Mare Nubium. Invading lava from Mare Nubium has partly destroyed the crater's walls and left only the vestige of a central peak. Note the rilles around its inner walls. A smaller, similarly flooded ring adjoining it to the east is Hesiodus, 27 miles (43 kilometers) in diameter.

Ptolemaeus Vast walled plain 96 miles (154 kilometers) in diameter, hexagonal in shape. Its ancient floor is heavily pockmarked with smaller craters, the most prominent being Ammonius, 6 miles (9 kilometers) wide.

Purbach Battered but still prominent large crater, 71 miles (115 kilometers) in diameter. Its floor contains ridges, and its north wall is interrupted by the oval crater Purbach G, 19 miles (30 kilometers) across, while its south wall intrudes into Regiomontanus.

Regiomontanus Flooded crater, 79 miles (127 kilometers) across. Its central peak has a summit craterlet. Forms a pair with Purbach, both noticeably hexagonal in outline.

Scheiner Crater 68 miles (110 kilometers) in diameter southwest of Clavius. The largest of three craterlets on its floor is Scheiner A, 7 miles (12 kilometers) wide.

Stöfler Large, flat-floored formation 81 miles (130 kilometers) wide, west of Maurolycus. Its eastern wall is destroyed by the intrusion of several craters, the largest being Faraday, diameter 43 miles (69 kilometers). The southern wall of Faraday is disturbed by Faraday C (17 miles [28 kilometers]), which itself intrudes into Stöfler P (25 miles [40 kilometers]).

Thebit Fascinating triple crater on south-eastern Mare Nubium. The main crater, 34 miles (55 kilometers) in diameter, is broken by the 12-mile (20-kilometers). Thebit A, which in turn is broken by the still-smaller Thebit L.

Tycho Magnificent crater in the Moon's southern uplands, 53 miles (85 kilometers) in diameter. Prominent at all angles of illumination, and brilliant under high lighting. Massively terraced walls rising up to 14,760 feet (4,500 meters), imposing central peak, and rough floor. Tycho is the major ray crater on the Moon. Rays from Tycho extend for 930 miles (1,500 kilometers) or more in all directions. Note the dark "collar" around Tycho under high illumination. Tycho is the youngest of the Moon's major features, some 100 million years old.

Walther Large crater 83 miles (134 kilometers) in diameter, considerably modified by landslips on the inner walls and by several interior craterlets. Appears almost square in outline.

Werner Prominent crater 44 miles (71 kilometers) in diameter with walls 13,780 feet (4,200 meters) high, notably sharper and more rounded than neighboring craters. The floor of Werner is dotted with several hills.

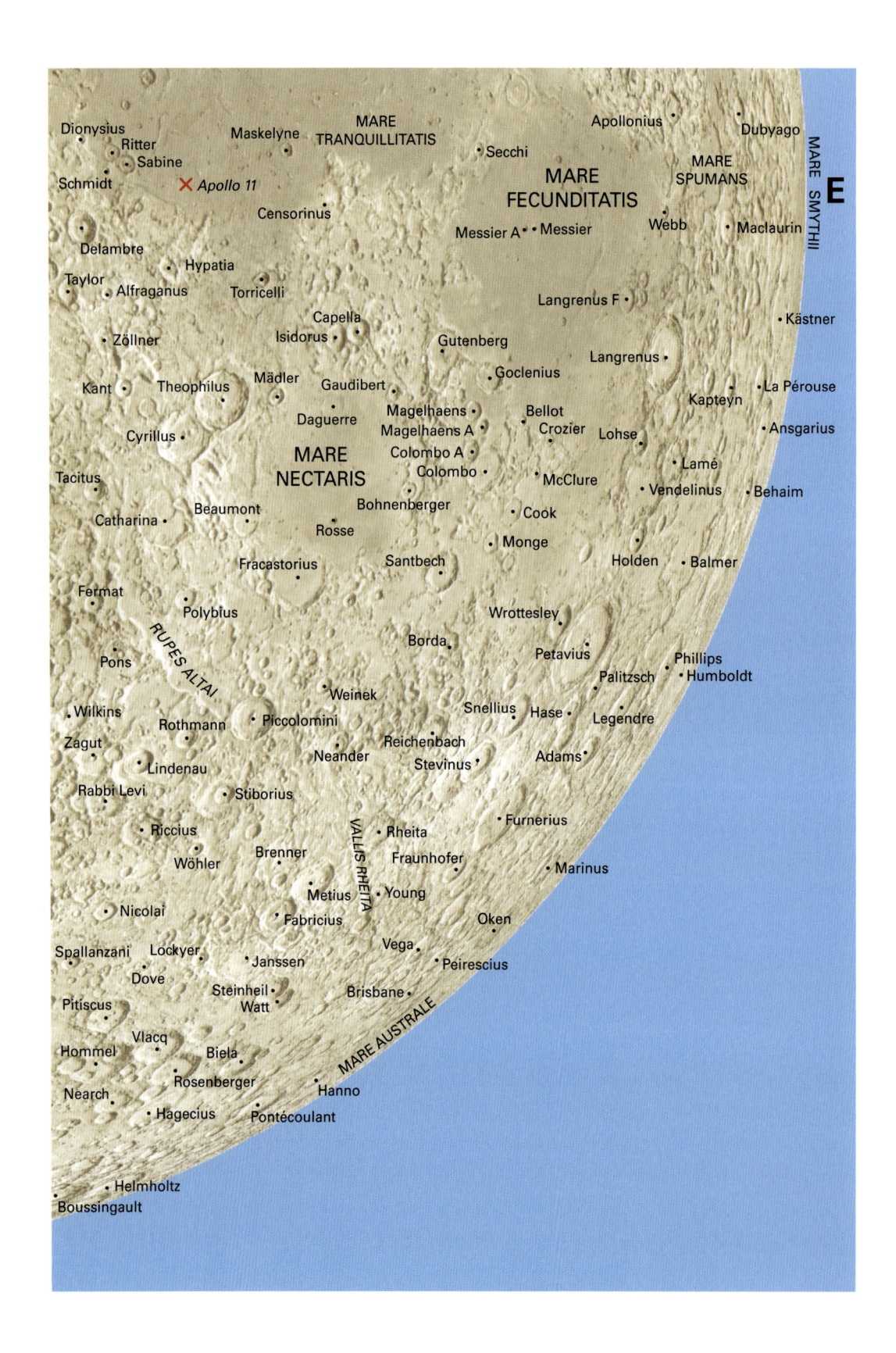

Moon map 6 (Southeast section)

Capella Prominent crater north of Mare Nectaris, 30 miles (48 kilometers) in diameter, deformed by a surface fault. Large central peak. Isidorus, 25 miles (41 kilometers) in diameter, adjoins it to the west.

Catharina One of a curving trio of craters around western Mare Nectaris. Diameter 62 miles (99 kkilometers). A faint ring, Catharina P, covers much of its northern floor.

Cyrillus Crater 61 miles (98 kilometers) in diameter, with complex terraced walls, multiple central peak, and rugged floor. Overlapped by Theophilus.

Fracastorius Horseshoe-shaped bay 75 miles (121 kilometers) in diameter on the southern shore of Mare Nectaris. Dark lava has breached its northern wall and flooded its interior. The crater Fracastorius D, 17 miles (27 kilometers) across, distorts its western wall.

Janssen Vast, irregular enclosure, 112 × 149 miles (180 × 240 kilometers) across, heavily bombarded. In the north it is interrupted by Fabricius, 49 miles (79 kilometers) wide, which has a central peak. On the west wall is a smaller (diameter 22 miles [35 kilometers]) sharp crater, Lockyer. Southeast of Janssen are the twin craters Steinheil and Watt (39 and 42 miles [63 kilometers and 67 kilometers] in diameter respectively). Farther north of Fabricius is Metius, diameter 52 miles (84 kilometers).

Langrenus Magnificent bright-walled plain on eastern Mare Fecunditatis with terraced walls, outer ridges, and a complex central peak. Its diameter is 82 miles (132 kilometers) and it is a ray center. Northwest of it, on Mare Fecunditatis, is a trio of smaller craters called Langrenus F, B, and K in order of decreasing size. South of Langrenus is the large, flooded formation Vendelinus, 88 miles (141 kilometers) across.

Mädler Prominent crater (diameter 17 miles [28 kilometers]) on northwest Mare Nectaris, with a central ridge.

Mare Fecunditatis Irregularly shaped 522 × 410 miles (840 × 660 kilometers) dark lowland area, connecting with Mare Tranquillitatis. On its western border it invades several craters, notably Gutenberg (diameter 44 miles [71 kilometers]) and Goclenius (diameter 34 × 45 miles [55 × 73 kilometers]). Note the numerous clefts in this area.

Mare Nectaris Rounded lowland plain 211 miles (340 kilometers) wide, bordered by several large craters, notably Theophilus, Cyrillus, Catharina, and Fracastorius. An outer mountain ring, Rupes Altai (the Altai Scarp), surrounds Mare Nectaris.

Messier and Messier A Elliptical pair of craters on Mare Fecunditatis, prominent despite their small sizes, 9 and 7 miles (14 and 11 kilometers). Two bright rays extend from the western member, Messier A. Both craters appear brilliant under high illumination. The pair may be the result of a glancing impact.

Palitzsch A crater and valley on the eastern side of Petavius. The crater Palitzsch itself, 26 miles (42 kilometers) wide, is at the southern end of the valley, Vallis Palitzsch, which is 68 miles (110 kilometers) long.

Petavius Magnificent walled enclosure 114 miles (184 kilometers) in diameter. A prominent rille runs across the floor from the massive, complex central peak to the terraced walls, which appear double in parts. Ridges radiate from its outer walls. West of Petavius is Wrottesley, 36 miles (58 kilometers) in diameter and with a central peak.

Piccolomini Beautiful crater on Rupes Altai (the Altai Scarp). Diameter 55 miles (88 kilometers), with a broad central peak and terraced walls.

Snellius Eroded crater 53 miles (86 kilometers) in diameter which straddles the shallow Vallis Snellius, running northwest toward Borda, 28 miles (45 kilometers) in diameter, and southeast past Adams, 39 miles (63 kilometers).

Theophilus Imposing crater 62 miles (99 kilometers) in diameter on the northwestern rim of Mare Nectaris, with a massive 7,215-feet (2,200-meter) central mountain. Terraced walls rise over 16,400 feet (5,000 meters) above the floor, with many external ridges.

Vallis Rheita (Rheita Valley) Crater chain northeast of Janssen and Fabricius. It can be traced for a total length of about 310 miles (500 kilometers). The crater Rheita itself, 44 miles (71 kilometers) in diameter and with a small central peak, lies at the northern end of the valley.

PLANETS AND DWARF PLANETS

Our Solar System is home to eight planets and five officially recognized dwarf planets, and there could be as many as 100 dwarf planets still awaiting discovery. Where did they all come from?

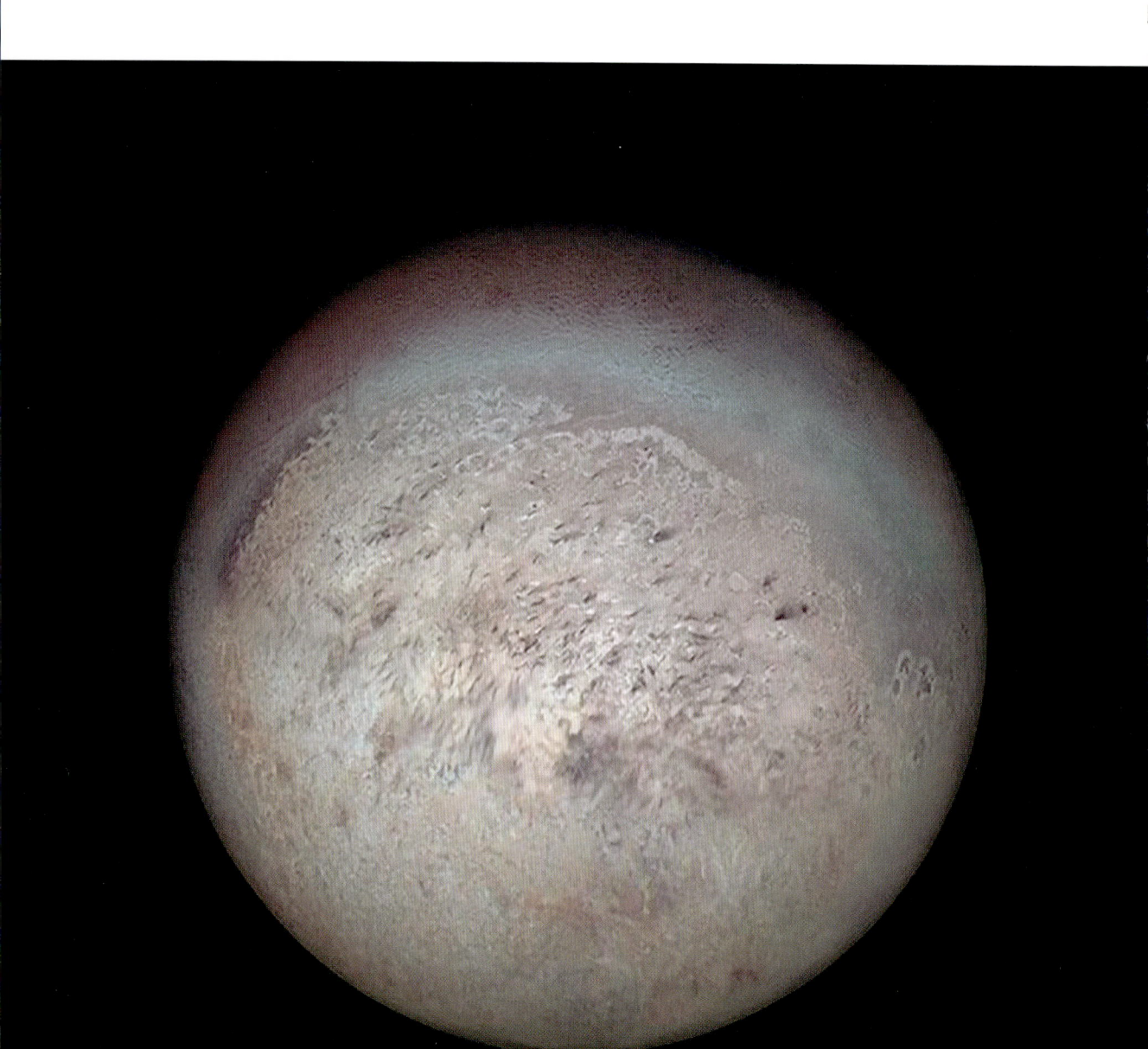

Our star, the Sun, was born 4.6 billion years ago from a massive spinning cloud of gas and dust called the solar nebula. Star formation within molecular clouds such as this is triggered by interstellar winds or shockwaves from supernovae, which cause a cloud to collapse under its own gravity. The collapse causes some of the material to form a star, and as this rotates rapidly, the remaining material spreads out into a rotating flat pancake of gas and dust called a protoplanetary disk, with the star at the center.

Over a few million years, the grains of material in the protoplanetary disk begin to clump together and these clumps then collide to form larger bodies called planetesimals—planetary precursors. These planetesimals continue to collide with each other and accrete more material until eventually they become large enough to form a series of planets in orbit around the star.

We once thought that we had a good idea of how our Solar System formed, but cutting-edge research into exoplanets (planets outside the Solar System) is giving us insights into what planetary systems in orbit around other stars look like, and most of them do not look like ours! While the exact mechanisms of how our Solar System formed are not fully understood, it's widely accepted that a combination of the composition of the molecular cloud coupled with temperature differences determined what the planets were made from and where they formed.

The inner Solar System was much hotter, due to its proximity to the newly born Sun, so volatile materials like water and methane couldn't condense there. Only materials such as metals and silicates could survive the heat, so all of the planets that formed within 4 astronomical units (AU) of the Sun were rocky planets. (1 AU is equivalent to 93 million miles [150 million kilometers]—the current distance between the Earth and the Sun.) One current model suggests that there were initially between 50 and 100 rocky bodies, or protoplanets. Over the next 100 million years, many of them collided and merged—one such collision is what formed the Earth–Moon system—and eventually we were left with the four rocky planets (Mercury, Venus, Earth, and Mars) and a band of leftover material called the asteroid belt—covered in more detail later in this chapter.

Farther than 4 AU from the Sun it was cold enough that volatile materials remained frozen, so the planets forming beyond this "frost line" were the gas giants. Jupiter and Saturn are primarily composed of hydrogen and helium, although they have denser materials in their cores. Uranus and Neptune formed even farther from the Sun, where it was colder still, so they are their own distinct class of planet: the ice giants.

Orbiting the Sun

The planets may not always have resided where they are now. Our new understanding of planetary formation suggests that they all underwent some form of migration in the years after they formed, until finally settling in their current orbits.

Because the planets formed from a flat protoplanetary disk, they all orbit in approximately the same plane and if we looked down on the Solar System from the Sun's north pole we'd see them all orbiting in an counterclockwise direction. The orbital path of each planet is not a perfect circle, but elliptical, so at some points it will pass closer to the Sun. The point where a planet is at its closest to the Sun is called *perihelion*, and when it's at its farthest from the Sun it's called *aphelion*. Earth's distance will vary by about 3.4 percent. Venus has the most circular orbit in the Solar System so its distance only varies by 1.4 percent, while Mercury has the most elliptical orbit of the main planets and at aphelion it is over 50 percent farther from the Sun than at perihelion.

When it comes to measuring each planet's distance from the Sun, the numbers start to get very big very quickly if we use miles or kilometers, which is why we use the astronomical unit (AU). The table below shows the distances for each plant in AU.

Opposite This NASA Voyager 2 image of Triton, Neptune's Moon, showing the southern polar region with streaking produced by geysers apparent on the icy surface. **Below** This table shows each planet's distance from the Sun, measured in AU (astronomical units).

Planet	Mercury	Venus	Earth	Mars	Jupiter	Saturn	Uranus	Neptune
Distance (AU)	0.4	0.7	1	1.5	5.2	9.6	19.2	30

With such huge distances between the planets and such a range of planet sizes, it's simply not possible to show the Solar System to scale in both size and distance at the same time in a single graphic, but the image right shows the relative sizes on one side and the relative distances on another. If you shrank the Sun down to the size of a medium-sized yoga ball, Earth would have a diameter of 0.2 inches (0.5 centimeters) and it would orbit at a distance of almost 213 feet (65 meters), which is larger than most people's backyard! At this scale, Neptune would have a diameter of just over 0.8 inches (2 centimeters) and orbit at a distance of nearly 1.2 miles (2 kilometers), and Pluto would have a diameter of 0.04 inches (1 millimeter) and orbit at a distance of 1.5 miles (2.5 kilometers).

Although we can place the distances from each planet to the Sun on a linear scale, the distances between the planets themselves at any one time is far more complex; they all orbit at different distances and at different speeds, so the relative positions are constantly changing. The distance between the Earth and Mercury, for example, can be anywhere between 48 million and 138 million miles (77 million and 222 million kilometers), and the distance between Earth and Venus can be anywhere between 23 million and 162 million miles (38 million and 261 million kilometers). Although Venus orbits closer to Earth than Mercury does, Venus has a much larger orbit, so Earth is actually closer to Mercury than Venus much of the time, which may seem surprising. Let's investigate why this is.

Mercury and Venus orbit closer to the Sun than Earth so their orbital mechanics are different than the outer planets. They're best viewed when they're at greatest *eastern or western elongation*, because this is the point where they are at their greatest angle away from the Sun when viewed from Earth. Both are constrained by their orbits so they are always a morning or evening object. When they pass between the Earth and Sun they're said to be at *inferior conjunction* and, although they're technically at their closest to Earth here, we can't see them because they're lost in the solar glare. Occasionally their orbits line up perfectly with the Sun so they can be seen transiting the Sun's disk. Mercury transits are more common and occur about 14 times per century but Venus transits are much rarer—the next one won't occur until 2117. When the inner planets pass around the back of the Sun, they're said to be at *superior conjunction*, so once again they're lost in the Sun's glare.

Let's return to the fact that Mercury is often closer to Earth than Venus. If all three planets happen to be on the same side of the Sun as shown in [fig 3a], then Earth is closer to Venus than Mercury. However, once Venus passes greatest western elongation, for the rest of its passage around the back of the Sun and until it is almost at greatest eastern elongation as shown in [fig 3b], it will always be farther away than Mercury regardless of where Mercury is in its orbit [fig 3c].

The planets that orbit farther out than Earth can never pass between the Earth and Sun like the inner planets do. They can only pass behind the Sun, when they are said to be in *solar conjunction* and are lost in the Sun's glare. The best time to view the outer planets is when they are at *opposition*, so called because they are on the opposite side of the Earth from the Sun. This is when they are at their closest to Earth; they will be at their biggest and brightest, and visible all night long, reaching their highest point at local midnight.

Above left This scale image shows the sizes of the planets to scale, compared with a section of the Sun. **Above right** This image shows the orbits of the planets to scale.

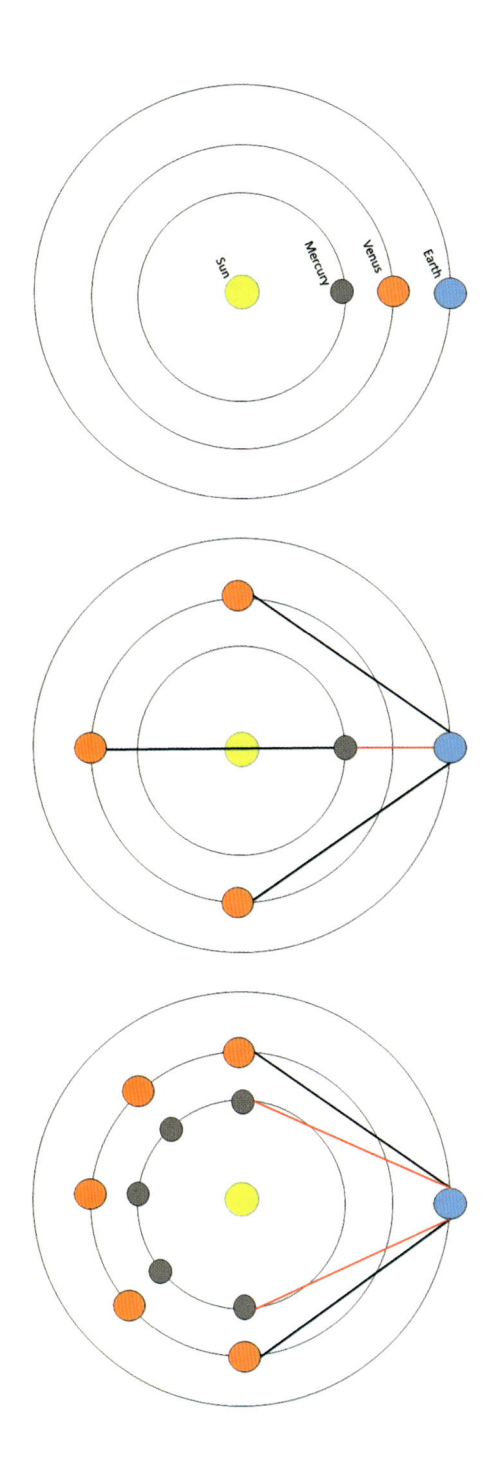

Because of the flat plane of the Solar System, when planets are observed from Earth they follow the path of the ecliptic—the invisible path in the sky that all Solar System bodies (apart from comets) follow. The ecliptic stretches from the east to the west, and if you're a northern hemisphere observer, objects that move along the ecliptic will do so in your southern sky. If you're a southern hemisphere observer, the ecliptic crosses your northern sky.

There are five planets that are bright enough to be seen with the naked eye, but without magnification they look like stars. However, their movements in the night sky are different to those of stars. Sometimes they move westward each day in *prograde motion*, sometimes they appear to slow down or come to a stop, and at other times they appear to be moving back toward the east—*retrograde motion*. This erratic behavior led ancient Greek astronomers to name them *asteres planetai* ("wandering stars," *planet* meaning "wanderer"). Additionally, planets are only visible to us because they reflect sunlight; unlike stars, they don't generate their own light. If you look carefully you'll notice that planets don't twinkle as much as stars do. Planets are disks rather than points of light; twinkling is caused by pockets of air with different temperatures rising and falling in the atmosphere. A pocket of air passing a star affects the entire point of light but when it passes a planet it only affects part of the disk of light at once, so its light appears steadier than that from a star.

Although the planets all orbit in the same direction, they move at different speeds, taking different times to complete one orbit of the Sun. The distances between each planet and Earth can therefore change significantly, leading to dramatic differences in apparent size and brightness—particularly noticeable when observing Venus, Mars, Jupiter, and, to an extent, Saturn.

Once the telescope was invented in the early 1600s, it allowed astronomers to observe the planets in more detail. One of the first prominent telescope users was the Italian astronomer Galileo Galilei (1564–1642), who noticed that when he observed Venus and Mercury at different times, they displayed phases in the same way the Moon does. This proved the geocentric model of the Universe (which had Earth at the center, and everything in orbit around it) was wrong. For the full range of phases to be visible on Mercury and Venus, they had to be in orbit around

Top (fig 3a) If all three planets of Mercury, Venus, and Earth all happen to be on the same side of the Sun as shown, then Earth is closer to Venus than Mercury. However, once Venus passes greatest western elongation, for the rest of its passage around the back of the Sun and until it is almost at greatest eastern elongation **Middle** (fig 3b), it will always be farther away than Mercury, regardless of where Mercury is in its orbit **Bottom** (fig 3c).

The planets and their rotational and orbital periods

Planet	Rotational period (in Earth hours)	Rotational period (in Earth days)	Orbital period (in Earth days)	Orbital period (in Earth days)
Mercury	1,408	58.67	88	0.2
Venus	5,832	243	224.7	0.6
Earth	24	1	365.2	1
Mars	24.6	1.04	687	1.9
Jupiter	10	0.42	4,331	11.9
Saturn	11	0.46	10,747	29.4
Uranus	17	0.71	30,589	84
Neptune	16	0.67	59,800	165

the Sun, not the Earth. Galileo's observations provided firm evidence to back up the heliocentric model put forward by Nicolaus Copernicus in the sixteenth century, which placed the Sun at the center of the Universe instead. Galileo also discovered the four largest moons of Jupiter; by plotting their movement night after night, he surmised that they were orbiting Jupiter, not Earth—yet more evidence that the geocentric model was incorrect.

As telescope design improved, it became possible to observe surface features on some planets and ascertain their rotational period. A day in this context is the time taken for a planet to complete one rotation on its axis (this includes a day and a night), whereas a year is the time taken for it to complete one orbit. These characteristics vary hugely from planet to planet.

The table above shows the rotational periods for each planet measured in Earth hours and days, along with the orbital periods measured in Earth days and years. You can see that a day on Mercury is nearly 60 Earth days, yet its year is only 88 days. A day on Mars is similar in length to an Earth day, but a Martian year is double that of Earth. The gas giants and ice giants have days that are a lot shorter than an Earth day but they have years that are significantly longer; in fact at the time of writing, only 1.07 years have passed on Neptune since its discovery in 1846!

Before we take a closer look at each planet in turn, there are some terms in our data tables listed with each planet that may need a brief explanation. Firstly, the mass and volume figures are listed as a comparison with Earth's mass and volume, and the density is measured relative to the density of water. The axial rotation period is how long it takes a planet to complete one rotation on its axis, relative to the stars. This will be slightly different than measuring relative to the Sun, so here on Earth, a day measured by how long it takes the Sun to return to its highest point is 24 hours, but when you measure how long it takes a star to return to the same place (a *sidereal day*), the value is 23 hours 56 minutes 4 seconds. There are two separate values for the orbital period; one is sideral and one is synodic. The *sidereal period* is the length of time it takes to return to the same position relative to the stars, whereas the *synodic period* is the length of time it takes a body to return to a particular position relative to the Earth. The two values can be very different. (In the data tables, "day" is abbreviated to "d.") The *axial tilt* is how much a planet is tilted on its axis, so a value of zero means no tilt; Earth is tilted by 23.5 degrees. *Orbital eccentricity* is a measure of how oval the orbit is, measured on a scale between 0 and 1: a completely circular orbit has an orbital eccentricity of 0, whereas an orbit so eccentric that it would escape the Sun's gravity is 1. Finally, *orbital inclination* is a measure of how tilted the planet's orbital path is.

Mercury

Diameter: 3,030 miles (4,879 kilometers)
Mass (Earth = 1): 0.06
Mean density (water = 1): 5.43
Volume (Earth = 1): 0.06
Axial rotation period (sidereal): 58.65 d
Orbital period (sidereal): 87.97 d
Orbital period (synodic): 115.88 d
Axial tilt: 0°
Mean distance from Sun: 35.98 x 65.8 miles (57.91 × 106 kilometers)
Orbital eccentricity: 0.206
Orbital inclination: 7°
Number of moons: 0

Below This photograph of Mercury was taken by the NASA's MESSENGER's wide angle camera (WAC) on January 14, 2008, as the space probe receded from the planet, recording a part of Mercury not previously seen. The camera was equipped with special color filters, allowing for a false-color image to display Mercury's details and color differences that would not have been possible with the human eye.

With a diameter of 3,030 miles (4,879 kilometers), Mercury is the smallest planet in the Solar System and this, plus the fact that it orbits so close to the Sun, means that it can be a difficult planet to observe from Earth. When Mercury lies too close to the Sun it's dangerous to look at it with binoculars or a telescope, so the best time is when it's at greatest eastern or western elongation. Even then, Mercury will always be a twilight object that requires an unobstructed horizon, but it's worth seeking out because when it's at its best it can be as bright as Sirius. If you observe Mercury through a larger telescope you'll see it changing phase as it moves through its orbit.

The difficulty of observing Mercury in any great detail previously led to some misconceptions about its rotational period. Being so close to the Sun, the Italian astronomer Giovanni Schiaparelli (1835–1910) postulated that it was tidally locked in the same way that our Moon is, so it must take 88 days to complete one rotation—the same time it takes to complete an orbit of the Sun. In 1965, astronomers at the Arecibo Observatory in Puerto Rico made a surprising discovery. They bounced radio waves off Mercury and were able to deduce that it spins once every 59 days; it is tidally locked but it has a 3:2 spin-orbit resonance, meaning that it rotates three times on its axis every two orbits of the Sun. A day on Mercury therefore last two-thirds of a year, with the Sun rising and setting very slowly. Due to its proximity, the Sun appears two and a half times larger in Mercury's sky than it does from Earth and this leads to some dramatic consequences. The intense heat and lethal doses of radiation from the Sun bake the rocks on Mercury's day-side to temperatures of over 750°F (400°C), but with no atmosphere to hold the heat, the night side of Mercury is a frigid −290°F (−180°C). It is too small and hot to retain a real atmosphere but it does possess a very tenuous surface exosphere that includes hydrogen, helium, oxygen, calcium, potassium, magnesium, and silicone. However, this exosphere is not stable—it is continuously lost and replenished by interaction with the solar wind and by the radioactive decay of elements found in Mercury's crust.

Our first close-up views of Mercury came from the Mariner 10 flyby in 1974. The planet is heavily cratered and bears such a striking resemblance to our Moon that at a glance you may struggle to tell them apart. Although the craters—formed from numerous impacts from asteroids, meteorites, and comets, particularly during the early days of the Solar System—may look the same as those on the Moon, there are some subtle differences. Firstly, Mercury's craters aren't as deep for a given diameter, and the ejecta blankets and rays haven't spread as widely as they have around lunar craters. This is due to Mercury having a stronger surface gravity—double that of the Moon, although still only 38 percent of that on Earth. There are, however, some craters near the poles that are so deep that their interiors are permanently shielded from sunlight and so any ice delivered by cometary impacts or gasses that may have seeped from the rocks are preserved in a deep freeze.

There are also other features that are unique to Mercury and not seen on our Moon. *Lobate scarps* are meandering cliffs that may be several hundred miles long and 0.6 miles (1 kilometer) tall, and they can even cut through old craters. These are believed to have been caused by Mercury shrinking slightly as its core cooled early in its history, causing compressions and faults to form in the rocks. Mercury is also home to *inter-crater plains*, another feature with no lunar counterpart. These are regions that lie in between the large craters, characterized by having small craters peppered across them. The craters here pre-date the larger ones and could possibly be the result of volcanic processes or caused by ejecta rocks from the larger impacts.

The most prominent feature on Mercury is the huge Caloris Basin, around 950 miles (1,500 kilometers) in diameter—larger than Mare Imbrium on the Moon (see page 220) and about a quarter of the planet's diameter. It contains concentric rings of mountains plus radial ridges and grooves that give it the appearance of a bull's-eye. The basin is thought to have been formed by a huge impact 3.9 billion years ago, and its interior and the area around it was subsequently flooded by molten rock. Geological activity died out 3 billion years ago and the only changes since then have been due to random meteorite strikes.

Although Mercury resembles the Moon, its surface is darker than the lunar surface and it has a much higher density. For its small diameter it has a surprisingly large mass—0.055 times that of Earth. This is probably due to an iron core that makes up three-quarters of its diameter. One explanation for this is that Mercury was once a much larger body with a differentiated core, but its outer layers were stripped off during collisions with one or more asteroid-sized bodies. This process could also have been responsible for knocking it into an elliptical orbit. Collisions from micrometeorites were responsible for another surprising feature that was discovered in 2001—Mercury's tail. Usually only comets or active asteroids have tails, but a combination of meteorite strikes and the effect of solar wind causes gases to be emitted from Mercury's surface and this is sculpted by the solar wind to form a tail. The tail is predominantly made from sodium, resulting in a yellow/orange color. It changes in brightness as Mercury orbits the Sun but it's not visible unless photographs are taken through a special filter tuned to the yellow sodium glow.

Sending probes to Mercury poses a huge technical challenge because they need to overcome the massive gravitational force of the Sun. Mariner 10 was the first to visit, completing two flybys in 1974 and a third in 1975, and it was able to map 45 percent of the planet's surface. During its first close approach it discovered that Mercury has a magnetic field. The probe ran out of fuel eight days after its final close approach, and it's believed to be still in orbit around the Sun, making close passes of Mercury every few months.

The MErcury Surface, Space ENvironment, GEochemistry and Ranging (MESSENGER) probe was launched in 2004 and it performed flybys of Earth and Venus to help it reach the correct trajectory to fall into orbit around Mercury. It performed three Mercury flybys in 2008 and 2009, then entered into an orbit around the planet in 2011 and continued to study and map the hemisphere that hadn't been observed by Mariner 10 until it was deliberately crashed onto the surface in 2015.

BepiColombo is a joint mission between the European Space Agency and the Japanese Aerospace Exploration Agency. It launched in 2018 and is due to arrive at Mercury in late 2025. It has a mapper probe that contains spectrometers to closely study the planet's surface composition, and a magnetometer probe to study its magnetic field. They are planned to operate for a year. The fact that Mercury has a magnetic field means that it has auroras, although they're only visible in X-ray wavelengths, unlike the auroras that occur on Earth.

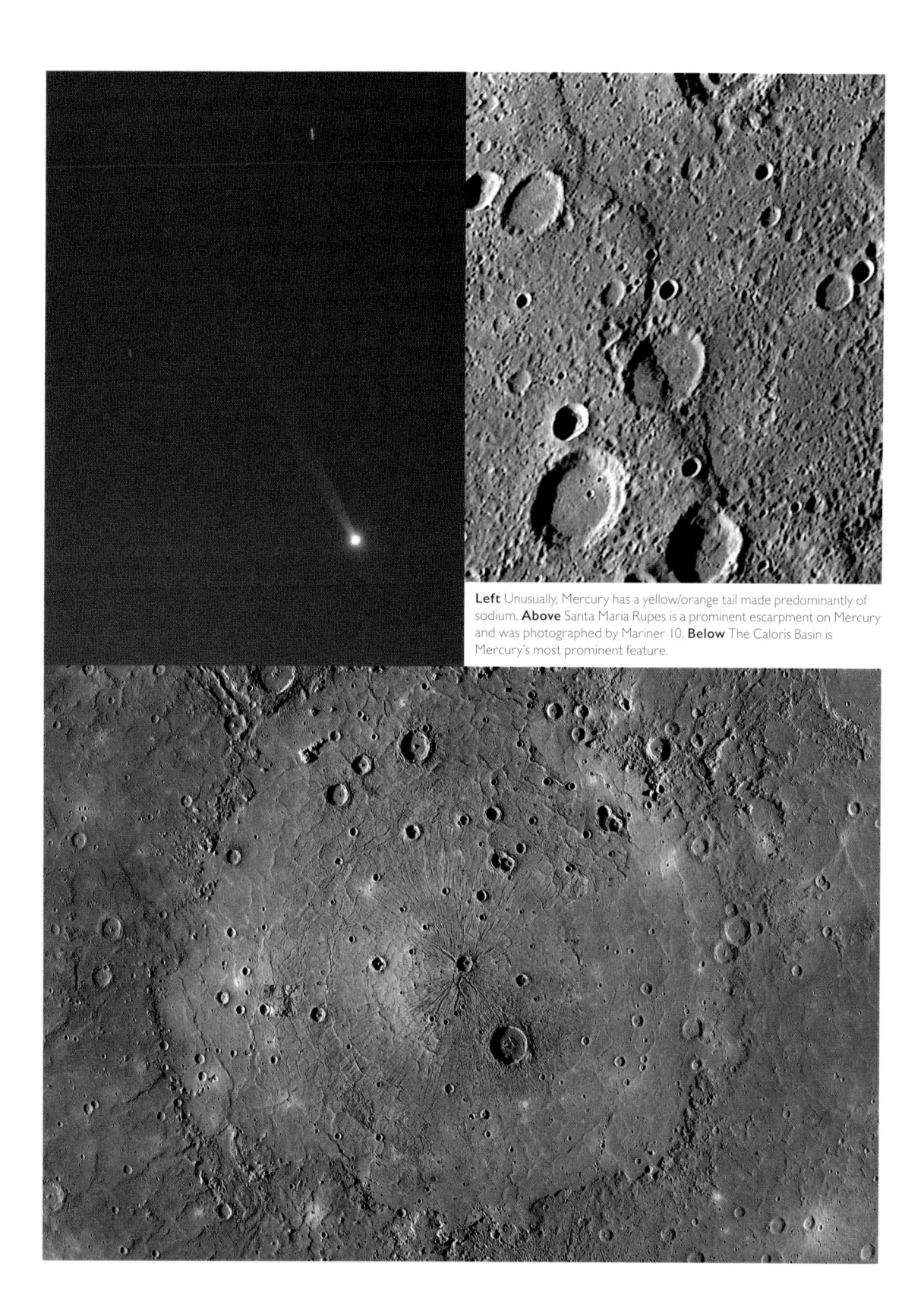

Left Unusually, Mercury has a yellow/orange tail made predominantly of sodium. **Above** Santa Maria Rupes is a prominent escarpment on Mercury and was photographed by Mariner 10. **Below** The Caloris Basin is Mercury's most prominent feature.

Venus

Diameter: 13,100 miles (21,104 kilometers)
Mass (Earth = 1): 0.82
Mean density (water = 1): 5.24
Volume (Earth = 1): 0.86
Axial rotation period (sidereal): 243.02 d
Orbital period (sidereal): 224.70 d
Orbital period (synodic): 583.92 d
Axial tilt: 2.6°
Mean distance from Sun: 67.2 x 65.8 miles (108.2 × 106 kilometers)
Orbital eccentricity: 0.007
Orbital inclination: 3.4°
Number of moons: 0

Below This stunning image of Venus was taken with the Akatsuki Ultraviolet Imager (UVI) in 2018. While the planet is obscured by a cloud layer, it shines brightly due to its sulfuric acid.

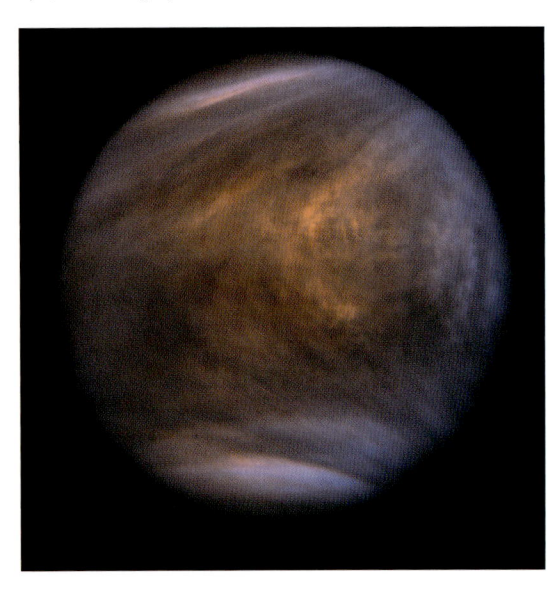

Venus is the brightest object in our night sky apart from the Moon—so bright, in fact, that it's frequently reported as a hovering UFO! Its orbit lies at 0.7 AU and, being an inner planet like Mercury, it's constrained by its orbit, so it always appears as either a morning or evening "star." Venus has a considerably wider orbit and can therefore appear much higher in the sky than Mercury, and at its best it can be visible for several hours before sunrise or after sunset. With a diameter of 13,100 miles (21,104 kilometers) it's very similar in size to Earth, but what makes it so bright is a dense blanket of unbroken clouds that reflect three-quarters of all the sunlight that hits it. This cloak of cloud prevents us from seeing the planet's surface.

You definitely don't need a telescope to view Venus, but if you have one, you'll see a bright white, almost featureless disk that may show a faint hint of some cloud markings, although you can easily see the changing phase of Venus even with low magnification. Despite the thick cloud layer that obstructs our view of its surface, the planet's brilliance is a real draw.

This cloud layer made it extremely difficult for astronomers to ascertain the time it took for Venus to rotate, and early measurements produced conflicting results. Radar observations were carried out in the 1960s and the results were surprising. Venus is the only planet that rotates in a clockwise, retrograde direction. One day on Venus is 243 Earth days—it has the slowest rotation of any of the planets—and a year is 224.7 days, making it the only planet in our Solar System where a year is shorter than a day. It takes 584 days for Venus to move through its phase cycle because this is the time taken from one superior conjunction to the next. This is slightly longer than a year because Earth is also moving relative to Venus.

When Venus is at its "full" phase it's at superior conjunction and cannot be viewed from Earth, but this is when it's also at its smallest apparent diameter. As Venus moves through its phases it is changing position relative to Earth and its apparent diameter varies dramatically. Its most favorable position in both distance and phase is when it's at 27 percent illuminated crescent and this occurs close to greatest eastern and western elongation, when it can reach a magnitude of −4.7—nearly seven times brighter than the second most prominent planet, Jupiter.

The Mariner 2 probe (launched in 1962) gathered information about Venus's atmosphere, while Venera 4

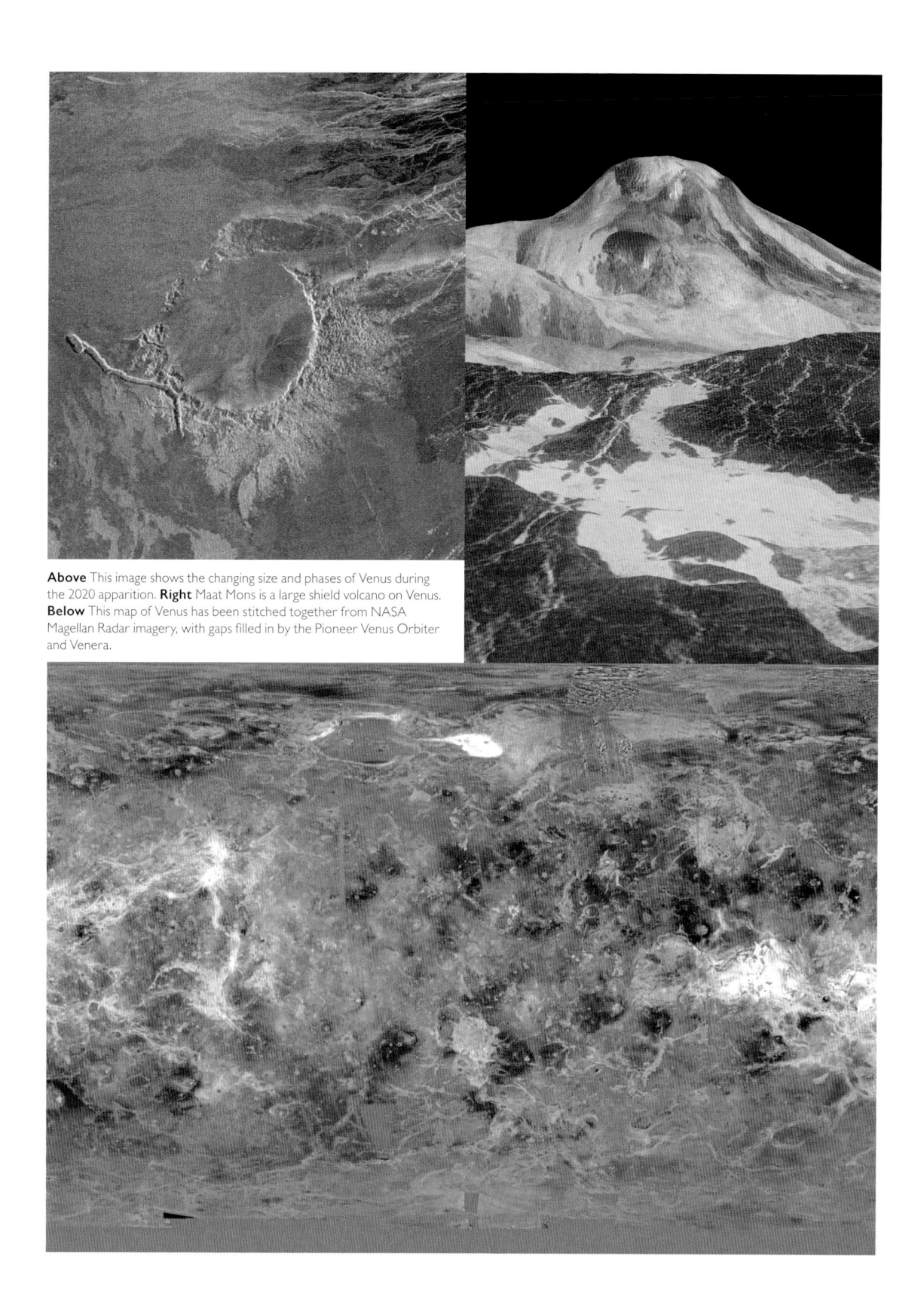

Above This image shows the changing size and phases of Venus during the 2020 apparition. **Right** Maat Mons is a large shield volcano on Venus.
Below This map of Venus has been stitched together from NASA Magellan Radar imagery, with gaps filled in by the Pioneer Venus Orbiter and Venera.

STARGAZING FOR BEGINNERS 245

(launched 1967) successfully deployed a set of science experiments in the Venusian atmosphere before it impacted the surface. The impenetrable layers that make Venus shine so brightly are a thick layer of sulfuric acid clouds, beneath which lies a dense carbon dioxide atmosphere at a crushing pressure 92 times that of Earth at sea level—combined with a temperature of almost 930°F (500°C), this is one of the most inhospitable environments in the Solar System. Over three-quarters of the sunlight hitting Venus is reflected by the clouds and just 1 percent actually penetrates the atmosphere, so you may be wondering why it's so hot on the surface.

The answer is the *greenhouse effect*. The 1 percent of the sunlight that does get through is absorbed by the atmosphere and re-radiated at longer wavelengths. Visible light can pass through the carbon dioxide atmosphere, but infrared radiation cannot, so the heat is trapped at the surface. The sulfuric acid clouds themselves are also stronger than car battery acid, and they absorb of infrared radiation, so the atmosphere and clouds together turn Venus into a giant heat trap. (Let this be a sobering lesson to us on Earth, contemplating the dangers posed by greenhouse gases in our atmosphere.)

If things didn't already sound grim enough, Mariner 10 discovered extremely high wind-speeds in the Venusian atmosphere. The densest cloud layer is located 30 miles (50 kilometers) above the surface and corrosive sulfuric acid rain falls from it. The sky below has the appearance of an overcast day here on Earth, but the gloom is broken by flashes of lightning. While Venus may be named for the heavenly goddess of love, the planet is actually an incarnation of hell. Nevertheless, despite its apparently inhospitable surface environment, it's not out of the question that extremophile microorganisms (microbes that can thrive in extremely challenging conditions) could survive in the clouds of Venus because all of the prerequisites for photosynthesis are plentiful—water, carbon dioxide, and sunlight.

In 2019, a group of astronomers reported changing patterns in the absorbance of light within Venus's atmosphere. This could be attributed to chemical changes we're as yet unaware of, but the absorbance was identical to that caused by microorganisms in Earth's atmosphere. In 2020, another group of astronomers reported the possible detection of phosphine in the upper levels of Venus's clouds.

The results are inconclusive at this stage, but if phosphine does exist, it could be formed by processes we don't yet fully understand or it could be formed by microbes. Rocket Lab is sending a probe to study the Venusian atmosphere and to look for organic molecules—the mission is due to launch in 2025.

There has been extensive exploration of Venus. In 1967 the Venera 4 probe impacted the suface and studied both the atmosphere and surface temperature. Later Venera probes also landed and sent back photos of the planet's surface. Venera 15 and 16 orbited between 1983 and 1984, mapping the terrain using radar technology, as did Magellan, between 1990 and 1994. From 2006 and 2014, the Venus Express orbited the planet, providing incredible observations of its atmosphere, and the space probe Akatsuki (also known as the Venus Climate Orbiter) has been in orbit since 2015. The surface is mostly rolling plains covered in a sulfurous yellow glow. There are three main continental areas—one being Ishtar Terra, equal in size to the United States and home to a mountain range, Maxwell Montes, that towers 7.5 miles (12 kilometers) above the mean surface level, higher than Mount Everest on Earth. The largest continental area is Aphrodite Terra, which is the size of South America and is cut by a system of rift valleys that extends for thousands of miles.

Magellan's radar also spotted impact craters with diameters ranging from 1 mile up to 60 miles (2 kilometers up to more than 100 kilometers), so asteroids and other meteoroids have clearly been able to fall through the atmosphere without burning up completely. And one of the most exciting discoveries on Venus's surface was active volcanism, with fresh-looking lava flowing from some mountains. One active volcano lies near the equator in Aphrodite Terra and, at a height of over 5 miles (8 kilometers), it is the second-highest mountain on the planet. Other volcanic features include pancake-like lava domes, and analysis from landers has revealed that the rocks are similar in composition to volcanic basalts on Earth.

Mars

Diameter (equatorial): 4220 miles (6,792 kilometers)
Mass (Earth = 1): 0.11
Mean density (water = 1): 3.94
Volume (Earth = 1): 0.15
Axial rotation period (sidereal): 24.62 h
Orbital period (sidereal): 686.98 d
Orbital period (synodic): 779.94 d
Axial tilt: 25.2°
Mean distance from Sun: 141.6 x 65.8 miles (227.9 × 106 kilometers)
Orbital eccentricity: 0.093
Orbital inclination: 1.9°
Number of moons: 2

Below This first true-color image of Mars was taken during the 2007 voyage of the ESA Rosetta spacecraft using OSIRIS color filters.

Mars is the fourth planet from the Sun and is approximately half of the diameter of Earth. It is often called the Red Planet because of its distinctive color, which to the naked eye appears much stronger than that of any of the surrounding stars. Its color is also why it's associated with the god of war. For an amateur astronomer, Mars is a very exciting planet to observe telescopically because, when it's well positioned and there are no any active dust storms, we can see through its thin atmosphere to its surface features and polar ice caps, all features that undergo seasonal changes.

A day on Mars is about 40 minutes longer than an Earth day but its year is nearly twice as long as ours, at 687 days. As with all of the outer planets, the best time to observe Mars is when it's at opposition, and this occurs every 26 months. Mars orbits at an average distance of 1.5 AU but because it has an elliptical orbit its distance can vary dramatically, so some oppositions are more favorable than others. If an opposition occurs when Mars is at perihelion, as it did in 2018 and will do again in 2035, it can come as close as 35 million miles (56 million kilometers) to us. During such times it is a brilliant red star that shines at magnitude −2.9 and this is the best time to observe it with a telescope. When opposition occurs when Mars is at aphelion, as it did in 2012 and will in 2027, it is nearly double the distance away at 62 million miles (100 million kilometers) and will therefore appear at half the size, so even a powerful telescope will struggle to pick out surface features. This is why astronomers never waste a chance to observe Mars during favorable oppositions. Camera technology has improved so much in recent years that amateur astrophotographers can capture incredibly detailed photographs, even with modest telescopes. Because Mars orbits farther out than Earth, it doesn't display crescent or half phases like the inner planets do, but if you observe it when it's at an elongation of 90° from the Sun, it will display a gibbous phase (see page 219).

The mystery of Martian canals

The first astronomer to observe Mars through a telescope was Galileo, but the first person to draw a sketch of its features was the Dutch astronomer Christiaan Huygens in the 1600s. During the nineteenth century, telescope technology improved so much that more detailed surface features could be observed. When Mars reached opposition during perihelion in 1877, the Italian astronomer Giovanni

Top This image shows the changing size of Mars during 2020 apparition, taken with a Celestron C11 telescope and ASI224MC camera.
Above Taken by the Hubble Space Telescope in 2001, this image shows Mars before and during a global dust storm. The two faces look dramatically different, with the surface almost completely obscured by the raging storm.

Schiaparelli observed it from Milan using a 8.7-inch (22-centimeter) telescope and produced some detailed maps. These recorded features that looked like long, straight lines, which Schiaparelli labeled as *canali* ("channels")—later mistranslated into English as "canals."

Although canals on Mars were later shown to be an optical illusion, other astronomers were influenced by the idea of them and, indeed, recorded seeing them, although not always in the same places. American astronomer Percival Lowell (1855–1916) proposed that these canals were human-made irrigation channels built by intelligent beings to channel melt water from the polar caps to the darker areas on the Martian surface. He believed them to be regions of vegetation such as moss or lichen because they appeared larger and darker in the summer. He published three books on this subject and maintained the belief that these were human-made features for the rest of his life. In the early 1900s, even with improvements in telescope design and the advent of astrophotography, canals did not appear in photographs and most serious astronomers did not see them, yet with the idea firmly seeded, others still claimed to have observed them—they were even included on the 1962 map of Mars produced by the US Aeronautical Chart and Information Center. Spectroscopic analysis showed that no water was present in the Martian atmosphere, disproving the whole idea, yet the idea of an advanced civilization on Mars would inspire many sci-fi stories. The canals would certainly not be the last Red Planet conspiracy theory!

Much of the Martian surface looks orange/red because it is covered in iron oxide. This is very obvious when viewed telescopically, as are the darker, grayish/brown areas that Lowell once thought were vegetation but in fact are rock and dust. These darker patches can change in appearance seasonally, as dust is blown around, so any map of Mars will only be an approximation of its true appearance. These changes do not take place at a geological level; they are simply changes in surface brightness as darker dust is relocated. (They are what are called *albedo features*—the term albedo referring to how reflective something is.) The most prominent dark albedo feature is Syrtis Major, a large triangular region that's visible in a modest telescope as long as it's on the Earth-facing side at the time of observation. Also achievable with a modest telescope is the huge circular lowland region known as Hellas. In stark contrast to these regions are the bright polar ice caps. Mars has an axial tilt very similar to the Earth's, so it has seasons that are similar to those we experience. The ice caps grow and shrink with the seasons.

While a perihelion opposition is the best time to observe Mars, it's also the time when increased temperatures mean Martian winds are stronger and more likely whip up dust storms. When they occur, surface features may appear diminished or completely obliterated. Sometimes only a small area is affected, but Mars is home to the largest dust storms in the Solar System, bringing winds of up to 100 miles (160 kilometers) per hour and dust that shrouds the entire planet.

Mariner 4 provided us with our first close-up photos of Mars in 1965 and there has subsequently been extensive exploration of the planet by space probes, orbiters, landers, and rovers, which together continue to make new discoveries. Mars's northern hemisphere has lower-lying land that has been flooded with lava from volcanoes and much of it is covered in wind-sculpted sand dunes. This contrasts with the southern hemisphere, which is comprised of heavily cratered highland regions that resemble the lunar highlands.

Martian volcanoes

One of these highland areas is the Tharsis Bulge, which straddles the equator and contains a chain of volcanoes, including Arsia Mons, Pavonis Mons, and Ascraeus Mons (*mons* meaning "mountain"). In March 2024 it was announced that a new giant volcano had been discovered in the same region. Spanning 280 miles (450 kilometers) and reaching height of just over almost 5 miles (9 kilometers), this volcano was so eroded and difficult to identify that it had been hiding in plain sight for years. It's thanks to years of repeatedly studying imagery from Mars orbiters that we continue to make discoveries like this. Northwest of this chain lies the biggest volcano in the Solar System, Olympus Mons. With a diameter of 370 miles (600 kilometers) and a height of 13.6 miles (21.9 kilometers), it is three times higher than Mount Everest and is so big that if you stood on the perimeter edge, the other edge would be lost over the horizon. The perimeter is visible from Earth in the form of a white ring. White clouds are often reported above the Tharsis area.

Martian canyons

Extending eastwards from the Tharsis Bulge is a vast network of canyons, Valles Marineris, which is up to 300 miles (500 kilometers) wide, 2.5 miles (4 kilometers) deep, and 2,500 miles (4,000 kilometers) long. It is the largest canyon in the Solar System and to say that it dwarfs the Grand Canyon would be an understatement because Valles Marineris would actually span the entire United States. What is visible from Earth is a feature called Coprates—darker-colored dust that collects on the canyon floor and provides a lower-albedo contrast to the surrounding landscape. We know from studying Mars that liquid water once flowed on its surface and it's widely believed that Valles Marineris was formed by water erosion, but it's recently been suggested that it began life as a tectonic rift on the surface and the crack that formed was subsequently eroded by water.

Martian craters

Not content with having record-breaking volcanoes and canyons, Mars is also home to the largest recognized impact basin in the Solar System, Utopia Planitia. It is located in the low-lying region in the northern hemisphere and it has an estimated diameter of 2,100 miles (3,300 kilometers). There are more than 43,000 craters greater than 3.1 miles (5 kilometers) on Mars, most of them in the southern hemisphere, and several of the large ones are visible from Earth. The Argyre Planitia impact basin is around 1,100 miles (1,800 kilometers) in diameter, with a depth of 3.2 miles (5.2 kilometers). The largest exposed crater is Hellas, with a diameter of 1,400 miles (2,300 kilometers) and a depth of 23,000 feet (7,000 meters); this is a similar size to Oceanus Procellarum on the Moon. Another notable crater is Isidis Planitia, which has a diameter of 930 miles (1,500 kilometers).

The Martian atmosphere

Mars has an atmosphere but because it lost its magnetosphere about 4 billion years ago, the solar wind is stripping it away. What remains is about 96 percent carbon dioxide, 1.93 percent argon, and 1.98 percent nitrogen, as well as traces of water vapor and lots of Martian dust particles. The average atmospheric pressure is less than 1 percent of Earth. The atmosphere is so thin that it cannot retain much heat, so temperatures vary from highs of up to 95°F (35°C) in the summer to a low of around −166°F (−110°C) in the winter, although the polar regions can drop to lows of −202°F (−130°C). Due to the unbreathable atmosphere and variations in temperature, when astronauts do step foot on Mars,

Top left Olympus Mons, as taken by the ESA Mars Express, is the largest volcano in the Solar System. **Top right** This image of Valles Marineris was created using a mosaic of images from the Viking 1 orbiter in 1980. It is the largest canyon in the Solar System. **Bottom left** Hellas Basin is one of the largest impact basins in the Solar System, believed to have been formed between 3.8 and 4.1 billion years ago, when a large asteroid hit Mars.

they'll need spacesuits similar to those worn by those who walked on the Moon—although with Mars's gravity about 38 percent of Earth's, it will be easier to take those steps than it was on the Moon.

When sunlight reaches Earth, the gases and particles in our atmosphere cause the light waves to scatter in all directions. Sunlight is made up of seven colors—red, orange, yellow, green, blue, and violet. Blue has a shorter wavelength than the other colors, so is scattered more easily, which is why we see a blue sky during the day. This is called *preferential scattering*. At sunset or sunrise, when the Sun is low in the sky, sunlight passes through a thick layer of atmosphere, dust, and pollutants, which allows predominantly red and yellow light to pass through. On Mars, a carbon dioxide atmosphere combined with iron dust in the air means that red light is scattered during the day, so the daytime sky always has a reddish hue. Then at sunset the red light disperses, leaving behind a bluish hue. Space probes and rovers have also photographed several kinds of cloud, crepuscular rays, and even atmospheric optical effects in the Martian atmosphere.

Each polar cap has a permanent core of water ice several feet thick, but in the winter this is augmented by carbon dioxide that freezes out of the atmosphere and produces a "hood" of frost that can extend far beyond the poles. As the seasons change, the carbon dioxide evaporates from one pole and migrates to the opposite hemisphere and freezes out again at the opposite pole. This process causes the atmospheric pressure to change by around 25 percent and it's thought that this movement can help to trigger seasonal dust storms. Dust devils are also a common occurrence on Mars and the Mars rovers have captured photos of them several times. Dust devils are small, isolated vortices of wind that can occur when warm air rises quickly through cooler air to form an updraft that begins to rotate. This then picks up dust from the ground and forms a rotating column of dust that moves across the landscape. Although similar in appearance to a tornado, dust devils are triggered from the ground upward, rather than from rotating clouds that come down to the ground.

Water on Mars

Although Lowell's canals don't exist, there's plenty of evidence that liquid water once existed on Mars. There are sinuous river deltas, dried-up river beds, river-tumbled smooth rocks and pebbles, and evidence of flash floods. Direct evidence has come from four of the Martian rovers. Spirit and Opportunity both landed in 2004—in Gusev Crater and Meridiani Planum respectively—and both found minerals that indicated that water had been present. The Curiosity rover has explored Gale Crater since 2012 and has photographed layers of sedimentary rocks that formed when water flowed inside the crater. The Perseverance rover ("Percy") landed in Jezero Crater in 2021 along with its companion, the Ingenuity helicopter. Ingenuity was sent to Mars purely as a technology demonstration with five planned flights. It was the first aircraft to achieve powered flight on another planet and it went on to

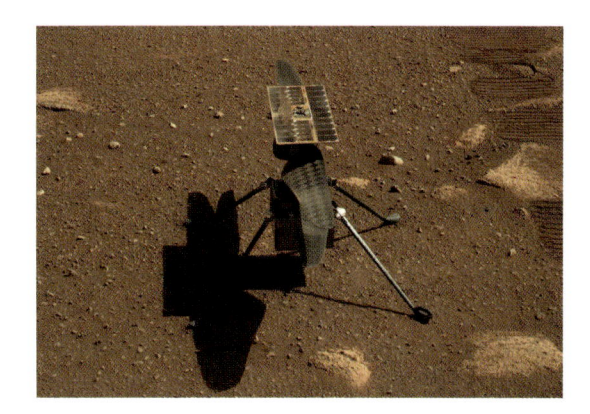

Top NASA's Ingenuity helicopter on Mars, as photographed by Mastcam-Z on Perseverance Rover in April 2021, was the first aircraft to achieve powered flight on another planet. **Bottom** Perseverance ("Percy") Rover even managed to take a selfie while on Mars.

STARGAZING FOR BEGINNERS 251

complete 72 flights, totaling almost 130 minutes of flight time and 10.5 miles (17 kilometers) in distance. Ingenuity's mission sadly ended in early 2023 when a rotor blade was damaged, but it nevertheless marked a huge milestone in planetary exploration and it helped to scout out areas for Percy to explore in future.

In the meantime, Percy has been collecting samples and caching them ready for a future mission to collect them and bring them back to Earth for analysis. An orbiter is due to launch in 2027 and a lander in 2028, and the samples are expected to arrive back on Earth in 2033. One day humans will walk on the Red Planet so it's a very exciting time for Mars science. Before he died (in 1996), American astronomer Carl Sagan recorded a message for the first people who'll step on Martian soil, and in 2008 NASA's Phoenix lander carried it to Mars: (Read more on page 391.)

> *Maybe we're on Mars because we have to be, because there's a deep nomadic impulse built into us… and for 99.9% of our tenure on Earth we've been wanderers.*

Is there life on Mars?

At present the atmospheric pressure is too low for liquid water to exist on Mars but there is still plenty of water there in ice form, the majority of it at the poles. If all of the water in the south pole region melted, it would create enough water to cover the entire planet to a depth of 36 feet (11 meters). There is also water ice in the surface soil in the form of hydrated minerals, ice in the bottoms of craters—and, in 2016, NASA reported the discovery of a large amount of underground ice in the Utopia Planitia region.

Water is a prerequisite for life as we know it, so if liquid water was present in the past on Mars, could microbial life have existed there? Are there any parts of Mars that still have microbial life today? Findings answer to these questions is one of the key goals of many Mars missions, but as yet no direct evidence has been found. The first search for life took place when the Viking 1 and 2 probes landed on Mars in 1976. The photographs were low resolution and one photo in particular led to yet another conspiracy theory.

Viking 1 released an image of the Cydonia region that included a lava dome measuring approximately 2.2 x 2 miles (3.6 x 3 kilometers)—photographed under a low Sun angle and at low resolution it resembled a human face and became known as the "Face on Mars." This was an example of pareiodolia—a phenomenon whereby the human brain sees likenesses (often faces) in random images. Perhaps with the story of Lowell's canals still lingering, some people believed that the face might be evidence of a city ruin once inhabited by an ancient civilization. Later high-resolution images taken by the Mars Global Surveyor proved that the feature looked nothing like a face (see right), but this region of Mars has inspired yet more sci-fi stories, and in 2006 the rock band Muse released a song called "Knights of Cydonia." There are conspiracy theorists who spend every day studying images from the Mars missions and among other things they have "found" an iguana, a woman, a mermaid, a rat, and even the face of British actor Les Dennis. Of course none of these things really exist; it's just pareiodolia at work.

Moons of Mars

Mars has two small moons, both believed to be asteroids that strayed too close to the planet and were captured its gravity. They were discovered in 1877, but observing them is beyond the capability of most amateur telescopes. Phobos is the larger of the two, measuring about 16 x 11 miles (26 x 18 kilometers). At a distance of just 3,700 miles (6,000 kilometers), it orbits closer to its parent than any other moon in the Solar System, and it does so speedily, orbiting Mars three times a day. Deimos measures about 10 x 6 miles (16 × 10 kilometers), orbits at a distance of almost 14,600 miles (23,500 kilometers) and takes just over 30 hours to complete one orbit. It's believed that there may be other smaller moons yet to be discovered. The Perseverance rover has captured both Phobos and Deimos transiting the disk of the Sun from down on the Martian surface. If the moons were any larger this would cause a total solar eclipse, but they are so small that they create a solar transit instead.

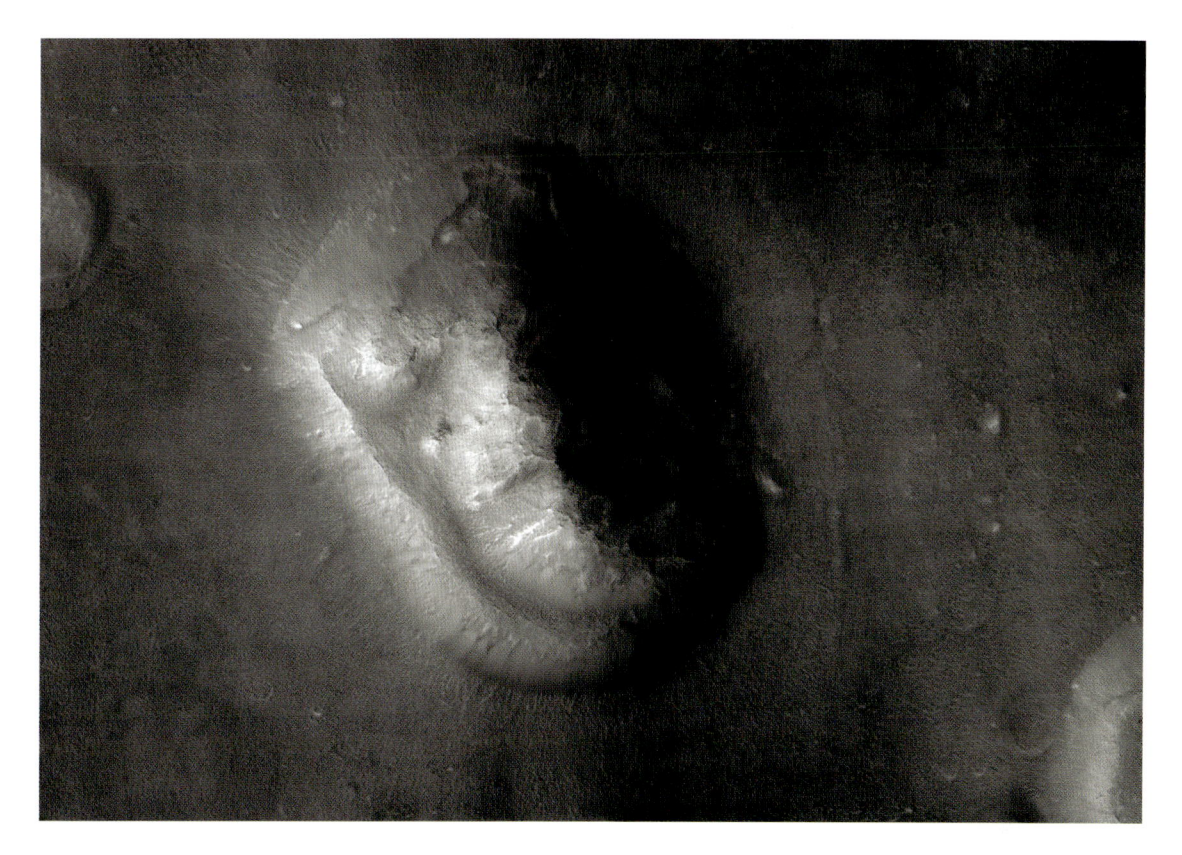

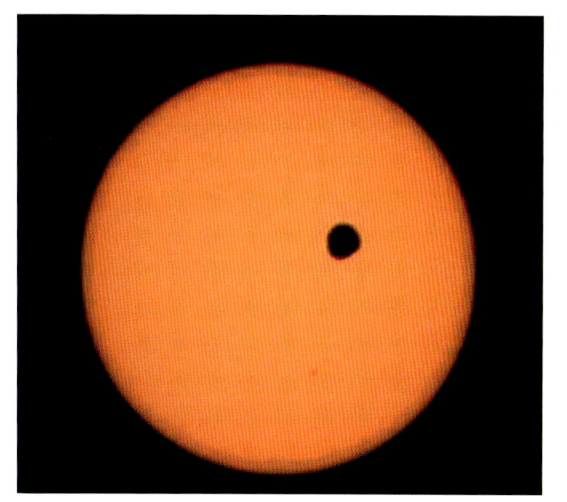

Top The High Resolution Imaging Science Experiment onboard the Mars Reconnaissance Orbiter captured this image of what was commonly known as the "Human Face." In greater detail, this image was able to prove that the feature does not resemble a face. Perseverance Rover captured both Phobos' **(Above left)** and Deimos' **(Above right)** transits across the Sun in April 2022.

Jupiter

Diameter (equatorial): 88,846 miles (142,984 kilometers)
Mass (Earth = 1): 317.83
Mean density (water = 1): 1.33
Volume (Earth = 1): 1,321
Axial rotation period (sidereal): 9.84 h
Orbital period (sidereal): 11.86 y
Orbital period (synodic): 398.88 d
Axial tilt: 3.1°
Mean distance from Sun: 483.6 x 65.8 miles (778.4 × 106 kilometers)
Orbital eccentricity: 0.048
Orbital inclination: 1.3°
Number of moons: 95+

Below This image of Jupiter was taken by the Hubble Space Telescope in April 2014 and shows the steadily shrinking Great Red Spot.

Jupiter, the first of the gas giants, is often referred to as "King of the Planets" because it has the largest diameter and the greatest mass of all the planets in our Solar System; in fact, its mass is two and a half times greater than all the other planets combined. Although its orbit lies beyond the asteroid belt at 5.2 AU, its diameter is nearly 93,000 miles (150,000 kilometers)—the equivalent of 11 Earth diameters—so it's large enough to be within the reach of most amateur telescopes. It reaches opposition every 13 months and during opposition it can reach magnitude −2.9, thanks to its highly reflective clouds, which make it the third-brightest natural object in the night sky, and very easy to spot with the naked eye.

Even with small binoculars or a small telescope you can see the four largest moons of Jupiter orbiting the planet—it's fascinating to watch their positions change over the course of a night or even a week. Sometimes you can see the shadow of one or more of the larger moons transiting Jupiter's disk. With a modest telescope you can begin to see the disk of Jupiter and you'll notice that it's not perfectly spherical—it bulges at the equator to form an oblate spheroid. You may also begin to see "surface" features (we use the term "surface" to refer to what we can see on the disk of Jupiter even though we're actually observing features in the top of Jupiter's cloud systems).

Jupiter's cloud belts

When observing Jupiter through a telescope, the first feature you may see is a pair of parallel dark bands that lie either side of its equator; they'll come in and out of focus as you look through layers of moisture in the atmosphere. You may also see the Great Red Spot, a huge anticyclonic storm system that was first definitively observed in 1831, but this or a similar object may have been observed as early as 1665. The storm lies about 5 miles (8 kilometers) above the cloud tops and is about the size of Earth but it has been steadily shrinking since it was first observed so it's now a fraction of the size it once was. Nevertheless, it's still a relatively easy feature to observe with modest equipment, and certainly within the reach of amateur astrophotographers.

Jupiter rotates extremely quickly, with the equatorial regions completing one full rotation in just under 10 hours, and the polar regions rotating about 5 minutes slower. Upon closer inspection you'll notice that it is criss-crossed with alternating bright and dark cloud belts, which rotate in alternating directions, and wind speeds of up to 310 miles (500 kilometers) per hour create lots of churning and swirling along the boundaries between the belts. If you're observing Jupiter with a large telescope or taking high-resolution photographs, it never looks exactly the same twice. It's both fascinating and important to record what you see during each session—either photographically or, even better, with sketches. In addition to the Great Red Spot there are numerous other storm systems that can be seen as white ovals. The behavior of these features tells us that Jupiter is not a solid object.

Because of the orbital plane of the Solar System, when viewed from Earth we can never observe the polar regions of Jupiter, but in August 2016 the Juno space probe sent back the first ever photographs of the northern polar region and it has subsequently photographed the north and south polar regions several more times. Jupiter is often used to give spacecraft a gravity assist, so over the history of spaceflight we have received observations of Jupiter from probes flying by as they headed for other worlds. The Galileo probe was the first to actually orbit the Jupiter system, in 1995, and it remained in orbit for seven years. The main spacecraft had a separate spacecraft—the Galileo entry probe—that was designed to withstand passing through Jupiter's atmosphere. It was released and survived for almost an hour, collecting data, before succumbing. The main probe was deliberately destroyed by crashing into Jupiter at the end of its life. At the time of writing, Juno is still in orbit around Jupiter and the mission has been extended to September 2025. In addition to photographic observations, Juno is studying Jupiter's magnetosphere.

Top left Jupiter's North Pole region, as imaged by NASA's Juno in 2016.
Top right This map of Jupiter's south pole region is the most detailed global color map of the planet ever produced, constructed by images taken by Cassini during a flyby of Jupiter in 2000. **Above** This image of Jupiter's Southern Storms at its south pole, as seen by NASA's Juno spacecraft from an altitude of 32,000 miles (52,000 kilometers).

Jupiter's composition

In terms of composition, Jupiter is predominantly hydrogen—the lightest element in the Universe—with some helium. Although these elements are light, when present in this quantity they end up forming an object with a high mass. The brighter cloud belts are comprised of ascending frozen ammonia crystals, but the darker belts are where the gases have descended and are a little warmer; this is, of course, a relative term because the temperature of the cloud tops is around −240°F (−150°C). The complex mixture of chemicals in Jupiter's atmosphere includes ammonia, phosphorus, sulfur, and possibly hydrocarbons. As they rise up through the atmosphere and into the clouds they change color as they interact with UV light from the Sun, so commonly seen colors include yellow, orange, brown, red, and purple.

Storm systems like the Great Red Spot, which can itself vary in color from red and pink to gray, are poorly understood, but they're believed to be upward-spiraling columns of gas similar to hurricanes on Earth. Jupiter emits twice as much heat as it receives from the Sun, probably because of ongoing contraction with its interior, and it is this that drives the turbulent cloud belts and storm systems.

The Shoemaker–Levy 9 impact

One of the most remarkable events to have occurred in the history of planetary observation was the impact of comet D/1993 F2, also known as Shoemaker–Levy 9, in 1994. The fragmented comet was discovered in 1993 by Carolyn and Eugene Shoemaker and David Levy. It had been captured by Jupiter's gravity and a close encounter with the planet in 1992 had caused the comet to fragment into 21 pieces; those fragments were now on a collision course with Jupiter. Between July 16 and 22, 1994, the world watched as each fragment hit Jupiter, leaving dark brown scars in the cloud tops that were visible from Earth for several weeks. The Hubble Space Telescope captured the impact scars and this event really brought home the importance of studying potentially hazardous near-Earth objects. There have been several impacts to Jupiter captured by amateur astronomers since then.

Jupiter's core

At around 0.6 mile (1 kilometer) below the clouds, the temperature and pressure increase so much that hydrogen is compressed into a liquid, forming liquid-hydrogen oceans around 12,400 miles (20,000 kilometers) deep. The pressure is the equivalent of 3 million times Earth's atmosphere, and under such compression the hydrogen is in such a dense state that it behaves like a metal; hydrogen in this state is called metallic hydrogen. The convection that takes place in this area is responsible for Jupiter's intense magnetic field, which is ten times stronger that Earth's magnetic field and extends out 50 Jupiter diameters in distance. This magnetic field means that extremely energetic auroras occur at Jupiter's poles and they have been photographed by space telescopes and probes. Jupiter is thought to have a rocky core beneath the metallic hydrogen ocean.

Jupiter's moons and rings

At the time of writing, Jupiter is home to 95 moons but this number is almost certainly going to increase as telescope and camera technology improves. As mentioned, Jupiter has four moons that are easily visible with binoculars and—for those with extremely good eyesight—with the naked eye. These are Io, Europa, Ganymede, and Callisto. Having been discovered by Galileo, they're often referred to as the Galilean moons. Watching them perform a celestial dance around Jupiter each day is rivetting, just as it was for Galileo when he did the same, and realized that they were orbiting Jupiter, not Earth (see page 240). These four worlds have the potential to be extremely exciting places from an astrobiology perspective.

Above This Hubble image shows the impact scars (appearing as brown spots) on Jupiter from fragments of comet Shoemaker-Levy 9.

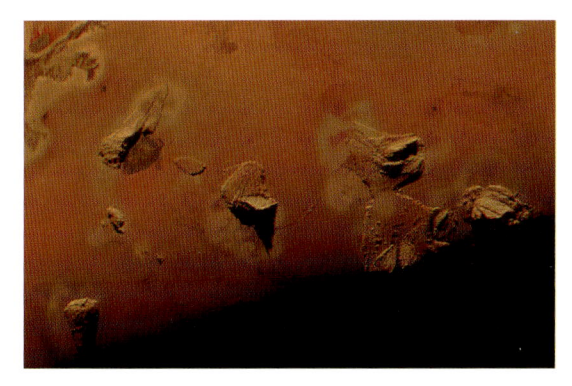

Io is the closest Galilean moon to Jupiter. With a diameter of about 2,260 miles (3,640 kilometers), it is slightly larger than our own Moon, and it takes 42.5 days to orbit Jupiter. It is the most volcanic body in the Solar System, and when the Voyager 1 probe did a flyby in 1979 it photographed eight volcanoes erupting simultaneously, with hundreds of other volcanic vents visible. They erupt lava and sulfur, and the latter solidifies on the surface, giving the moon a garish yellow color. The reason there is enough heat in such a small body this far from the Sun is because of tidal heating as a result of Jupiter's gravity. This creates a body that is constantly erupting fresh lava. The surface of Io is therefore continuously being resurfaced, so very few impact craters are visible.

Some of the sulfur also escapes and adds a yellow coating to Jupiter's innermost moon Amalthea. Close to the orbit of Amalthea is a faint, tenuous ring system that's located just 18,600 miles (30,000 kilometers) above Jupiter's cloud tops. Although too faint to see from Earth without the largest of telescopes, the rings were discovered by the Voyager space probes in 1979 and were subsequently investigated by the Galileo probe, the Hubble Space Telescope, and the James Webb Space Telescope, the latter being responsible for our most recent image of them, in July 2022. Unlike Saturn's rings, which are made of ice, Jupiter's rings are dusty and are thought to be the remains of one or more tiny moons that have broken up.

Moving out from Io, the next Galilean moon is Europa, the smallest of the four with a diameter of around 1,940 miles (3,100 kilometers), making it slightly smaller than our own Moon. Its surface is a bright and smooth layer of ice, 60 miles (100 kilometers) thick, and the low number of impact craters suggests that the surface is quite young, so it's thought that a liquid-water ocean beneath this icy mantle is contributing to resurfacing. The heat that keeps this water liquid is once again due to tidal heating caused by gravitational forces from Jupiter and is thought to be the cause of the prominent criss-crossed markings all over the surface. The crust is constantly flexing and cracking, which allows some of the slightly melted slush to seep through the cracks before refreezing. The Galileo space probe studied this so-called *chaos terrain* in a region called Conamara Chaos and was able to photograph it under a low Sun angle—it revealed a landscape full of huge ice rafts and ice cliffs. The Juno space probe has photographed Europa more recently. The presence of warm, liquid water beneath the surface makes Europa a prime candidate for the potential existence of subsurface microbial life. The Europa Clipper mission is due to launch in October 2024 and arrive at Europa in April

Top This false-color composite image shows Jupiter's four largest moons, also known as the Galilean satellites, first discovered by Galileo. Shown from left to right (in order of increasing distance from Jupiter): Io, Europa, Ganymede, and Callisto. **Above** This NASA image shows the shadows created by Io's volcanoes.

2030, and one of its key goals is to study habitability. The American Poet Laureate Ada Limón was also asked to pen a poem that will be engraved on the spacecraft:

Arching under the night sky inky
with black expansiveness, we point
to the planets we know, we
pin quick wishes on stars.
From "In Praise of Mystery:
A Poem for Europa"—Ada Limón

The next Galilean moon is Ganymede, the largest of the four, and with a diameter of around 3,270 miles (5,270 kilometers), the largest moon in the Solar System. It is even larger than Mercury in size but only half its mass. Ganymede's surface is comprised of silicate rock and water ice, but it's thought to also have a liquid-saltwater ocean about 120 miles (200 kilometers) below the surface. It has a metallic core and is the only moon in the Solar System to have its own magnetosphere. Ganymede's surface is a mix of two types of terrain: some is older, darker, and heavily cratered, while other areas are lighter in color with fewer craters, where ice has formed a new surface over the top of the old. The lighter regions often have visible grooves, thought to be due to stresses on the ice.

The final Galilean moon is Callisto, the second largest of the four with a diameter of around 3,000 miles (4,800 kilometers). This moon is saturated with craters of all sizes, the largest being Valhalla, with a diameter of 1,860 miles (3,000 kilometers). It is comprised of equal parts of rock and ice, so is the least dense of the Galilean moons. It doesn't undergo the same tidal heating that the other three do, but it's still thought to have a liquid-water ocean at a depth of around 185 miles (300 kilometers), so once again there's a chance that it could harbor microbial life.

Top Jupiter has incredible rings and auroras, as displayed in this stunning James Webb Space Telescope composite image.

Saturn

Diameter (equatorial): 74,900 miles (120,536 kilometers)
Mass (Earth = 1): 95.16
Mean density (water = 1): 0.69
Volume (Earth = 1): 764
Axial rotation period (sidereal): 10.23 h
Orbital period (sidereal): 29.45 y
Orbital period (synodic): 378.09 d
Axial tilt: 26.7°
Mean distance from Sun: 886.5 x 65.8 miles (1426.7 × 106 kilometers)
Orbital eccentricity: 0.054
Orbital inclination: 2.5°
Number of moons: 146+

Below This NASA/ESA Hubble Space Telescope's image of Saturn was taken in June 2019, as the planet made its closest approach to Earth that year, at approximately 0.85 billion miles (1.36 billion kilometers) away.

With its beautifully bright ring system around its middle, there's no doubt that Saturn is the most beautiful of the planets. It is the second gas giant and orbits at a distance of 9.6 AU, so a year on Saturn is equal to 29.4 Earth years. Although with its diameter of 74,900 miles (120,536 kilometers), it is the second largest planet in terms of size, it only has about a third of Jupiter's mass and a density that is about 70 percent of water; if you had an ocean large enough, Saturn would float on it! Like Jupiter it has a very fast rotation, with a day lasting just 11 hours.

Saturn's rings can reflect more light than the planet itself, so when at opposition approximately every 14 months, it can reach magnitude –0.3; it would be less than half of this magnitude without the rings. As Saturn orbits the Sun, the tilt of its rings changes between 0 and 27° over an approximately 15-year cycle. The rings have a diameter of 170,900 miles (275,000 kilometers) but for the most part they are only about 320 feet (100 meters) high, so this axial tilt plays an important role in how bright Saturn appears in our skies. In 2025 the rings will completely disappear from view as we will be looking them edge-on.

To the naked eye, Saturn appears as a yellow-colored star. Through a good pair of binoculars that are mounted to keep them stable, it will look oval-shaped because of the rings. When you look through a telescope with higher magnification, though, you will see that the rings are distinct from the planet—as long as the axial tilt is favorable, of course. And as you increase the magnification you will begin to resolve the gaps in the rings—in particular, the region known as the Cassini Division (see page 260).

However, there is more to see on Saturn than just the ring system. Like Jupiter, Saturn is comprised mostly of hydrogen with some helium, and it is also subjected to tidal heating, which means that the planet emits more heat than it absorbs from the Sun. Although Saturn is about 86°F (30°C) cooler than Jupiter, this tidal heating drives weather systems in Saturn's clouds. The bands lack the stark contrast seen on Jupiter because they're obscured by haze, but they are present, and approximately every 30 years a large white storm system erupts and remains visible for several weeks or months. The wind speeds can reach 1,100 miles (1,800 kilometers) per hour.

When the Voyager probes flew by Saturn, they caught a glimpse of the north pole region and captured part of a very intriguing feature, and when the Cassini probe was in orbit around Saturn it gave us a clearer look at it. The cloud belts here have formed an enormous hexagonal high-speed hurricane that is twice the diameter of Earth. The hexagon is spinning around a tight vortex at the north pole with winds of about 300 miles (500 kilometers) per hour. This symmetrical feature had scientists perplexed for a while because we don't see this kind of pattern with storms on Earth, but the composition of Saturn's atmosphere is more uniform and there is no solid surface below. This allows Saturn's jet stream to settle into a symmetrical shape. Saturn also experiences auroras at the poles, photographed by the Hubble Space Telescope several times.

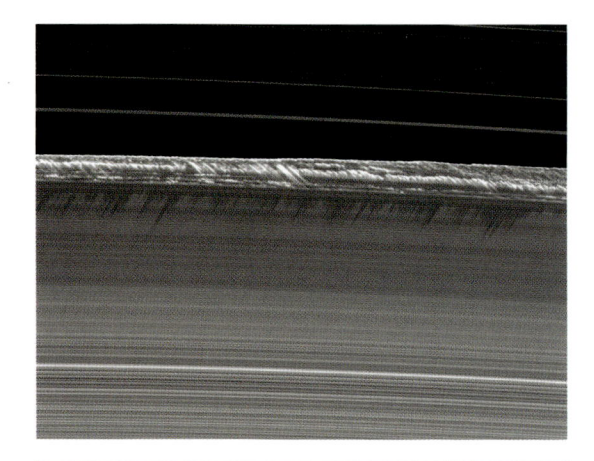

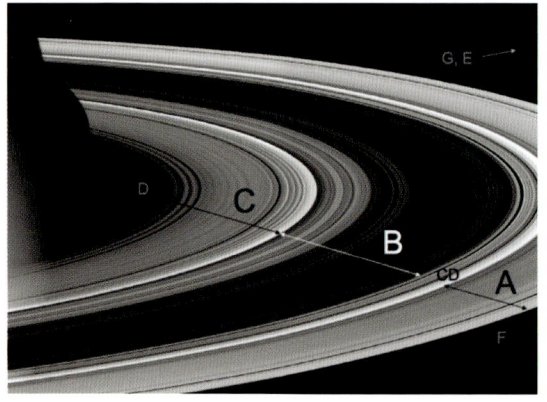

Top left This Hubble image shows the change in Saturn's axial tilt, from 1996 to 2000. **Top right** This image was taken by NASA's Cassini in 2009, showing Saturn's B Ring Peaks as vertical structures as high as 1.6 miles (2.5 kilometers), casting long shadows above the plane of the rings. **Bottom left** Saturn's South Pole, as photographed in 2017 by the Cassini space probe. **Bottom right** Observed by the Cassini space probe, this image shows Saturn's rings, here classified as A through G.

Saturn's rings

It's impossible not to be in awe of Saturn's ring system when you observe the planet directly or see it in photographs. Because they're so bright, the rings can look like a solid object at low magnification, but in 1655 Christiaan Huygens discovered that they are not solid but, rather, a swarm of smaller particles orbiting Saturn. The seven rings are named in the order in which they were discovered, from A to G, but that isn't the order in which they lie (see image above). From Saturn outwards they are D, C, B, A, F, G, and E. The D ring is very faint and closest to Saturn. The main rings are A—which is the outermost ring and most easily seen in a telescope—B and C. The Cassini Division is the largest gap in the rings and separates the A and B rings. The outer part of Ring A itself also has a narrow gap called the Enke Gap, which can be observed with large telescopes. Just outside the A ring is the narrow F ring, shepherded by tiny moons, Pandora and Prometheus. Beyond that are two much fainter rings named G and E. Saturn's diffuse E ring is the largest planetary ring in the Solar System, extending from Mimas' orbit to Titan's orbit, about 621,500 miles (1 million kilometers). Observers with access to larger telescopes observe ripples within the rings—a sign that the density of the material varies within them.

We got our first close-up look at Saturn when Pioneer 11 did a flyby in 1979, followed by Voyager 1 and 2 flybys in 1980 and 1981, and the Cassini probe began orbiting Saturn in 2004. Nothing could prepare astronomers for the detail that was captured by these probes: the rings were actually a system of thousands of narrow ringlets and gaps, with thread-like ringlets even occuring in the gaps. For the most part the rings are extremely thin in relation to their diameter; at the same ratio, a compact disk would need to have a diameter of 3 miles (5 kilometers). The bodies that make up the rings are mostly water ice and dust, ranging from the size of a dust speck right up a house. The Cassini probe was able to photograph the rings under a low Sun angle and it made an astonishing discovery—some fragments along the edges of the rings are the size of the Rocky Mountains! The fragments within the rings are continuously colliding with each other and this exposes brighter, fresh ice from below the surface, giving the rings their high albedo. Many of Saturn's small moons play a role in the structure of the ring system.

Saturn's moons

There are currently 146 moons around Saturn but, as with Jupiter, new moons are being discovered constantly. So far, 131 of them are less than 30 miles (50 kilometers) in diameter and there are probably an immeasurable number of small moonlets that reside within the rings and act as shepherd moons—so called because they prevent the rings from spreading out farther, or they clear out small particles from the gap in which they orbit. Saturn's major moons in order of their orbital position are Mimas, Enceladus, Tethys, Dione, Rhea, Titan, and Iapetus. These moons are bright enough to be observed through larger telescopes and can be captured photographically.

Titan is Saturn's largest moon, with a diameter of 3,200 miles (5,150 kilometers), and it takes 16 days to orbit the planet. Bigger than Mercury, it's the only moon to have a substantial atmosphere—in fact, the atmospheric pressure on the surface of Titan is 50 percent greater than at sea level on Earth. Being so far from the Sun it is very cold on Titan's surface, with temperatures of about −292°F (−180°C). Titan's atmosphere is 90 percent nitrogen, with methane making up most of the rest. Clouds of orange smog prevent direct observations of its surface but it has been mapped by probes using infrared technology. In January 2005 the Huygens probe parachuted down to the surface of Titan and discovered a landscape that at a glance looked Earth-like, with lakes, rivers, rocks, and pebbles, together with weather systems, winds, and rain. However, there is one major difference—it rains methane, and those rivers and lakes are liquid-hydrocarbon lakes of methane and ethane. Such a cold and inhospitable-sounding world doesn't at first seem like a candidate for microbial life, but there is thought to be a subsurface ocean. Titan is also rich in complex organic molecules and is a prebiotic environment thought to be similar to primordial Earth. While life may not exist there right now, there is a chance that it could evolve in the future.

Iapetus has a diameter of 910 miles (1,470 kilometers) and is Saturn's third-largest moon. It takes 79 days to orbit Saturn and is notable because one side of it is five times darker than the other. This is thought to be due to dust from a darker, more distant moon, Phoebe, being swept up by the leading side of Iapetus while its trailing side remains lighter.

Enceladus is the second smallest of the major moons, and with a diameter of 310 miles (500 kilometers) it is just a tenth of the size of Titan. It takes just under 1.5 Earth days to complete one orbit of Saturn. Despite being small, it's an incredibly interesting world because it's covered in fresh, clean ice, making it one of the most reflective bodies in the Solar System. It was discovered by William Herschel in 1789 but we knew very little about it until the Voyager and Cassini space probes were able to provide more detail. Enceladus displays very active cryovolcanism, which causes geysers of salty water vapor and other organic molecules to vent out into space. These geysers look stunning in photos taken by the Cassini probe. Some of this water falls back to the surface as snow, which is why the surface is young, with few visible impact features. Some of the material erupted from Enceladus forms Saturn's E Ring. The source of the geysers is a subsurface liquid-water ocean but the source of the heat is not currently fully understood—there is too much to have been generated by tidal heating alone. The presence of liquid water plus organic molecules make Enceladus a prime candidate for subsurface microbial life.

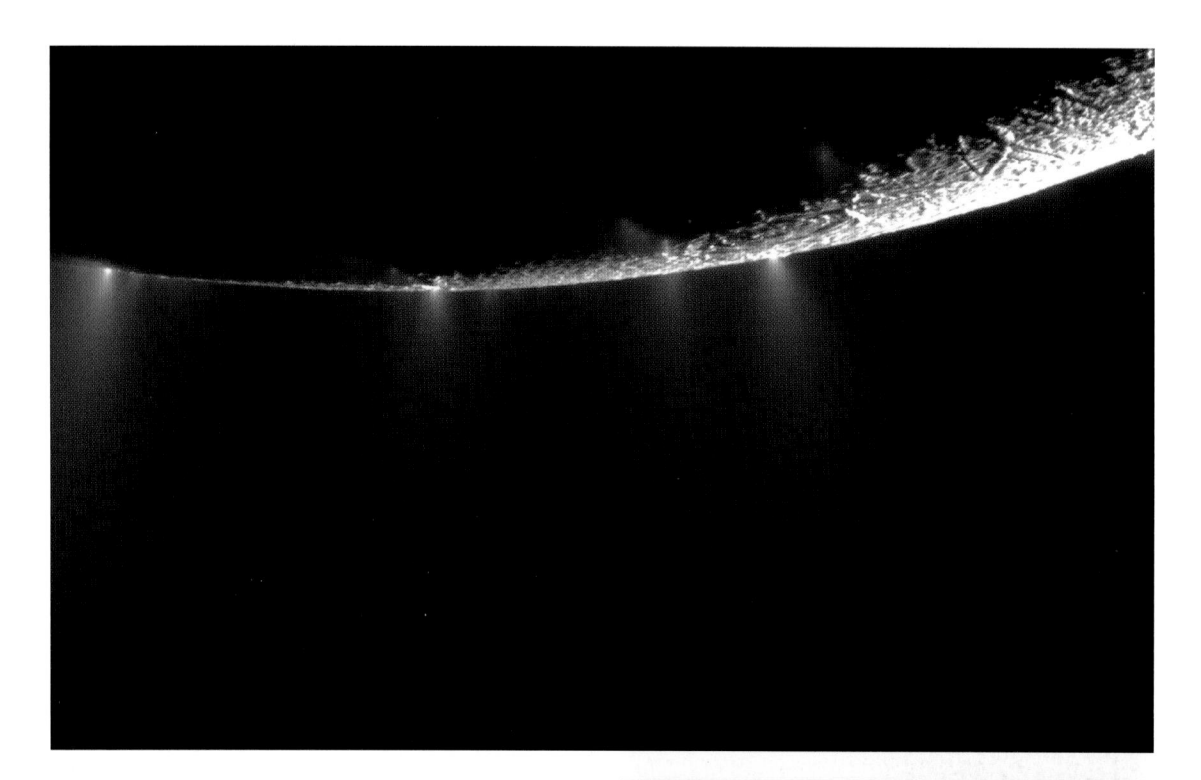

Mimas is the smallest major moon of Saturn and has a diameter of 246 miles (396 kilometers). It takes just under one Earth day to complete an orbit of Saturn. It is saturated with craters but is notable for one particularly huge crater (86 miles [139 kilometers]) with a large central peak, Herschel—named after William Herschel, who discovered Mimas in 1789. This impact feature gives Mimas a striking resemblance to the Death Star from the *Star Wars* movies, although Mimas is actually three times larger.

Top Saturn's moon Enceladus features geysers of salty water vapor and particles. More than 30 individual jets can be seen in this image, taken by the Cassini spacecraft during a 2011 flyby. **Above right** Mimas is dominated by the large Herschel Crater (seen here on the right of the photo), in this Cassini 2010 image.

Uranus

Diameter (equatorial): 31,763 miles (51,118 kilometers)
Mass (Earth = 1): 14.54
Mean density (water = 1): 1.27
Volume (Earth = 1): 63
Axial rotation period (sidereal): 17.24 h
Orbital period (sidereal): 84.02 y
Orbital period (synodic): 369.66 d
Axial tilt: 82.2°
Mean distance from Sun: 1784 x 65.8 miles (2871.0 × 106 kilometers)
Orbital eccentricity: 0.047
Orbital inclination: 0.8°
Moons: 28+

Below Hubble observations have allowed for the discovery of larger rings and previously unknown moons, in addition to the faint rings seen here in this 2003 Hubble image.

Uranus is the seventh planet from the Sun and it has a diameter of 31,763 miles (51,118 kilometers), making it the third-largest planet in terms of diameter. It orbits at 19.2 AU and, because of its location so far out in the Solar System, it is classed as an ice giant. When at opposition it reaches magnitude +5.5, so it's technically just on the boundary of naked-eye visibility, but without magnification its movement day to day is so small that it was not recognized as a planet by early astronomers. It had in fact been observed and included in star catalogs many times prior to its official discovery by William Herschel in 1781, and even he at first identified it as a comet.

If you familiarize yourself with the nearby starfield around Uranus ahead of time, you can easily spot it just with binoculars. Once the position of Uranus is known, you can then watch it slowly move against the background stars as the days pass. It takes 84 years to complete one orbit of the Sun. If you take a simple photograph with a DSLR or a camera phone, you'll be able to see Uranus looking like a star. With magnification, its cyan color can be seen, but even with large telescopes it's extremely difficult to see any kind of surface features visually because of its distance and its very small *angular diameter*. Angular diameter refers to how large a spherical object appears when observed from Earth, and it's measured as a fraction of the full 360-degree circle of the sky. The more distant a planet is, the smaller its angular diameter; the outer planets are so small that their angular diameters are measured in arcseconds (there are 3,600 arcseconds in 1 degree). Uranus's angular diameter is between 3.4 and 3.7 arcseconds—about a tenth that of Jupiter. (Neptune's is even smaller—between 2.2 and 2.4 arcseconds.) That said, camera technology now allows the backyard astronomer to begin to capture some of Uranus's surface features if they have a large telescope.

Uranus is thought to have an iron core, surrounded by an outer icy mantle in the middle, and an outer layer that is a gaseous envelope of hydrogen and helium with about 2 percent methane, which gives it is color. It only takes Uranus 17 hours to complete one rotation, which is retrograde, but it has the most extreme axial tilt of any of the Solar System planets as it is tilted by 98°; something large likely impacted it in the past and knocked it over. This means that the polar regions point toward the Sun, so despite its rapid rotation, one pole gets 43 years of direct Sun, followed by 43 years of night. This results in some pretty extreme weather, with the wind speeds of 540 miles (860 kilometers) per hour. It's also the coldest of the planets, with average temperatures of −370°F (−224°C).

Like the other giant planets, Uranus has banding and dark spots on its surface, and it has a faint ring system and a magnetic field, so auroras have been photographed by the Hubble Space Telescope. There are currently 28 known moons, the major moons being Miranda, Ariel, Umbriel, Titania, and Oberon. The outer Solar System is less well explored so the only spacecraft to have visited Uranus was Voyager 2, in 1986. The James Webb Space Telescope has also photographed it.

Below Not much is known about the auroras of the ice giant Uranus. In 2011, Hubble became the first Earth-based telescope to capture an image of these auroras. This image is a composite image of the planet by Voyager 2 and two different Hubble observations.

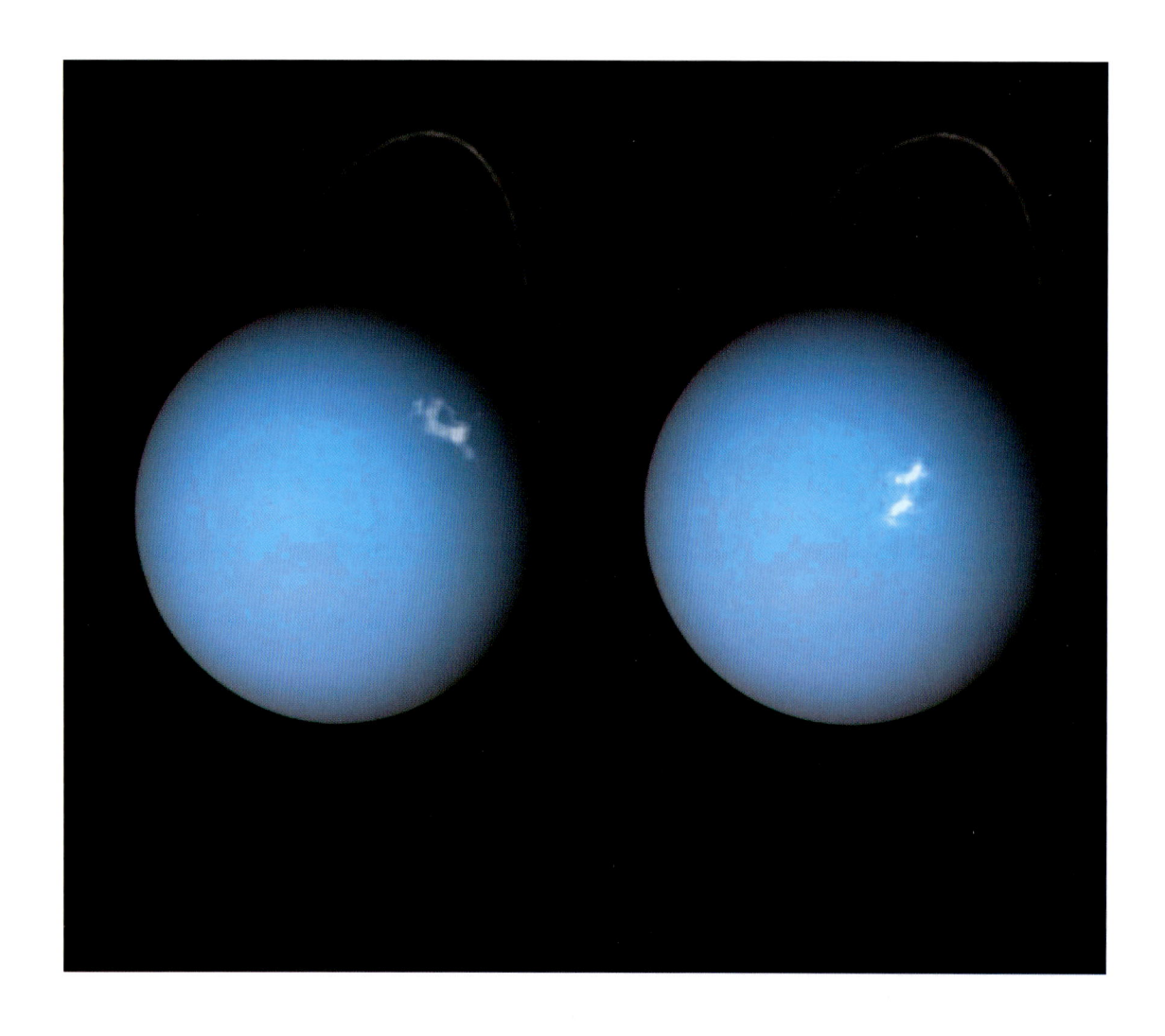

Neptune

Diameter (equatorial): 30,775 miles (49,528 kilometers)
Mass (Earth = 1): 17.15
Mean density (water = 1): 1.64
Volume (Earth = 1): 58
Axial rotation period (sidereal): 16.11 h
Orbital period (sidereal): 164.8 y
Orbital period (synodic): 367.49 d
Axial tilt: 28.3°
Mean distance from Sun: 2795 x 65.8 miles (4498.3 × 106 kilometers)
Orbital eccentricity: 0.009
Orbital inclination: 1.8°
Number of moons: 16+

Below This high-resolution image of Neptune shows the planet in close to its natural color, with the Great Dark Spot, smaller Dark Spot Jr, and white cirrus clouds in view. The photograph was taken by Voyager 2 in August 1989.

Neptune is the most distant planet, orbiting at 30 AU. It has a diameter of 30,775 miles (49,528 kilometers), so it is slightly smaller than Uranus, but has very similar features and characteristics. It is a little warmer than Uranus, but it is still frigid at −328°F (−200°C). With an average magnitude of +7.78, observing this object definitely requires binoculars or a telescope. As occurred with Uranus, Neptune had been recorded as a star by several astronomers prior to its official discovery, but its movement was so slight that it went unnoticed. In 1821 the French astronomer Alexis Bouvard published orbit tables for Uranus, but subsequent observations showed that it deviated from its predicted path. Bouvard believed that the gravitational effect of another planet was causing this variation so the hunt for a new planet in the predicted location began; it was finally discovered in 1846. Neptune only takes 16 hours to rotate but it takes 165 years to orbit the Sun, so its movement against the background stars is very small. As stated earlier, since its discovery in 1846, only just over a year has passed on Neptune!

Neptune is similar in composition to Uranus but it displays more activity and variation in weather patterns, and dark-color storm spots have appeared a few times. In 2018 one such storm spot produced sustained wind speeds of 1,300 miles (2,100 kilometers) per hour—the fastest recorded wind on any of the planets. Historically, images of Neptune have shown it to be a much darker blue than Uranus, but in 2023 some color recalibration was done that revealed them to be more similar to each other. From Earth it's extremely difficult to observe any kind of detail on Neptune, but advanced imaging techniques with large telescopes can bring out some features.

Neptune has a faint ring system and to date 16 moons have been identified. Probably the most interesting is Triton, a moon that has a substantial atmosphere. Despite being extremely cold at −390°F (−235°C), there are active geysers within the polar caps that demonstrate a subsurface liquid-water ocean, which makes it another candidate for microbial life to potentially exist and definitely worthy of further exploration by future space probes. Voyager 2 is the only spacecraft to have visited Neptune (in 1989), and it verified the presence of a magnetic field as well as imaging its active weather systems. The Hubble Space Telescope and James Webb Space Telescopes have both imaged Neptune.

Dwarf planets

Older astronomy textbooks will state that the Solar System has nine planets, with the outermost one being Pluto, so why do we now only have eight? In 2005 a tenth planet was discovered, one that was very similar in size to Pluto and, with telescope and camera technology improving, it was obvious that we were on the brink of discoveries that would lead to a huge increase in the number of planets in our Solar System. In 2006, the International Astronomical Union (IAU) published *Resolution B5: Definition of a Planet in the Solar System*, which included the following definitions:

A planet is a celestial body that (a) is in orbit around the Sun, (b) has sufficient mass for its self-gravity to overcome rigid body forces so that it assumes a hydrostatic equilibrium (nearly round) shape, and (c) has cleared the neighborhood around its orbit.

A "dwarf planet" is a celestial body that (a) is in orbit around the Sun, (b) has sufficient mass for its self-gravity to overcome rigid body forces so that it assumes a hydrostatic equilibrium (nearly round) shape, (c) has not cleared the neighborhood around its orbit, and (d) is not a satellite.

According to these new definitions, Pluto was demoted to dwarf planet status and our Solar System became an eight-planet system. This is a decision that some people feel very strongly about, but whether Pluto is a planet or dwarf planet doesn't take anything away from its unexpected beauty.

Until the New Horizons mission visited Pluto in 2015, our best photo of the planet was little more than a few blurred pixels. Orbiting at a distance of between 29.6 and 49.3 AU and having a diameter of less than 1,500 miles (2,400 kilometers), it only ever reaches magnitude +15, so it's a very difficult object to observe—just capturing it as a star-like dot requires a large telescope. It takes nearly 248 years to complete one orbit of the Sun so its movement against the background stars is extremely small, but with a large telescope you should notice movement every couple of days. The other interesting thing about Pluto's orbit is that is has an eccentricity of 0.25, which is a quarter of the way to being removed from the Solar System! Its orbit means that the difference between aphelion and perihelion is almost 20 AU, and this brings Pluto to within the orbit of Neptune, but they are in stable orbital resonances, which prevents them from colliding. Pluto has five moons, the largest being Charon, which is just over half the size of Pluto itself. The two are tidally locked and because of their orbital dynamics they are treated as a binary pair.

The New Horizons mission launched in 2006 and arrived at the Pluto system in 2015. Some people were skeptical that anything of interest would be found on such a cold rock so far from the Sun, but it was full of surprises. The surface displays great variety in terms of color and brightness—most notably its "heart" (otherwise known as Tambaugh Regio), which is a heart-shaped basin, 620 miles (1,000 kilometers) wide. This bright surface feature displays active cryovolcanism in the form of churning, polygonal nitrogen ice cells, covering a region called Sputnik Planitia on one lobe of the heart. The whole region is being regenerated due to convection; there are no visible impact features and there are signs of glacial flows into and out of the area. There are also peaks on Pluto's surface that look to be ice volcano mergers. All of this points to there being a subsurface liquid-water ocean. There are now calls for a Pluto orbiter mission so we can get a greater understanding of this fascinating world.

The definition of what constitutes a dwarf planet has blurred some of the lines with regards to what is simply a trans-Neptunian object (with an orbit beyond Neptune) and what is a dwarf planet, with some objects technically being both. Some may not meet the hydrostatic equilibrium requirement, but at the time of writing the other most likely proposed dwarf planets are Ceres, Eris, Haumea, and Makemake. However, some astronomers also include Quaoar, Sedna, and Gonggong, and yet more bodies are being proposed. Additionally, there will be more dwarf planet discoveries when the Vera C. Rubin Observatory comes online in 2024/25. This is still a rapidly evolving area of astronomical research, which we'll touch on again later in this chapter.

Opposite top Until NASA's New Horizons spacecraft captured this image of Pluto in 2015, only very blurred images were available. This enhanced-color view from July 2015 shows a range of colors and landforms. **Opposite bottom** This close-up image from New Horizons shows the frozen plains of Pluto's Sputnik Planitia.

20 miles

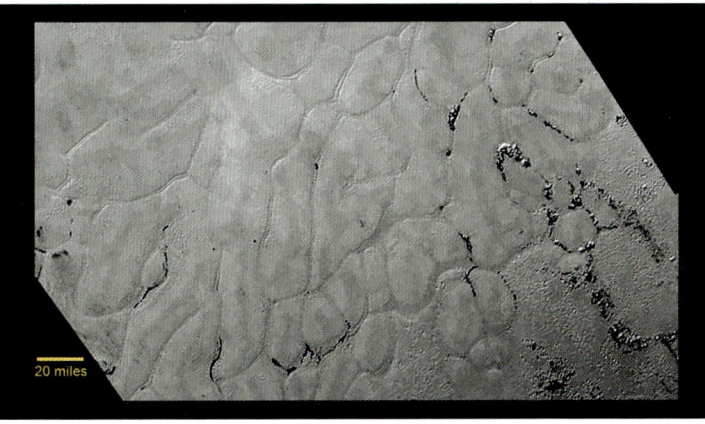

ECLIPSES AND OCCULTATIONS

Around 4.6 billion years ago our star, the Sun, was born from a massive spinning cloud of gas and dust called the solar nebula. Star formation within molecular clouds such as this is triggered by interstellar winds or shockwaves from supernovae, which cause the molecular cloud to begin to collapse under its own gravity. The collapse causes some of the material to form the star and, as the star rapidly rotates, the remaining material spreads out into a rotating flat pancake of gas and dust, with the star at the center. This pancake is called a protoplanetary disk, and this is the material that over time will become planets.

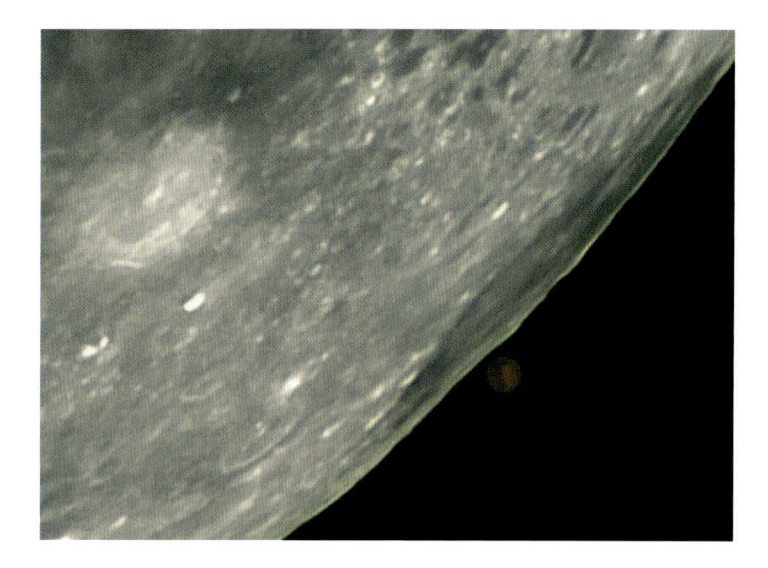

Over millions of years, the material begins to clump together and form planetesimals, which are the precursors of planets. The exact mechanisms of planet formation are poorly understood, but it's widely accepted that planetesimals continue to accrete material within their respective orbits. The composition of the molecular cloud determines what the planets are made from. In our Solar System, the inner four planets are rocky and the outer planets are gas giants. The boundary between the rocky planets and gas giants lies between Mars and Jupiter and it contains an orbiting disk of leftover chunks of material called the asteroid belt (see page 272).

Because they were formed from a flat disk, all planets and asteroids orbit in an almost flat plane, and when they are viewed from Earth, all Solar System bodies (excluding comets) appear to travel along an imaginary band called the ecliptic. Earth rotates on its axis toward the east, so the Sun, Moon, and planets are all seen to rise in the east and set in the west. The planets move across our sky at different speeds than the stars, sometimes even appearing to move backward day by day (what's known as retrograde motion). The Moon and Sun also move across the sky at different speeds, so a planetary game of cat and mouse ensues. The Moon and planets may catch up with each other and appear in the same part of the sky, when they are said to be in conjunction. Of course, they're not actually near each other; this is just a line-of-sight effect when viewed from Earth. An asteroid may pass in front of a star, the Moon may pass in front of some bright stars or a planet, or in fact any Solar System body whose orbit is closer to Earth can pass in front of a more distant body, and this is known as an *occultation*. Mercury and Venus orbit closer to the Sun than Earth so they can sometimes pass in front of the Sun—the silhouette of the planet is then seen transiting the solar disk. Mercury transits the Sun about 14 times per century but Venus transits are rarer. The last Venus transit was in June 2012 but the next one won't be until 2117. The Moon is our closest neighbor and is in orbit around the Earth; when it passes in front of the Sun it causes an eclipse.

Opposite This arresting photograph by Mary McIntyre clearly shows the Sun's corona during a totality of the solar eclipse on August 21, 2017.
Top Mars emerging from behind the Moon following the lunar occultation on December 8, 2022 in this photo from Oxfordshire, U.K.

Anatomy of a solar eclipse

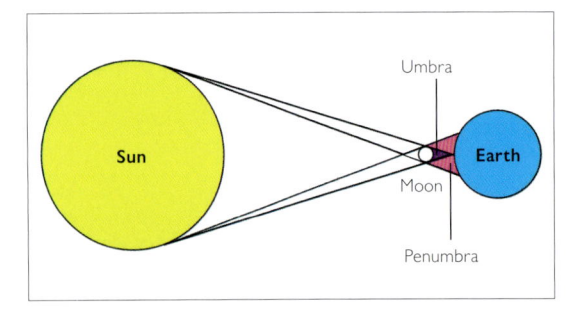

Solar eclipses

The Moon is 400 times smaller than the Sun, but the Sun is 400 times farther away, so when viewed from Earth, the Sun and Moon are almost the same size. The size of the Moon varies throughout the month because its orbit is not perfectly circular; when the Moon is at perigee—at its closest point—it will appear larger than it does at apogee, when it's at its farthest point. We're the only planet in the Solar System that has this chance ratio, so we're the only planet lucky enough to see one of nature's most impressive sights: a total solar eclipse.

As the Moon orbits the Earth, the portion of its surface that's illuminated by the Sun changes, leading to the constantly changing phases seen during a lunar cycle. During a New Moon, the Moon is in the same part of the sky as the Sun, so the side of the Moon facing Earth gets no visible sunlight and we can't see it. Sometimes the Sun and Moon align perfectly, and the Moon blocks out all of the Sun's visible surface, the photosphere; this allows us to see the Sun's beautiful tenuous outer atmosphere, known as the corona. This part of an eclipse is called totality, and it's the only time that the corona can be seen from Earth.

If you viewed a solar eclipse from the lunar nearside, you'd see the small shadow of the Moon being cast down onto the Earth. This shadow moves across the Earth's surface at the speed of the Moon's orbit around the Earth—2,288 miles (3,683 kilometers) per hour—although the rotation and curvature of the Earth will cause variations in the speed from eclipse to eclipse. The shadow has a darker region in the center called the umbra, and a less dark region around the outside called the penumbra. To see a total solar eclipse you need to be in the line of the umbral shadow—known as the path of totality.

This is only about 100 miles (160 kilometers) across and will typically travel a path of around 10,000 miles (16,000 kilometers) across the globe, but the shadow doesn't always travel over well-populated areas; it very often passes over oceans or unpopulated Arctic/Antarctic regions or deserts. Totality only lasts from a few seconds up to a maximum of 7 minutes 32 seconds. If you're outside the umbral shadow but in the penumbral shadow, you'll only witness a partial eclipse.

If a solar eclipse occurs when the Moon is at apogee, its apparent diameter is too small to completely cover the Sun, so it leaves a border around the outside often referred to as the "ring of fire." This is an annular eclipse.

There's a New Moon every month so you'd be forgiven for thinking that a total eclipse should also happen every month, but this isn't the case. For a solar eclipse to occur, the Sun, Moon, and Earth need to be in a perfectly straight line, but the Moon's orbit is slightly tilted, so most months the Moon does not eclipse the Sun. Approximately twice a year there is a partial eclipse, but as mentioned above, these don't always occur in populated areas. Total solar eclipses occur approximately every 18 months.

Lunar eclipses

A lunar eclipse occurs when the Moon passes through the shadow of the Earth being cast out into space. Lunar eclipses can only occur during a Full Moon, when the Moon is on the opposite side of the Earth from the Sun and the lunar nearside is fully illuminated. As the Moon moves through Earth's shadow, the shadow starts to block out the light received by the Moon, and a curved disk with a blurred edge is seen moving across the lunar surface. Earth's shadow has an umbral and penumbral region and if the Moon passes through the penumbral shadow only, a penumbral eclipse occurs, during which there is a subtle dimming effect on the edge of the Moon. If the Moon moves through part of the umbral shadow, a partial lunar eclipse occurs, but if it enters the umbral shadow fully, a total lunar eclipse occurs and the lunar surface turns a shade of brown, red, or orange—what's sometimes referred to as a "Blood Moon." Light refracting through Earth's atmosphere is thought to cause this effect, but the exact color is determined by local conditions, such as how much dust or moisture is in the air, so it can't be predicted accurately. If Earth had no atmosphere, the Moon would remain completely black during totality. The shadow of the Earth is much larger than the lunar shadow seen on the Earth during a solar eclipse, so the much smaller Moon takes a long time to pass through it. The full duration of a lunar eclipse is several hours, with totality lasting between 30 minutes and an hour.

As with solar eclipses, the Moon's tilted orbit means that lunar eclipses only occur a couple of times a year. When they do occur, they're visible from a much larger area of the Earth than solar eclipses.

Anatomy of a lunar eclipse

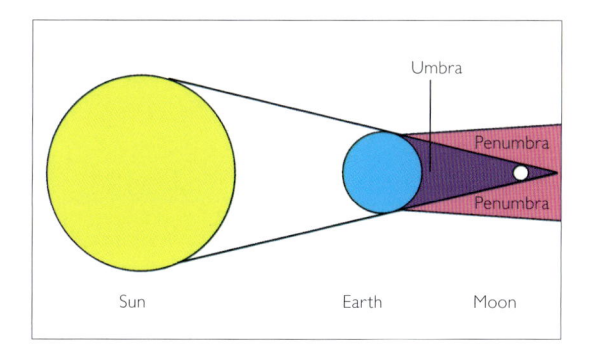

Saros cycles

The alignment of the Sun, Earth, and Moon to form total solar or lunar eclipses is cyclical in nature. The synodic month is the exact time it takes for the Moon to cycle through its phases: 29 days 12 hours 44 minutes and 2.9 seconds. A Saros measures the time taken for 223 synodic months (approximately 18 years) to pass. If there's an eclipse on a particular date, one Saros period later an identical eclipse will occur, but because the Saros isn't an exact number of days in length, each successive eclipse will be about eight hours later in the day than the previous one. For solar eclipses this means the path of totality will move westward about a third of the way around the globe, so although the two solar eclipses will be identical in orbital geometry, they'll not be visible from the same place. There are also multiple Saros cycles operating at the same time. In the 18-year period of one Saros cycle there'll be approximately 40 other lunar or solar eclipses taking place, each of them with their own Saros. There's also a relationship between solar and lunar eclipses; exactly half of a Saros—a "Sar"—after a lunar eclipse, a solar eclipse will occur, and vice versa. Knowledge of Saros cycles has helped astronomers to predict eclipses for millennia, with records dating right back to the Babylonians.

Below This montage demonstrates the visual stages of a lunar eclipse, created from images in Mill Valley, California, U.S.A, on August 29, 2007.

ASTEROIDS AND COMETS

As discussed previously, when the Solar System formed, material within the protoplanetary disk began to clump together, coalescing over time into planetesimals and, eventually, planets. Material within the orbit of each body was eventually swept up by that planet, but some material was prevented from forming into planets because Jupiter's gravity disrupted the process. The result was a torus of leftover material. This torus orbits the Sun in between Mars and Jupiter and is known as the asteroid belt.

If you've ever seen a movie where a spacecraft flies through this belt, ducking and diving through a dense stream of asteroids, this is not a realistic representation. Most of the asteroids are an average of 600,000 miles (1 million kilometers) apart, so if a spacecraft encountered an asteroid, the chances of it flying past another would be very low!

By 2023 it was estimated that there were over a million objects in the asteroid belt. (The number is constantly growing for two reasons. Firstly, asteroids are constantly colliding and breaking into smaller pieces, and comets can also fragment. Secondly, telescope/camera technology and detection methods are always improving, allowing us to see smaller and fainter objects.) While a million is a huge number, most of these are very small. The total mass of the asteroid belt is only about a thousandth of Earth's mass, and about 60 percent of the total asteroid belt mass belongs to its four largest asteroids: Ceres, Vesta, Pallas, and Hygiea. Ceres has a diameter of about 590 miles (950 kilometers), which is about the length of the United Kingdom, making it large enough to be classed as a dwarf planet. The average diameter of Vesta, Pallas, and Hygiea is less than 370 miles (600 kilometers). The remaining objects range in size, going down to the size of a dust particle.

Despite the huge number of objects in the asteroid belt, these are spread out over such a vast area that it's more accurate to think of it as a very diffuse swarm of different-sized orbiting bodies. The nearest edge of the band is 1.86 billion miles (3 billion kilometers) from the Sun, and the outer edge is nearly 3 billion miles (5 billion kilometers) from the Sun.

Early astronomers thought the asteroid belt was the remains of a single body that had disintegrated but we now believe that it's actually fragments of many different small bodies. Although the mass of the asteroid belt is only a thousandth of Earth's mass, a few Earth masses of material are thought to have originally been available in this part of the protoplanetary disk. The effect of Jupiter's gravity was to essentially "stir up" this mix and cause collisions that continuously broke pieces apart rather than allowing them to coalesce into a single body. Collisions within the asteroid belt are one source of the interplanetary dust that exists in the space between the planets. Sunlight that's scattered off this dust is responsible for the phenomenon known as zodiacal light, a very diffuse triangle of light extending from the ecliptic toward the direction of the Sun.

Jupiter's powerful gravity is responsible for organizing certain classes of asteroid into particular orbits within the belt, with many of them having orbital resonances with that planet. If something orbits the Sun twice for every time Jupiter orbits it, the object is said to have a 2:1 orbital resonance; if something orbits the Sun five times for every time Jupiter orbits it three times, it has a 5:3 resonance, and so on. Asteroids reside in different parts of the asteroid belt, but there are some areas where very few asteroids exist; these are called Kirkwood gaps, after the American astronomer Daniel Kirkwood, who discovered their existence in 1866. Kirkwood realized that these gaps were not random, but instead corresponded to regions that had once had specific orbital resonances with Jupiter, occurring at 3:1, 5:2, 7:3, and 2:1 resonances. When a small body has an orbital resonance with a much larger body, the gravitational forces upon the smaller body can cause it to be accelerated or decelerated, and this can push it away from a particular orbit and into a more stable one. Most of the bodies that used to reside in the Kirkwood gaps have been cleared out by Jupiter's gravitational force. Jupiter's gravity also has the effect of locking asteroids in particular stable orbital resonances.

The movement of asteroids between different resonances is known as orbital evolution and understanding this behavior is very important when it comes to identifying near-Earth objects (NEOs). These are asteroids or comets that have orbits passing near Earth. Potentially hazardous asteroids (PHAs), whose orbits bring them onto potential collision courses with planets have, in astronomical terms, quite short life spans, yet new PHAs are constantly being discovered. The supply of PHAs is being replenished by gravitational forces from Jupiter and Mars, pushing them into the inner Solar System—a process that can also work in reverse; Jupiter pulling PHAs back out of the inner Solar System has protected Earth from collisions. Occasionally, asteroids are captured by a planet and orbit it instead. We have evidence of this in the Mars system—Phobos and Deimos, the two moons of Mars, are believed to be captured asteroids.

Opposite The relative size comparison between Ceres (one of Earth's largest asteroids, also classed as a dwarf planet), the Earth, and the Moon are displayed here.

Asteroid classification

There are three main categories of asteroids—C-type carbonaceous asteroids, S-type silicate asteroids, and X-type hybrid asteroids—but it's difficult to definitively assign all asteroids to one of these categories because they include many subclasses with overlapping characteristics.

C-type asteroids are primarily found in the outer region of the asteroid belt and make up 75 percent of all visible asteroids. They're rich in carbon so therefore have a low surface brightness (albedo)—because carbon is dark it doesn't reflect light as well—and spectral analysis shows that their composition is very similar to that of the early Solar System, making them a potentially important source of information. Spectral analysis involves splitting the light from astronomical objects and studying its component parts. (If you've ever seen a spectrum of light created when sunlight shines through something acting as a prism, such as a rainbow, you've seen this in action.) Astronomers study these components to gain information on temperature, density, mass, and more.

S-type asteroids are mostly found in the inner part of the asteroid belt and make up about 17 percent of the total asteroid population. They're rich in silicates, with some metals present, and a lack of any carbonaceous compounds suggests that they've undergone changes since the early days of the Solar System, with tidal heating and surface melting resulting in their current composition. They are brighter than C-type asteroids.

X-type asteroids are primarily found in the middle of the asteroid belt and make up most of the remaining asteroid population. They can be partially split into subgroups based on their albedo: M-type metallic, P-type primitive, and E-type enstatite. Spectral analysis suggests that many of these asteroids consist of nickel–iron, and they are thought to have come from the cores of differentiated bodies (bodies that were large enough for their gravity to have begun to pull the heavier elements into the center to form a core) that were then broken apart by collisions.

Some asteroids don't fit within these classes and it's not certain if all X-class asteroids actually have the same composition, or if they're just bodies that don't fit well into the C- or S-type classes. Many asteroids are not a solid chunk of any one material but are comprised of lots of different-sized fragments of material that have accreted over time and are held together by gravity—essentially, gravitationally bound "rubble piles."

In addition to the asteroids that reside within the asteroid belt, there's another class of asteroids known as the Jupiter trojans. These share Jupiter's orbit around the Sun and are located in one of the planet's two stable Lagrange points—points where there is gravitational equilibrium on a small body. One group is located 60° ahead of Jupiter in its orbit, and the other is 60° behind. The first Jupiter trojan was discovered in 1906 but we now know that there are about 10,000 of them. Most are D-type—so named because they are very dark and tend to have spectra that are very red. However, some trojans are P-types and C-types.

Some asteroids in the outer asteroid belt contain a lot of ice, and small impacts can cause this ice to turn directly from solid to gas in a process called sublimation, causing the asteroids to display similar characteristics to comets. These are classed as active asteroids, but again, the exact boundary between what's a comet and what's an active asteroid can be blurred. On September 26, 2022, the Double Asteroid Redirection Test (DART) mission impacted the asteroid Dimorphos, a moonlet in orbit around the asteroid Didymos. This mission was the first to test if an asteroid could be deflected by impacting it. Following the impact, photographs taken by the Hubble Space Telescope show Dimorphos looking exactly like a comet.

Until recently, much of our asteroid knowledge came from spectral analysis. This can be unreliable because some asteroids display spectral features usually associated with a completely different class. For example, some S-type asteroids can show similar spectral characteristics to M-class asteroids, even though there's no metal present in them. Asteroid sample-return missions have allowed the direct study of samples, and in December 2020 the JAXA Hayabusa2 spacecraft returned samples from C-type asteroid Ryugu. One sample was taken from the surface, and another from the subsurface material that was expelled after an impactor was launched; they were found to contain organic compounds and water. In September 2023, NASA's OSIRIS-REx spacecraft returned samples from another C-type asteroid, Bennu. The first analyses revealed a high water content and a high percentage of carbonaceous

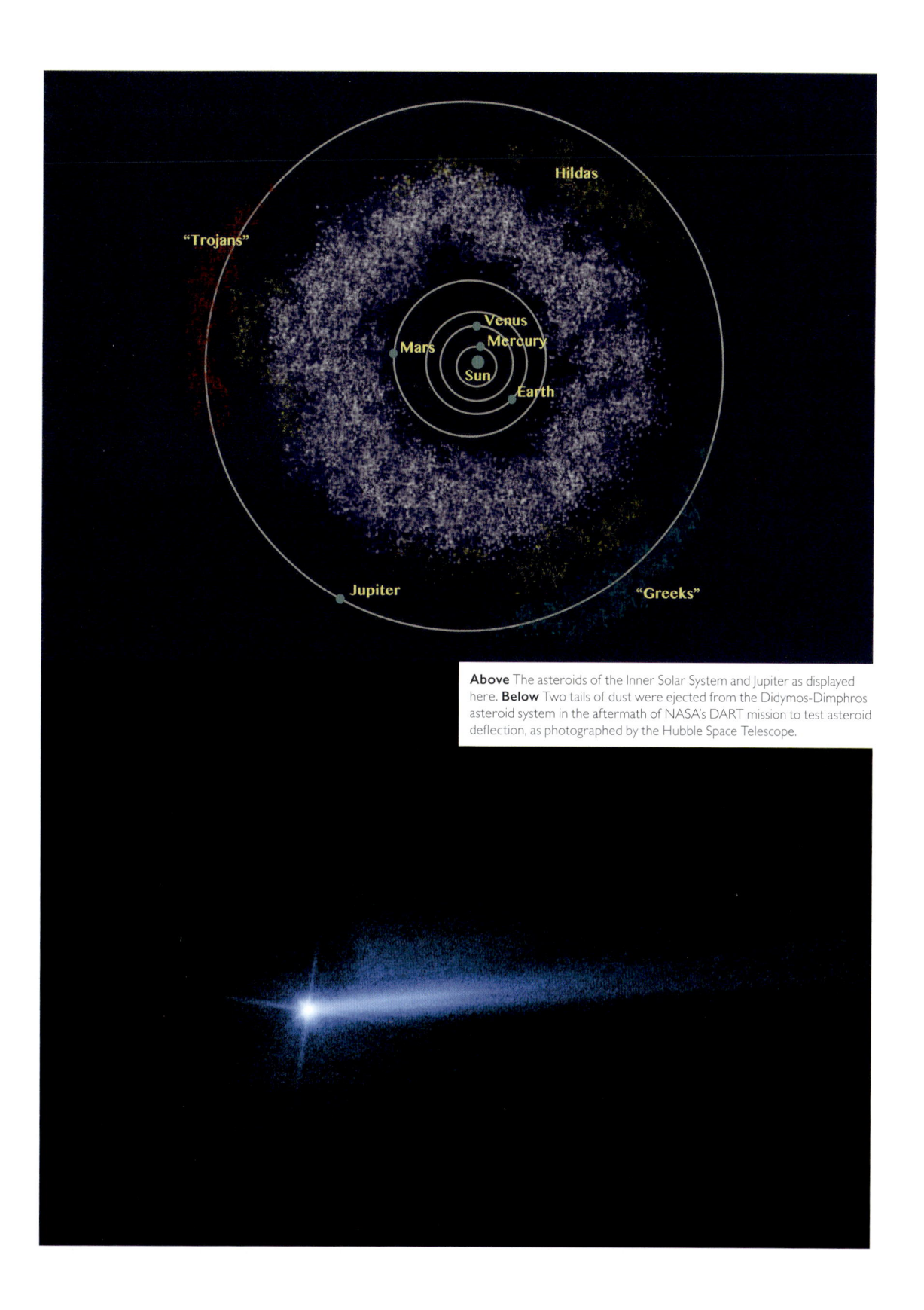

Above The asteroids of the Inner Solar System and Jupiter as displayed here. **Below** Two tails of dust were ejected from the Didymos-Dimphros asteroid system in the aftermath of NASA's DART mission to test asteroid deflection, as photographed by the Hubble Space Telescope.

material. Samples from both of these missions will give us a better understanding of the early Solar System and help to answer the question of where Earth's water came from. One other way we can learn more about asteroids is by studying meteorites, because there are some fragments within carbonaceous chondrites that contain primordial Solar System material, and some have inclusions that pre-date the birth of our Sun. Being able to locate and analyze these is vitally important for investigating the composition of the solar nebula—something we'll return to on page 286.

The Kuiper Belt

For many years astronomers believed that our Solar System extended beyond the orbits of Uranus and Neptune, but existing technology made it difficult to discover small objects so far from the Sun. The American astronomer Percival Lowell started to search for a planet beyond Neptune in 1906 and his observatory unknowingly captured two faint images of Pluto in 1915, but it wasn't until fellow American Clive Tombaugh began to use a blink comparator on pairs of photographs taken at different times that Pluto was finally discovered. This was not as easy as it sounds because Pluto is extremely small and faint, and it takes almost 248 Earth years to complete one orbit around the Sun, so its daily motion is very small. It took Tombaugh nearly a year of searching, but he was finally able to confirm Pluto's discovery on February 18, 1930. Our Solar System now had a ninth planet but it was very small compared to Neptune; astronomers therefore believed that other small bodies must exist. In the United States, Frederick C. Leonard was the first astronomer to suggest the existence of these bodies, and in 1943 Kenneth Edgeworth published a paper in the *Journal of the British Astronomical Association* in which he hypothesized that the outer region of the Solar System was populated by a large number of comparatively small bodies.

The Dutch-American astronomer Gerard Kuiper did a lot of work on this in the 1950s, but the technology to find such small bodies that move such a small amount each day did not yet exist. In 1977, asteroid (2060) Chiron was discovered between Jupiter and Uranus, but its orbit was found to cross the path of Saturn's orbit. Subsequently, asteroid (5145) Pholus was discovered and this also crossed the orbits of the gas giants. This was a new class of asteroids called centaurs. As mentioned earlier, asteroids with orbits that bring them into the path of larger planets are not long-lived because they'll be subjected to gravitational changes and possibly even impacts, so the discovery of these two asteroids indicated a steady supply of centaurs coming from somewhere. It was postulated that there was another, much larger belt of bodies that extended way beyond the orbit of Neptune. It was named the Kuiper Belt in honor of Gerard Kuiper, but others refer to it as the Edgeworth–Kuiper Belt to acknowledge Edgeworth's contribution.

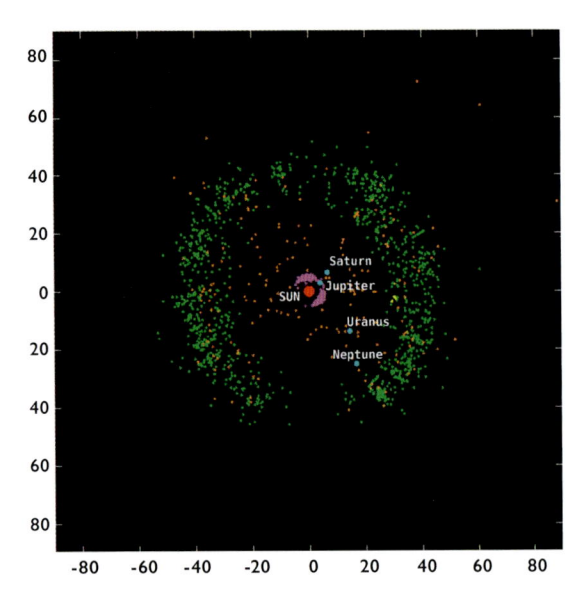

In 1992, the minor planet Albion was the first Kuiper Belt Object (KBO) to be discovered after Pluto and, as technology has improved, the number of KBOs has reached in excess of 100,000. There are three distinct classes of KBO—plutinos, classical, and scattered disk—and these have particular orbital resonances with Neptune. As the name suggests, Pluto belongs to the plutino group, and these objects are in a stable 3:2 resonance with Neptune. As more and more plutino objects were discovered, it became necessary to reclassify Pluto as a dwarf planet (defined as a body that's large enough that its own gravity has formed an almost spherical shape), although a great many astronomers remain angry about its "demotion." The largest asteroid, Ceres, is large enough to be

Top The Outer Solar System's Object Positions are charted here, with lime green for Kuiper Belt Objects, orange for scattered disk objects or centaur, and magenta for Jupiter's "trojans."

considered a dwarf planet, even though it resides in the asteroid belt. Some of the scattered-disk objects exhibit centaur-like behavior, so once again there's an overlap between subgroups.

The Kuiper Belt is 20 times wider and up to 200 times more massive than the asteroid belt, and the orbits of some of the scattered-disk objects extend out to nearly twice the distance of Neptune's orbit. It's now believed that the scattered disk region is the source of short-period comets—comets that have an orbital period of under 200 years (see below). We still don't fully understand the Kuiper Belt, but as technology improves we'll continue to learn more. The Vera C. Rubin Observatory in Chile houses the Large Synoptic Survey Telescope, which is due to come online by 2025. This has a primary mirror measuring 27.5 feet (8.4 meters), and the entire sky will be imaged every few nights using a 3200-megapixel camera, generating a vast amount of data over a wide field. One of the goals is to map NEOs and KBOs, and it's estimated that the number of small Solar System bodies could increase by a factor as high as 100. This will undoubtedly help us to learn more about the populations of these objects but there'll be a need for citizen science projects to help with data classification.

The Oort Cloud

The Oort Cloud is believed to be a vast cloud of many billions of orbiting bodies that never formed planets. It is made up of two regions. The disk-shaped inner region starts at about 2,000 AU (1 AU, or astronomical unit, is the distance from the Earth to the Sun, equating to 93 million miles [150 million kilometers]). The outer region is spherical and it encases the entire Solar System, extending out to as far as 200,000 AU, which equates to 3.2 light years. For context, the Voyager 1 space probe was launched in 1977 to study the outer Solar System. It's traveling at 38,210 miles (61,493 kilometers) per hour and will take 300 years to reach the inner edge of the Oort Cloud; once there it will take a further 30,000 years to pass through to the outer edge. The Oort Cloud is so far away that it's only loosely bound to the Solar System and is therefore prone to gravitational effects from passing stars. If a passing star gives one of the icy bodies a gravitational kick, it can send it off on a very long journey to the inner Solar System.

Comets

Comets are low-density, icy bodies that never made it past the planetesimal stage. They formed in a region of the solar nebula that was cold enough for ice to form—the same region of space where Uranus and Neptune formed. Many comets are probably similar in structure to an icy version of C-type asteroids, but they are mostly water ice and also contain frozen volatile compounds such as carbon dioxide, carbon monoxide, ammonia, and methane. Some comets also contain organic molecules, hydrocarbons, and amino acids. This ice also contains dust grains, so the ices are more like fluffy, dirty snow than the kind of solid ice you find in an ice-cube tray in your freezer.

The main body of a comet is called the nucleus, which is usually dry and dusty, with a very low albedo; they are among the least reflective bodies in the Solar System so it's very difficult to see them until they approach the inner Solar System. Comet nuclei are usually less than 38 miles (60 kilometers) in diameter, but measuring their size accurately is difficult because we often can't see them until they begin to form a coma—a fuzzy cloud that surrounds and obscures the nucleus (see below). They're usually irregular in shape because they've never accreted enough material to have the amount of gravity required to form an almost spherical shape.

Comets fall into two main categories: periodic (short period) and non-periodic (long period). The orbital period of a comet is how many years it takes for it to complete one orbit into the inner Solar System. Short-period comets have an orbital period of under 200 years, so Halley's Comet (1P/Halley), for example, which has an orbital period of about 76 years, is a short-period comet. Long-period comets have orbital periods longer than 200 years, some even reaching into millions of years. The scattered-disk objects within the Kuiper Belt is thought to be the source of short-period comets, whereas long-period comets come from another source of icy planetesimals called the Oort Cloud, or Öpik–Oort Cloud. Many comets with orbital periods of less than 20 years are Jupiter-family comets; although they originated in the Kuiper Belt, their orbits are controlled by Jupiter's gravity.

Comets have been observed for millennia, and throughout history they've often been considered bad omens, associated with natural disasters or the deaths of famous people, even if such deaths occurred several years before or after the comet's apparition. Comets don't follow the path of the ecliptic as other Solar System bodies do; they can appear anywhere in the sky, sometimes coming in at very steep angles and changing in brightness as they move across the sky. This unpredictable behavior made people very suspicious of them.

In the days before telescopes were widely used by amateurs, the number of comets that were readily visible was quite small, so they tended to be named with the year of their apparition, and perhaps the name of someone whose death had become linked to it. Examples include Caesar's Comet of 44 BCE, which was visible soon after Julius Caesar was assassinated, and the Great Comet of 1680. Later, comets often took the name of the person who discovered them or calculated their orbit. In the twentieth century, up to 1994, another further system was used, with comets named according to the year of discovery plus a lower-case letter referring to the order of discovery within that year. Once the comet had passed through perihelion the name was changed to the year of perihelion, followed by a Roman numeral referring to the order in which those comets reached perihelion. (Comets are said to be at perihelion when they reach the point in their orbit that's closest to the Sun.) This was an extremely impractical system once technological improvements dramatically increased the number of comet discoveries, so the modern naming convention was brought into effect.

If a comet has a period of less than 200 years, the name starts with a P. If it's longer than 200 years, or non-periodic, then it stars with the letter C. If a comet name begins with D it means it has disintegrated or been lost, while X is used for comets for which we have no reliable orbit—these are usually historic comets that visited before orbit mathematics were refined. More recently, the I class was added for interstellar objects that didn't originate in the Solar System. After this prefix letter comes the year of discovery and then a letter that relates to when in the year a comet was discovered. The year is broken up into half months, so a comet discovered between January 1 and 15 would be labeled A, the second half B, the first half of February would be C, the second half D, and so on. The number after that relates to the number of comets discovered in that half month. As an example, a non-periodic comet discovered on January 7, 2024 that's the fourth discovered during that half month would have the name C/2024 A4. Finally, the name of the discoverer is included in parentheses at the end. This could be the sky survey (carried out by robotic and space telescopes) or up to three named individuals responsible for the discovery. This system is complicated but necessary now that we're discovering around 20 to 30 new comets every month. Historic comets have been given new designations but people often still refer to them by their old name—for example, 1P/Halley is so named because it was the first periodic comet identified, but it's frequently referred to simply as Halley's Comet.

As comets move toward the inner Solar System and get closer to the Sun, their ices sublimate (turn directly from a solid to a gas), resulting in jets of gas and dust leaving the comet. When the Rosetta spacecraft followed 67P/Churyumov-Gerasimenko as it passed through perihelion, it photographed some incredible jets of material being ejected from fissures in the comet's nucleus. The ejected material begins to form a very tenuous atmosphere around the comet nucleus, and this is called a coma. The coma is usually made up of water and dust, and it will start to form when the comet is within 3 to 4 AU from the Sun. Although a comet nucleus may be relatively small, the coma can become huge; diameters of thousands or even millions of miles have been observed. The Great Comet of 1811 had a coma that exceeded the diameter of our Sun, at almost 864,000 miles (1.4 million kilometers). Due to sunlight heating up the volatile gases and forming dicarbon (two carbon atoms joined together), comas often have a green glow. Dicarbon is not very stable so it breaks down into single carbon atoms again before the dicarbon reaches the tails, so the tails are often a different color than the coma. When a comet reaches a distance the equivalent of the orbit of Mars (approximately 1.5 AU) it's close enough to the Sun that it can be affected by solar wind. This blows some of the gas and dust away and the coma can reduce in size, but the sublimated material instead contributes to the size of the tails.

Comets move at speeds of between 12 and 44 miles (20 and 70 kilometers) per second, so you'd be forgiven for thinking that its tails are the trail of material it leaves as it whizzes through space. This is not exactly the case. Although this material does leave the comet as it moves along its orbit, its tails are always sculpted by the solar wind, which pushes the tails away from the Sun. Once a comet has gone past perihelion and is on its way back out of the Solar System, the comet is usually traveling tails first.

The solar wind pushing gas and dust away from the comet causes two visible tails to form. The dust tail is illuminated by the Sun and usually looks white or yellow. Although sculpted by the solar wind, the dust particles can be slower to spread out, and because the comet is moving in a curved orbital path, the dust tail may also look curved. A dust tail can reach lengths of 6 million miles (10 million kilometers). Occasionally a comet also displays an anti-tail—a tail that points in the opposite direction to the other tails. This is a line-of-sight effect caused by dust being left in the comet's orbital plane as the comet moves.

The gas that leaves the comet becomes ionized by UV radiation from the Sun and this forms the ion tail. It is usually blue because it contains carbon monoxide ions. The ionized gas is also swept away from the comet in the direction that the solar wind is traveling, but it does so more readily than the dust tail because the ions are smaller, so the ion tail always points away from the Sun in a straight line. Ion tails can be even longer than dust tails, with some examples having been seen to reach lengths of 93 million miles (150 million kilometers).

Top Comet 67P/Churymov-Gerasimenko, as taken by the NavCam on the Rosetta spacecraft on July 7, 2015, showing jets of material coming from the comet's nucleus.

Comets are at their brightest when they are at perihelion. That said, they can form fissures that lead to increased outgassing and jets that cause the comet to suddenly increase in brightness; this is when a comet is said to be in outburst. Although we detect many new comets a month and can calculate their orbital parameters very well, it's extremely difficult to predict exactly how bright a comet will be as it approaches perihelion, or what its tails will look like. The orbital parameters may also be changed by outgassing from fissures that can cause the comet to change course. As the great Canadian astronomer (and discoverer of 23 comets) David H. Levy once said, "Comets are like cats: they have tails, and they do precisely what they want!"

Occasionally there are comets that are easy to see with the naked eye that would be spotted by a non-astronomer casually looking at the night sky. These are known as "great comets" and occur on average once a decade. One of the best recent great comets for northern-hemisphere observers was C/1995 O1 (Hale–Bopp), which became clearly visible to the naked eye in January 1997 and remained visible for several months. It was brighter than most stars and even visible in the evening twilight. In January and February 2007, southern-hemisphere observers were treated to C/2006 P1 (McNaught), which was so bright that it was visible in daylight and had a spectacular dust tail. Although nowhere near as impressive as Hale–Bopp, C/2020 F3 (NEOWISE) put on a nice show in the northern hemisphere in June and July 2020. It was large and bright enough to be photographed easily with a smartphone camera or budget DSLR camera, so rapidly became the most photographed comet in history.

On its journey into the inner Solar System, a comet will potentially undergo significant surface changes and shed a large percentage of its material. Comets with periods of only a few years will, in astronomical timescales, be very short-lived because all of their volatiles will be stripped away, leaving a "dead comet," although the line between active asteroid and dead comet can be very blurred.

Because comets leave behind a trail of dust, when Earth passes through that debris field every year, it gives rise to meteor showers (which we'll look at in the next section). Some short-period comets visit us every few years, where others may only visit once in our lifetime. Comets that have made repeated perihelion passages will become significantly transformed and some are not able to survive the journey, with exposure to the Sun's radiation causing them to fragment. Comets that pass very close to the Sun are called sungrazers. About 90 percent of sungrazing comets belong to a subcategory called the Kreutz Group; all comets in this group are believed to be fragments of one giant comet that broke into many pieces. There are other sungrazer subcategories too—the Kracht, Kracht 2a, Marsden, and Meyer Groups—but these are still poorly understood.

Long-period comets have orbital periods ranging from 200 years to thousands or even millions of years, and some will only ever come into the inner Solar System once. Very long-period or non-periodic comets have not been subjected to as many transformative processes, so these give us the chance to study a potentially pristine piece of the very early Solar System.

As of 2023, there have only been two confirmed visitors that originated from outside of our Solar System. The first was 1I/2017 U1 'Oumuamua, a long, thin object that was discovered in October 2017. It didn't show visible cometary activity but changes to its trajectory suggested there was active outgassing of some kind taking place. The second object was comet 2I/Borisov, discovered in August 2019, and this did display cometary characteristics with a visible coma, and a tail that was 14 times larger than the Earth. Its composition was found to be very depleted in water and dicarbon, but rich in carbon monoxide, giving the comet more of a blue color. While this is less common in comets originating in our Solar System, a similar composition was observed with C/2016 R2 (PanSTARRS). We've now also discovered at least 30 exocomets—comets that are orbiting stars other than our Sun or are rogue comets not tied to a star—and that number will continue to rise as discovery methods improve.

Numerous comets are visible at any one time from Earth, but they're usually so faint that a moderately sized telescope is required to see them. The improvements in CCD and CMOS camera technology for more affordable prices has allowed the amateur astronomy community to image comets more successfully than ever before. Every amateur observation of a comet is valuable to the professional comet-observing community, especially for long-period or non-periodic comets that we only get one chance to observe.

The subject of water within comets is an interesting one. The exact mechanisms involved with early planet formation are not fully understood, and we still don't have definitive answers for where all of Earth's water came from; a huge amount of the Earth's surface is covered in oceans, and terrestrial processes cannot account for it all. One popular theory is that water and other organic compounds were delivered to Earth by impacts from a huge number of comets and asteroids during a period of heavy bombardment 4 billion years ago. Several space missions have sent probes to comets, and one of the things they've studied is the isotopic composition of the water contained within them—some samples have matched Earth's water, but others haven't. We also know that comets contain molecules such as amino acids and other important building blocks of life, so some scientists believe that comets may be in part responsible for the origins of life on Earth.

Top Comet C/1995 (Hale-Bopp), as photographed in Croatia on March 29, 1997, was one of the best recent comets for northern-hemisphere observers, as it stayed visible to the naked eye for several months.

Above Comet C/2006 P1 (McNaught), as seen on January 23, 2007 over Victoria, Australia, was a treat for southern-hemisphere viewers and so bright that it was visible in daylight. **Below** The main body of a comet is called its nucleus, which is dry and dusty. As comets get closer to the Sun, their ice turns directly from a solid to a gas, resulting in jets of gas and dust leaving the comet to form a tail.

METEORS AND METEORITES

Interplanetary dust is the name given to the dust in between the planets. It comes from impacts on rocky planets, rocky or icy moons, asteroid impacts, comet outgassing, and more, so the supply of it is constantly being replenished. The size and structure of the dust particles is extremely varied but larger fragments are usually referred to as meteoroids. This term applies to larger asteroid or comet fragments as well, but there's no official size cut-off for what constitutes a meteoroid rather than dust particle.

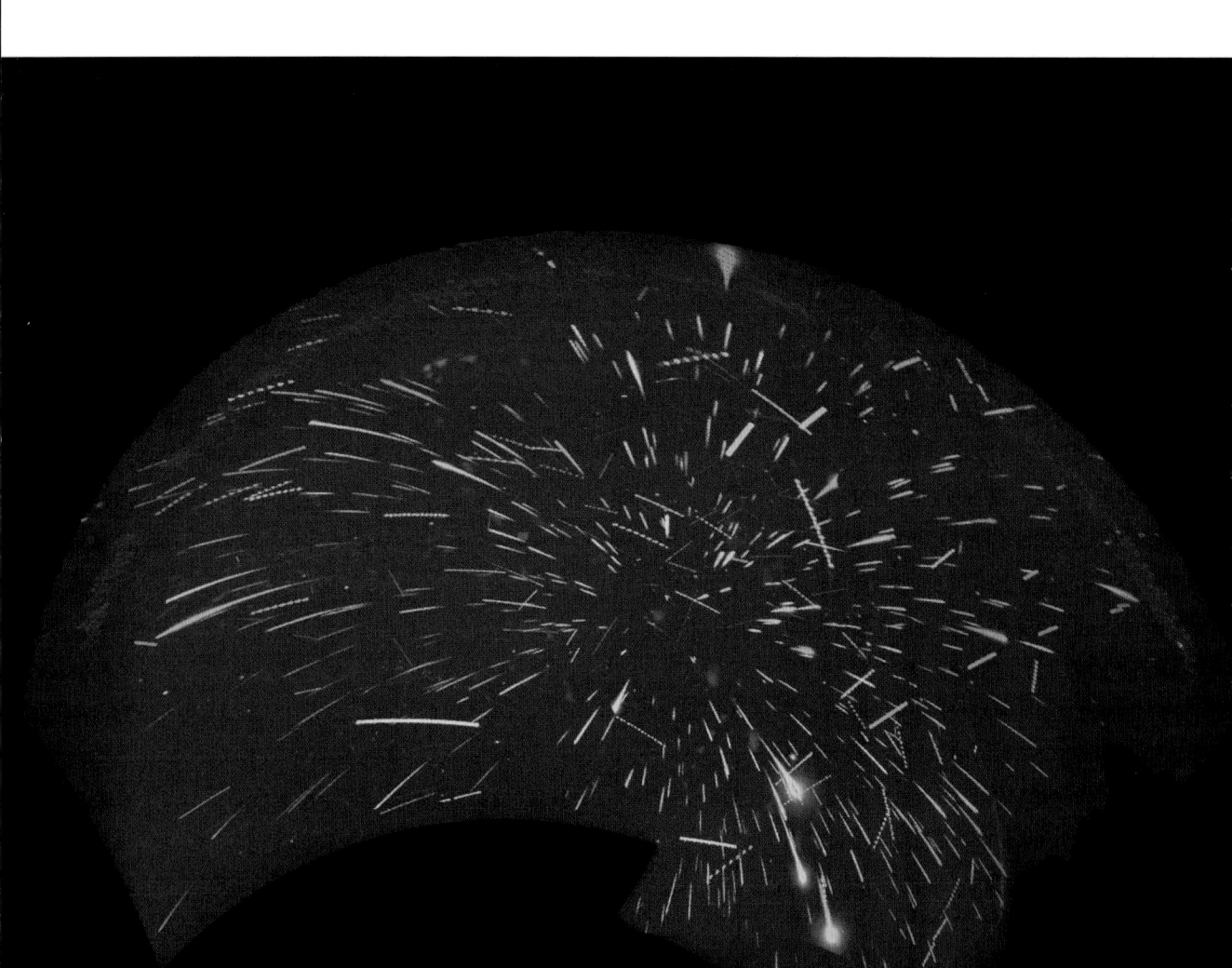

When a meteoroid passes through Earth's upper atmosphere, the friction causes the meteoroid and the air around it to heat up. The fragment burns up and leaves a visible streak of light called a meteor, or "shooting star." Meteors move at a speed of between 50 and 230 feet (15 and 70 meters) per second, so the streak of light is only visible for a second or two. Most meteors result from very small dust grains and they burn up at altitudes of 35–50 miles (60–80 kilometers) or higher. Because they burn up so high in the atmosphere they can be observed from a great distance. A meteor seen near the horizon may actually be more than 100 miles (160 kilometers) away, while those closest to us are those directly above our heads. On a clear night you can expect to see two or three meteors per hour; these are called sporadics. Across the whole sky there'll be about 50 meteors per night, although there'll be fewer during summer when the nights are shorter. Approximately 100 tons of interplanetary dust hits Earth's atmosphere every day, but most small fragments are completely vaporized. The larger the fragment, the brighter the meteor will be, and meteors that exceed magnitude -4 are classed are fireball meteors, or bolides.

Meteor showers

The trail left behind by a comet is called a meteoroid debris stream, and when Earth passes through it we experience a meteor shower. We cannot directly observe a comet's debris stream, but observing the number of meteors and how bright each event is can tell us a lot about how the stream is evolving. A comet can actually produce more than one meteor shower; comet 1P/Halley is responsible for both the Eta Aquariids and the Orionids. Most meteor showers have parent bodies that are comets, and even those whose parent body is not known are still believed to be comets, with one exception—the parent body of the Geminids is asteroid 3200 Phaethon and it's not certain if this is an asteroid that displays cometary activity or if it's an old, dead comet that has lost most of its volatiles.

Meteor showers are named after the constellation that the shower radiant lies in. The radiant is the point in space that the meteors appear to come from. (The meteors don't actually come from that constellation, though—this is just a line-of-sight effect.) The meteors belonging to a particular shower can appear anywhere in the sky, but if you trace the line of their path back, they'll appear to originate from that radiant. The radiant of the Perseids meteor shower lies in the constellation of Perseus, the Geminids lies in Gemini, and so on, but the radiant is not actually fixed in space—it can drift during the duration of a meteor shower because of Earth's movement through the stream. Many news articles promoting meteor showers will quote a figure called the zenith hourly rate (ZHR) and state that this is how many meteors you'll see, but this is a common misrepresentation of the facts used to provide a click bait headline, and the press frequently embellish this number even further.

The ZHR of a shower is a theoretical figure that makes several assumptions: that the shower radiant is at the zenith (the point directly above the observer's head), that the observer can look at the entire sky all at once, and that there is no moonlight or other light pollution. This is very unlikely! Almost no meteor shower radiant gets even close to the zenith, but the higher the radiant is, the more meteors you'll see. If the radiant is low to the horizon, it straight away halves the ZHR figure. Even if you have perfect vision and have allowed your eyes to fully adapt to the dark, humans cannot survey the entire sky at once; at best you'll see about a third of it. And if the Moon is bleaching the sky, or if you live in an urban area, the light pollution will obliterate any faint meteors. The ZHR has a cut-off of magnitude +6, which is the limit of the human eye, but even the most experienced observer will struggle to see a meteor that faint. So, a shower with a quoted ZHR of 120 meteors per hour, when viewed from a semi-rural location, with no moonlight and fully dark-adapted vision, may translate to around 30 meteors per hour —fewer if there is a bright Moon. One other thing to consider is that even though we're passing through a predicted debris stream, meteors are totally random events. You may be sitting outside on a freezing December night observing the Geminids meteor shower and see absolutely nothing for 20 minutes, then suddenly see many meteors in a short space of time. It's impossible to predict exactly when or where a meteor will occur, or how bright it will be. We do know, however, that there are more meteors in the pre-dawn hours; this is

Opposite Perseid Meteors Captured during the 2022 Perseids Meteor Shower. This a combined image from four Global Meteor Network meteor cameras in Oxfordshire, U.K. and clearly shows that all of the meteors appear to be coming from the same spot in the sky. This is the Perseids radiant.

because the dawn side of the Earth is facing directly into the meteoroid stream. It's akin to walking forwards in a rain shower—the front of you gets wetter than the back, so a late night (or very early start) is recommended.

A meteor shower is usually active for a few weeks, with a peak of activity focused over a couple of days, but it depends on how compact the debris stream is. Some showers caused by a recent comet visitor have a peak of just a few hours, with little activity either side of that, but showers like the Perseids may be active for over a month. If a parent comet has a short orbital period, the debris stream is being replenished regularly and will yield more meteors, but one from a long-period comet that rarely visits will have a more diffuse debris stream. Occasionally Earth passes through a denser part of the debris stream and this leads to a dramatic increase in meteor rates known as a meteor storm. The Leonids have produced several meteor storms throughout history, and in 1833 it was estimated that over 200,000 meteors were observed every hour for a nine-hour period from North America; this must have been an incredible sight to behold, but of course it was interpreted as a bad omen. Between 1997 and 2002 the Leonids again put on a great show, and were said to have produced several thousand meteors per hour.

The debris streams from different parent bodies are moving at different speeds, so different meteor showers have different characteristics—some have slow-moving meteors, and others have faster meteors. The composition of the parent body also affects the appearance of the meteor as it burns up, which is why some meteors have green trails, whereas others are more likely to be orange. Some showers don't have a high ZHR but they're more likely to produce fireballs. Every meteor shower is unique and worth observing, and it can be done with no special equipment; all you need is a comfortable recliner and dark adapted eyes. And remember that even if you see far fewer meteors than expected, that's still valuable information for scientists.

Below Perseid Fireball photographed by Mary McIntyre from Oxfordshire, U.K. on August 13, 2022. Despite the bright moonlight, the fireball clearly showed the green trail that is characteristic of Perseid meteors.

Meteor cameras

For many years meteor observations were made visually and, like comets, meteor showers were viewed with superstition or interpreted as a bad omen. That said, early astronomers had a good understanding of the relationship between comets and meteor showers, and in 1587 a book based on an anonymous treatise about comets called *Kometenbuch* ("The Comet Book") was published. It featured 13 watercolor paintings of comets and meteors, and one of these depicted a meteor shower, with the meteors appearing from a radiant. Modern-day technology has revolutionized meteor science and there are now global networks of cameras that are recording the sky every night from dusk until dawn, monitoring meteor activity. Most of them are recording short video clips, some using all-sky cameras to monitor fireball meteors, and others using low-cost CCTV cameras that are better at capturing faint meteors. The Global Meteor Network (GMN) uses open-source software that runs on a Raspberry Pi computer—Raspberry Pi Meteor Station (RMS) software—and is able to analyze all of the meteor events captured on each camera. If a meteor is captured on one camera, the software can make an educated guess as to whether it belongs to a meteor shower or if it's a sporadic, and can estimate the magnitude. However, the real science can begin if the same meteor is captured on cameras in different locations because RMS can then more accurately confirm if it was a shower meteor or sporadic. Several new meteor showers have been discovered, thanks to the work of the GMN project, and space agencies use the data to keep their launching spacecraft safe.

Any meteor that's captured on three or more stations can be triangulated and RMS will determine the ground track, calculate how fast the object was moving, what angle it entered the atmosphere, calculate a probable orbit trajectory for it, and determine if it was asteroidal or cometary in origin. Based on how bright the event was, it can also estimate the mass of the fragment and determine if any material survived and fell to the ground to become a meteorite. In the United Kingdom there are over 200 active GMN cameras and the resulting data has been used to create a vast, searchable archive, with a webpage for every triangulated event, together with all the orbital data. The cameras have also been great at recording "Earth-grazing" meteors—those objects that come in at too shallow an angle to make it all the way through the atmosphere, so instead skip off it, leaving a fast, long trail before heading back out into space. Monochrome cameras are more sensitive than color cameras so most of the images from the camera networks are black and white, but they still produce some incredible imagery.

Meteorites

Meteorites are not inherently rare; in fact, micrometeorites are raining down on Earth constantly. We learned earlier that about 100 tons of interplanetary dust hit our atmosphere every day, and it's been estimated that about 5,100 tons of micrometeorites fall to Earth every year. They are the tiniest fraction of an inch in diameter and therefore so small that rather than vaporize, the friction heats them up. They begin to melt into a roughly spheroid shape, then cool again as they fall through the rest of the atmosphere. They can be found in silt on rooftops and gutters, and have even been isolated from the sea bed. It takes skill and practice to separate them from human-made spherical objects—from angle-grinders, brake disks, fireworks, and gunshot residue—and from naturally occurring grit and silt, but it's a really fun project.

The study of larger meteorite fragments is crucially important to our understanding of the early Solar System because it's impossible to send spacecraft to all of the existing asteroids or comets. It is estimated that 20,000 meteorites with a mass of 3.5 ounces (100 grams) or larger fall to Earth each year and they can be scientifically very valuable. In principle they can be found anywhere on dry land, but in areas that are full of vegetation, or heavily populated, this poses a challenge because they often resemble terrestrial rocks, animal droppings, charred remains from a barbecue, or leftover debris from industrial processes in foundries, and so on. They're much easier to find in barren landscapes such as deserts, where they really stand out against the sand, or in Antarctica, where glacial processes can concentrate meteorites into specific areas—and again, they are very obvious sitting on top of ice and snow.

Meteorites that are discovered on the ground are called "finds." Scientists can analyze them to establish their classification and can usually tell how long they've been present on Earth, but we usually have no other information. If a fireball has been observed and meteorites from that event subsequently recovered, these are called "falls." It's worth noting that a meteorite is cold to the touch when it first lands because, after it has burned up in the upper atmosphere, it has then fallen though many miles of cold atmosphere on its way to the ground. Even though some meteorites get so hot that they have a fusion crust around the outside, witnessed meteorite falls have even been observed with frost on their surface.

If the event has been observed visually then we have a little more information about it, based on knowing roughly the direction of travel and how bright it was. Until recently, meteorite falls were much less common than finds, but there have still been many recorded throughout history. In recent years there's been an increase in the number of recovered falls, thanks to the work of the meteor camera networks. If a bright event has been captured on multiple cameras, the camera networks work together with meteor scientists to calculate the potential fall zone, and trained experts can then go hunting for meteorite fragments. If recovered, not only do we have a new meteorite to analyze, we also have its orbit and trajectory so we know exactly where in the Solar System it came from.

Larger meteorites that survive are usually asteroidal in origin. Some are from comets, but cometary material is much more delicate and therefore less likely to survive. We occasionally get meteorites that come from the Moon or Mars. These are a result of a large impact that flung pieces of surface material up into space, from where it has made its way to Earth, so these are quite rare.

There are three main meteorite classifications. Iron meteorites are composed mostly of iron, stony meteorites are composed of silicate material, and stony-iron meteorites are a mixture of silicates and metals. Iron meteorites are believed to be cores of differentiated planetesimals that were smashed apart by a violent impact. Stony meteorites are from the crust, and stony-iron meteorites are believed to have come from the boundary between the core and crust.

Stony meteorites are those that fall to Earth most frequently, but iron meteorites are found most often because they are less likely to be eroded by the weather. Ninety percent of stony meteorites are chondrites, which means they contain small spherical pieces of silicate material called chondrules. They're interesting to study because they're believed to have

been formed by local heating and cooling processes within the solar nebula. They may also have been formed by surface processes on planetesimals or by impacts. Most chondrites are further classified as ordinary chondrites, but there's another type called carbonaceous chondrites, which are carbon-rich meteorites containing organic compounds such as amino acids. Chondrites are of great interest to planetary scientists because they have primordial chemical compositions, so they tell us a huge amount about the early solar nebula—in particular, the carbonaceous chondrites that contain inclusions that are actually older than the Solar System; studying these inclusions give us a snapshot of what the solar nebula looked like before our Sun was born.

Chondrites are classified further into groups, according to their chemical and mineral composition. Class C is the carbonaceous chondrites and they in turn have eight subcategories based on studies of particular meteorites, with the main ones being CI, CM, and CV; these have even further subcategories, based on how well preserved the chondrules (grains) are. E chondrites contain a form of pyroxene called enstatite. The ordinary chondrites, class O are subdivided based on how much iron they contain. H chondrites have relative high iron content; L chondrites have a low iron content(between 5 and 10 percent); and LL chondrites have a very low iron content of less than 2 percent.

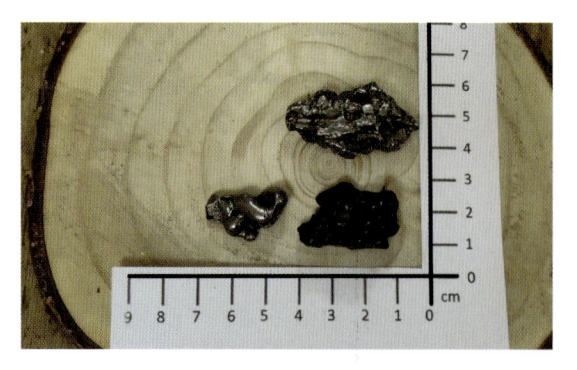

Above From the author's private collection, these specimens show iron, stony-iron, and carbonaceous chondrite meteorites.

Notable meteorite falls

There have been many notable meteorite falls throughout history. At 5 p.m. on April 9, 1628 there was bright fireball with associated sonic boom, and subsequent 22-pound (10-kilogram) meteorite fall in Hatford, Berkshire, in the U.K. A woodcut illustration of the event features an army of celestial bodies above the clouds firing huge canons! In October 1992, many people witnessed a meteorite strike and damage a car in Peekskill, New York, and the car has since been displayed around the world. The largest event in recent history was the Tunguska event in Russia. A piece of stony asteroid, about 160–200 feet (50–60 meters) in diameter, entered the atmosphere and exploded at an altitude of 3–6 miles (5–10 kilometers), causing a meteor air burst. The explosion, estimated to be the equivalent of 10 to 15 megatons of TNT, thankfully took place over a sparsely populated area of East Siberian taiga, but it flattened about 80 million trees across 830 square miles (2,000 square kilometers). Three people were reported to have lost their lives, but had this occurred over a more populated area, the casualties would have been significant. Because the meteorite exploded above ground, it left no visible crater.

At 9.20 a.m. on February 15, 2013, people in Chelyabinsk were driving to work when numerous dashboard cameras captured a huge daytime fireball. This was an 59-foot (18-meter) diameter, 8,950-ton near-Earth asteroid that had entered the atmosphere at over 12 miles (19 kilometers) per second. It exploded at a height of about 18.5 miles (30 kilometers), with the energy of about 430 kilotons of TNT. It created a powerful shockwave that caused extensive damage over an irregular oval area of about 62 miles (100 kilometers) across and about 18.5 miles (30 kilometers) wide, and nearly 1,500 people needed medical attention. The event showered the area with LL chondrite meteorite fragments.

The meteor camera networks have played a big role in helping to find the fall zone of meteorites following bright meteor events. On February 28, 2021, a fireball witnessed across the United Kingdom resulted in about 11 ounces (300 grams) of carbonaceous chondrite meteorite being recovered from Winchcombe, Gloucestershire. The samples were almost identical in composition to the samples returned from asteroid Ryugu. On February 13, 2023, meteor cameras also helped with the recovery of fragments of the 3-foot (1-meter) diameter near-Earth asteroid 2023 CX1, which traveled eastward across the British Channel and landed in the Normandy region of France. (It's worth noting here that in recent years, meteorites have become very collectable and this has led to unscrupulous sellers selling fake meteorites for huge sums of money across the internet. Make sure you do your research and only buy from trusted sources.)

Impacts from comets and asteroids are not uncommon so it's vitally important that the big sky-survey telescopes are constantly monitoring the sky for the next big impactor. The DART mission has shown that it's possible to slightly change the course of an asteroid by impacting it, but there's still more work needed to help protect Earth from potentially hazardous objects.

Top Perseids meteor shower, as seen here over the Red Rocks of Sedona, Arizona, U.S.A.

CHAPTER THREE

STARGAZING TOOLS AND TECHNIQUES

STARGAZING BASICS

Sitting outside under a beautifully dark, star-filled sky is an awe-inspiring experience, no matter how many times you do it. Looking at the night sky can give us a sense of calm and make us feel insignificant compared to the vastness of space, putting any problems we're dealing with into perspective. In essence, stargazing is easy—go outside at night and look up and you become a stargazer! However, there are several things to think about as a beginner that will help you get the best out of your stargazing sessions.

Keeping comfortable and warm

It sounds really obvious, but if you're cold or your neck hurts, you aren't going to stay outside observing for long, and if you then spend all of the next day in pain, you won't be motivated to go out stargazing again anytime soon. It's essential that you develop good habits from the start so that you can observe for longer and get more enjoyment out of it.

Firstly you need to get into a comfortable position. Tipping your head backward from standing can put a lot of pressure on your neck, leading to pain and even injury to your neck and back, so it's far more ergonomic to sit in a garden recliner that will support your entire spine. Zero-gravity garden recliners make excellent observing chairs, because you can tip yourself back at different angles to observe different parts of the sky with minimal impact on your posture.

What you wear is also very important, because even in summer it can get really cold at night, so choose lots of thin layers and make sure you have an extra layer of clothing and a blanket. You'll lose heat from any body part that isn't covered, so remember your hat and gloves and slip some reusable hand warmers into your coat pockets to warm up your fingers when needed. Remember to keep your legs and feet warm, too, by wearing thick socks, as well as waterproof shoes or boots to keep you dry when there's dew or frost. In cold or wet conditions you'll also need good soles on your footwear to avoid slipping. Lastly, having a thermos of a hot drink with you is a lovely treat if you start to feel cold.

Be aware of the Moon phase

Lunar observing is fun and rewarding, but if you're wanting to observe fainter objects that require a dark sky, the presence of a bright Moon will ruin your plans. There are smartphone apps and many free resources online that will help you keep track of the lunar cycle and give you the moonrise and moonset times each day. This information can also be found in monthly astronomy magazines as well as night-sky guide books and almanacs.

Previous pages Whether stargazing with high-tech equipment or just the naked eye, the night sky offers incredible wonders to behold. **Opposite** Everything from a full total solar eclipse (seen here) to a delicate scattering of stars in a peaceful sky can be seen with enough patience. **Above** Be sure you are comfortable, dry, and warm for any forays into stargazing. Don't forget that nights can get cold even on a warm day.

Let your eyes get dark adapted

Have you ever been out walking at night and been dazzled by the bright headlights of a car driving toward you? If so, you'll remember that even after the car has passed, you still can't see very well for a while. The pupil of the eye is an opening that allows light to pass through to the retina. Bright lights cause the pupil to contract, reducing the amount of light entering the eye, but when it's dark, the pupil dilates, allowing in more light, so it takes time to transition from one extreme to the other. When you first go outside at night after being in a well-lit room, you'll not see very much. You'll notice that over time your eyes get accustomed to the dark and eventually become fully dark adapted, but this process takes about 30 minutes, so you need to be patient. It's a fun exercise to count the number of stars in a particular area of the sky when you first go outside, then repeat the count every ten minutes to see how many more you can see. While you're waiting for dark adaptation, make sure you avoid looking at your cellphone or tablet, and if you need to use a flashlight, make sure it uses red light, which has a less damaging effect on dark adaptation than other colors. You can even buy sheets of red film that can be cut to size and used as a filter to cover screens, but it's still better to avoid exposure to screens altogether.

Another useful trick to learn is something called "averted vision." You may already know that there are two kinds of cells in your eyes: rods and cones. Rods don't detect color but they do detect dim light, whereas cones are adapted to detect bright light and color. Cones are located in the center of your eye, and rods are located around the edges, so in low light at night, your peripheral vision is more sensitive than your central vision. Although it feels completely counterintuitive, if you look slightly off to one side of the object you want to observe, you'll notice that the object becomes a lot brighter than it is if you look directly at it. If you keep moving your eyes to and from the object, it will appear to brighten and fade, or blink in and out of view. There is even a planetary nebula—NGC 6826, a supernova remnant—that was named the "Blinking Planetary" because of the way the layers of ionized gas that were thrown out when the star exploded appear to blink into view when you use averted vision, then disappear when you look directly at the central star. It's quite a strange feeling the first time you experience it, but averted vision is an essential part of astronomy, both when observing with the naked eye and when looking through binoculars and telescopes.

Plan your targets and keep an observing diary

You'll have a much more productive stargazing session if you research and plan what you want to see beforehand because some constellations, naked-eye star clusters, and deep-sky objects are only visible at certain times of year, or are best viewed at a certain time of night. You can do your research well ahead of time and produce a list of objects you'd like to look for; this will then allow you to maximize your observing time while outside.

It's also really useful to keep an observing diary, recording exactly what you've seen—for example, how many stars you were able to see in a cluster, or how clear a particular deep-sky object was. You can also record the lunar phase, temperature, wind conditions, any wildlife you saw or heard, and the sky conditions (which we'll discuss again later). This will make your observing diary much more interesting to read again at a later date—you'll be recording a much fuller story of how a session goes, even if things don't go as well as you hoped.

Every season is unique and has a new set of things to observe, so planning and recording will help you cram as much into each observing session as possible and give you a real sense of achievement.

Above Make sure you allow your eyes to adapt before attempting to stargaze. The bright lights of a car, flashlight, or streetlamps all cause the pupil to contract, and it takes time for your eyes to transition to a dilated pupil necessary for seeing in the dark.

Learn the terminology

When you first pick up a stargazing guide there can be so much new terminology to learn that it can feel a bit overwhelming, but it really doesn't take long before those new words and concepts become second nature. While it's really easy to just download an app and point your smartphone or tablet at the sky and wait for it to tell you what you're looking at, this isn't ideal because screens will wreck your night vision. It's far better, as you become a more experienced observer, to master looking at a star map or planisphere, then learn to understand how the celestial coordinate systems work and how to find objects for yourself. A planisphere is a simple analog star-chart computing instrument made from two overlapping flat pieces that are joined at the center and can rotate independently. The bottom piece is a circular star chart with constellations marked on it and monthly calendar dates printed around the outside. The top layer has an elliptical cut-out viewing window that allows only a portion of the star chart to be seen, and the 24-hour time cycle printed around the rim, so when you line up the time and date to match the time of observing and hold the planisphere above your head, what you see in the viewing window corresponds to what you see in the night sky. These are produced for certain bands of latitude in the northern and southern hemispheres and the star chart will remain relevant for a lifetime. Some planispheres have planetary visibility tables printed on the back of them, covering a specific number of years. Because planets don't follow the same regular annual changes that the stars do, once the planet tables have expired they're no longer any use, but you can still use them for observing stars.

When you're observing bodies within our Solar System, this is somewhat easier because they all travel within an invisible band in the sky called the ecliptic (see page 71). Because the Solar System was formed from a flat disc of gas and dust, all of the planets and dwarf planets orbit within an almost flat plane; Solar System bodies always move along that same band of sky when viewed from Earth, rising in the east and setting in the west. All Solar System objects rise roughly in the east and set roughly in the west, but if you're in the northern hemisphere, they transit across the southern sky, so it's impossible to see any of these objects in the northern sky. If you're in the southern hemisphere, the objects transit across the northern sky, so from Australia, for example, you'll never see Solar System objects in your southern sky.

There are eight planets in the Solar System, all in orbit around the Sun. From our vantage point on Earth, the planets can behave in unexpected ways, which is why early astronomers called them "wandering stars." Planets can appear in the same part of the sky as each other and this gives the impression that they're really close together, but in reality there are vast distances between them. When two bodies are visible in the same part of the sky they're said to be "in conjunction" with each other; planets may be seen in conjunction with each other, with the Moon, with bright stars, or with star clusters.

Mercury and Venus are inner planets and their orbits are closer to the Sun than Earth's orbit. When its orbit takes one of these planets behind the Sun, it's said to be at "superior conjunction"; being in the same part of the sky as the Sun, we cannot observe it from Earth. The planet will then emerge as an evening object until it reaches greatest eastern elongation—this is the farthest easterly point the planet can reach when viewed from Earth. The planet is constrained by its own orbit, so Mercury, being the innermost planet, can never lie that far from the Sun. Once the planet passes greatest eastern elongation it once again moves toward the Sun but this time it passes between the Earth and the Sun and into "inferior conjunction" (where once again it cannot be seen) before emerging once more as a morning object until it reaches greatest western elongation. Occasionally the inner planets line up perfectly with the Sun and we can see them transiting the solar disc. Mercury transits occur about 14 times per century but Venus transits are much rarer (see page 269).

Mars, Jupiter, Saturn, Uranus, and Neptune are the outer planets, so it's not possible for them to pass in between Earth and the Sun to inferior conjunction. When these planets pass around the back of the Sun they're said to be at "solar conjunction" and we cannot observe them. Our best view of the outer planets is when they reach "opposition"—so called because they're on the opposite side of the Earth to the Sun. This is when the planets are fully illuminated by sunlight; they appear at their biggest and brightest, and are visible all night. The apparent diameter of Mars can change dramatically during an apparition, so it's always best to observe it at opposition. The location of the planets in our sky is also affected by our movement along our own orbit, so some apparitions are not as favorable for viewing as others.

The planets are all traveling in their orbits at different speeds; the closer to the Sun, the faster the planet will move, so there are times when a planet orbiting closer to the Sun will overtake one farther away from the Sun; when viewed from Earth this can look like the planet is moving backward. If you've ever spent time on a busy highway, you'll be familiar with this effect. If the traffic in your lane is moving slightly faster than an adjoining lane, the cars in the other lane can appear to be moving backward, when in reality they're still moving forward, just at a slower speed than you. The same thing happens with the planets—if you're viewing the apparent movement of the planets from Earth, most of the time the planets are moving farther toward the west each day in "prograde" motion, but an undertaking maneuver causes this movement to slow, stop, then switch direction, sending the planet into "retrograde" motion. Just like highway traffic, the planet hasn't actually started moving backward, it's simply a line-of-sight effect.

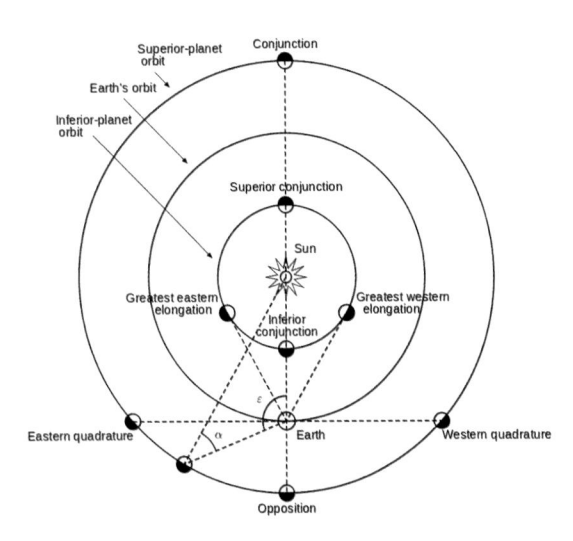

Above This diagram shows the positions of the inner and outer planets at different points in their orbit. The inner planets are best viewed when they are at greatest elongation, as that is when they have the greatest separation from the Sun when viewed from Earth. The outer planets are best viewed when they are in opposition. When they are at conjunction, they are too close to the Sun to be visible from Earth.

Finding things in the night sky

We know that humans have studied the night sky for millennia because there are cave paintings dating back around 17,000 years in Lascaux, France, that likely depict the stars in the region around Taurus and the Summer Triangle, as well as what seems to be a representation of stars around the Corona Borealis in a cave in Pico del Castillo, Spain. To date, these are the oldest known recordings of night-sky stars. Many Indigenous civilizations divided the sky into constellations, with star patterns depicting mythological objects or characters. The seasonal changes in star patterns and the apparent rotation of the sky due to Earth's rotation on its axis featured in stories that were then passed down through generations. There was particular importance placed on the constellations that lie along the ecliptic. This became known as the Zodiac and the earliest known depiction of it was on the ceiling of the Hathor temple at Dendera, Egypt, which dates back to around 50 BCE.

Many of the constellation names that we use today are based on ancient Greek and Roman mythology, recalling epic battles, stories of the afterlife, and more than a handful of unsavory characters. Historically, constellation names referred specifically to these star groups, but later a constellation came to refer to a defined section of the sky. In the process of officially dividing the sky into sections it became necessary to assign names to groups of stars that hadn't been included in previous constellations, leading to additions such as Camelopardalis the Giraffe, and Monoceros the Unicorn (see pages 109 and 154).

So how do we know where to find things in the sky? Think of the night sky as a sphere around the Earth. That sphere has been divided up into coordinates in the same way that our terrestrial surface has been divided up into a latitude and longitude grid. There are two different coordinate systems in use— the *altitude-azimuth (Alt-AZ) system* and the *equatorial (EQ) system*—and it's essential to understand the basics of both. Both are grid systems, with the whole sphere of the sky divided up into vertical segments a bit like segments of an orange, then further divided into concentric horizontal slices.

The Alt-AZ system is specific to the observer's location, with the center of the grid located at the zenith—the point directly above your head. From there, the divisions extend out like a dartboard. The altitude lines are horizontal and refer to how high above the horizon an object is, and the azimuth lines are vertical and refer to how far round from north the object is. This is a spherical grid system with the divisions measured in degrees, so with reference to altitude, something that sits at your horizon is at 0°, something directly overhead and at the center of the grid will be at +90° degrees, and something at the nadir (the point directly below your feet in the other hemisphere) is at −90°. The azimuth lines are also measured in degrees and refer to how far away from north the object is and this incorporates the compass points with north at 0°, east at 90°, south at 180°, and west at 270°. This grid system is useful because it will tell you in real terms where an object lies in the sky from your location, so for example, on a particular date at 7 P.M., the Moon may be located 35° above the horizon and 110° round from north, or Jupiter could be located 40° high and 180° from north.

There is a big problem, however. If you observe any Solar System or celestial objects for any length of time, you'll observe them moving because of Earth's rotation, so their Alt-AZ coordinates are constantly changing. While it's really useful having a coordinate system specific to your location, the downside is that the coordinates are only correct for you; if somebody is located 100 miles (160 kilometers) away, they can usually still see the same objects but the Alt-AZ coordinates of those objects will be different for them. There will also be differences in the rise and set times from different locations. On a random day in November, the Sun rises 27 minutes earlier in London than it does in Edinburgh, and sets 3 minutes earlier in Edinburgh than London. There are similar differences seen in the rise and set times of the Moon between the two cities. The differences are even greater if you compare the Alt-AZ coordinates between two different continents, so clearly this system is not ideal when communicating the location of objects to people who are not local to you. Luckily another system exists that overcomes this issue and that's the *equatorial (EQ) coordinate system*.

The equatorial coordinate system also breaks the sky up into segments and slices, but the central point of the grid is located at the north celestial pole rather than at the observer's zenith. The north celestial pole is located above Earth's North Pole and the south celestial pole is above the Earth's South Pole, so the center of the equatorial coordinate grid will vary with your latitude. In the northern hemisphere we currently have the star Polaris located very close to the north celestial pole, so this makes it easy to find the centre point of the equatorial coordinate system; if you're located at 52° latitude, Polaris is 52° above the northern horizon, but if you're at 35° latitude, Polaris will be 35° above the northern horizon, and so on. Interestingly, if you're located exactly at the North Pole, your Alt-AZ grid would actually match your equatorial coordinate grid, but the poles are the only places on Earth where this is the case. The rotational axis of Earth actually changes by about 1° every 72 years, and over a period of about 26,000 years the axial poles will paint out a circle around the celestial poles, causing the stars to slowly change position. This is known as "precession" and the movement is often likened to that of a spinning top that begins to wobble as it slows down, but this movement is so slow that it has little effect on observations within our lifetime.

The two measurements used in the equatorial coordinate system are declination (DEC) and right ascension (RA). Declination is a measure of the angle of an object relative to the celestial equator, so it basically tells you the altitude of the object; the north celestial pole is +90°, the celestial equator is 0°, and the south celestial pole is −90°. Right ascension is the equivalent of terrestrial longitude and is a measure of the angular distance eastward from the vernal equinox. It's measured in sidereal hours, minutes, and seconds (see page 28) because it's a measure of the time it takes for a celestial object to complete one full rotation of the celestial equator; in one hour, objects move 15° of RA, and in 24 hours they move 360°. This grid system is fixed on the sky, so as Earth rotates and the stars and Solar System bodies move, so does the grid. This means that the RA and DEC values of most stellar and deep-sky objects don't change much—at least not within our lifetime—and they're the same regardless of the observer's location, so this is a more useful system. The Sun, Moon, planets, dwarf planets, asteroids, and comets are not fixed in space like stars, so their RA and DEC values do change. If you're located anywhere other than the North or South Pole and you overlay the Alt-AZ grid onto the equatorial coordinate grid you'll see they look completely different. However, if you're at the poles, the two different grid systems line up—but remember that the Alt-AZ coordinates of an object will be changing as

296 THE BACKYARD STARGAZER'S BIBLE

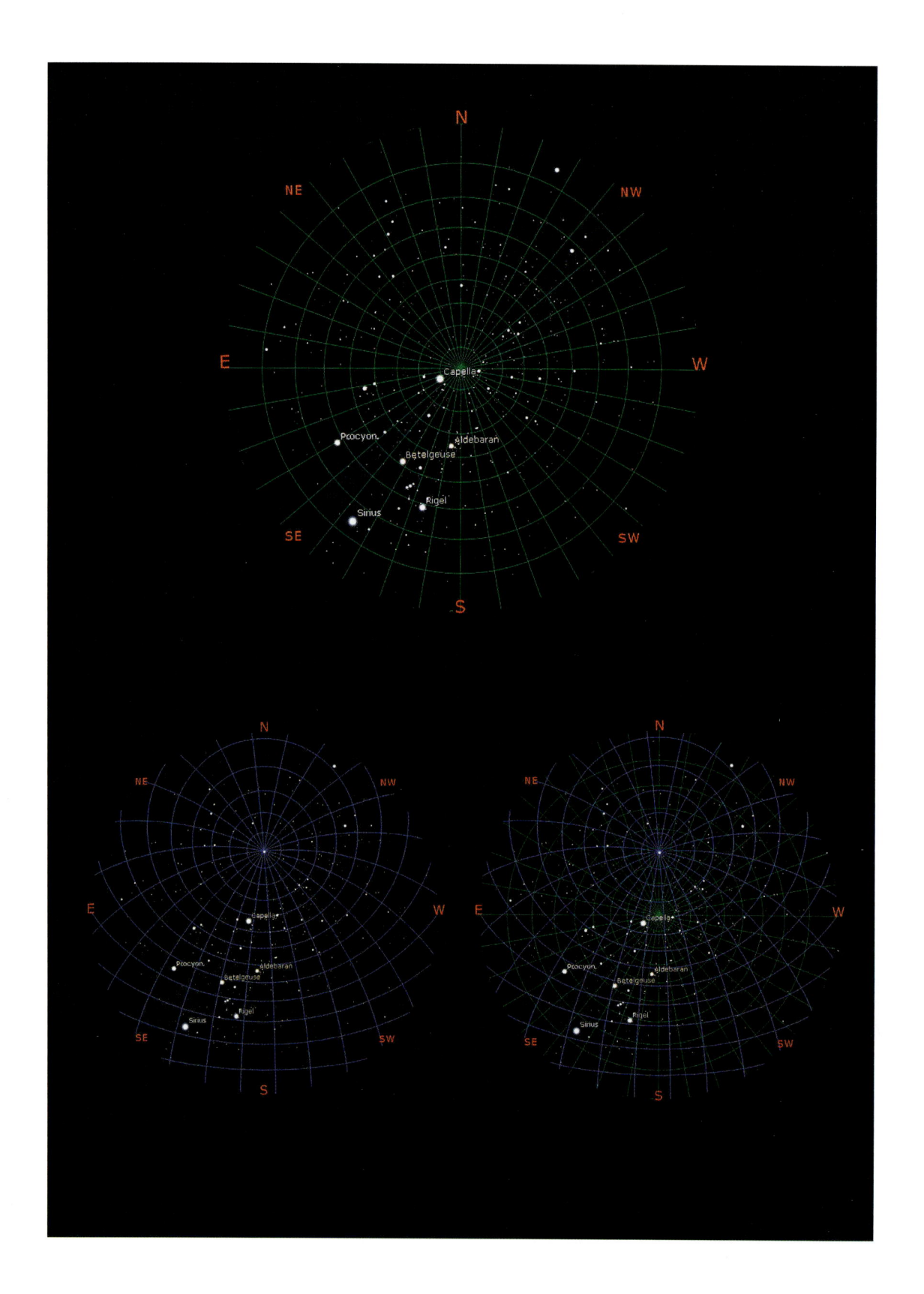

the object moves across the sky, whereas the equatorial coordinates remain fixed, for reasons explained earlier.

You may be wondering at this point why the stars are fixed on the equatorial coordinate grid. Well in reality they're not; stars are actually all moving relative to each other, but most of them are so distant that their movement is imperceptible within short time periods, and in astronomical terms a human lifetime is incredibly short. There are some nearby stars whose movement can be detected over time, and if you observe them several years apart you can see their proper motion, but these are rare; for the most part, stars remain fixed within our observational lifetime.

Rotation of the sky

If you take a long-exposure photograph of the stars at night using a static tripod, you'll see them appear to move around the sky, creating star trails. The movement we capture in these images is actually the movement of Earth rotating, not the stars moving. Because of Earth's counterclockwise rotation, stars rise in the east and set in the west, and if you point your camera at the celestial poles, the stars trace out beautiful concentric circles. As mentioned in the previous section, in the northern hemisphere the star Polaris sits close to, but not exactly at, the celestial pole, so in long-exposure photographs you can see that it paints out its own tiny circle. If you point your camera toward the celestial equator, the star trails take on a flatter appearance, then form tighter circles again

Opposite top This Altitude-Azimuth (Alt-AZ) grid was created using Stellarium. This coordinate system is specific to the observer's location, with the center of the grid above the observer's head. The stars do not follow these grid lines as Earth rotates. **Opposite bottom left** This Equatorial (EQ) grid was created using Stellarium. This grid system has its center at the celestial pole and is fixed on the sky, so as Earth rotates, the grid rotates with it. This means the coordinates are the same regardless of the observer's location. **Opposite bottom right** This Alt-AZ grid has been overlaid onto the EQ grid to show how different the two grid systems are. During the night, stars will follow the lines on the EQ grid, not the AZ grid. **Below** Long-exposure photographs capture the rotation of the night sky around the celestial poles—the point of the sky directly above the North and South Poles—in the form of star trails. It takes just under 24 hours for the stars to paint out full circles as shown in this image, which was created by Mary McIntyre by combining images taken in the U.K. during a full year and blending them together.

towards the opposite pole. If you live in the polar regions where, in the winter time, you have 24 hours of darkness, you can take star-trails photographs continuously for 24 hours and the stars form complete circles. It's also possible to create this effect artificially by taking star-trails images several times a year with your camera in exactly the same place and pointing in exactly the same direction, then merging the images digitally. In the photograph below you can see the tiny circle created by Polaris in the center. This rotation is what causes objects to change orientation as they move across the sky over an entire night.

The definition of a "day" is the time taken for a body to complete one rotation on its axis. Here on Earth, a day is 24 hours long—or is it? If we use the stars to measure how long it takes for Earth to complete one rotation on its axis, we get a different figure. You can test this yourself if you go out every clear night for a week and stand in exactly the same spot, then note the time that a bright star disappears behind a chimney or lines up with a telegraph pole. You'll see that a star returns to that same spot nearly four minutes earlier each day—3 minutes 56 seconds earlier, to be precise. This day lasting 23 hours 56 minutes 4 seconds is called the sidereal day.

So why do clocks tell us that a day is 24 hours long? It's because if we do the same thing but measure how long it takes the Sun to reach its highest point in the sky each day, we get a value of 24 hours. This is a solar day, and the reason it's slightly longer than the sidereal day is because in the time it takes Earth to complete one rotation on its axis it has also moved along its orbit around the Sun (by nearly 1° because one complete orbit is 360° and there are 365 days in an Earth year), so the Earth has to rotate a tiny bit farther each day before the Sun returns to its previous position.

With regards to studying constellations, the 3 minutes 56 seconds makes very little difference day to day, but month to month you start to see that the constellations are moving westward. Constellations close to the celestial pole are still moving westward each day but they remain visible all year round; these are known as circumpolar constellations. If you look for the Plough (Big Dipper), a seven-star asterism (the name given to a general group of stars) within the constellation of Ursa Major (the Great Bear), you can see its orientation changing with the seasons. The constellations closer to the celestial equator display the greatest seasonal variations. For example, in the northern hemisphere we only see Orion during the winter months, and we only see Sagittarius during the summer months. If you're planning to study a particular constellation you need to know if it's circumpolar or whether you'll need to wait for the right season to observe it. While it's great having access to the circumpolar constellations all year round, it's also good having the variation in the non-circumpolar parts of the sky because each season then brings its own unique observing targets.

Seeing and transparency

If you look at the Moon through a telescope with high magnification you'll notice that its surface looks as if it's covered with a transparent shimmering layer that causes its features to come in and out of focus. This is because from most back-garden locations we're observing the sky through a thick, hazy, turbulent layer of atmosphere that contains moisture, dust, pollen, volcanic ash, or other pollutants. Most professional observatories are located high up on mountains because this gets their telescopes up above as much of that atmosphere as possible, providing better viewing conditions. There are two aspects of the atmosphere that affect observing—seeing and transparency—and it's important to note both of these in your observing diary.

Seeing is a measure of how much turbulence there is in the atmosphere, and this can be made

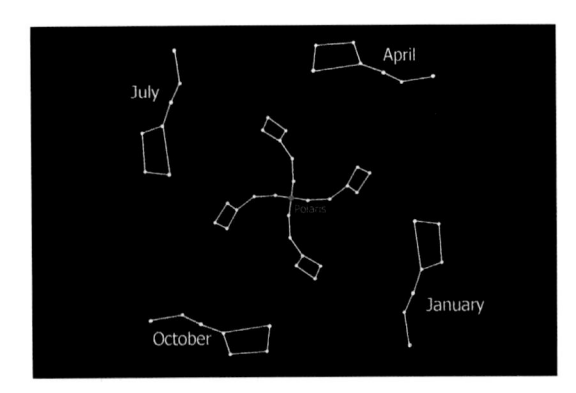

Above The Plough, or Big Dipper, is a seven-star asterism in the constellation of Ursa Major (the Great Bear), and it can be used to locate Polaris (the Pole Star), which lies at the end of the tail of Ursa Minor (the Little Bear). Stars appear to rotate around Polaris, but because stars rise approximately four minutes earlier each day, the constellations shift positions as the weeks pass. This means the Plough is oriented differently during different seasons. Credit: Mary McIntyre

worse by local weather conditions, haze from light pollution, humidity, or even hot air rising from buildings. The effect of poor seeing is that you lose resolution, making it difficult to resolve double stars or to pick out surface features on the Moon and planets. Seeing is measured on a scale of 1 to 5, with 1 being perfect, 2 good, 3 moderate, 4 poor, and 5 very bad. Transparency is a measure of sky clarity; under good transparency you can observe faint objects that are not easily visible under poor transparency. Good seeing and transparency don't always go hand in hand. It's possible to have good transparency with poor seeing, as is common after a rainstorm has just passed over, or you can have very good seeing conditions on a misty night when the transparency is poor. One thing they both have in common is that the higher up in the sky you look, the better the conditions will be. When objects are close to the horizon, conditions will always be more challenging than if you're observing something up near to the zenith, so always try to observe the more challenging targets when they're higher in the sky. When you're filling in your diary, make a note of the limiting magnitude up at the zenith and record the seeing and transparency conditions at the time you made your observations.

Star hopping and using the stars as pointers

Trying to learn all 88 constellations at once can feel like an overwhelming task, so one approach is to become familiar with a small number of brighter, easier-to-spot constellations, and use these as signposts to other star groups or deep-sky objects. This is where Stellarium or a planisphere (see page 305) are helpful because you can not only check the star patterns but also the orientation of the constellation at the time you're observing—it's much more difficult to find a new constellation if you're expecting to see it a different way up.

In the northern hemisphere we can begin by using the Plough, or Big Dipper, because it's a very prominent feature of Ursa Major. This group of seven stars looks like a saucepan, and if you follow the line of the two stars on the outer edge of the "bowl," the next star you see with the naked eye is Polaris. You've now found the north celestial pole and you only need to learn this once because Polaris will not move much within our lifetime. Polaris is part of the constellation

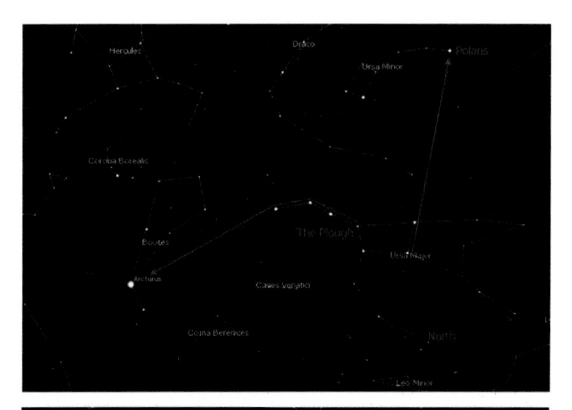

Top The Plough, or Big Dipper, within Ursa Major (the Great Bear) can used as a signpost to find other constellations. The "tail" is pointing to the Red Giant star Arcturus. The two stars on the outside of the "bowl" point toward Polaris and Ursa Minor (the Little Bear). This chart was created using Stellarium. **Above** This large asterism, called the Summer Triangle, contains the three brightest stars in the northern hemisphere summer sky; Vega in Lyra, Deneb in Cygnus, and Altair in Aquila. This chart was created using Stellarium.

300 THE BACKYARD STARGAZER'S BIBLE

of Ursa Minor (the Little Bear) and this also has a bowl and handle, with Polaris lying at the end of the handle. The stars of Ursa Minor are quite faint, but from a semirural location you should be able to see some of the bowl stars.

From mid-February to mid-October, we can use the Big Dipper to locate the large, kite-shaped constellation of Boötes (the Herdsman). This time, follow the line down from the handle of the Big Dipper and it points to Arcturus, a red giant star that is 25 times larger than our Sun and sits at the bottom of Boötes. From April to October, three bright stars, Vega, Deneb, and Altair, form a large triangle-shaped asterism known as the Summer Triangle. Vega is part of the small constellation of Lyra and it sits just outside the band of the Milky Way.

Deneb is the brightest star of Cygnus the Swan, a big cross-shaped constellation that lies along the line of the Milky Way. If you observe Cygnus from a semirural location you should be able to see the band of the Milky Way passing through it. If you can't see the Milky Way with the naked eye, pointing binoculars in that direction will reveal countless numbers of stars that form part of it. At the head of Cygnus, Albireo is one of the most beautiful examples of a color-contrasting double star, with the primary star a lovely yellow, and the smaller star a contrasting blue. If you point your binoculars just under halfway between Altair and Vega, you'll see a very endearing little asterism called Brocchi's Cluster—also known as the Coat Hanger because it resembles a perfect upside-down coat hanger. This group of stars was once thought to be a gravitationally bound cluster, but we now know this shape is a chance line-of-sight effect.

The winter sky in the northern hemisphere from October to January is dominated by one of the brightest and most easily recognizable constellations—Orion, the Hunter. Many constellations bear little resemblance to the character they're named after, but Orion does sort of resemble a human figure, or alternatively an old-fashioned sand egg-timer. Just a quick naked-eye glance at Orion reveals a beautiful pair of contrasting-color stars. The red giant star Betelgeuse, in the top-left corner, is a reddish-orange star almost 700 times larger than our Sun. In stark contrast, Rigel, in the bottom-right corner, is a blue giant star, around 79 times larger than the Sun. At the center of the egg-timer, a row of

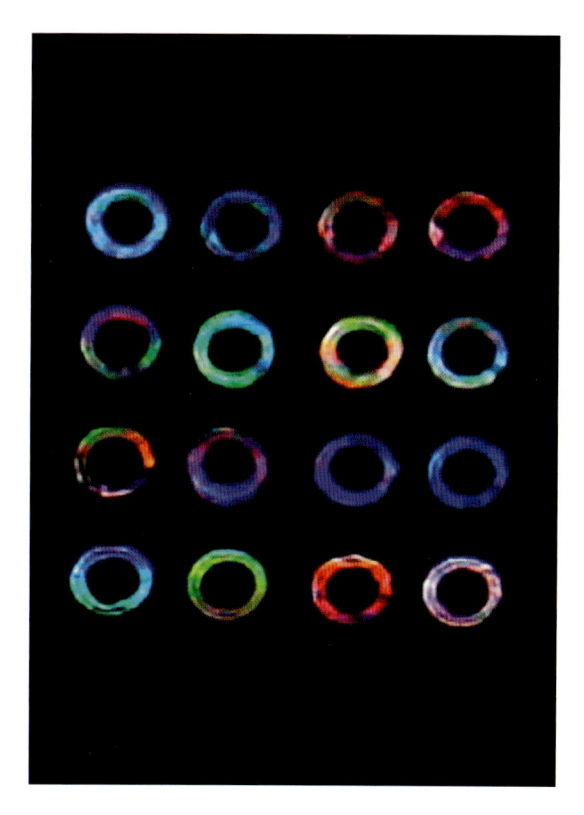

Opposite top One of the finest color-contrasting double stars, Albireo, shown here as photographed through a 10" Dobsonian telescope. Albireo A is a golden color and is located about 430 light years away. Albireo B is a blue star located about 400 light years away. Credit: Mary McIntyre **Opposite bottom** Brocchi's Cluster is an asterism also known as the Coat Hanger because of its strong resemblance to an upside-down coat hanger. The differently colored stars are not actually gravitationally bound in a true cluster; it is actually a chance line of sight effect that gives it its shape. Credit: Mary McIntyre **Above** Created by Mary McIntyre, this image shows the many scintillating colors of the brightest star in the northern hemisphere, Sirius. It was created by defocusing the camera to better show up the different colors caused by atmospheric turbulence deflecting the starlight. A mirror lens was used and, much like mirror telescopes, these lenses cause stars to form rings when out of focus.

302 THE BACKYARD STARGAZER'S BIBLE

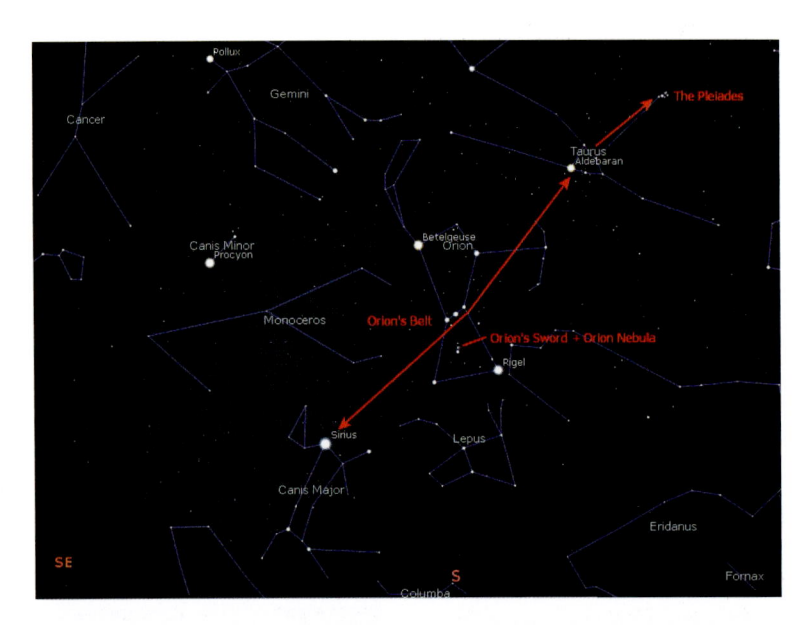

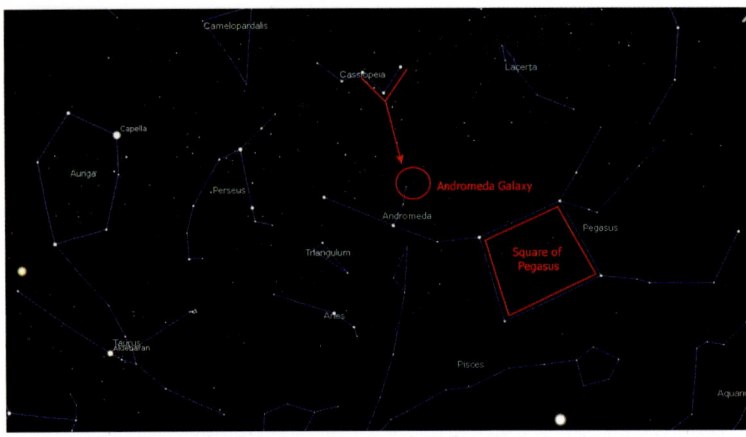

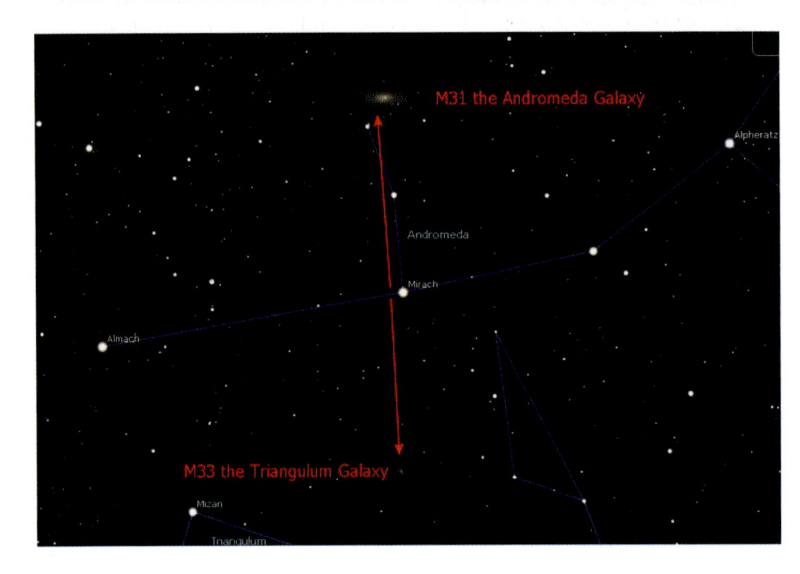

three stars form Orion's Belt. Look below that and you'll see a line
of three much smaller stars in a vertical line—this is Orion's Sword. If you look carefully at the sword, you'll notice that the center star looks more fuzzy and diffuse than the other two. That's because it's not actually a star at all, it's the great Orion Nebula; if you look at it through binoculars, the nebula, with its winged-shape, becomes more apparent.

We can then use Orion's Belt as pointers to some other objects. If you follow the belt stars down to the lower left, you'll see a twinkly star called Sirius (the Dog Star), one of the brightest stars visible in the northern hemisphere, and its scintillation creates rapidly changing colors. If you follow Orion's Belt to the upper right, you come to another red giant star that's 40 times bigger than our Sun, Aldebaran. It's part of the constellation of Taurus (the Bull) and lies in the V-shaped Hyades star cluster, which contains over 100 stars.

If you continue the line up from Orion's Belt and Aldebaran you'll come to a really gorgeous little cluster, the Pleiades—also known as the Seven Sisters, even though there are significantly more than seven stars in the cluster. The stars can be difficult to resolve with the naked eye, but through binoculars you'll see signs of the blue gas and dust that surrounds the cluster. It was once thought that this cluster was a stellar nursery, with all the stars having been formed from the surrounding material, but it's now thought to be a line-of-sight effect. Nevertheless, it's one of the jewels of the winter night sky and an extremely popular target for astrophotographers.

The W-shaped constellation of Cassiopeia (the Queen) is circumpolar and it can be used as a pointer to find the second most distant object visible to the naked eye—the Andromeda Galaxy, a spiral galaxy that is a whopping 2.5 million light years away. Although the diameter of the galaxy is the same as six Full Moons, it has a low surface brightness and only the brighter, core region is visible to the naked eye. Technically this galaxy is visible all year round from the northern hemisphere, but you'll have a better chance of spotting it when it's higher in the sky in the autumn months. Once you've found Cassiopeia, look at the point of the right-hand side of the W and think of it as an arrow pointing downward; follow that line down about three times the length of one of the Vs making up the W, and there lies the galaxy. Below the Andromeda Galaxy is a line of three stars; the brightest one at the bottom is Mirach. Take a note of the distance between the Andromeda Galaxy and Mirach, then travel that same distance in the opposite direction and you'll find an even more distant galaxy—the Triangulum Galaxy, 2.72 million light years away. It has a lower surface brightness than the Andromeda Galaxy so it can be more challenging to spot, even with binoculars, but it's technically visible to the naked eye if you have really good sky conditions.

Once you've found some of these brighter constellations and deep-sky objects you'll begin to get a mental picture of the sky map. You'll then be able to use the star patterns that you're familiar with to star-hop to adjacent constellations. The technique of star hopping is an essential skill for telescope users because most deep-sky objects are too small and faint to be visible in the telescope's finderscope. Instead you begin by pointing at a bright star near the object of interest, then slowly hop from star to star in the right direction until you land on the target. Most people use computerized telescopes these days so this is a skill that many people don't bother to learn, but it will really help you to memorize the sky.

Opposite top The three stars in Orion's Belt can be used as pointers. Follow the line down to find Sirius and then follow it up to find the red giant star Aldebaran in Taurus. Continue that line upward to find M45 the Pleiades. This chart was created with Stellarium. **Opposite middle** Cassiopeia, or the Square of Pegasus, can be used to locate M31, the Andromeda Galaxy. This chart was created with Stellarium. **Opposite bottom** From the Andromeda Galaxy, follow the slightly kinked line of three stars to find the star Mirach, then continue in that direction the same distance again to find M33, the Triangulum Galaxy. This chart was created with Stellarium.

STARGAZING FOR FREE

When we think of astronomers we often conjure images of learned professors sitting in purpose-built observatory domes, surrounded by expensive telescopes and other equipment. While it's certainly helpful to have those things, you don't need expensive instruments; all you need is your eyes, a notepad, and a pencil. There are several things to observe that don't require a telescope, and even your naked-eye observations can contribute to science. We touched on smartphone and tablet apps here, and while the light from your phone can affect your night vision, you can still use an app to plan your observing session before you go outside and let your eyes adapt to the dark.

Apps

Stellarium is a must-have free planetarium app that you can download on desktop computers, laptops, tablets, or smartphones. Once downloaded, you provide your location and it will then show you what the sky looks like at any time on any date—past, present, or future. It's an essential tool for looking at where and when objects rise and set, seeing where the planets are relative to the constellations, checking where the band of the Milky Way is, or noting if there are any interesting conjunctions coming up. When faced with an entire sky full of stars it can be daunting trying to find all of the constellations. Stellarium allows you to turn on constellation lines and labels to help you figure out the size of each one and how it's oriented at the time of observing. You can zoom out to get your bearings around the sky, or zoom in to get a closer look at a particular area. If you already have a pair of binoculars for birdwatching or terrestrial use, you can configure the "ocular" view so that Stellarium will show you what an object will look like through those binoculars. You'll then know what to expect. On smartphones and tablets you can open the app and point it at the sky and it will show you what you're looking at. It's definitely not ideal to have that screen shining for your whole stargazing session, but it can certainly help when getting started.

Stellarium shows you everything in your day and night sky, but you can also get apps that will specifically track the lunar cycle, monitor and alert you if there's an aurora alert, or track the International Space Station or other satellites that are going to be making a visible pass. The internet is full of great resources too, including NASA's amazing Scientific Visualization Studio. This is a suite of outreach tools available for the general public, science communicators, and news outlets to access free of charge. SVS was founded in 1990 by Jim Strong from NASA's Goddard Space Flight Center and was initially designed to create near real-time visualizations from satellites for the public to view. Around the end of the 1990s, it moved toward pre-rendered visualizations of Earth science topics, including ocean currents, sea levels, weather and climate changes, and air quality. It also provides tools to view lunar phases and libration for each month using satellite imagery from lunar orbiters, in order to study helio

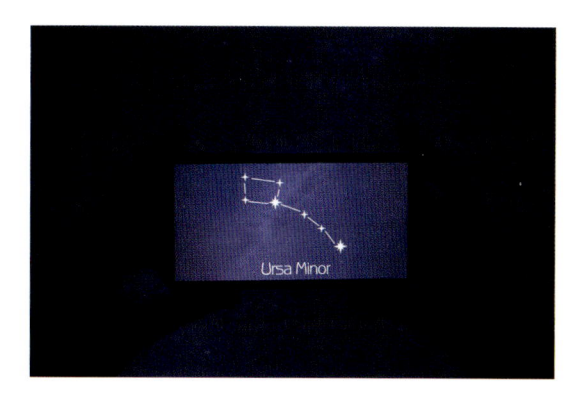

Opposite The sky has much to offer stargazers, and equipment is not always necessary. Here, a group is watching Comet Neowise by the Ottawa River, Canada. **Top** This photography by Mary McIntyre shows the full arch of the Milky Way photographed from the top of a hill above a rural Oxfordshire village, U.K., with a budget DSLR camera and 10 mm wide angled lens. Seven images were stitched together to create this panorama. **Right** The growth of starwatching apps can be a great addition to free or low-cost technology in the pursuit of night sky identification. Seen here, the Ursa Minor constellation (the Little Bear) can be identified using an app on a smartphone.

physics, cosmology, exoplanets, and much more more. And finally, it's really helpful to buy a star map or planisphere and a red flashlight, all of which are pretty inexpensive and are less likely to interfere with your dark adaptation. Once you own these items they can be used for many years.

Naked-eye observation

As mentioned already, there are many things you can observe without any optical aids. Constellations are always fun to observe and it's really interesting diving into Greek and Roman mythology and trying to see if you can make out the objects or characters being represented. The Milky Way is another excellent target for naked-eye observation. From North America, the best views are from late February through October, depending on where you are based, when the part of the Milky Way that's nearer the galactic core is visible low in our southern sky, but the band of the Milky Way spans the whole sky and, although not as striking in winter, it's still visible all year round. The Milky Way really suffers if there's light pollution from the Moon or artificial lights, though, so make sure you choose a dark site, and check that the Moon won't interfere.

There are several meteor showers throughout the year and observing them is a really worthwhile and fulfilling pastime. The best way to do this is to lie on a garden recliner with a blanket, and a pencil and pad to record how many meteors you see. Moonlight definitely has an impact because it renders the fainter meteors invisible, but brighter meteors will be clearly visible so it's still worth observing under these conditions. Each meteor shower is named after the constellation that the shower radiant lies in—the Geminids radiant lies in Gemini, the Perseids radiant lies in Perseus, and so on—but meteors can be seen anywhere in the sky, not just near the radiant. Not all meteors you see will belong to the main shower you're observing, because there are often multiple minor showers active at the same time, plus about three sporadic meteors per hour that don't belong to any shower. If you trace back along the path that a meteor has taken, you'll be able to tell if it originated from the radiant. If you keep a record of how many shower meteors you see, even if significantly fewer than expected, that information is useful for the scientific community, so do submit your observations to bodies such as the American Meteor Society (AMS) and the Society for Popular Astronomy (SPA), to be stored in their data archives.

We learned in Chapter 2 that lots of tiny fragments of interplanetary dust are scattered throughout our Solar System, and at certain times of year you can see this dust illuminated by sunlight. This is called zodiacal light and if you fancy a bit of a challenge, see if you can spot the faint cone of light extending up from the horizon and along the ecliptic. This phenomenon is completely invisible if there's any moonlight or too much light pollution, but if you're somewhere dark enough, it's easier to see when the ecliptic is higher. Look west after sunset during spring, or east before dawn in autumn, and try to spot the triangle of light. As with all astronomical objects it shows up better in photographs but it's possible to see it visually if you let your eyes get dark adapted and you use averted vision.

There aren't many things as impressive as a display of the aurora borealis (northern lights) or aurora australis (southern lights), with their shimmering, dancing curtains of ethereal light moving across the sky, so it's no wonder that these are items on many bucket lists. If you've ever seen an aurora you'll understand why many of the stories associated with them involve spirits of the departed. It may surprise you to know that it's not uncommon for an aurora to be seen at midlatitude, especially when the Sun is at its most active, because it's caused by charged particles from the Sun interacting with Earth's magnetic field; the more active the Sun is, the more charged particles will come our way, which increases auroral activity.

Photographs of auroras are often extremely colorful—for several reasons. Cameras are taking exposures of several seconds, camera sensors are more sensitive to color than the human eye in low light, and photographers often boost the color saturation of their images to help an aurora stand out against the sky. When you see an aurora first-hand, especially at midlatitudes, it's more likely to be a muted grayish-green—or no color at all—so it can be easy to miss,

Opposite top Aurora displays are much more subtle in brightness and color to the naked eye compared to what cameras can capture. This photograph was taken in June 2015 from Oxfordshire, U.K. during a very strong geomagnetic storm, then edited to make it look the way it looked visually. Credit: Mary McIntyre **Opposite bottom** This is the same photo presented as the camera "saw" it. Cameras take long exposures and are more sensitive to color in low light than the human eye, so the colors stand out much more. Credit: Mary McIntyre

STARGAZING TOOLS AND TECHNIQUES 307

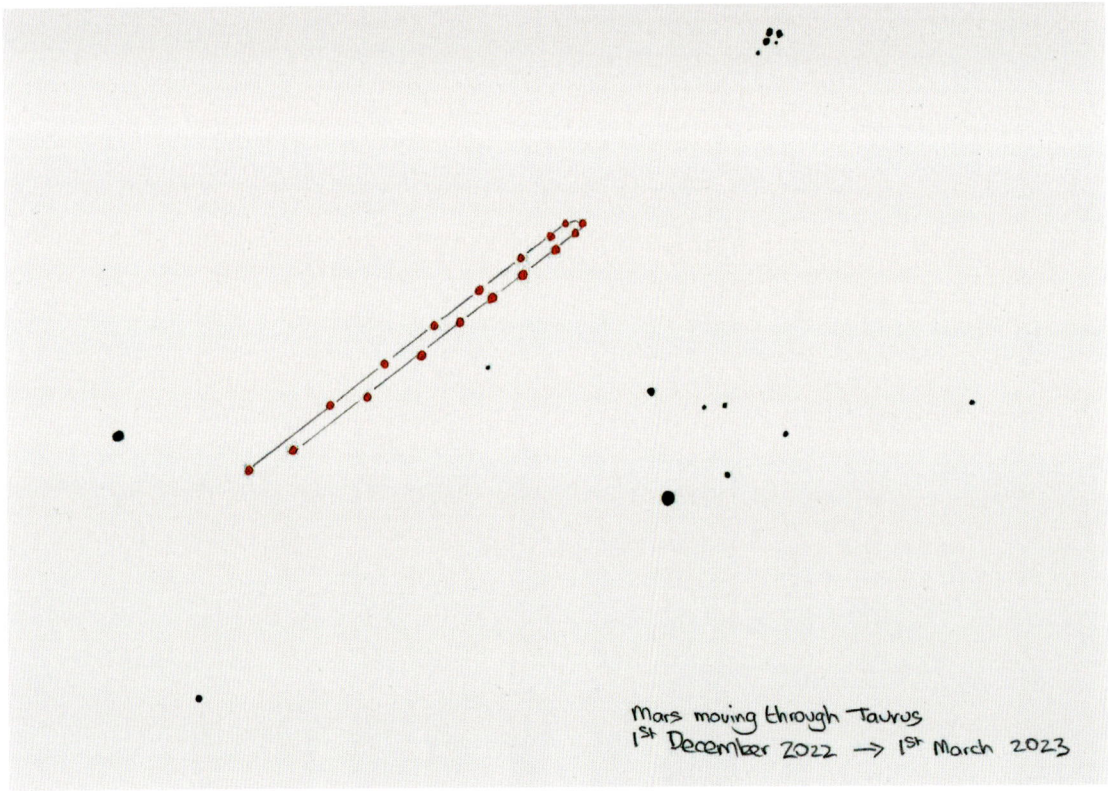

Mars moving through Taurus
1st December 2022 → 1st March 2023

especially if there's moonlight. It's helpful to take photographs at the same as observing visually because it can help to confirm if what you're seeing is an aurora or just thin clouds or car headlights in the distance. Auroras can extend over huge areas, so you don't need any kind of magnification to see them, just your eyes. Keep a note of exactly what you're seeing, and pay particular attention to shape, color, speed of movement, and the height of the display. The AMS and SPA also appreciate receiving visual reports of aurora displays, especially those seen from farther south, so once again your observations will contribute to their archives.

Although it might seem that we've been very critical of the Moon because it can affect your ability to observe faint objects, don't fall into the trap of demonizing the Moon because in itself it's a wonderful thing to observe. Seeing it waxing and waning through its cycle is fascinating, and the gentle blue light that bathes the landscape when the Moon is gibbous, or nearly full, is so restful. It's also interesting seeing how the lunar landscape changes with different amounts of illumination, as lunar "maria" (seas) become easier to discern and ejecta ray patterns from its larger craters become visible. The Moon by itself is lovely, but when you have the brighter planets in conjunction with it, it makes for a very pretty view. Because the Moon and planets all move along the ecliptic, it's possible to see Mercury, Venus, Mars, Jupiter, and Saturn near the Moon, and sometimes you even get multiple planets near the Moon at the same time. Another special event is when the Moon occults a planet or bright star (as discussed in Chapter 2), and this is also something you can observe without binoculars.

Before the invention of photography, astronomers had to draw what they saw to keep a record of their observations. These days we have smartphones that are capable of taking extremely good photographs, but there is a tendency to take a photo, share it to social media, then never look at it again, so you don't actually learn that much about your subject. Please, please put your phone away and try drawing what you observe. You don't need an art degree, or to create scientifically valuable drawings; something that's a record of what you're seeing is good enough. You'll find that in order to draw something, you need to look at it much more carefully and for longer, with the result being that you'll notice features you missed to begin with. Spending that extra time will not only make you feel a connection to the object, it will also add to your knowledge. You can draw anything you observe, whether it's constellations, the movement of a planet through a constellation over a month or two, how the Moon changes each day, or what you witness during a meteor shower. Even if you're only recording what you're seeing with your naked eye, you'll benefit from it immensely and you'll become a better observer as a result. Absolutely everybody has the ability to create useful astronomy drawings, so do try it. You don't even need any fancy art supplies—just some printer paper, a pencil, and an eraser.

Opposite top This photo by Mary McIntyre shows a Waning Crescent Moon with earthshine and Venus as seen from Oxfordshire, U.K. in February 2014. Because planets and the Moon all move along the ecliptic, they can appear in the same part of the sky. This is a line of slight effect and the bodies are not actually close to each other. The photo was exposed for long enough to capture earthshine on the non-illuminated side of the Moon. **Opposite bottom** This sketch shows the constellation Taurus with the movement of Mars over a three-month period, from December 1, 2022 to March 1, 2023 plotted onto it. Credit: Mary McIntyre

STARGAZING EQUIPMENT

When you've been enjoying your views of the night sky unaided for a while, you may decide that the time has come to buy your first telescope, but what should you buy? In actual fact, you're better off starting out with binoculars. The view they offer has the same orientation as what you see with the naked eye, so they can help you find your way around and learn to star-hop before being faced with an instrument that inverts the view. Using even a basic telescope requires some knowledge of the sky, so it's worth investing some time in observing with binoculars first.

Binoculars

Binoculars utilize both eyes, so the views through them can have more dimension, which is very pleasing for an observer. When you look with two eyes you have more of a sense of depth than when you look with one eye; although you aren't technically seeing in 3D, it seems as if that's the case. Additionally, most people have better vision in one eye than the other, so by using both eyes you achieve a "binocular advantage," which can allow you to see better in low light than if using just one eye. Another point is that there is a small blind spot in each eye where the optic nerve enters the retina, so if you're using both eyes at once, the blind spot for each eye is in a slightly different place in your field of view, thus helping to negate the blind spot. Binoculars also have the advantage of being extremely portable. If they are 10 x 50 or smaller, you should be able to hold them without any support, but keep your elbows tucked in so that the weight is supported by the entire arm to help prevent fatigue. If you're observing for long periods, or your binoculars are larger or heavier, you'll find it easier to mount them.

The cheapest and simplest way to support the weight is to turn a broom upside down and rest the binoculars on the brush; this isn't elegant, but it's surprisingly effective! If you mount a pair of binoculars on a standard camera tripod you can observe objects that are quite low in the sky, but the tripod legs can get in the way and you're limited by how impractical it is to bend and tip your head back to look through the binoculars when they're pointing at objects higher in the sky. A parallelogram mount gives more stability and a far greater range of movement. These consist of two longer pairs of parallel pieces of wood or metal that are joined using two shorter, vertical pieces, one of which is attached to a tripod. The pieces are able to pivot at the joins in such a way that you have a greater range of movement, plus you can adjust the height of the mount while keeping the binoculars pointing at the target. This is invaluable at star parties or outreach sessions because the binoculars can quickly be adjusted for observers of different heights with minimal fuss. They can be used seated or standing, or even adapted to work while you are lying on a recliner. You can buy readymade parallelogram mounts, although they can be expensive. There are also many free resources available online, providing instructions on how to build one yourself using whatever materials you have spare in your garage.

Telescopes

Upgrading to a telescope gives you the advantage of greater magnification and more flexibility with regards to changing eyepieces that allow for a variety of different magnifications. Telescopes are also easier to use for astrophotography. Once you're sure that you're ready to begin your telescope journey, it's time to make that very important decision about which one is right for you. There are three basic telescope types: refractors, reflectors, and compound telescopes. There are many other types, too, but we'll just cover the most popular options here.

Refractors

The first telescopes invented were refractors, and the simplest design is an achromatic refractor, which is a hollow tube with a concave lens, or "dioptric," at one end and a concave (or convex) eyepiece at the other end. The lens refracts parallel rays of light and focuses them to a focal point inside the tube; the distance from the lens to the focal point is the telescope's "focal length." The rays of light cross over at the focal point and are brought back to parallel rays again by the eyepiece, and the crossing over of the rays at the focal point mean the image you see in the eyepiece is upside down. This type of refractor is usually more affordable but it suffers from what's known as chromatic aberration—because red, green, and blue light don't focus at exactly the same point through a simple lens, objects will have color fringes around them both visually and in photographs. If you don't mind that imperfection, they're still fun instruments to use. Historically, one way around this was to make the tubes longer—which is why many vintage refractors are extremely long—but today we have a better solution, and that is to use different materials to produce the telescope lens.

Apochromatic refractors have lenses that are made with extra-low dispersion materials and consist of two or three layers sandwiched together. This doublet or triplet design corrects chromatic aberration, but these telescopes are more expensive to produce and are

Opposite Telescopes and binoculars can provide increased viewing ability. This large Newtonian reflector telescope provides an excellent support for seeing night sky phenomena over the bright lights of a city.

therefore more expensive to buy. If you're using a refractor for visual work, make sure that there is a diagonal at the eyepiece end so that the eyepiece sits at 90° to the telescope body. This is a much more ergonomic way to observe, particularly if you're pointing the telescope up high. Smaller refractors give a wide field of view, which is excellent for imaging of larger galaxies and nebulae but they can also give great views of the Moon and can reveal some detail on the larger planets, although the view will be quite small. They're very portable instruments and require almost no maintenance, so they're great all-rounders. The aperture of a refractor is limited to 39 inches (100 centimeters) because the weight of the glass lens can cause distortions and larger-aperture refractors require a longer tube length. Refractors are usually mounted on a tripod.

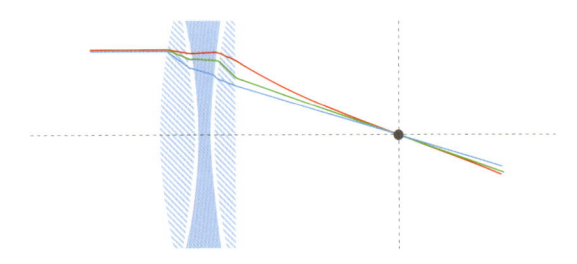

Above Extra low dispersion materials sandwiched together form the front lens of modern apochromatic refractors, and this corrects the chromatic aberration seen in budget refractors, caused by the three colors of light not coming into focus at exactly the same focal point.

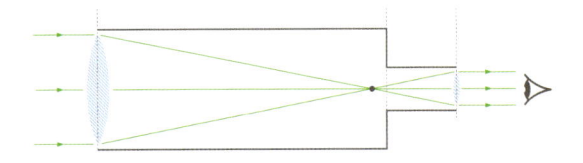

Above Refracting telescopes work by refracting parallel rays of light and focusing them on a focal point inside a tube.

Reflectors

Newtonian reflectors—a type of reflecting telescope invented by Isaac Newton—use an open tube with a primary mirror or "catoptric" at the bottom end. Light rays enter the tube, then the mirror focuses the light onto a much smaller secondary mirror, which is suspended on a structure called the "spider" near the top of the tube. The light is then directed to the eyepiece, which is on the side and near the top of the tube. Because mirrors are involved, the object will be a mirror image of how it would look to the naked eye. The primary and secondary mirrors on reflecting telescopes can be a variety of shapes, but the Newtonian reflector usually has a parabolic primary mirror and a flat secondary mirror that intercepts the light rays before they reach the focal point at the end of the tube. Because only one surface of a telescope mirror needs to be perfectly polished, these telescopes are cheaper to produce than refractors, so you can get a larger-aperture instrument for your budget, in sizes ranging from 74 mm right up to 500 mm. This mirror design can suffer with coma aberration, though, which can make stars take on a non-rounded shape around the edges. They're really good for observing faint objects because no light is absorbed by lenses, so they're affectionately referred to as "light buckets" by astronomers. Smaller reflectors can be mounted on tripods but larger telescopes are often put onto a simple rotating base that sits on the floor. These are called Dobsonian mounts and, when paired with a reflector, are one of the easiest kinds of telescopes to use because you just rotate and tilt the telescope until it's pointing at the object you want to look at. You can even buy small tabletop Dobsonians that are very quick and easy to use. Because of the open-tube design, the air inside the telescope needs to cool down to the outside temperature before use, otherwise warmer air currents in the tube will affect your views.

Giving such great bang for your buck, it's very tempting to go for the largest aperture you can afford, but keep in mind that the larger the aperture, the longer the tube, so you can end up needing to stand on a ladder to look through the eyepiece. Larger-aperture mirror telescopes can be big and heavy and therefore difficult to transport, too. You'll also need to think about storage when the telescope is not in use. Because the mirror is open to the elements, it may need to be cleaned periodically, and the mirrors need to be collimated to keep the image sharp, especially after transportation. Collimation is the process whereby all of the optical and mechanical components of the telescope are aligned perfectly so that the light can be brought to good focus. If a telescope is out of collimation it's impossible to get it to focus properly. Some mirror telescopes require collimation every time they're transported, but this is relatively simple for the end user to do themselves once they've learned how

Below Design and light path of a Newtonian reflector. The primary mirror is parabolic and the secondary mirror is flat. **Bottom left** Orion Telescopes 254 mm / 10 inch diameter Newtonian reflector on a Dobsonian mount. **Bottom right** Skywatcher Heritage 76 mm / 3 inch diameter table-top mini Newtonian reflector on a Dobsonian mount.

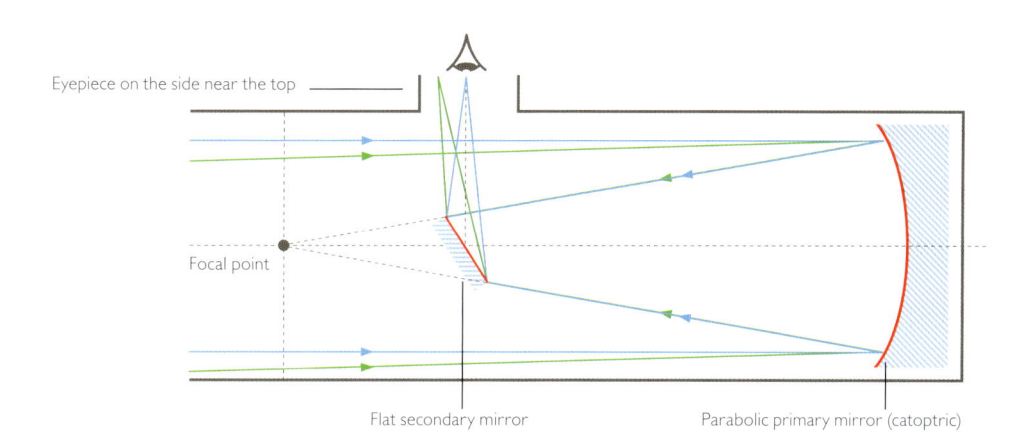

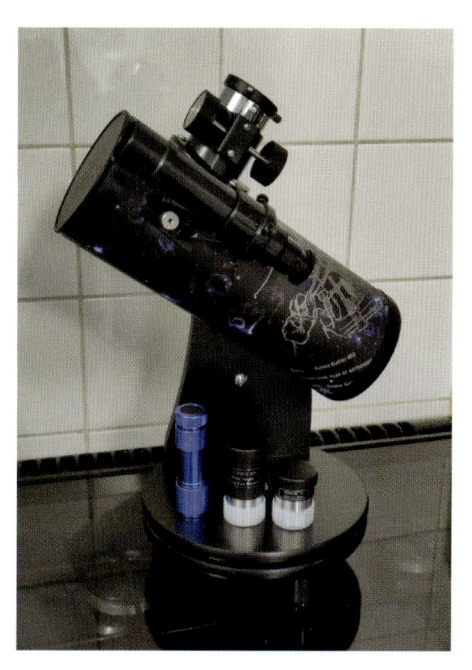

Eyepiece on the side near the top

Focal point

Flat secondary mirror

Parabolic primary mirror (catoptric)

STARGAZING TOOLS AND TECHNIQUES 313

to do it correctly. One exception is Maksutov telescopes, which have difficult-to-access primary mirrors, so they may need to be sent to a professional. If you're thinking of using a reflector telescope for astrophotography, it's worth noting that many of them cannot achieve focus without the use of a 2x magnification Barlow lens. The Barlow lens (named after Peter Barlow) is a lens that can increase the effective focal length of a telescope when placed before the eyepiece or camera, so it increases the magnification of the object you're observing. They are commonly available in 2x, 3x, and 5x magnification. Every doubling of magnification will quarter the amount of light coming through, so this will reduce the apparent brightness of the object. This can be an advantage for bright objects, but a definite disadvantage for faint objects.

One additional consideration for imaging is that the spider assembly obscures some of the light as it enters the instrument and produces four-pointed star spikes on photographs. Star spikes are viewed as an advantage by some because they make your photographs more closely resemble images taken with the Hubble Space Telescope—in fact they're so popular among some astrophotographers that they buy software that artificially adds star spikes to images taken through a refractor—but they're loathed by others, so this is really down to personal choice.

Compound telescopes

Compound telescopes use a "folded mirror" design that is comprised of an open tube with a primary mirror at the bottom and a secondary mirror that directs the light rays through a hole in the center of the primary mirror and onto an eyepiece that sits at the bottom of the tube. The focal point of this kind of telescope sits way outside the tube, so they pack a really long focal length into a very short tube. A classical Cassegrain has a parabolic primary mirror and a hyperbolic secondary mirror, but a specialist kind of Cassegrain known as a Ritchey–Chrétien is extremely popular in professional observatories; these use hyperbolic primary and secondary mirrors designed to avoid coma or spherical aberration. The open-tube design suffers with the same disadvantages as Newtonian reflectors and they need to be collimated. Most Ritchey–Chrétien telescopes are optimized for imaging rather than visual work but because the secondary mirror is suspended on a spider, you'll get star spikes on your images.

One final mirror telescope that has become very popular in recent years is the Schmidt-Cassegrain (SCT). This is a "catadioptric" design—using both mirrors and lenses. Like other Cassegrains, these pack a long focal length into a short tube. The primary mirror is spherical, which is easier and cheaper to manufacture, but they usually suffer with spherical aberration—an issue that's addressed by a transparent corrector plate on the front of the telescope, which corrects the aberration before the light hits the primary mirror. There's no need for a spider because the corrector plate holds in place a convex secondary mirror that also acts as a field-flattener. The light then goes through a hole in the center of the primary mirror and into an eyepiece. This brilliant design allows large-aperture, compact instruments to be produced for an affordable price. They're also very versatile, with the advantage of a sealed tube that keeps the mirror clean and dry. They still need to be collimated, though, and in order to achieve focus on an SCT, the whole mirror needs to move back and forth, which increases the risk of "mirror flop" when slewing the telescope.

Mounts

If choosing a telecope wasn't already a complicated enough decision to make, you also need to consider how you're going to mount it. The two basic mount types are Alt-AZ (Dobsonian mounts come under this category) and equatorial. For visual observing you can get away with a less sophisticated mount and accept that you'll have to keep manually nudging the telescope along as the object moves out of view; the larger the aperture of your telescope, the faster objects move out of shot. You can also get computerized mounts that have a "GoTo" function—you just select the object you want to view and the telescope will slew to it and track it so it stays in the field of view of your eyepiece, which can save a lot of time. This is very useful for sketching and it allows you to maximize your observing time rather than wasting any of a clear night trying to manually track down faint objects. However, in order to set up these mounts you first need to align your telescope on some bright stars in different parts of the sky so that the mount knows where it's pointing—and you can't do this if you have

Classical Cassegrain

Focal point

Hyperbolic secondary mirror

Parabolic primary mirror (catoptric)

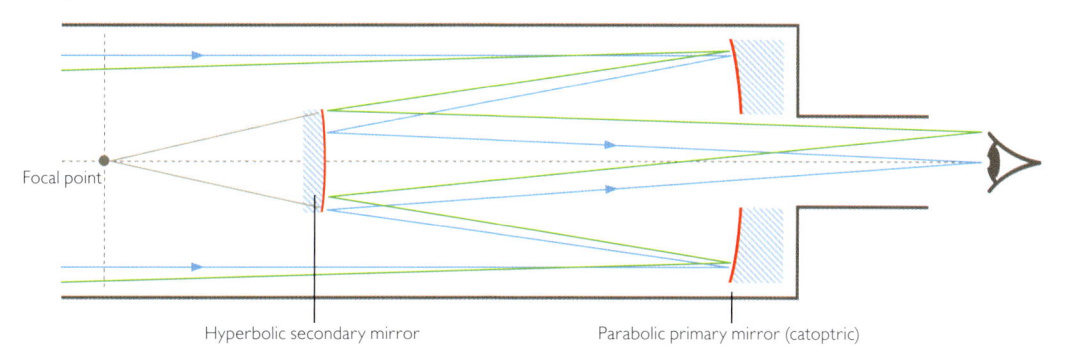

Ritchey-Chrétien (specialized Cassegrain)

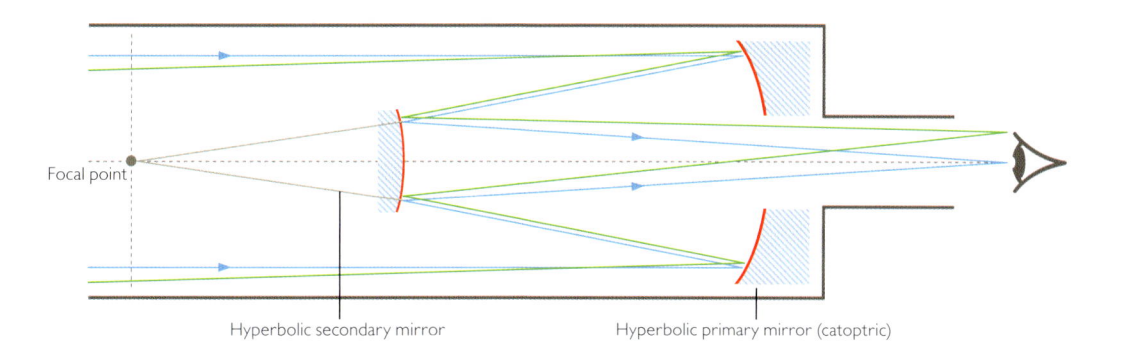

Focal point

Hyperbolic secondary mirror

Hyperbolic primary mirror (catoptric)

Schmidt-Cassegrain

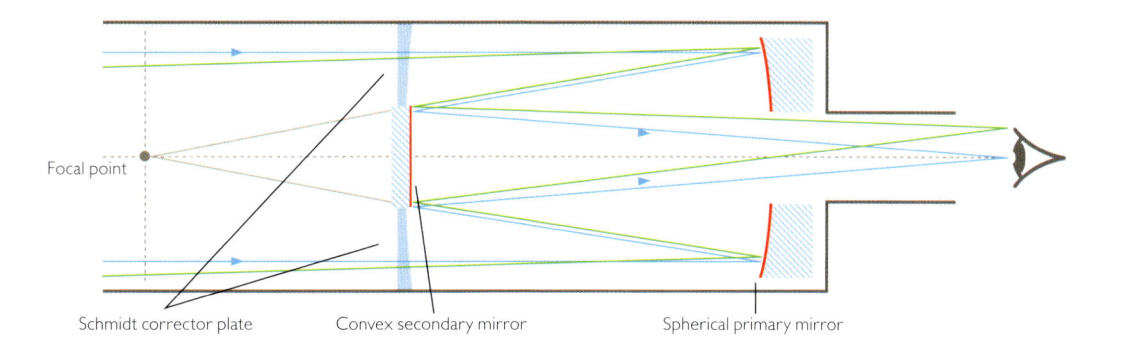

Focal point

Schmidt corrector plate

Convex secondary mirror

Spherical primary mirror

Top Design and light path of a classical Cassegrain telescope. The primary mirror is parabolic and the secondary mirror is hyperbolic. The secondary mirror bounces light back through a hole in the center of the primary mirror and into the eyepiece. **Middle** Design and light path of Ritchey-Chrétien telescope. This telescope is a specialist Cassegrain, and both the primary and secondary mirrors are parabolic. Almost all professional observatories now use Ritchey-Chrétien telescopes. **Bottom** Design and light path of a Schmidt-Cassegrain telescope, which has a spherical primary mirror and convex secondary mirror. Spherical mirrors are less expensive to produce but they suffer from spherical aberration. However, the corrector plate on the front of the telescope corrects this before the light hits the primary mirror.

STARGAZING TOOLS AND TECHNIQUES   315

no knowledge of the stars. This is why learning the sky with your binoculars first is so important.

We discussed the Alt-AZ and equatorial coordinate systems on page 294. Telescope mounts that move in Alt-AZ are easier to set up because they don't need to be polar aligned, but they will only track along the altitude and azimuth grid lines—remember that the stars do not follow those lines, unless you are observing from the North or South Pole! For visual work this is usually adequate, but for photography you start to suffer with field rotation in your photographs once you go past a certain exposure time. Relating to astrophotography, field rotation is the circular star-trailing pattern seen in long exposures taken with a mount that is tracking in Alt-AZ rather than an equatorial mount. This is due to the mount following the Alt-AZ coordinate lines when the stars are actually tracking the equatorial coordinates. Equatorial mounts will track along the equatorial coordinate grid, so they're the best option for serious astrophotography. However, they need to have their polar axis aligned to the celestial pole and they require counterweights that are set up to perfectly balance the weight of the telescope, so they're cumbersome to transport—and it takes time to get everything set up. Once set up, the mount needs its polar axis to be accurately aligned, then the mount needs to be aligned on some bright stars before use. Set-up time can be greatly reduced if you establish a permanent pier in your garden, or even build a home observatory shed.

Digital telescopes

In recent years there's been an exciting new development in telescope technology in the form of smart digital telescopes. These amazing gadgets are very compact and require almost no set-up—they automatically take a photo of the sky and compare that with a built-in star map, then figure out where they're pointing using a technique called plate-solving. Once it has figured out where it's pointing, you can select targets from the library and the telescope will slew to the object and begin taking images to produce a live-stack that, over time, builds up a longer exposure and therefore a more detailed image. (In astrophotography, image stacking is the process of taking multiple identical photographs of an object and digitally stacking them together to produce a single, more detailed image. Live stacking is the process whereby computer software does this in real time and displays the image on the screen as it does so. The longer the live stack goes on, the more detailed the image gets.) These telescopes allow you to take digital photographs of objects that would otherwise be invisible from a particular location, or to see a more detailed image of a brighter object than what you'd see through an eyepiece. The technology is improving all the time and newer releases of this type of telescope are becoming more affordable. They're not a substitute for actually looking through an eyepiece,

Top left An EQ5 Pro equatorial mount. The axis of this type of mount is lined up with the celestial pole and has counterweights on a bar to balance the weight of the telescope. Once properly set up and aligned, it will track the stars accurately. **Top right** Skywatcher AZ_GTi mount. Although this type of motorized mount does track, it only tracks in Alt-AZ so it will not follow the stars as accurately as an equatorial mount does. However, it has the advantage of a faster set-up and no counterweights.

but they're a very appealing option for many people because they're so easy to use. They're very compact so don't take up a lot of storage room, yet they pack a real punch.

Choosing and taking care of your equipment

Choosing your equipment is an important decision to get right because too many telescopes end up sitting unloved and gathering dust in a shed or attic because they were not the right match for the observer. In addition to the various pros and cons of the different telescope types, it's worth reiterating that most telescopes can be multipurpose instruments and, with the exception of astrographs, which are specifically made for imaging, most telescopes can be used for visual work as well as photography. If you're planning to be a purely visual astronomer, your budget would be better spent on some good-quality eyepieces—you can even get specialist eyepieces that are suited to people with astigmatism or who need to wear glasses while observing. If imaging is going to be more your thing, then investing in a good equatorial mount is a better use of your money. Your telescope needs to be easily accessible and you need to be able to carry it to your observing location, so consider both size and weight and think seriously about the practicalities of lifting and carrying it.

It's extremely important that you avoid getting your telescope wet because this can corrode the moving parts, so you need a safe, dry place to store it when not in use. A cold, damp environment can cause mildew or algae to grow on mirrors or lenses, and spiders love to make new homes inside open-tubed instruments, so always keep the dust covers on the front of the telescope, on the eyepiece holder, and on both sides of your eyepieces. If you're observing outside but store your equipment inside, put any covers back on before you bring it into the house—unless your telescope or binoculars have fogged up, in which case bring them in and let them fully demist before putting the covers on, otherwise the moisture will be trapped on the lenses. Try to avoid rubbing the lenses or eyepieces because you could damage the coatings on the glass if there are any abrasive particles on the surface; it's better to use a rocket blower or a soft brush to gently brush the dust away.

We talked earlier about keeping warm and dry, but it isn't just the observer who needs protection. If the temperature at night falls below the dew point, your equipment will start to get wet as the dew settles. This can cause lenses and mirrors to fog up, which is not ideal. Although it can be quickly rectified with a quick warm blast from a hair dryer, it can easily be avoided in the first place by using a dew heater. Affordable dew heaters are available to buy but you can create your own makeshift version by using some reusable hand warmers and an old sock that's big enough to fit over the end of your telescope or binocular lenses. Simply cut off the toe end of the sock to leave you with a tube, slide it over the end of your instrument, then activate the hand warmers and tuck them in so they lie close to the lens or primary mirror; this should keep them above the dew point and prevent them from fogging up. This solution is easy to transport and doesn't require external power so there's no need to worry about batteries running; when the hand warmers get cold, just replace them with a new pair. If you're using a larger-aperture telescope you can still use the hand warmers but you'll need a sock that's long enough to wrap around the tube rather than slide over it. Tuck the hand warmers inside the sock spaced out evenly and tie the ends of the sock around the telescope. When you're observing on a cold night, try to avoid breathing on your eyepieces because they will fog up and cloud your view.

Astronomy equipment can be an expensive investment but if you choose wisely and look after it properly it will last you for many years and bring you many hours of joy.

Right A low-cost alternative to buying a dew heater for your telescope is to cut the end of an old sock, wrap it over the end of the telescope tube then insert some reusable hand warmers into it. This will help prevent dew from forming on the telescope.

URBAN AND SUBURBAN SKIES

Astronomical observations can be challenging in locations that have higher levels of light pollution and sadly the amount of artificial light at night (ALAN) is becoming a significant problem. If you look at satellite photos of Earth taken from space, you can see the ever-increasing amount of light pollution extending out from towns and into suburban and rural areas, despite the fact that the extremely detrimental effect it has on biodiversity and human health is now well documented.

Light pollution

Most species on Earth have spent millions of years evolving to function with a cycle of day and night, but the increase in ALAN—particularly the recent influx of blue-rich LED lighting that mimics daylight—is affecting insects, birds, fish, mammals, and trees. In 2000 it was reported that over 4 million birds die every year in the United States alone as a direct result of ALAN, and this was probably an underestimate. Bats are going hungry because the insects they usually feed on in-flight are so busy being attracted to bright lights at night that their breeding cycles are affected. Some species of turtle require a dark, moonless night in order to successfully hatch their young, but bright lights are having an impact on this process. Bright lights at night near water encourage algal blooms that are damaging to ecosystems. The list goes on and on.

It isn't just wildlife that's negatively impacted by ALAN, either. Human circadian rhythms are severely impacted by exposure to blue-rich white light at times when it should be dark because this affects the production of the hormone melatonin. It's also worth remembering that in addition to regulating sleep cycles, melatonin is essential for well-being and plays a role in the immune response to cancer cells. The introduction of blue-rich LED lighting has been rolled out globally without proper assessments of its impact on human health and the environment, and without regard to its effect on the night sky.

In addition, the energy used to power all of these lights is a considerable problem, with the world now facing a climate emergency. We need to make more of an effort to use the right amount of light at night—only using illumination where it's absolutely necessary. Because LEDs are energy efficient and cheaper to run, there's a tendency to install lights that are far too bright and are left on all night. Even if they are shielded, when these overly bright white lights shine down onto a wet, frosty, or snow-covered surface, the light reflects straight back up into the sky, causing even worse levels of light pollution than was seen with traditional unshielded sodium lights. Using a dimmer, warmer-toned LED that's well shielded to prevent unnecessary light spill makes a big difference. Bright security floodlights fitted on domestic and commercial properties are relatively inexpensive,

Opposite This spectacular Pink Supermoon rises over the New York skyline on April 26, 2021. **Top** This photograph by Christoph Geisler shows light pillars over Laramie, Wyoming. Light pillars form when ice crystals higher up in the atmosphere sink down and reflect light from an unshielded light source to form what look like tall, thin pillars.

320 THE BACKYARD STARGAZER'S BIBLE

but their casings often have a wide-angled opening, so light spills out far beyond where it's needed. These lights often remain illuminated all night—or if they're motion triggered, they're so sensitive that they flick on constantly. Ironically, they shine straight into the faces of passing motorists and pedestrians, which actually reduces their effectiveness when it comes to exposing criminal activity taking place.

There is never a need for lights to shine outward or upward into the sky, where they affect wildlife and ruin the dark sky. And interestingly, if external lights spill over into the windows of a neighboring property, it counts as light trespass in the U.K. and is considered an offence. If a security light is needed, select one that's not aggressively bright and preferably angle it straight downward, or at least under 22°. Several years ago, France brought in strict measures reducing ALAN, so hopefully other countries will follow suit.

Unshielded lights reflect off clouds, so the negative effect of badly designed lighting is amplified when thin cloud is present; we've all seen the huge domes of light that extend up over a town when you approach them at night. If you live in a major city you'll be familiar with the ever-present orange glow at night; the majority of city dwellers have never seen a truly dark sky. This can all make you feel very disheartened about taking up astronomy, but there are still things worth observing, such as the Moon, brighter planets, brighter constellations, and conjunctions. You may even consider investing in the correct equipment for safe observations of the Sun because that's one area of astronomy that's definitely not affected by light pollution!

In very cold weather the frozen ice crystals that are usually located higher up in the atmosphere can sink down to lower levels and form "diamond dust"—crystals that are suspended in the air. Unshielded light sources can then reflect off the ice crystals and form light pillars. These look like a beam of light shining straight upward but the pillar is not above the light source, or anywhere else—like rainbows and all ice haloes, it's simply an atmospheric optical effect. While colored light pillars can be very pretty, resembling pillars of an aurora, every one of these beams is created by an outside light source that's not properly shielded.

Improving your views

If you live in an urban area, just getting yourself into a position by a wall or shed, where you are shielded from lights shining directly at you, will make a big difference to what you can see above you. If you lie on a recliner and cup your hands around the outside of your eyes as a shield, you'll notice the sky directly above you looks much darker. You can also buy binocular and telescope eyepiece shields that help to block out distracting stray light and they work extremely well. If you are into astrophotography and most of your light pollution comes from old-fashioned sodium lamps, you can buy light-pollution clip filters that filter out these wavelengths of light. Sadly new LED lights are broad spectrum so cannot be filtered out, but many towns now have better shielding on LED lights, so you may find that your skies have improved—at least, when road surfaces are dry.

If you still find light pollution is affecting your photography it's worth investigating astrophotography of deep-sky nebulae using narrowband filters such as hydrogen alpha, sulfur II, and oxygen III with a monochrome camera. Because they're capturing light at a very specific wavelength—beyond what the human eye can see—the images are not affected by light pollution in the same way. You can also buy tri-band and quad-band filters that allow you to image at multiple narrow wavelengths of light at the same time, which reduces imaging time. The downside of narrowband imaging is that you need to take much longer exposures and lots of them, so you may need a higher-specification mount to help you achieve this.

Opposite top This photo by Mary McIntyre shows a huge cone of light coming from a badly angled security light on a misty night. This kind of light spill is extremely damaging to both human health and nocturnal animals. **Opposite bottom** This photo shows an example of better designed directional shielding on streetlights on a misty night.

The Bortle scale

In 2001, *Sky & Telescope* magazine published an article by amateur astronomer John E. Bortle in which he described his new numerical scale. The Bortle scale measures how much light pollution there is in a particular location from 1 to 9, with 1 being the darkest sky possible and 9 being the worst possible inner-city light pollution. It measures visibility criteria of several popular targets and includes the limiting visual magnitude for each step of the scale. The limiting visual magnitude is the faintest object that is visible using a 12.5-inch (32-centimeter) reflector telescope, in your current sky conditions. When creating the different steps of the Bortle scale there was an attempt to clarify what the limiting magnitude was within each step—for example, a Bortle 1 sky has a limiting visual magnitude of 17.5, a Bortle 5 sky has a limiting visual magnitude of 15, and Bortle 8 sky has a limiting visual magnitude of 13.

Dark sky reserves now exist where artificial lights are restricted at night in order to protect the dark skies in that location, and visiting these sites is an amazing experience. However, you don't necessarily need to seek out a Bortle 1 location in order to enjoy stargazing as there's a stark difference in what is visible just moving from a Bortle 5 or 6 to a Bortle 4 location. If you're a complete beginner trying to learn your way around the sky you may find that a little bit of light pollution can actually help you to identify the constellations; there are many experienced amateur astronomers who've struggled to pick out star patterns they thought they knew well when faced with the many additional stars visible from a very dark location. Scientifically, the accuracy of the Bortle scale is not perfect, but it's still a useful measure that can help you to plan your stargazing sessions.

Above The Bortle scale shows how light pollution affects the night sky, from the worst (City/Inner City Sky) to the best (Excellent Dark Sky Site). Even visiting a location just one point darker on the scale can make a big difference to your skies. **Opposite** This wooded are shows poor light pollution and how it can affect visibility of any detail in the night sky.

The stargazing experience

When you have access to binoculars or a telescope, an entirely new world of observations opens up to you, but as a new visual astronomer it's vital that you manage your expectations. We've all been wowed by the incredible images we see online and in magazines, but those images were created by taking multiple exposures over many hours, then digitally stacking them (see page 367) and processing the stacked image to bring out the detail. There is no small or moderate-sized telescope that will give you views anything like those photos, so people can sometimes be underwhelmed when they first look through a telescope. However, there is still so much beauty in the night sky waiting to greet you at the eyepiece. Yes, many deep-sky objects look like indistinct, gray smudges, but keep in mind how far away those objects are. If you're looking at the Andromeda Galaxy, the photons of light left that galaxy 2.5 million years ago and they're now being focused on your eye. It's impossible not to feel a real connection with an object when you think in these terms. Visual astronomy and astrophotography are two very different sides to astronomy but there's room for both in every astronomer's life.

If you're using a telescope, you can completely change the magnification by swapping eyepieces. You can increase the focal length by adding a 2x, 3x, or 5x Barlow lens, for example, or you can decrease the focal length by adding a focal reducer—there are many different viewing options using the same telescope. You can also attach filters to improve your viewing experience. An ultra high contrast (UHC) filter, for example, will help bring out more detail in nebulae and galaxies while suppressing light pollution.

When creating sketches of your observations, use a clipboard and a dim red light to provide gentle illumination on the page. You can use a clip-on LED reading light on the dimmest setting and wrap some red film around the light several times to diffuse it down a bit. Even a dim red light will affect your night vision a little, so choose one eye to look through the eyepiece and the other to look at what you're drawing. As you look through the telescope, keep your drawing eye closed, and as you draw, keep your observing eye closed so that it doesn't get affected by the light. As you open and close each eye in turn, pirate noises are optional!

DARK SKY SITES

The International Dark-Sky Association is a non-profit founded in 1988 by an astronomer–physician team. Headquartered in Arizona, the group aims to preserve and restore optimum conditions for viewing the night sky through advocacy and conservation efforts. Much of their work is done through more than 70 local chapters worldwide, listed on their website.

Light pollution gained attention in the 1970s as astronomers noted the effect of artificial light on their ability to observe the night sky. At present, around a third of the world can no longer make out the Milky Way; in the United States the percentage is closer to 80 percent of the population. Almost everyone in Europe and the U.S.A. has a view of the night sky undermined by light pollution. The Cooperative Institute for Research in Environmental Sciences offers an Atlas of Artificial Sky Brightness at https://cires.colorado.edu/Artificial-light.

Light pollution not only affects the study of astronomy and enjoyment of stargazing, it affects the Earth and its animals as well, leading to ecosystem imbalances and hormone dysregulation, along with other adverse health impacts.

The International Astronomical Union champions the Dark and Quiet Skies Global Outreach Project to advocate for the importance of sky conservation focusing on both dark and quiet. Their outreach efforts address the risks posed by satellite radio interference as well as visual impacts. Their site offers the opportunity to take a pledge to "become a dark and quiet sky protector," as well as certificates from the Office for Astronomy Outreach that document this commitment.

The International Dark-Sky Association mentioned above hosts the International Dark Sky Places program, which certifies sky conditions around the world. There are currently more than 200 places representing 22 countries in their catalog, including communities, parks, and sanctuaries that are certified as offering protected views of the night sky. We have provided a selection of each below.

Dark Sky Communities

Dark Sky Communities implement measures to curb light pollution and raise awareness about the value of preserving the night skies.

Bee Cave, Texas, U.S.A.
Bisei Town, Ibara City, Japan
Bon Accord, Alberta, Canada
Coll, Scotland
Fulda, Hesse, Germany
Jelsa, Croatia
Sark, Channel Islands
Sopotnia Wielka, Poland
Xichong, Shenzen, China

Dark Sky Parks

Aenos National Park, Greece
Albanyà, Spain
Appalachian Mountain Club Maine Woods, Maine, U.S.A.
Arches National Park, Utah, U.S.A.
Big Bend National Park, Texas, U.S.A.
Bodmin Moor Dark Sky Landscape, Cornwall, England
Cherry Springs State Park, Pennsylvania, U.S.A.
De Boschplaat, the Netherlands
Death Valley National Park, California, U.S.A.
Desengano State Park, Brazil
Eifel National Park, Germany
Elan Valley Estate, Wales
Great Basin National Park, Nevada, U.S.A.
Hehuan Mountain, Tawain
Iriomote-Ishigaki National Park, Japan
Mayo Dark Sky Park, Ireland
Naturpark Attersee-Traunsee, Austria

ASTRONOMY CLUBS AND ORGANIZATIONS

The United States is home to more than 600 amateur astronomy clubs. The Astronomical League publishes a quarterly magazine called *Reflector* and a directory that catalogs amateur astronomical groups by region in the U.S.A., from the Northwest through the Great Lakes region to the Southeast. The league is committed to combating light pollution and encouraging young people to participate in amateur astronomy, among other goals.

Observing Programs include the Bright Nebula Observing Program in Washington, the Carbon Star Observing Program in Arkansas, and the Youth Astronomer Observing Program in Florida.

Sky & Telescope magazine features a club finder on their site, with offerings that range from a public high-school planetarium in Maryland to an Astronomy Club in Moscow that hosts an annual event every New Year's Eve.

In the U.K., the *BBC Sky at Night* magazine lists astronomy societies located throughout the U.K. and Ireland. There are close to 300 clubs and societies in the region. The clubs' intentions differ, but many encourage observing the night sky in groups and sharing equipment. Some clubs, like the Isle of Man Astronomical Society, have their own observatories available for tours.

The International Meteor Organization located in Belgium has initiated a coordinated effort of amateur work in the area of meteor phenomena.

Go Astronomy (at go-astronomy.com) lists clubs and organizations worldwide and includes a global astrotourism directory, featuring planetariums, space museums, dark sky sanctuaries, and star parties.

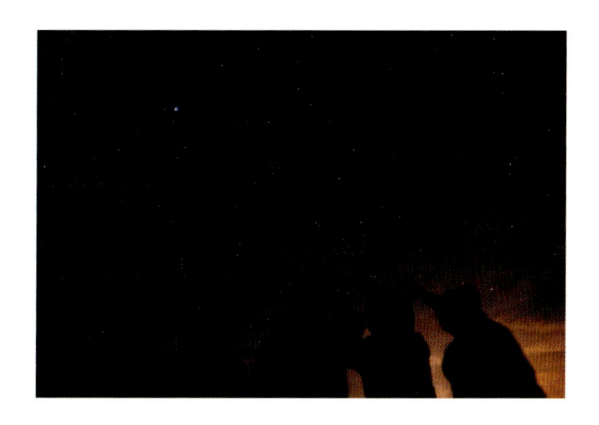

Northumberland National Park and Kielder Water & Forest Park, England
Parc National du Mont-Tremblant, Canada
Ramon Crater Nature Reserve, Israel
Sky Meadows State Park, Virginia, U.S.A.
Tomintoul and Glenlivet-Cairngorms, Scotland
Waterton–Glacier International Peace Park, Canada and Montana, U.S.A.

Dark Sky Sanctuaries

!Ae!Hai Kalahari Heritage Park, South Africa
Aotea Great Barrier Island, New Zealand
Bardsley Island (Ynys Enlli), Wales
Boundary Waters Canoe Area Wilderness, Minnesota, U.S.A.
Gabriela Mistral, Chile
Pitcairn Islands, Pacific Ocean

Dark Sky Reserves

Alpes Azur Mercantour, France
Bannau Brycheiniog National Park, Wales
Cranborn Chase, England
Greater Big Bend International Dark Sky Reserve, Mexico
Kerry International Dark Sky Reserve, Ireland
Mont-Mégantic, Montreal, Canada (the world's first Dark Sky Reserve)
NamibRand Nature Reserve, Namibia
River Murray, Australia

At the time of print there was only one DarkSky-approved Lodging site. Under Canvas Lake Powell—Grand Staircase in Utah, U.S.A.

Above Clubs can be a great way to have support and guidance in viewing the night sky, such as with this group observing the Milky Way from a dark sky location.

The Milky Way Galaxy, as seen here in four different Dark Sky Parks. **Above** Set against the dramatic foreground of the Dark Sky Park of Arches National Park, U.S.A., the Milky Way provides a truly stunning background. **Below** The Milky Way can also be seen here, at the Dark Sky Park of Balanced Rock, Big Bend National Park, Texas, U.S.A. The silhouetted stargazer with headlamp highlights the incredible scale and vastness of the night sky.

Above The Milky Way can be seen here at the Dark Sky Park of Mt Hehuan in Taiwan. **Below** Santa Elena Canyon can be seen here under the Milky Way, at the Dark Sky Park of Big Bend National Park.

CHAPTER FOUR

THE ART OF THE SKY

THE NIGHT SKY IN ART

Our early ancestors had a real connection to the night sky. With no clocks or digital calendars to provide reminders, the movement of Solar System bodies marked the passage of time in a regular and reliable way. It's impossible not to notice the daily rising and setting of the Sun, but to mark longer intervals, people studied the Moon, whose 29.5-day phase cycle provided an easily observable monthly metronome; the word "month" is even a derivation of the word "Moon."

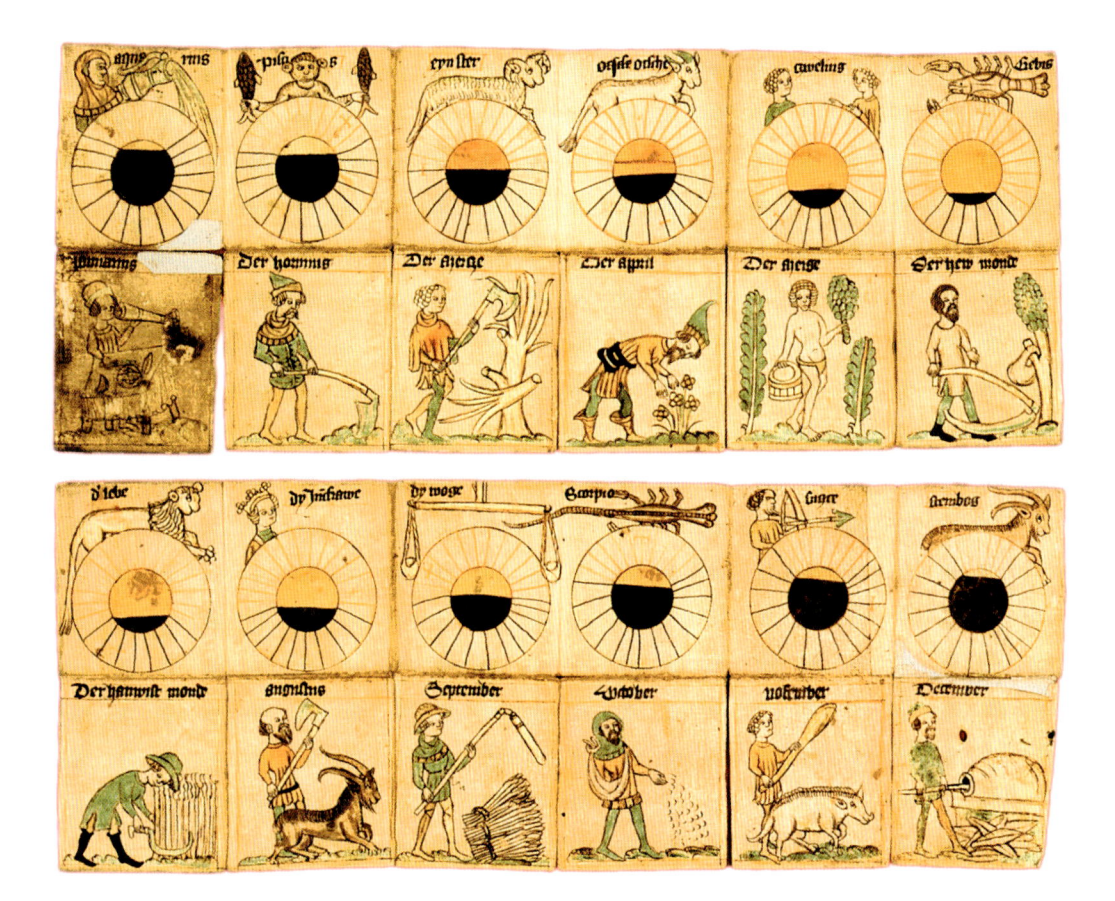

Indigenous peoples used the Moon as a calendar and assigned names to each Full Moon during the year. These names then served as reminders to undertake certain farming tasks, for example, or to prepare for a particular migratory animal as a food source.

Early philosophers had detailed knowledge of lunar cycles, as well as the lunar and solar eclipse Saros cycles, and lunar and lunisolar calendars were produced. One example is the calendar monument that was built at Warren Field, in Scotland, which dates back to around 8,000 BCE. This had pits that corresponded to lunar phases, together with a point that corresponded to sunrise on the winter solstice. This feature marked the exact passage of a year and it was necessary because the lunar cycle is not perfectly aligned with the solar calendar. Another example is the Antikythera mechanism, an ancient Greek hand-powered orrery dating back to around the second century BCE and said to be the oldest known example of an analog computer, featured cogs that represented the lunar phase and Saros cycle, as well as many others serving different astronomical functions.

The ever-turning wheel of the sky and the shifting positions of the constellations throughout the year were also a way to mark the passing of the seasons, as was the knowledge of the equinoxes, and summer and winter solstices. Wood or stone circles that align to the solstices—Stonehenge being one example— were built all over the world; more than 4,000 of these were built in Northern Europe alone, and

Previous pages Vincent Van Gogh's *Starry Night Over the Rhone* (1888, oil on canvas). **Opposite** The Nebra sky disk is a bronze and gold disk of around 12 inches (30 centimeters) diameter with a weight of 5 pounds (2.2 kilograms), having a blue-green patina and inlaid with gold symbols. These symbols are interpreted generally as the Sun or full moon, a lunar crescent, and stars. Since 2002, it has been in the State Museum of Prehistory Halle (Saale), Germany. **Above** Circa 1400, Faltkalender mit Monatsbildern is an example of a folding pocket calendar produced for farmers during the Middle Ages.

around 1,000 still exist in the British Isles. Portable sky-measuring instruments such as the Nebra sky disc discovered in Germany in 1999, and the "Divine Couple Ring" discovered in Knossos in 2002, both date back to around 1600 BCE and feature night-sky objects, so they're believed to have been used as portable calendars. Folding pocket calendars were produced for farmers in the Middle Ages, and in one example from around 1400, the *Faltkalender mit Monatsbildern* ("Folding Calendar with Monthly Pictures"), each month featured a zodiac sign—the constellation that the Sun lies in for that month—as well as a picture that corresponded to the agricultural tasks to be completed that month, and a little illustration indicating approximate daylight hours.

The wonder of the night sky has inspired art all over the world since the beginning of human civilization. Early astronomers had to record their observations by sketching what they saw with the naked eye, and they brought constellations to life by producing stunning artwork to illustrate the mythological stories that inspired their names, but capturing the night sky in art is something that can be traced back way further than that.

In 1940, a network of caves was discovered near the village of Montignac, in France's Dordogne region, by an 18-year-old dog walker called Marcel Ravidat. The Lascaux Cave is now a UNESCO World Heritage Site. There are several chambers within the cave network, and the walls and ceilings are covered in paintings that are thought to date back 17,000 years. They feature large animals, including horses, deer, and aurochs (an extinct species thought to be the ancestor of today's domestic cattle), as well as local fauna that corresponds to the Palaeolithic fossil record. In 2000, some of the paintings were studied by Dr. Michael Rappenglueck of the University of Munich, who identified two paintings that he believed depicted constellations. In one part of the cave, he noted a painting of a bull that appears to have dots that correspond with Taurus, and another that corresponds to the three stars of the Summer Triangle. Rappenglueck also discovered another example of an apparent night-sky depiction, dating back 14,000 years, on the walls of the Cueva di El Castillo cave in the mountains of Pico del Castillo, Spain—this time covering the area around Corona Borealis. To date, the Lascaux paintings remain the oldest artistic depictions of the night sky, but there are hopefully others out there waiting to be discovered. There are also petroglyphs (rock carvings) all over the world depicting constellations and astronomical events, some of these dating back more than 5,000 years.

As humans began to try to make sense of the world around them, whether through philosophy or religious writings, artists were inspired to create art that provided a visual representation of their theories and discoveries. Then, once the telescope was invented in the early 1600s, it led to numerous new discoveries, and astronomers sketched, painted, or engraved images to record what they saw at the eyepiece. In addition to a wealth of astronomy sketches produced by both amateur and professional astronomers, depictions of the night sky have inspired other forms of art throughout history. From astronomical objects depicted as deities with faces, to bible illustrations and accurate recordings by astronomers, artists have taken inspiration from the beauty of space—and continue to do so.

Opposite The golden Divine Couple Ring from Poros showcase a seated figure and a standing one, surrounded by rocks, flanked by two birds, and below a sun and moon image. The figure is said to be a Minoan solar proto-diety. The photographs are by Giorgos Rethemiotakis, and the ring can be found at the Heraklion Museum, Greece.

THE ART OF THE SKY

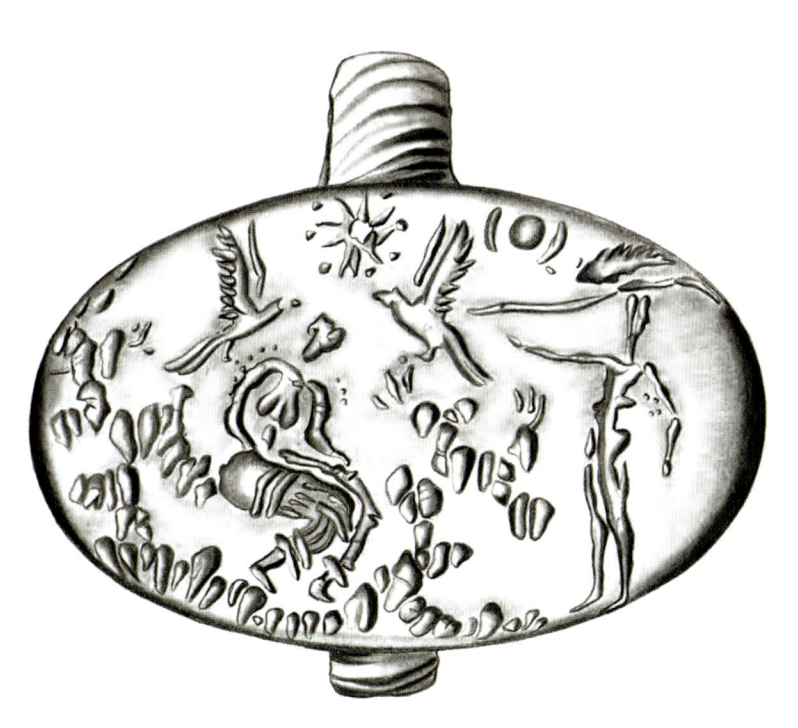

Michael Wolgemut
(1434–1519)

In 1493, the German physician and cartographer Hartmann Schedel wrote the illustrated encyclopedia the *Nuremberg Chronicle*, which included human history as told in historical accounts from around the world as well as through biblical passages. It was one of the first books of its kind to incorporate illustrations and text, and the German painter and printmaker Michael Wolgemut was commissioned to produce the illustrations.

Wolgemut was an incredibly talented artist and he ran a large workshop where many artists trained as apprentices. The bulk of the work carried out involved taking Wolgemut's designs and transferring them to woodcut blocks, which were used to provide book illustrations for many Nuremberg publishers, including those responsible for the *Nuremberg Chronicle*.

The *Nuremberg Chronicle* features a series of woodcuts that depict the six days of Creation, followed by the seventh day of rest, as told in the Book of Genesis. The first illustration (day one) is a simple design that depicts the creation of night and day, then each illustration increases in complexity to include: the creation of the sky and sea on day two; land and vegetation on day three; the stars, Sun, and Moon on day four; and sea creatures and birds on day five. In all illustrations, God is represented by a disembodied hand in the top left corner, but on day six, when animals and humans are created, God takes human form, creating Adam from a lump of clay with his left hand as he blesses him with his right hand. Day seven, the day of rest, shows God and nine orders of angels sitting above the newly created Universe. Inspired by the Ptolemaic geocentric model of the Universe, Earth sits at the center and is surrounded by concentric spheres of water, air, fire, and the ten celestial spheres. In the corners, the wind blows from the four cardinal points and, as is common with allegorical works, is depicted as deities with faces.

Prior to Gutenberg's invention of the printing press in around 1440, books were hand-written, so they were expensive and rare. Hartmann Schedel was one of the first cartographers to use the printing press to reproduce text and illustrations for his books, although Wolgemut often hand-colored his illustrations after printing. When the *Nuremberg Chronicle* was being produced, the teenage Albrecht Dürer was an apprentice in Wolgemut's workshop, so there's a good chance he was involved in its illustrations. Dürer went on to improve the production methods used by Wolgemut, and became a remarkable artist in his own right.

Above This portrait of Michael Wolgemut was painted by Albrecht Dürer in 1516. **Opposite** Image 78 (top) of the *Nuremberg Chronicle*, by Michael Wolgemut, is one of 1809 woodcuts produced from 645 blocks. *The Nuremberg Chronicle* is a universal history compiled from older and contemporary sources and is one of the most densely illustrated and technically advanced works of early printing. It contains 1809 woodcuts produced from 645 blocks. The Nuremberg entrepreneur Sebald Schreyer and his brother-in-law, Sebastian Kammermeister, financed the production of the book. Michael Wolgemut and his son-in-law Wilhelm Pleydenwurff executed the illustrations in around 1490. The views of towns, some authentic, some invented or copied from older models, are of both artistic and topographical interest. This brilliantly colored copy, owned by Schedel, contains valuable additional matter, such as Erhard Etzlaub's map of the road to Rome. Along with the rest of Schedel's library, the book became part of the library of Johann Jacob Fugger, which in 1571 came into the possession of Duke Albrecht V of Bavaria. Image 79 (bottom) of *The Nuremberg Chronicle*, with folios depicting days five and six of Creation. On day five God created the sea creatures and birds and on day six God created animals and humans.

THE ART OF THE SKY 335

Above Folio 92 (Comet, 1506) of *The Augsburg Book of Miracles* shows a comet streaking through the sky in the year 1506. *The Book of Miracles* is an illuminated manuscript made in Augsburg in Germany in the 16th century.

The Augsburg Book of Miracles

(Augsburger Wunderzeichenbuch)
1552 (illustrator unknown)

Throughout history, many regular astronomical apparitions have been interpreted as portentous signs—signifying bad luck, or occasionally good. In 1552, a bound collection of 167 watercolor and gouache paintings of "miraculous" events was published, vaguely following the timeline laid out in the Bible. Now known as *The Augsburg Book of Miracles* (after the German city where it was produced), it contained 60 folios that featured perfectly ordinary events, many of them astronomical in nature, and many of which had been deemed responsible for some kind of natural disaster or catastrophe.

Comets have long been associated with the deaths (or births) of historical characters, or linked with famine, war, natural disasters, or pretty much any other unfortunate event that might have occurred within living memory of a comet being sighted, so they have featured extensively in art. One of the earliest depictions was that of Halley's Comet, on the eleventh-century Bayeux Tapestry—foretelling the death of English King Harold II at the Battle of Hastings. In his 1305 painting *Adoration of the Magi*, Giotto di Bondone depicted the Star of Bethlehem as a comet, but astronomers have not found any evidence that a bright comet was visible at the time Jesus was said to have been born.

Many comet appearances prior to 1552 are discussed in *The Augsburg Book of Miracles*, accompanied by some of the most striking comet paintings ever created. These are extremely stylized, and deliberately dramatic, to represent the disastrous events they were alleged to have caused. Folio 92 features a comet that was visible in 1506, and the text below it reads,

> *In the year 1506, a comet appeared for several nights and turned its tail towards Spain. In this year, a lot of fruit grew and was completely destroyed by caterpillars or rats. This was followed eight and nine years later, in this country and in Italy, by an earthquake, so great and violent that in Constantinople a great many buildings were knocked down and people perished.*

In other words, a comet was considered responsible for an infestation of caterpillars, and then an earthquake, almost a decade after its passage through the inner Solar System! Meteors, eclipses, and many other apparently portentous events are presented in the same stylized manner throughout *The Augsburg Book of Miracles*, providing a stark contrast to the way these objects would be being painted by the late 1800s, by artists such as Frederic Edwin Church (see page 350).

Galileo Galilei
(1564–1642)

Astronomers and scholars took a more scientific approach to astronomy art, creating accurate drawings and maps from their observations. In England, Queen Elizabeth I's astronomer William Gilbert (1544–1603) produced the earliest known lunar map in around 1600 from naked eye observations, and this was published in 1651.

The Italian polymath Galileo Galilei is often credited as the first astronomer to create a sketch of the Moon as seen through a telescope, but that honor actually goes to the English astronomer Thomas Harriot, who created a lunar drawing using his new "Dutch Trunke" telescope on July 26, 1609. Harriot's telescope did not have the same resolving power as Galileo's, but he was still able to see that the lunar surface had a "strange spottedness."

Galileo originally enrolled at the University of Pisa to study medicine but after attending a geometry lecture he switched to mathematics and natural philosophy. During his career he made many discoveries in fundamental science, but he's best known for his work in the field of astronomy.

Galileo's first telescope only had a magnification of 3x, but by making modifications to the design he was able to increase that to 30x. At that time, the heavens were believed to be perfect and unchanging, but Galileo was able to demonstrate that the Moon was not the translucent, perfect sphere it was believed to be; rather, it was covered in mountains and craters.

In his early sketches of the Moon, Galileo wasn't trying to achieve accuracy—instead he was demonstrating how light and shade played over the uneven lunar landscape to create the shapes and textures he saw. He also observed dark patches on the Sun—sunspots—that moved over time. So just like the Moon, the Sun was not a perfect sphere created by God.

Galileo also observed the planets. He discovered the four brightest moons of Jupiter and, after studying their movement, realized they were actually orbiting Jupiter—a fact that went against the contemporary belief that everything was in orbit around the Earth. Galileo also noted that Venus showed the same changing phases that we see on the Moon; this could only be possible if Venus was orbiting the Sun.

Galileo published over a dozen written works during his career, including the 1610 treatise *Sidereus Nuncius* ("Starry Messenger"). This was the first publication based on telescopic observations and it featured many of his drawings. His discoveries challenged the beliefs of the time and in 1633 he was found to be "vehemently suspect of heresy" for declaring that the Sun was at the center of the Universe and that the Earth and planets were in orbit around it. He was placed under house arrest, which continued until his death in 1642. Many since, including Albert Einstein and Stephen Hawking, have considered Galileo responsible for the birth of modern science.

Above Portrait of Galileo Galilei (1564-1642) by Justus Sustermans, circa 1640, on display at the National Maritime Museum, Greenwich, U.K. **Opposite top** Galileo's 1613 sketch of the Sun, including several sunspots, that he was able to see as moving over time. **Opposite bottom** Galileo's lunar sketches, published in 1610 in *Sidereus Nuncius*, were designed to demonstrate how sunlight on the lunar terrain created areas of light and shade rather than trying to create topographically accurate drawings.

THE ART OF THE SKY

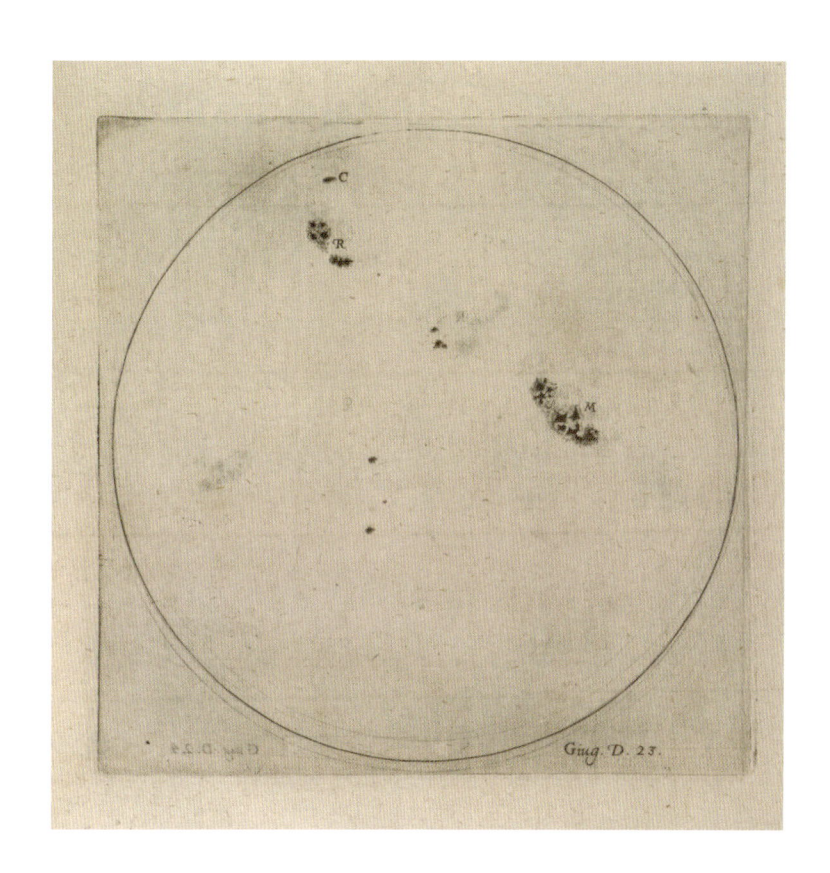

DE MACROCOSMI PRINCIPIIS. 41

Ut in mundi primordio, ubi tenebræ cujusque cœli cum partibus lucidis, quas viscositas spirituum in illis conclusorum, informationisque aviditorum am-

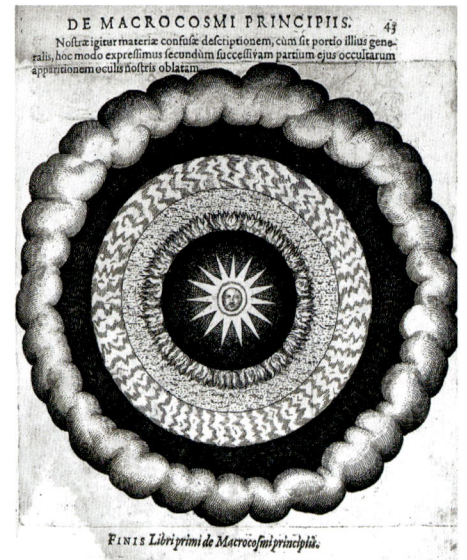

DE MACROCOSMI PRINCIPIIS. 43

Nostræ igitur materiæ confusæ descriptionem, cùm sit portio illius generalis, hoc modo expressimus secundùm successivam partium ejus occultarum apparitionem oculis nostris oblatam

FINIS *Libri primi de Macrocosmi principiis.*

Et sic in infinitum.

Et sic in infinitum.

Et sic in infinitum.

Et sic in infinitum.

Robert Fludd
(1574–1637)

The seven pictures in the Creation series from the *Nuremberg Chronicle* (see page 335) are colorful and charming, and a stark contrast to the Creation series created by the English physician, mathematician, and cosmologist Robert Fludd. Fludd can certainly be considered a polymath—he obtained several degrees from Oxford University, including one in medicine, but he also studied some less-conventional subjects, including theurgy (the practice of rituals), and paracelsianism, an early modern medical movement based on the theories and therapies of Paracelsus.

Fludd primarily followed the teachings of Paracelsus with regards to his medical career, but he also had an interest in the occult and believed there was much wisdom to be found in the teaching of natural magicians. In 1617, Fludd began working on *Utriusque cosmi, maioris scilicet et minoris metaphysica, physica atque technica historia* ("The Metaphysical, Physical, and Technical History of Two Worlds, the Macrocosm and the Microcosm"), with which his aim was to explain all of the visible and immaterial Universe while still connecting humans with divine beings. He produced some striking illustrations to accompany his text, and this included his Creation series.

The first painting in the series is a black square on a white background with the text *Et sic in infinitum* ("And so on to infinity") around the sides; this represented the "nothingness" from which everything was created. Aristotle couldn't accept that vacuums existed and proposed that "ether" filled the space in between everything, but Fludd's abstract black square is a suggestion of a vacuum, and it pre-dates the famous *Black Square* painted by the Russian artist Kazimir Malevich by nearly three centuries.

Subsequent paintings in Fludd's series show chaotic fires that subside and form a star-like body in the center that's surrounded by rings of smoke. This has some similarities with our current models of the Solar System formation, even though Fludd believed in the geocentric model. In a painting called *Fiat luc* ("Let There Be Light"), the black square is again present but the Holy Spirit is represented as a dove that has carried a ring of divine light across the center of it. The final painting in Fludd's series features the geocentric model of the Universe, with Earth at the center, surrounded by the 11 rotating spheres. Earth itself contains drawings of Adam and Eve, with Eve reaching for the tree. Angels keep watch from around the outside, while the light-bringing dove now faces away from Earth because man is acting of his own free will.

Fludd produced numerous works during his career but many of them were controversial. He debated with some of the great philosophers of his time, including Johannes Kepler and the French philosopher Pierre Gassendi, and he was condemned by many of his contemporaries for his interest in the occult.

Opposite top The folded engraving "De metaphysico macrosmi…ortu" (1617) is one of Fludd's Creation series and (left image) depicts chaotic fires that formed during the early stages of creation. (Right) The fires have now subsided and a star-like body resides at the center of concentric rings of smoke. **Opposite bottom** "Et sic in infinitum" (1624) is the first folio in Fludd's Creation series. It shows a black square with the translated words "And so on to infinity" around the edges. This is Fludd's suggestion of a vacuum; the nothingness from which the universe was created.
Above Robert Fludd was an English Paracelsian physician, astrologer, mathematician, cosmologist, Qabalist and Rosicrucian apologist. Robertus Fludd copperplate engraving by Johann Theodore de Bry (1650).

Andreas Cellarius

(c. 1596–1665)

Pareidolia is the tendency to see meaningful patterns in something that is completely random, and the multitude of mythological stories that have been assigned to star patterns, the Milky Way, meteor showers, and even aurorae are examples of this. Breaking the sky up into sections is helpful for a cartographer, but peoples from around the world have projected their own stories onto these patterns, bringing the stars to life for their specific culture. To breathe even more life into those stories, astronomers and artists have produced images of the constellation myths, each depicting their own interpretation. Constellations that lie along the ecliptic and are therefore considered to be part of the zodiac have always gained more attention, with the earliest depiction being a bas-relief on the ceiling of a temple in Dendera, Egypt, dating to around 50 BCE.

In the geocentric model of the Universe, the outermost rotating celestial sphere was assumed to contain all of the stars, so it was common to see constellations depicted back to front because that's how they would have looked if viewed from outside the sphere. Books that contained celestial maps were illustrated with copper and steel engravings that were then often colored by hand. In 1660, the Dutch–German cartographer Andreas Cellarius published a star atlas known as the *Harmonia macrocosmica* ("Cosmic Harmony"), which contained some of the most striking cosmic charts ever produced, featured across 29 double-pages.

Some of the charts were incredibly innovative because the engraver responsible had the idea to show not just the constellations as they would look from the outside, but also a map of the Earth sitting at the center; behind the northern or southern hemisphere stars, a map of the respective hemisphere of Earth could be glimpsed through the constellations. The complexity of the work was astounding. Cellarius also included charts with the Sun at the center, and planispheres that had the classic mythological figures replaced with biblical figures, based on a Christian constellation map produced by Julius Schiller in 1627. The biblical constellations idea didn't really catch on...

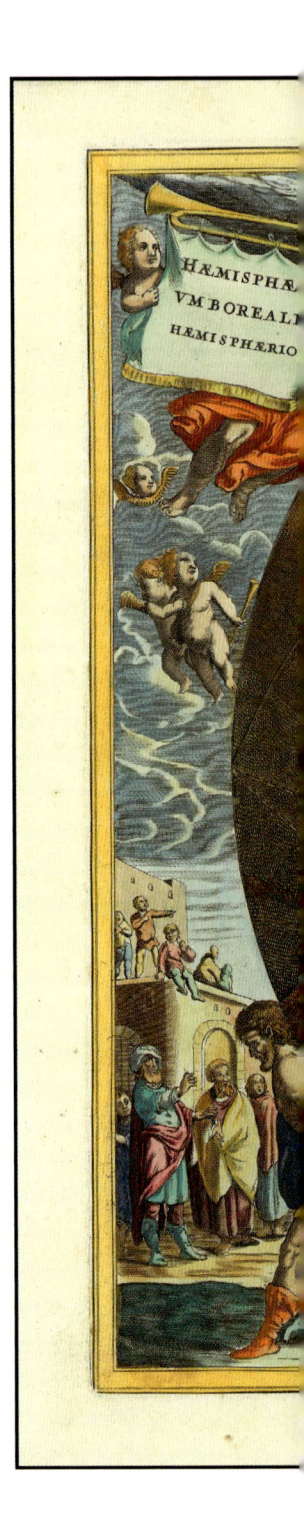

Above *Haemisphaerium Stellatum Boreale Cum Subiecto Hemisphaerio Terrestri*, by Andreas Cellarius, is a chart featuring illustrations of the constellations superimposed over the top of a map of the world. The constellations are depicted as they would look from outside of the Earth's celestial sphere, so they are back to front from how they appear to us from Earth. From Cellarius' *Harmonia Macrocosmica*, first published in 1660.

THE ART OF THE SKY 343

Claude Mellan

(1598–1688)

Claude Mellan was a French draftsman, engraver, and painter. He was much sought after as a portrait artist, always drawing his subjects from life, then producing the final work as an engraving—a method that he used to create a self-portrait in 1635. Although he primarily created portraits and religious works, Mellan was commissioned that same year by the French philosopher and astronomer Pierre Gassendi to produce some engravings of the Moon. At the time it was commonplace to use cross-hatching for shading dark areas, but Mellan opted for the approach of using evenly spaced horizontal lines instead, resulting in some of the finest, most distinctive lunar art ever created. Mellan finished his three engravings in 1637, but when copies were sent to Galileo he was said to have been irritated by them, saying rather waspishly, "They couldn't have been created by anyone who had seen the Moon with his own eyes." This couldn't have been further from the truth—Mellan had in fact based what he drew on his own observations through a telescope!

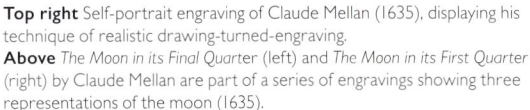

Top right Self-portrait engraving of Claude Mellan (1635), displaying his technique of realistic drawing-turned-engraving.
Above *The Moon in its Final Quarter* (left) and *The Moon in its First Quarter* (right) by Claude Mellan are part of a series of engravings showing three representations of the moon (1635).

Maria Clara Eimmart
(1676–1707)

The German astronomer Maria Clara Eimmart was the daughter of the painter, engraver, and amateur astronomer Georg Christoph Eimmart. She became her father's apprentice and he taught her how to sketch and engrave. Then, with the profits from their work, they were able to fund the building of their own private observatory.

Combining her love of drawing and astronomy, Maria soon gained recognition for her incredibly detailed eyepiece pastel sketches and engravings; during her lifetime she created more than 350 pastel drawings of the Moon alone—a collection entitled *Micrographia stellarum phases lunae ultra 300* ("Detailed Illustrations of Over 300 Lunar Phases"). So detailed were Eimmart's sketches that they were used to create a new lunar map.

In addition to her lunar work, Eimmart drew comets, planets, eclipses, and atmospheric optics. Many of her sketches were based on her own observations but she also collated observations from other astronomers and created drawings from them—for example, a panel showing the phases of Mercury, based on observations by the German astronomer Johannes Hevelius. Eimmart's pastel drawings were highly distinctive because they were created with pale pastels on dark blue cardboard. In addition to her astronomical work, she produced drawings of flowers, birds, and classical subjects, but these have since been lost. Eimmart died in childbirth in 1707.

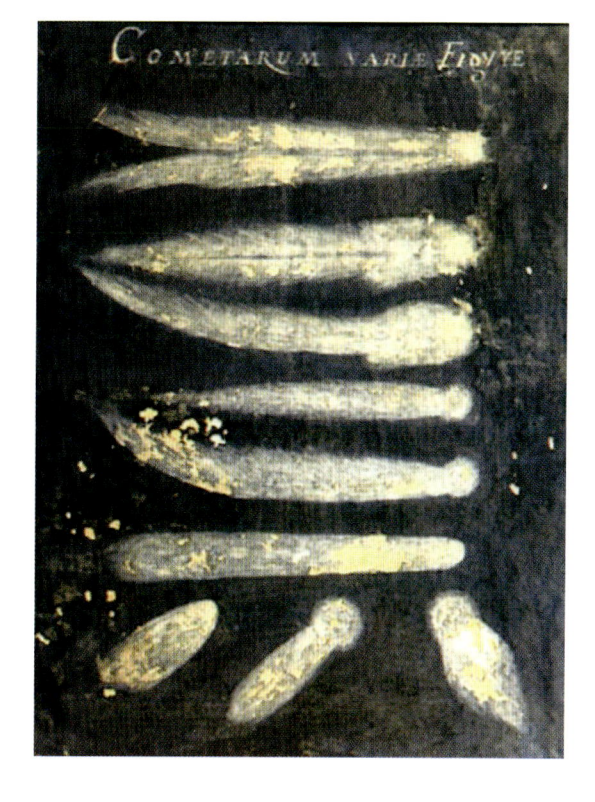

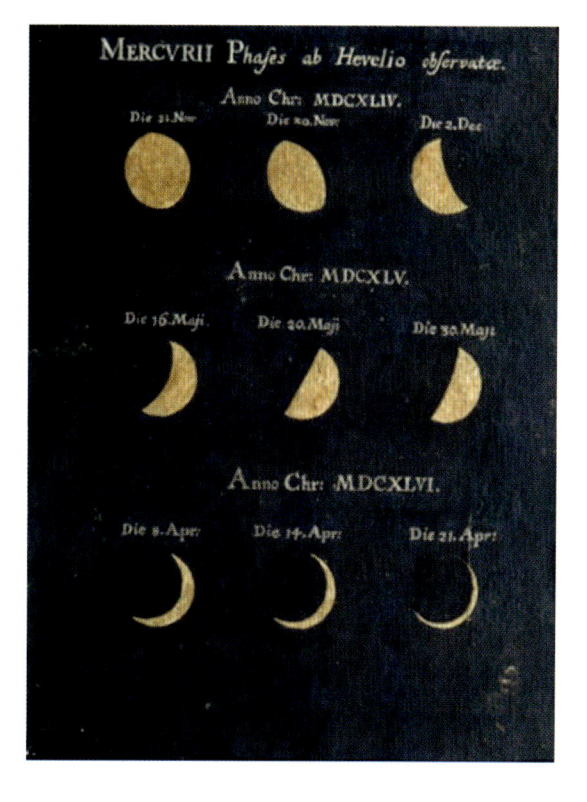

Top "Cometarum varie figyre" (comets in various ways) is part of a series of sketches created with pastels on blue cardboard by Maria Clara Eimmart during the 17th century. **Right** "Phases of Mercury" sketched with pastels on blue cardboard during the same period. The sketches were based on observations from the observatory of Johannes Hevelius on November 21, November 30, and December 2, 1664; May 16, 20, and 30, 1665; and April 8, 14, and 21, 1666.

Henry Pether
(1800–80)

The Moon has not just inspired astronomers over the years. Many painters have created nocturnes during their careers and, even if the Moon itself hasn't always been present, moonlight on clouds, or landscapes bathed in moonlight, have featured extensively. Henry Pether was an English landscape painter who specialized in moonlit scenes in Britain, Paris, and Venice—with the Moon often taking a central position within the composition. Pether was best known for painting serene moonlit views of the River Thames near Greenwich, many of which were produced between 1850 and 1865. One example depicts Greenwich Hospital in the foreground with a Full Moon low in the sky, reflecting off the water. It's beautifully painted, with photorealistic architecture. Interestingly, moonlight has not always been used to create restful scenes like those painted by Pether—in stark contrast, many artists have used moonlight breaking through cloud to illuminate stormy seascapes, evoking great drama rather than serenity.

Below *The Thames and Greenwich Hospital by Moonlight* was painted by Henry Pether circa 1854-65. It hangs in the National Maritime Museum, Greenwich, U.K. The moon was a favorite subject of Pether's and he gave it prime position within many of his paintings.

James Nasmyth and James Carpenter
(1808–90 and 1840–99 respectively)

Some of the most incredible lunar artwork ever produced featured in the heavily illustrated 1847 book *The Moon: Considered as a Planet, a World and a Satellite*, by the Scottish inventor and amateur astronomer James Nasmyth and the British astronomer James Carpenter. Photographic techniques at the time were not producing the sharp images Nasmyth and Carpenter wanted for their book illustrations, so in addition to creating stunning drawings and paintings, the pair made extremely detailed plaster models of the lunar surface that they then photographed under studio lighting. This allowed them to simulate the low angles that produce the sharp lunar shadows we see through a telescope's eyepiece as the Sun rises and sets over a crater or mountain range. The fine detail in some of these models is so incredible that you'd be forgiven for thinking they're photographs taken from a present-day lunar orbiter.

The lunar mountains in Nasmyth and Carpenter's book are depicted as extremely steep, rugged ranges. Before up-close photographs of the Moon existed, it was widely assumed that the landscape must be as rugged as their images implied, but the stark contrast between light and shade is due to the lack of an atmosphere on the Moon combined with a low Sun angle, not due to the landscape itself. Most of the lunar highland peaks have been eroded by 4 billion years of micrometeorite impacts and so lack the razor-sharp, steep-sided peaks the shadows seem to indicate.

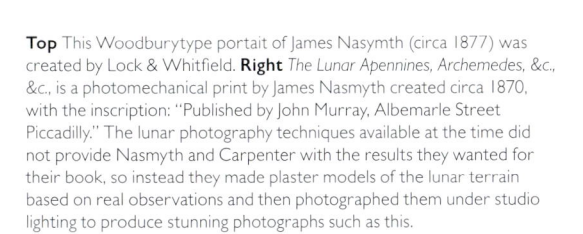

Top This Woodburytype portait of James Nasymth (circa 1877) was created by Lock & Whitfield. **Right** *The Lunar Apennines, Archemedes, &c., &c.*, is a photomechanical print by James Nasmyth created circa 1870, with the inscription: "Published by John Murray, Albemarle Street Piccadilly." The lunar photography techniques available at the time did not provide Nasmyth and Carpenter with the results they wanted for their book, so instead they made plaster models of the lunar terrain based on real observations and then photographed them under studio lighting to produce stunning photographs such as this.

Above First sketch of a spiral "nebula" (or galaxy), as published in 1850 by Lord Rosse, in *Observations on the Nebulae, Philosophical Transactions of the Royal Society*. **Opposite** Portrait of William Parsons, 3rd Earl of Rosse (before 1867), creator unknown.

William Parsons, 3rd Earl of Rosse
(1800–67)

William Parsons was an Anglo-Irish astronomer, naturalist, and engineer, and president of the Royal Society. He was born in York but after his father's death in 1841 he became Lord Rosse and inherited a large estate in Parsonstown, in present-day County Offlay, Ireland. There, at his home Birr Castle, Rosse had a 72-inch (1.8-meter) reflecting telescope built. Known as the Leviathan of Parsonstown, the size of the telescope was unprecedented—it was the largest telescope in the world at the time, and the speculum mirror alone weighed 3 tons! Lacking information about how other telescope makers had constructed their instruments, Rosse had also had to devise many of his own methods. Construction began in 1842 and it was first used in 1845, although the Irish Famine delayed its regular use for a couple of years.

Using the Leviathan, Rosse cataloged many nebulae and observed that some of them displayed a spiral structure; these were in fact galaxies. One observation in 1845 led him to create his famous eyepiece sketch of the Whirlpool Nebula—what we now know as Messier 51, the Whirlpool Galaxy. Rosse created a second, more detailed sketch of this galaxy in 1850. For many years this sketch has been recognized as the first recorded observation of the spiral structure within the Whirlpool Galaxy, and the drawing does bear a striking resemblance to modern-day photographs.

The French physicist and telescope maker Léon Foucault has long been known as the "father of the reflecting telescope" in its modern form. In the late 1850s, he built his largest telescope, a Newtonian reflector with an 31.5-inch (80-centimeter)-diameter mirror. Before the telescope was moved to its home in Marseille it was used by astronomers at the Paris Observatory, and one of the recorded observations was of the Whirlpool Nebula. The eyepiece drawing created of it was thought to be lost—a loss lamented by the astronomer William Tobin in a paper in 1987, and again in his biography of Foucault in 2003, explaining that it would have been useful to compare the view seen through Foucault's telescope with that seen through Rosse's Leviathan. Then in 2008, an article was published in the *Journal of Astronomical History and Heritage*, announcing that Foucault's Whirlpool Nebula sketch had been found in the Paris Observatory archives. It had been created in April 1862 by Jean Chacornac, and when compared to modern CCD images of the galaxy, it showed that Chacornac's drawing of the overall structure was actually more accurate than Rosse's sketches, though of course this wasn't known until high-resolution photos were taken of M51 years later.

Aside from the overall structure not being as accurate, there's no doubt that Rosse's sketches were much more detailed, probably due to the larger telescope aperture. Irrespective of who "did it better," both sketches are significantly more detailed than any other eyepiece sketches created by their peers, and Rosse's famous drawings went on to inspire one of the most famous artists in history—Vincent van Gogh (see page 353).

Frederic Edwin Church
(1826–1900)

Born in Connecticut in the United States, Frederic Edwin Church belonged to the Hudson River School of landscape painters, and was best known for creating stunning, photorealistic paintings of sunsets, mountains, and waterfalls. He also created paintings that included rainbows and astronomical phenomena.

On June 20, 1860 a very bright cluster of meteors was visible from the United States, and they were observed and painted by Church. He used a bold color palette that's somewhat reminiscent of Vincent van Gogh's paintings of the night sky (see page 353), but Church's painting establishes a serene scene with a highly realistic landscape and reflections of the meteors in the water below. It's in stark contrast to the depictions of such events in *The Augsburg Book of Miracles* (see page 337).

The aurora borealis and aurora australis (northern and southern lights) are among nature's most beautiful sights. Just like constellations, aurorae have inspired mythological stories, even in mid-latitude regions where displays are rare, and their beauty has inspired numerous works by artists and meteorologists who have undertaken expeditions to Arctic regions to study them in person. Church created a painting of the aurora borealis in 1865, presented as the backdrop to an expedition made by the Arctic explorer Isaac Hayes, his friend and pupil. Church had in fact based his painting on a sketch and description of the event provided by Hayes.

Aurora Borealis features a sky full of color, illuminating Hayes' schooner in the foreground. The really interesting thing about this particular painting is the straight, colorless beam of light shown at bottom right. This is actually Strong Thermal Emission Velocity Enhancement (STEVE), a phenomenon that wasn't officially named until 2016, when it began to be investigated properly, even though it might have been reported as far back as 1705, and aurora photographers had been capturing the phenomenon for decades. An aurora is caused by charged particles exciting molecules in the atmosphere, which results in photons of light being emitted. STEVE is only ever seen during a strong aurora display, but it's actually caused by a ribbon of hot plasma.

Above *Aurora Borealis* by Frederic Edwin Church (1865) was painted based on observations made during an Arctic expedition by Arctic explorer and pupil of Church, Isaac Hayes. The painting now hangs in the Smithsonian American Art Museum.

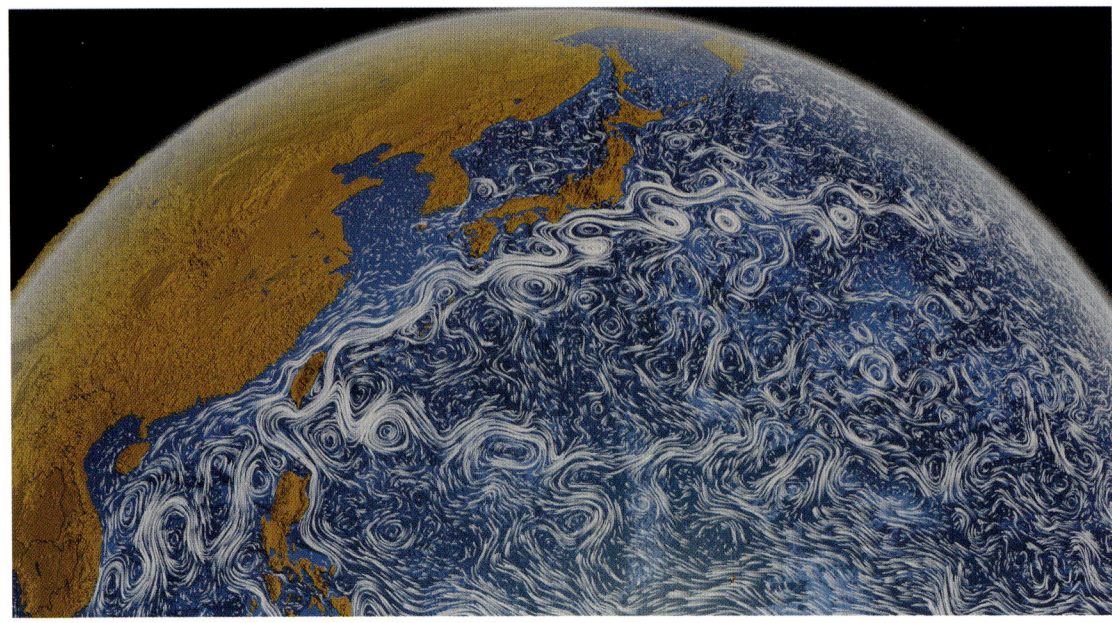

Vincent van Gogh
(1853–90)

The Dutch Post-Impressionist Vincent van Gogh remains one of the most influential artists in the history of Western art. He painted landscapes, still lifes, portraits, and self-portraits. He was prolific, too, creating around 2,100 works during his lifetime. His output was characterized by highly expressive brushstrokes and bold colors—a style that was greatly underappreciated at the time.

Nocturnes were not Van Gogh's usual subject matter, but in 1888 he painted *Café Terrace at Night*. While the main focus of this work was the artificially illuminated café, the background featured an abstract starry sky—the first of his paintings to do so. Later that year, Van Gogh painted *Starry Night Over the Rhône*, from a vantage point that enabled him to capture the reflections of the gas lights in the town of Arles. The sky above features the well-known seven star asterism the Plough, or Big Dipper, in Ursa Major, with the stars depicted in paler shades than the reflected lights below. Although this group of stars would not actually have been visible in that part of the sky when viewed from his location.

In 1889, Van Gogh created one of the world's most widely recognized pieces of art, *The Starry Night*. At the time he was a patient at the Saint-Paul asylum in Saint-Rémy-de-Provence, and the foreground features some of the landscape that he could see from his room. He then combined this with abstract night-sky objects, including the constellation Aries, the planet Venus, and a waning Crescent Moon, although this was not the lunar phase at the time of painting. What makes this painting so distinctive are the swirling patterns that fill the sky—a pattern thought to have been inspired by seeing Lord Rosse's M51 sketches (see page 348). What's also striking about this painting is the remarkable similarity it has to modern-day satellite imagery, which reveals the movement of global wind patterns and ocean currents (see bottom image). While these are patterns that Van Gogh could not have foreseen, it's still fascinating to view his work in this light.

After suffering from mental ill-health for many years, Van Gogh took his own life in 1890. He's only known to have sold a single named painting during his lifetime and he died in poverty, but after his death, his work began to receive the recognition it deserved. Today his paintings rank among the most expensive ever sold.

Opposite top In *The Starry Night* (1889, oil on canvas), Vincent Van Gogh created a foreground based on what he could see from his hospital room in Saint-Rémy-de-Provence and combined it with an abstract night sky that features the constellation Aries, Venus, and a Waning Crescent Moon. The swirling patterns in the sky are believed to have been inspired by the Whirlpool Galaxy sketches created by William Parsons.
Opposite bottom The similarities between Van Gogh's post-Impressionist painting and data visualisation model are striking. This print resolution still of the Kuroshio Current shows the ocean surface currents around the world during the period from June 2005 through December 2007. The visualization does not include a narration or annotations; the goal was to use ocean flow data to create a simple, visceral experience. This visualization was produced using model output from the joint MIT/JPL project to estimate the Circulation and Climate of the Ocean, Phase II or ECCO2. ECCO2 uses the MIT general circulation model (MITgcm) to synthesize satellite and in-situ data of the global ocean and sea-ice at resolutions that begin to resolve ocean eddies and other narrow current systems, which transport heat and carbon in the oceans. ECCO2 provides ocean flows at all depths, but only surface flows are used in this visualization. The dark patterns under the ocean represent the undersea bathymetry. Topographic land exaggeration is 20x and bathymetric exaggeration is 40x. This visualization was shown at the SIGGRAPH Asia 2012 Computer Animation Festival. **Above** Simply called *Self-Portrait* (1887), painted by Vincent Van Gogh in oil on artist's board.

Joan Miró
(1893–1983)

A great many artists have been inspired to create abstract art in response to stars and constellations, with Pablo Picasso and Georgia O'Keefe being notable among them, but a particularly striking series of paintings in this vein was created by the Spanish Surrealist Joan Miró between January 1940 and September 1941. Miró's "Constellations" series (the name was given to the series retrospectively) comprised 23 paintings—gouache on paper. He considered it one of his most important works, and art historians likewise view the series as a high point in his career.

Begun just after the turbulence of the Spanish Civil War in his home country (Miró had previously settled in France, although he would return to Spain as he was completing the series), "Constellations" explores the themes of wonder, love, and escape—escape from war, brutality, and genocide. Miró explained in an interview that the paintings had evolved during a period in which he had become increasingly drawn to music, stars, and the night sky, and they employ his characteristic dreamlike style, with a play of abstract shapes suggesting stars in constellations. The crescent Moon that's present in so much of his work also occurs regularly throughout the series.

Below Joan Miró's Surrealist painting *Ciphers and Constellations in Love with a Woman* was painted in 1941. This was created as part of Miró's "Constellations" series of 23 works of gouache on paper, and he considered this series as among the most important of his works. The abstract shapes in the work alludes to constellations in the night sky.

David A. Hardy
(1936–)

Space art as a genre is still very new compared to other art movements. One could argue that some of the lunar landscapes featured in James Nasmyth and James Carpenter's 1847 book *The Moon: Considered as a Planet, a World, and a Satellite* (see page 347) belong to the category of space art, but strictly the genre has its roots in the 1940s, with the work of the French illustrator and astronomer Lucien Rudaux and the American painter Chesley Bonestell. Rudaux and Bonestell created paintings depicting the landscapes of other planets and moons within our Solar System long before any planetary flybys had taken place. Because they pre-dated direct flyby imaging, some of these early works are now known to be inaccurate depictions, but their widespread publication in books and magazines captured the imagination of an entire generation of space enthusiasts who saw these alien worlds brought to life. One of the most famous was Bonestell's painting of Saturn as it would look from its moon Titan, which was published in *Time* magazine.

The British space artist David A. Hardy has been creating stunning space landscapes since 1952, making him the longest-established artist in his field, with his work having graced the pages of astronomy and space books for decades. A common theme is planets as viewed from their moons, but Hardy has created art based on almost every aspect of space, including highly accurate depictions of existing planets and their moons, and the lunar landscape, as well as imaginary alien landscapes. Most of his early work was done in watercolor and gouache, although he's worked with oils and pastels, too, and since 1991 he has also created digital art in Photoshop. Hardy has a very successful webstore selling prints of his popular works, in addition to the commissions he undertakes for book illustrations and custom artworks. Hardy's work is the benchmark for many budding space artists today.

Above In the gouache painting *Saturn from Enceladus*, Saturn and its rings are seen from the icy surface of its moon Enceladus. Whether painting existing planets or alien landscapes, David Hardy's work is both highly realistic and visually stunning.

Cathrin Machin
(1986–)

The British–Australian space artist Cathrin Machin has smashed records as the world's highest-crowdfunded painter and she has built an impressive list of clients around the world, which includes space billionaires, scientists, film stars, and tens of thousands of others. Having been fascinated with space since she was a child, her mission is to use her awe-inspiring, photorealistic work to encourage people to take a pause from their busy lives and look up at the stars in order to contemplate the vastness and beauty of the cosmos.

Machin paints in oils, a medium often favored by artists creating this kind of work, because the way the paints blend together allows her to capture the flowing motion of nebula clouds. She is also an astrophotographer and digital artist. Machin is fully aware that many people are losing the ability to see and share the night sky due to light pollution and is therefore determined to make her art as accessible as possible. She shares her visions of the cosmos with half a million followers across multiple social media platforms, and has recently opened the Fabric of Space store, where she sells her art in the form of everyday items such as blankets, desk mats, tapestries, and pet clothing.

Above Cathrin Machin's *When Two Souls Meet*, created in oil on board. Many space artists love to use oil paints because of how well they can blend. A thin layer of clear medium, such as liquin, is spread over a pre-prepared canvas or board covered in black gesso. Translucent oil colors are painted over the top of the clear layer, then small dots of opaque titanium white are blended on top of that using a soft brush. This picks up the paints below to create subtle color gradients and soft nebula clouds. Cathrin Machin uses this technique to produce imaginary nebula scapes as well as replicating photographs of real nebulae.
Right Cathrin Machin's *The Passage*, created in oil on board. Aurora Borealis/Aurora Australis has inspired artists for centuries. Using the same technique described above, here Cathrin Machin uses translucent greens in her base layer, then blends upward with titanium white to produce arcs of auroral activity. A more subtle version is reproduced underneath to create subtle foreground water reflections.

Olga Soby
(1980–)

Acrylic fluid painting is a relatively young technique. The first artist known to use fluid art effects in his work was the Mexican painter David Alfaro Siqueiros, in the 1930s, but it really gained popularity at the start of the 2020s. Acrylic paint is mixed with a variety of pouring mediums and/or water to achieve a fluid consistency, then it's poured onto a canvas, board, or other object. After that, the paint is manipulated in various ways—by tilting the canvas, or by swiping, splashing, or even blowing it around using a hair dryer or straw. It's a satisfying form of art to create and to watch being created, so it's extremely popular with content creators on YouTube and Instagram.

One of the world's finest fluid acrylic artists is Olga Soby. Soby is based in Canada, but her roots go back to Ukraine. She began her art journey in her twenties, with traditional acrylic painting, but once she discovered fluid art it became her biggest passion—a medium for self-expression, and a pivotal point in her artistic career.

Soby likes to say that instead of traditional brushes she chooses to paint with "elements": water, fire, airflow, and gravity are employed to both manipulate the paint and impact the viewer's emotions through color and movement. Her "Dancing Universe" series has pushed the boundaries of what can be created with these techniques. Using a variety of unconventional tools, such as giant cookie cutters, funnels, and a hair dryer, she has used multiple layers to create some stunning abstract art inspired by the magnificent beauty of space.

Soby's work now hangs in private collections worldwide, and through her successful YouTube channel and online art courses, she shares her skill and passion for fluid art to help aspiring artists around the globe to express themselves creatively.

Soby's work is just one example of a vast array of space-inspired art, and these visual representations of our Universe make space and astronomy accessible for people of all ages and experience levels. Thanks to the work of NASA and SpaceX, and the incredible images we now see via professional observatories and space telescopes, interest in space is at its highest since the Apollo era, and there's no doubt that it will continue to inspire artists for generations to come.

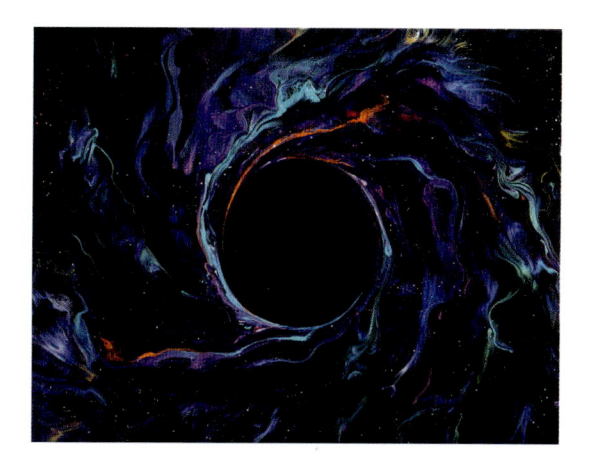

Top *Black Hole* by Olga Soby from her "Dancing Universe" abstract fluid art series. Soby mixes acrylic paint with water to create a fluid consistency. Black paint was laid down first before using a large cookie cutter to mask off a circular area at the center to represent the black hole. A variety of paint colors were added around the outside then manipulated using a hairdryer to form ribbons of material spiraling into the black hole. White stars were added next, including those close to the black hole with trails (again using a hairdryer) made to look like they're being pulled in, followed by iridescent glitter on all but the black hole for effect. Epoxy resin was added, building dimension and contrast, making the black hole darker and the spiraling colors brighter. **Bottom** *Dancing Planet* by Olga Soby, also from her "Dancing Universe" series. The planet at the center was created with a bright color palette to represent beauty within, her secret message for this painting. Colorful paints were added to the center and manipulated with a hairdryer as the canvas was rotated to create a planet with movement and energy.

ASTROPHOTOGRAPHY AND PROCESSING

Art inspired by the heavens has not been the only thing that has attracted new enthusiasts to the world of astronomy—the beauty of the night sky as captured via astrophotography has also played an important role. Put simply, astrophotography is the process of photographing an object in space, whether the Moon or Sun, galaxies, constellations, or other astronomical objects, including comets, eclipses, or meteorite showers. Astrophotographs can be taken on anything from a Smartphone camera to the Hubble Space Telescope—and everything in between—using a wide range of techniques and tools.

Early photography

Before the advent of photography, astronomers could only record what they saw with the naked eye, looking through a telescope, and the only way to do so was by creating a drawing, painting, or engraving. Photography, then, proved revolutionary, because exposures of several seconds or minutes brought an entirely new part of the cosmos into view.

The Frenchman Louis-Jacques-Mandé Daguerre pioneered the first commercially successful photographic process in 1839, which allowed a clear, sharp, detailed image to be developed on a sheet of silver-plated copper—a daguerreotype. When it came to the night sky, this process only worked for bright objects like the Moon; however, the first attempt resulted in a fuzzy blob, due to telescope tracking errors. This highlights the issue with long-exposure photography: objects in the sky are not stationary, so a telescope needs to track at the same speed as the object being imaged. In 1840, the American scientist and photographer John William Draper produced a more successful daguerreotype of the Moon, which was a 20-minute exposure taken through a 5-inch (13-centimeter) reflecting telescope, and in 1850 the first daguerreotype of a star, Vega, was produced.

Wet plate photography, invented by the American Frederick Scott Archer in 1851, used a glass plate coated with a silver nitrate solution to make it sensitive to light. From this, paper prints could be produced, which were much cheaper than daguerreotypes, but the limitation was that an image could only be exposed for as long as the plate stayed wet. The development of dry plate photography in the late nineteenth century allowed longer exposures to be captured and astrophotography became a serious research tool for professional astronomers. Dry plate photography was prohibitively expensive for most hobby astronomers but by the twentieth century, film photography had become more accessible, allowing enthusiasts to dabble with capturing nightscapes through their home telescopes.

The advent of digital photography brought about the biggest revolution in astronomy since the invention of the telescope. The invention of the CCD (charge-coupled device) chip in the 1970s allowed the first digital photographs to be captured, and by the 1990s, digital cameras had become widely available. In the early 2000s, CMOS (complementary metal-oxide semiconductor) sensor technology was developed and rapidly began to outperform CCD chips. CMOS chips are now widely used in both high-end and budget cameras, and in mobile phone cameras, being cheaper to produce and more light-sensitive, so able to capture faint objects in a shorter exposure time.

The basics of astrophotography

There are a lot of misconceptions about astrophotography—particularly the belief that it requires you to spend thousands of pounds on equipment, which just isn't the case. While it's certainly true that some equipment will achieve better results, it's still possible to get satisfactory results using budget options.

Simple wide-field photographs can be taken with most compact cameras, bridge cameras, and mobile phones, and this will allow you to photograph constellations (and the planets, if they're visible), the Moon, and conjunctions between various Solar System objects. Some newer phone cameras will even allow you to capture the Milky Way, and single exposures of a few seconds are all that's required.

Star trails photography is one of the easiest nightscape techniques to master because it's the one case where you actually want to capture the apparent movement of the stars. They are best captured at focal lengths of between 10 mm and 50 mm, and because you actually want your stars to trail, you only need a basic, static tripod. While it's tempting to put the camera into bulb setting and just leave it to capture one very long-exposure photograph, this doesn't yield great results. It's far better to take lots of 30-second shots—making sure the tripod doesn't move—and then use star trails software to combine these individual photographs into a single final image; StarStaX and Sequator are both free of charge and very simple to use. Keeping the ISO down to around 800 will preserve the star color, which looks pretty in star trails images. If you're pointing at the north celestial pole you'll get your stars forming concentric rings around Polaris, but pointing east, south, or west will give a very different result. And if you choose an attractive foreground object you can get a totally different look by simply changing your vantage point.

Opposite This example of astrophotography shows a timelapse photo of star trails in the night sky.

Focal lengths and ISO settings

The primary goal of all other types of astronomy imaging using a DSLR and lenses, or a camera attached to a telescope, is to capture as much light as you can without the stars trailing in your image. The Earth is rotating on its axis, and the stars and Solar System objects move in relation to it, so if your camera tripod or telescope mount are static, you'll need to set your exposure times accordingly: the longer your focal length is, the more apparent this movement will be. There's a simple calculation you can use to find a starting point for testing your exposure times—the "Rule of 500." Start with the number 500 then divide it by the focal length of the camera lens or telescope you're using; the result is the approximate number of seconds you can expose for before your stars will start to trail.

Example
1–18-mm focal length:
500 / 18 = 27.7 seconds

The results of repeating this calculation for several different focal lengths are shown in the table below.

Focal length	Calculated number of seconds
10 mm	50
18 mm	27.7
50 mm	10
300 mm	1.6
600 mm	0.83

There are a couple of reasons, however, why these figures might not always be the best exposure time. Firstly, the stars close to the celestial pole move a far smaller distance in a given time period than the stars near the celestial equator, so if you have an 18-mm focal length, you can probably do a 30-second exposure pointing at Polaris without any issues, but you'll almost certainly get trailing if you're pointing at the southern part of the Milky Way. Additionally, camera sensor size varies. If your camera has a cropped sensor, the exact focal length you get with a particular lens will be different than if you use the same lens on a camera that has a full-frame sensor.

It's helpful to get into the habit of doing test shots before every imaging session, so use the Rule of 500 calculation to get a starting point, then take some test photos with a series of exposure times around that length and check the image for trailing. Then try a couple of different ISO settings at your best shutter speed and review the images again. Once you've done this a few times you'll remember what works for your location and will be able to skip the test shots. When shooting in low light, you want the lens aperture to be wide open to let in as much light as possible, so set your f/stop to the lowest it will go. On some budget cameras, setting the f/stop to below 2 can introduce some noise, so again, do some test shots to check how the background looks.

Choosing a higher ISO will increase the amount of light getting to the sensor, but as well as the light you *do* want, it may also include too much of the light you *don't* want, so you need an ISO setting that works for your local sky conditions. If there's a lot of light pollution, haze, humidity, or moonlight, you'll need to take the ISO down to around 800. If you're somewhere darker you can set the ISO higher, but older, budget-range cameras may suffer from an unacceptable amount of noise if you set it to the maximum the camera is capable of, so in these cases, stick to a maximum of around 1600. Newer, more expensive cameras, however, may not suffer with the same noise issues. Again, do your test shots!

If you have a DSLR camera you can use lenses of different focal lengths to capture a variety of astronomical objects. A focal length of between 10 mm and 50 mm will allow you to photograph constellations, planetary conjunctions, the Milky Way, aurorae, zodiacal light, and meteor showers. An ISO of 800 or 1600 and f/3.5 is a good starting point for your camera settings for these objects. Tracking mounts for cameras are widely available and because they're tracking star movement, they allow you to take longer exposures. This is an advantage if you're in a dark sky area and can take longer exposures without too much light pollution intrusion.

Using a 300-mm zoom lens, or the digital zoom on your bridge camera, will allow you to get closer photographs of the Moon. Very short exposures will capture its surface features, while a slightly longer exposure of a Crescent Moon will capture Earthshine—the phenomenon where reflected light from Earth gently illuminates the shadow side of the lunar surface. When photographing the Moon, always keep a lookout for aircraft, because if you live under a busy airspace, there'll be many opportunities to capture an aircraft lunar transit!

If a bright planet is in conjunction with the Moon it makes for a wonderful photo opportunity, and a zoom lens can also be used to photograph Jupiter and the four Galilean moons, some star clusters, and some larger and brighter deep-sky objects. If you have a static tripod your exposures will be limited to just 1.6 seconds, which is why you're restricted to the brighter objects, but there are lots of objects that can be captured with shorter exposures.

Above Photographing an almost Full Moon in twilight gives you more foreground definition than photographing it in a dark sky and foreground objects add context and scale to the image.

Photographing the Sun

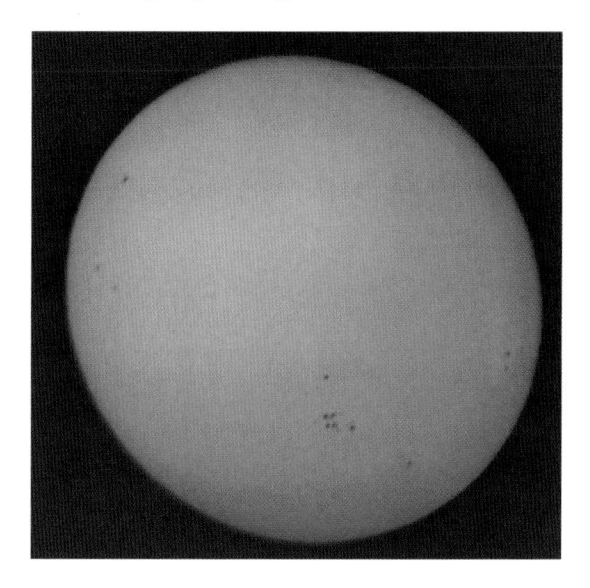

Safety is your number one concern when photographing the Sun. Never look directly at the Sun with the naked eye or through anything that magnifies, including the finderscope of your telescope and the viewfinder of your camera. It will lead to permanent eye damage, or even blindness. Do not attempt to photograph the Sun directly without a filter in place either because it will damage your camera sensor.

If you have a DSLR with a zoom lens, or a bridge camera with digital zoom, you can buy solar filters that sit on the front of the camera and will filter out 99.99 percent of the Sun's light. This will allow you to capture photographs of any sunspots present on the Sun's "surface"—but make sure you use the live view screen to focus and set your exposures and not the unfiltered viewfinder.

If you have binoculars or a telescope you can use the projection method as follows to project an image of the Sun onto a piece of paper, but never use this technique if any lenses or eyepieces in your telescope are plastic because the Sun will melt them. Point the telescope at the Sun, keeping the covers on the finderscope to prevent any accidental eye damage. Once you're pointing at the Sun, place a clipboard with a piece of white paper in front of the eyepiece so that you have a projected image of the solar surface. Use the telescope focuser to get the image as sharp as possible; you'll get a better result if you place the clipboard farther away from the eyepiece.

Once you have a sharp image, take a photograph of the paper using a mobile phone, bridge camera, or DSLR. Then put the cover back on the front of the telescope to avoid prolonged exposure to the Sun's heat. It goes without saying that you should never leave the telescope unattended while pointing at the Sun—people cannot resist looking through a telescope if they see one, but they WILL go blind if they do. The projection technique is not suitable for large reflecting telescopes because the internal parts of the telescope can get too hot.

You can also buy solar filters that fit on the front of a telescope and these will allow direct observation and photography of the Sun, but always check the filter carefully before each use—if there are any signs of damage, don't risk it. Keep the covers on the finderscope at all times to prevent accidental eye damage. Once the filter is in place, you can image the Sun using the same techniques described below, and the results will have better resolution than when using the projection method.

Opposite top The southern Milky Way from North Oxfordshire, U.K. photographed with a budget DSLR and 10-18 mm wide angled lens on August 14, 2023 by Mary McIntyre. The core of our Milky Way galaxy is not visible from the United Kingdom, but during the summer months it is possible to catch the region near to the core, low in the south. This photograph was taken from a semi-rural area so as to compensate for light pollution levels, 25 identical 25-second exposures were captured and then combined using image stacking (see page 367), which increases the overall total exposure while reducing noise. **Opposite bottom** Boeing 747 flying across a 33% illuminated Waxing Crescent Moon, photographed by Mary McIntyre from North Oxfordshire, U.K. on May 28, 2020 using a budget DSLR camera and 300 mm zoom lens. Lunar transits by aircraft are very common in countries where there is busy airspace. This aircraft was traveling from Chicago to Luxembourg. **Above** Cellphone photograph of sunspots on November 30, 2023 from a projected image of the Sun using a 102mm aperture refractor telescope.

Other photography through a telescope

If you have a telescope, the simplest way to take photos of the Moon is to use your cellphone camera. Universal cell-phone holders are available to buy and they allow you to align your phone's camera lens with the telescope eyepiece, then hold it in place while you tap the screen to take photos. You can even use your phone camera to zoom in and get closer photos of lunar craters or mountains. The CMOS sensors used in modern cellphones are of high quality and it's possible to capture excellent lunar photographs in this way—known as *afocal* photography. Some phone cameras are even able to take slightly longer exposures, which allows you to take photos of the brighter deep-sky objects. If you want to image the Sun with a solar filter in place, you can use exactly the same method with your phone camera.

The best way to capture high-resolution photographs through a telescope is to attach a DSLR directly to the telescope. The camera lens is removed and replaced with a T-ring that's specific to your camera; screwed into the T-ring is an extension tube. This tube needs to be the same diameter as your telescope eyepiece—the most common size is 1.25 inches (3-centimeters), but 2-inch (5-centimeter) tubes are also available. You simply remove the eyepiece of your telescope, attach the camera and T-ring assembly in its place, then use the telescope focuser to achieve the correct focus. You have now turned your telescope into a giant zoom lens! This method is called *prime focus* photography because you're focusing the light from the telescope directly onto the camera sensor, which gives a cleaner, sharper result.

If you're using a refracting telescope you can actually remove the whole diagonal assembly and attach the camera using a 2-inch (5-centimeter) tube. If you're using certain types of reflecting telescope you may find that you need to use a 2x magnification Barlow lens in order to achieve focus. A Barlow lens is really useful for imaging bright objects such as the Moon and planets because it effectively increases the focal length of your telescope, allowing you to get

Below Prime focus photography, where the DSLR camera is attached to the telescope with the diagonal still in place. The T-ring attaches to the camera in place of a lens and the 1.25-inch (3-centimeter) extension tube slots into the telescope instead of an eyepiece.

Left A single 90-second photograph of M42 the Orion Nebula through a 70 mm refractor telescope that is mounted on an equatorial tracking mount. This nebula has a bright core so the region at the center of the nebula is overexposed. **Above** An iPhone photograph of a Waxing Gibbous Moon taken through an 8' Dobsonian telescope. **Below** M31 the Andromeda Galaxy photographed with an apochromatic refractor with lenses that correct chromatic aberration so blue haloes are not present around the stars. Images by Mary McIntyre.

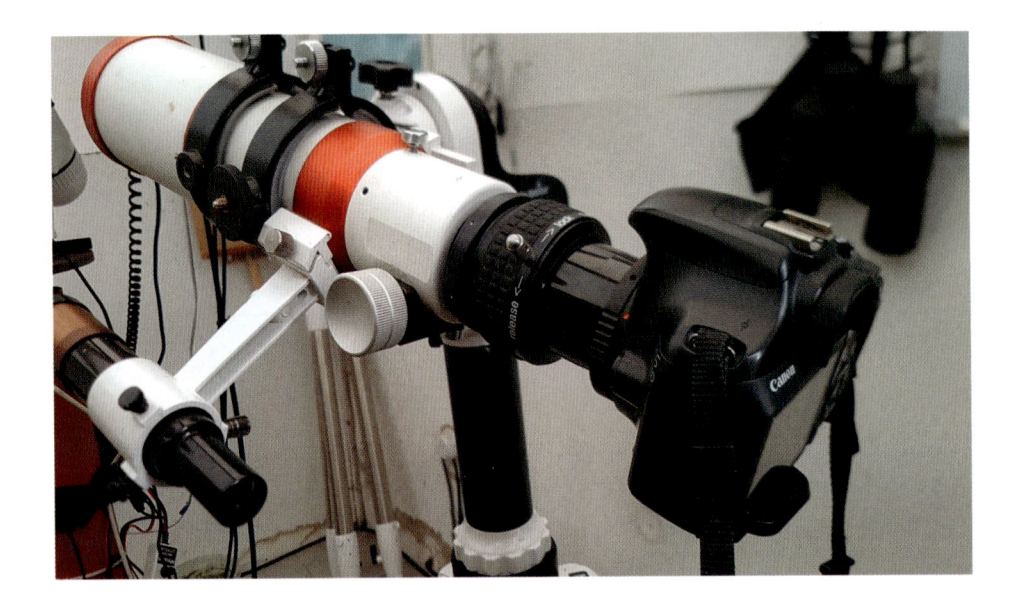

close-up shots of lunar craters, for example, or the surface features of Jupiter. It also adds contrast. Barlow lenses are available up to 5x magnification; however, they're not recommended for use on deep-sky objects because every doubling in magnification quarters the amount of light getting through. In these cases it's better to avoid Barlow lenses if possible—and, in fact, if you have a telescope with a long focal length but want to image deep-sky objects, it can be helpful to put in a focal reducer, which reduces the effective focal length of your instrument. This increases the light gathering as well as giving you a larger field of view.

In Chapter 3 we learned that most budget refracting telescopes are achromatic, so they suffer from chromatic aberration, which shows up as blue/violet fringes around the stars and Moon. This doesn't mean that you can't take photographs through an achromatic refractor—you will be able to capture some beautiful images—but your stars will have visible haloes. The chromatic aberration will be worse near the edges so keeping your object in the center of the field of view will help, and any blue haloes still present can be reduced by post-processing. You can think about upgrading to an apochromatic refractor at a later date if you find you're really enjoying astrophotography, but when you first start out, just use whatever equipment you have to practice your techniques.

With your DSLR camera attached to a telescope, exactly the same settings apply with regards to ISO—in other words, go with what works for your local sky conditions and the age of your camera sensor, and again the focal length will determine how long you can expose for before you get trailing, so the bigger the focal length of your telescope, the less time you can expose for. Here is where having a static mount can make things more difficult because very short-exposure photographs will not capture very many photons of light from your deep-sky object, although that you can still photograph some of the brighter deep-sky objects. If you take lots of identical photographs—called *light frames*—of your deep-sky object you can use computer software to do what is known as "image stacking."

Top Prime focus photography, with the camera attached to the telescope without the telescope diagonal in place. A T-ring is attached to the camera in place of a camera lens and the 2-inch (5-centimeter) extension tube slots into the back of the telescope.

Stacking

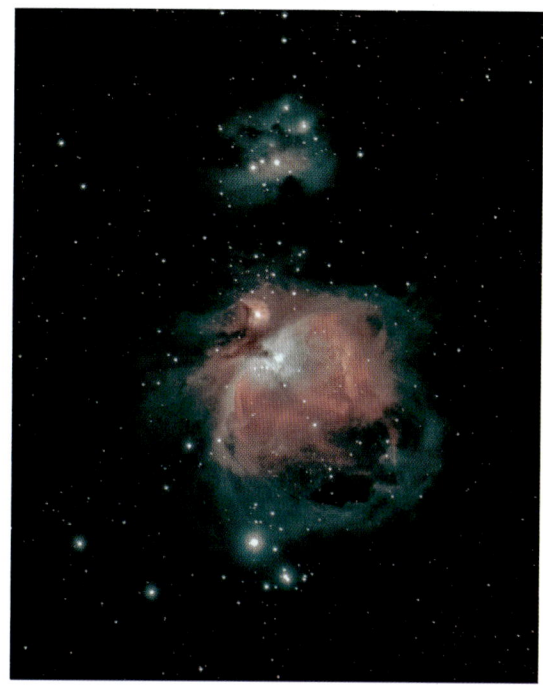

Stacking is a commonly used process in astrophotography, employed to increase the "signal to noise ratio." In real terms that means your signal—the light from your deep-sky object, lunar surface, or planet—gets brighter and more detailed while at the same time the noise from your camera sensor starts to average out. You can take hundreds of short exposures if you want to, remembering that you'll need to keep moving the telescope to keep your object in the frame, then the software will align each of the images using the star field to create a more detailed final result. There is a point of diminishing returns with lunar and planetary imaging, and the rotation of Jupiter is actually so fast that there's a limit on how many images you can stack in one go, but with deep-sky objects, the more photos you stack, the more detail you'll capture. You can actually photograph the same object every night for several weeks—or even months—then stack them all to produce an extremely detailed final result, but a few hours is usually more than enough.

The most commonly used stacking software for deep-sky photos is DeepSkyStacker, which is free.

If your images are very short exposures you won't get a lot of signal in each frame, and if it's a very faint object there'll be almost no signal, so the final stacked result won't give you results as detailed as those created with longer exposures. That said, you can still get pleasing results with shorter exposures, but it's definitely better to stick to the brighter objects, such as M42, the Orion Nebula, or M31, the Andromeda Galaxy.

Stacking is routinely used by photographers who have tracking mounts as well as those who work with static mounts. In Chapter 3 we learned about the celestial coordinate systems. Some budget-friendly mounts track in Alt-AZ, which means the mount

Top left The results of stacking 200 x 2-second exposures of M42 the Orion Nebula through a refractor telescope mounted on a static mount. Although there was very little signal in each individual exposure, image stacking produced a more detailed final image. **Top right** The results of stacking 103 x 90-second exposures for the outer regions and 134 x 10-second exposures for the core of M42 the Orion Nebula. There is significantly more detail in this version because each exposure had so much more signal, and this means the overall noise is also less. To compensate for the overexposed core, the shorter exposure image was digitally merged with the longer exposure to preserve the details in that region. Images by Mary McIntyre.

won't track the stars perfectly, but it will keep them in shot for longer than no tracking at all. When you're imaging the Moon or planets, an Alt-AZ mount is great for keeping the object in shot for long enough to take a series of photos to stack, but you can start to run into issues with longer exposures of deep-sky objects. Because the Alt-AZ tracking system won't compensate for the rotation of the sky, longer exposures will show up a circular star trailing pattern; this is called *field rotation*.

The way to get around this is to shorten your exposure times to a point where the stars don't trail, then take multiple images to stack. The stacking software can correct for the field rotation and will give you a nicely aligned image. You won't have as much "signal" from your deep-sky object in each photograph as you'd get with an equatorial tracking mount, but you'll have more than if you'd used a static mount. If you're using a large reflecting telescope, such as those available to buy on a Dobsonian mount, these have such good light-gathering capability that short exposures can yield really nice results. There are those who say you can't image with a Dobsonian telescope, but there are many astrophotographers who take amazing photos with them; again, just use what you have to develop your skills.

If you have an equatorial tracking mount, as long as your set-up and alignments have been done correctly you can start to really push your exposure times, but the mount by itself will struggle to keep things perfectly aligned for longer than three minutes. In any event, most locations now suffer with so much light pollution that three-minute exposures have a washed-out background anyway, so when you do your test shots you may find that between 90 and 120 seconds is long enough. Additionally, with the advent of satellite mega-constellations it's now almost impossible to get a photo that doesn't have satellite trails running through it, but thankfully some of the stacking algorithms will remove them.

You can buy light-pollution clip filters that sit inside a DSLR camera and these will remove sodium wavelengths from your image. This was really helpful in the days when all streetlights were orange, but now many places have moved to broad-spectrum LED lights, which can't be filtered out so easily. A combination of light pollution and moonlight can sometimes make you feel like it isn't worth going out imaging but there is another option: narrowband imaging.

Top Field rotation showing up in a 60-second exposure of M45 the Pleiades that was taken through a 71 mm refractor on an Alt-AZ tracking mount by Mary McIntyre. Although the mount does track, it doesn't perfectly follow the rotation of the sky, so longer exposures will produce stars with a circular trailing pattern.

There are filters available that you can screw into the imaging train that capture photons of light in wavelengths not visible to the human eye; you can then capture those photons using a monochrome camera to produce a grayscale image of part of a nebula. The most commonly used filters are hydrogen-alpha, oxygen III, and sulfur II. These will pick up detail in nebulae that we wouldn't otherwise be able to see; the monochrome channels are then merged using Photoshop or Affinity to produce a false-color image. They are sometimes also combined with images taken through red, green, and blue filters. This is a technique that was made popular by the Hubble Space Telescope and it gives you the advantage of being able to capture detail in a wavelength that isn't as affected by light pollution. Monochrome cameras have a higher resolution than color cameras so you get a higher-resolution final image, but you can still image through narrowband filters with a color camera. There are now filters available—such as tri-band or quad-band filters—that remove light pollution and allow you to image in two or three narrowband channels at the same time using a color camera. The results are improved further if your DSLR has been astronomy-modded, a process that removes the IR-block filter in a camera, which makes the camera more sensitive to colors near the red end of the spectrum.

There is one downside to narrowband imaging, though, and that's the necessary exposure times. When you're imaging just one wavelength of light, you need to take much longer exposures in order to capture enough photons of light from the nebula. Most equatorial mounts can't track accurately by themselves for the length of time needed, so autoguiding is required. This is a process where a smaller telescope plus an additional camera is mounted on top of your main telescope, making sure the smaller telescope is pointing at a bright star. Alternatively, you can use an off-axis guider, which attaches to the eyepiece end of the telescope and sends the light out to a second camera there instead. Using guiding software, the computer will aim to keep that star perfectly in the center, so if it drifts off due to the mount not tracking perfectly, it will feed a correction to bring the mount back into alignment, thus ensuring your main telescope tracks much more accurately. You can then take significantly longer exposures, even upward of ten minutes. This is an advanced technique, so something you should build up to, but once mastered will allow you to create some top-class astronomy photographs.

Most high-resolution images created today are not taken using a DSLR camera but a dedicated astronomy CCD or CMOS camera. These are high-frame-rate cameras that have lower noise and more sensitive sensors. They are small and lightweight so are less likely to cause your focusing tube to sag, or to add too much weight to your telescope mount. They are available for a variety of budgets—you'll find the lower-end options cost less than a mid-range DSLR but will give significantly better results.

For lunar, planetary, and solar images it's common to shoot a video of around 2,000 frames. Because these cameras are capturing at a high frame rate there's more chance of the camera grabbing frames during those crucial moments when the seeing conditions allow the image to be perfectly in focus. Even bright objects like the Sun and Moon benefit from stacking, so the software will sort the images by quality, then only the best-quality frames are stacked, giving a crisp final result. The software choices for stacking these kinds of images are AutoStakkert! or RegiStax.

Dedicated astronomy cameras are also used for deep-sky objects, but rather than a video they capture long-exposure still frames, then the best of those are stacked. Barlow lenses and narrowband filters can be used on them in the same way that they are used with DSLR cameras. Some of the higher-end astronomy CMOS cameras have cooling systems that keep the camera sensor cold, and this reduces noise. Cameras of this kind can be used on telescopes and microscopes, or sometimes fitted with an all sky lens, but they can't be used for other photography applications.

When you create deep-sky astronomy images, even though you've increased the signal-to-noise ratio by stacking, you'll still sometimes experience noise in your images, especially if you've had to stack short exposures. You may also see amp-glow from the camera electronics, or optical imperfections such as vignetting or bits of dust on the camera sensor (affectionately known as "dust bunnies"). The way to improve the final result is to shoot calibration frames. *Dark frames* are used to reduce noise, amp glow, and any hot or cold pixels that show up as red and blue dots on your image, and the way to shoot them is to put the lens cap back on the telescope or camera and take 20 to 30 shots with the identical camera settings

used for the photos. You then add these frames to the stacking software and it will create an average frame called a master dark and subtract it from all of your light frames.

Darks are often all that's required, but if you do have dust bunnies or vignetting you can help to remove them by shooting *flat frames*. These are slightly more difficult to shoot because you need to take a photo of an evenly illuminated surface. You can do this by stretching a white T-shirt over the end of your telescope on a cloudy day, or you can use a flats box that contains diffused white LED lights. Keep the ISO the same as you used for your images, then do some test shots using different exposures and check the histogram of the image on the back of your camera using the viewing options—you're looking for the image to be about 50 to 60 percent saturated, and you'll need about 20 identical shots. Again, the software will create a master flat and subtract it from your images; if it's been done correctly, the dust bunnies and vignetting will disappear. If you have marks on the telescope lens that are showing up in your image, you'll need to make sure you don't rotate the camera at all before shooting flats, otherwise the marks will be deleted from the wrong part of your image.

As a beginner don't get hung up on calibration frames; there are a great many astrophotographers who never use them. Most budget cameras will benefit from dark subtraction, but newer models give good results without, so if you're not seeing obvious imperfections in your stacked image, don't worry about it!

When you first get into astronomy, there's a risk of "aperture fever"—the urge to buy bigger and bigger telescopes, but bigger is not always better for astrophotography. Think about what you want to image; if large, sprawling nebulae or large spiral galaxies are what excite you, a small refractor will be a far better instrument because you'll only see a tiny part of the object with a larger-aperture telescope. If, however, you want to get into high-resolution planetary imaging, then a larger-aperture Schmidt–Cassegrain is probably a better option. Most telescopes can be multipurpose because you have the ability to change the effective focal length using Barlow lenses and focal reducers, but it's better to have your astrophotography goals clear in your mind before investing in expensive equipment that may not be right.

Opposite IC 1805 the Heart Nebula by Rachael and Jonathan Wood imaged with an ED80 refractor and ZWO294 CMOS camera, total exposure 12.5 hours, made up of 3 hours in H-alpha, 5 hours in OIII and 4.5 hours in SII. These three "narrow band" filters image details that are not visible to the human eye and capture the detail in a monochrome image which is then digitally combined to produce a highly detailed color image. Narrow band imaging is not affected by light pollution or moonlight, but much longer exposures are required.

Image processing

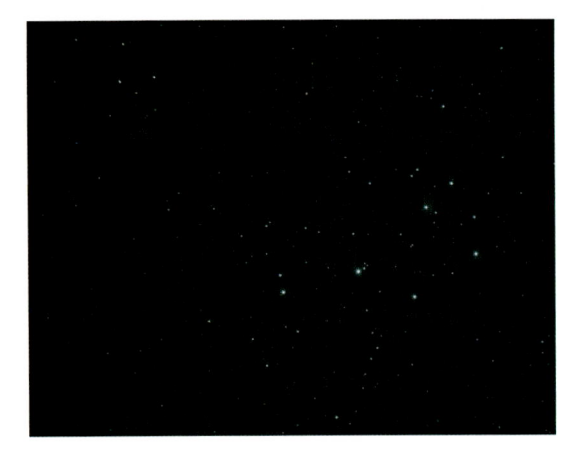

Once you've mastered the art of capturing your astronomy photographs you may think that's job done, but just as important as image capture is image processing. Single-shot wide-field images and star trails photographs often need very little attention, but almost all other images will require processing of some kind. If you experiment with stacking deep-sky images you may well be disappointed when the unprocessed output file looks completely dark, but don't worry—your data is in there just waiting to be teased out. Most astronomy objects have a huge dynamic range, meaning there are very faint parts, very bright parts, and lots in between, which can be a challenge to process. The goal is to enhance the faint areas while not over-exposing the brighter areas. Using basic levels adjustments will correct contrast differences, but curves will allow you to anchor one region of brightness while increasing or decreasing the brightness in other areas. This is commonly referred to as "stretching the image."

Many images will also benefit from some noise reduction and sharpening. Basic image processing can be done using Lightroom, Photoshop, or Affinity Photo, and advanced imagers like to use PixInsight, but there's also a free piece of software called FastStone Image Viewer that has some brilliant features and is far more intuitive for beginners to use. Every image you work on will be different, so try to approach this process organically rather than trying to follow a specific workflow that somebody else has applied to an image that was probably taken with totally different equipment. It's better to apply multiple small adjustments to the levels and curves rather than doing it in one big stretch, and less is always more—over-processed images don't look natural. You'll also need to crop out the stacking artefacts from around the outside edges, where your stacked images have overlapped. If you don't like the finished result, just start again from scratch and reprocess it—reprocessing images is a popular cloudy night activity for most astrophotographers!

Top left A stacked image of M45 the Pleiades straight out of Deep Sky Stacker with no processing yet applied. The image looks underwhelming and has lots of stacking artefacts around the edges, but the detail is hidden in there, waiting to be brought out by processing. **Top right** The same image of M45 after processing, with all the detail brought out by "stretching" the image using processing software. Images by Mary McIntyre.

Smart telescopes

In 2017, the first prototype of an entirely new generation of telescopes was unveiled at the CES trade show in Las Vegas. Combining the technology used in telescopes, telescope mounts, planetarium software, digital astrophotography, image stacking, and smartphone apps, the Unistellar eVscope is fully automated and can be operated by users who have absolutely no prior astronomy knowledge or telescope experience to create stunning high-resolution astronomy photos. It does the alignment fully autonomously using *plate-solving*—the process by which an image is taken, then compared to a star map so that the telescope knows exactly where it's pointing without any polar or star alignment required by the user. An object is then selected from an app on a smartphone or tablet, and the telescope will slew to the object and begin taking photos. The telescope will then create a *live stack* displayed on the app screen and, as each exposure is added, more detail will appear in the photograph.

Since this first release, several other companies have now released their own versions—some are more affordable, some have deep-sky filters and white-light solar filters, and some are small enough to fit into a rucksack, making them easy to transport. The advantages these telescopes offer are self-evident; absolutely anyone now has the ability to create beautiful astronomy photos within a matter of minutes, even if they've never owned a telescope before. Being able to share the beauty of space so quickly in an outreach situation is invaluable, but the one downside is that it may take away the opportunity for people to actually look through a telescope eyepiece and experience that real connection with an object that so many of us love. There's definitely still a place for both types of telescope.

Above M33 the Triangulum Galaxy taken by Cath Adams using a Seestar 50 smart telescope. This type of telescope is fully automated and allows people who have no prior astronomy or astrophotography experience to produce incredible night sky images with minimum fuss.

TIMELAPSE SKY PHOTOGRAPHY TIPS

Once you've mastered capturing a single photograph, whether it be wide-field or with a zoom lens, you can go on to take lots of images and create a timelapse video. Great timelapse subjects include capturing the general movement of the night sky, the Milky Way, aurorae, star trails, moonrise/moonset, noctilucent (night shining) clouds, daytime clouds, and sunsets.

It's vital that your tripod is kept completely still between exposures, so either hang something heavy from the hook underneath it, or use a bungee cord to secure it to the ground. If you're going to be taking photos for a long period to create a timelapse video, it's definitely advisable to make a makeshift dew heater out of an old sock and reusable hand warmers. This will prevent your camera lens from fogging up (see page 317).

Using a remote shutter cable not only lets you take a photograph without any camera shake, but when the camera is set to continuous mode you can lock the shutter cable in place and the camera will continue to take photos without you having to touch anything, therefore reducing the chance of you accidentally moving the tripod. The maximum exposure with a basic shutter cable is 30 seconds, because anything higher will go into bulb setting and the camera will just take one long exposure until you stop it. An intervalometer is a slightly more sophisticated type of shutter cable that allows you to not only set a shutter speed longer than 30 seconds, but can also leave a specified gap between your exposures. This is useful for objects that only need a very short exposure—for example, the rising Moon—otherwise you end up with thousands of images when one every couple of minutes is adequate.

Once you have all of your images, you can batch-process them if necessary to

Top Timelapse subjects that work well include the general movement of the night sky, Milky way, aurorae, star trails, moonrise and set, clouds, and sunsets.

correct any issues such as light pollution or haze. Lightroom makes this a quick and easy job because you make changes to one image, then paste those settings into an entire batch of images in a couple of clicks. If you don't have Lightroom, don't worry; if you got your exposures right you'll be able to make a lovely video without any processing.

You can use any video- or timelapse-creation software to make a video from the images. A popular choice for astronomy photos is Planetary Imaging Preprocessor (PIPP) because it allows you to choose your own frame rate, pause the final frame so the video doesn't end abruptly, and resize individual images to control the file size of the final video. The frame rate you choose will depend on how many images you've taken; if you only have a small set of images and you set a high frame rate, your video will be over in a few seconds. For example, if you've taken 60 photographs and you set the frame rate to 60 frames per second, your video will only be 1 second long, whereas 10 frames per second will give you a video that's 6 seconds long. Once you have the basic timelapse video you can use a video editor to add annotation and royalty-free music.

PIPP can also be used to center an object in each frame. This is a really useful tool if, for example, you want to create an animation of the apparent movement of sunspots as the Sun rotates. Lightroom or Photoshop can be used to manually crop and align images of the lunar surface so that you can use the images to create a timelapse of a lunar crater or a mountain sunrise or sunset. DeepSkyStacker can be used to align photographs of a starfield so that you can capture the movement of an asteroid, dwarf planet, or comet moving against the background stars. There are a great many subjects that work well in timelapse form.

If you want to create a star trails timelapse you'll first need to create a stacked star trails image using the free software StarStaX and tick the "Save cumulative files" box. This will save an image each time a new photograph is added to the stack, so when you scroll through the cumulative pictures you can see the star trails getting longer in each frame. If you don't follow this step you'll just get a simple night-sky timelapse with no star trails. Once you have the set of cumulative images, you can create the video in same way you would with your other photographs.

Timelapse videos are a great way to present your photography on social media, where video content is more favorable to the algorithm than still images. Sharing short video clips will engage your online followers, even if they're just a simple photo slideshow in video form, so it's a really useful skill to learn.

SOLAR ART PHOTOGRAPHY PROJECT

Observing and/or photographing the Sun on a daily basis is fun and fascinating in equal measure, so that's the subject of this first project. As we saw in the astrophotography section, you can safely observe the Sun's photosphere with a telescope or binoculars by projecting it onto a piece of white paper or cardboard, then take a photo of the projection with a smartphone camera, or draw what you see. If you don't have access to a telescope you can still complete this project because the Solar Dynamics Observatory (SDO) has a huge database of daily Sun photos captured by their space telescope, which has been studying the Sun since its launch in 2010. It includes the Sun imaged through multiple different wavelengths of light, but this project uses white light images because that's what you'd see if you projected the Sun onto paper or used a white light filter on the front of your telescope.

Although the Sun's photosphere is not as dynamic as the chromosphere, if you observe the Sun daily you'll see any visible sunspots appear to move across the Sun's disc. The sunspots themselves aren't actually moving—what you're seeing is the Sun's rotation. You can count how many days it takes the Sun to rotate by watching when the sunspots emerge from the eastern limb and disappear again over the western limb. In addition to the rotation, you'll also see the sunspots evolving over time; they may change size or shape, split into two distinct sunspots, or new sunspots might appear.

Materials needed
- Piece of white paper
- 2B Graphite pencil
- Eraser
- Blending stump or cotton bud (Q-tip)

While it's really quick and easy to take a photo of the projected image or to download the SDO photograph, taking the time to draw the sunspots will mean you have to study them in more detail, so that's the focus here. Sunspots are among the easiest things to draw; all you need are the simple materials listed above. In terms of paper, printer paper is fine to begin with, but paper designed for drawing is a heavier weight and has a smoother surface for blending. A 2B graphite pencil is preferable, but an HB pencil will work too. If you're creating astronomical drawings for scientific purposes then obviously it's essential that they're accurate, but if you're new to sketching or are just doing it for fun, try not to get too hung up on achieving perfect results straight away because you'll get better with practice.

In this example, photos taken every other day from July 6 until July 16, 2023 were downloaded from the SDO website, but you can of course follow the same steps using a live projected image of the Sun every other day.

Sunspots drawn with pencil in the first circle for day one of the project. Umbra and penumbra regions of the sunspots were included. Credit Mary McIntyre

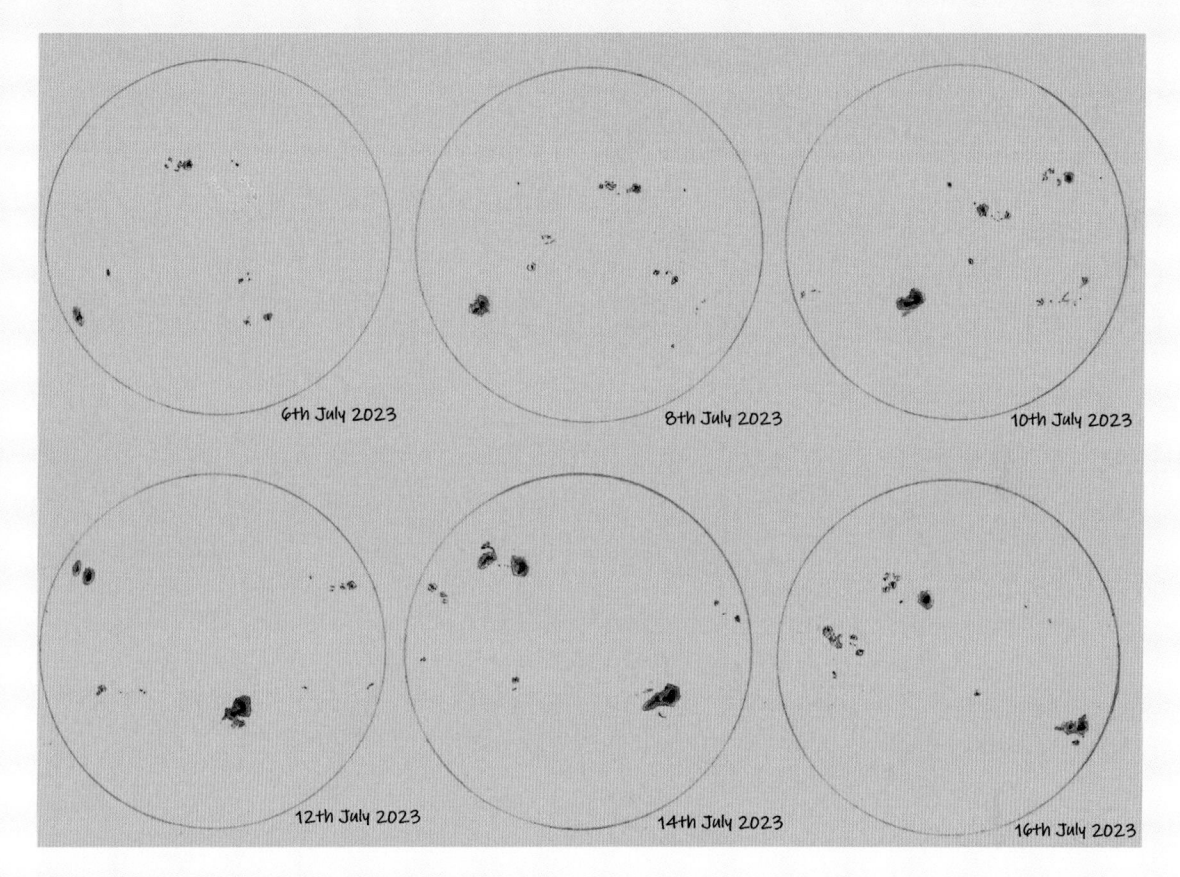

Pencil drawings of the sunspots on all six days completed. The changing position of the sunspots is due to the Sun's rotation and the sunspots have evolved during the six-day period. Credit Mary McIntyre

1. Draw six identical circles on your piece of paper.

2. In the first circle, mark the positions of the sunspots on your first day. Think of your circle like a clock face and try your best to be accurate with their placement as well as their shape and size.

3. You'll notice that the larger sunspots have a darker center (umbra), so fill that in first. The region around the outside is slightly paler; this is the penumbra. Apply less pressure with your pencil and draw the outline of this region. Lightly shade in the penumbra, then gently blend both regions with the blending stump or cotton bud.

4. Repeat this process for the other days until you have all six circles filled in, remembering to check the changing positions and any alterations in morphology. How have your sunspots moved during the given time period? Can you figure out how long it takes a sunspot to travel from limb to limb?

5. If you fancy a challenge, carefully photograph your sketches, then align them and create a timelapse animation that shows the movement of the sunspot across the Sun's surface. (There are tips for timelapse photography on pages 374-5.)

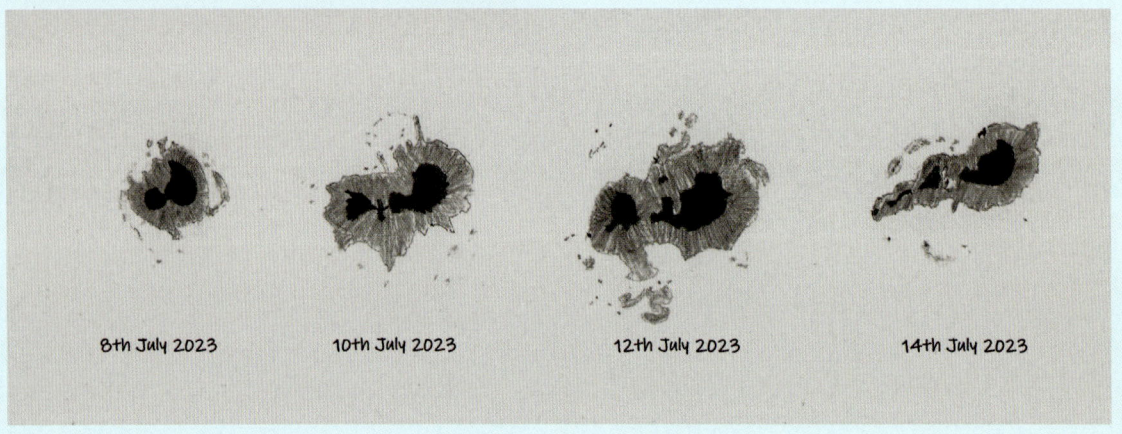

8th July 2023   10th July 2023   12th July 2023   14th July 2023

Another fun activity is to take a closer look at a larger sunspot region to see how its morphology changes over time. When a sunspot is near the limb, its shape is affected by foreshortening, so it's better to wait until it has fully rotated into view for your first drawing. With this is mind, here we just used four reference images, taken on July 8, 10, 12, and 14, 2023. In our example we're studying the evolution of sunspot AR13383, but you can start by zooming into a large sunspot in your own photographs, or by getting a closer look at your projected image.

1. Draw the outline of the umbra and fill it in.

2. Now outline the shape of the penumbra, but when you shade it in, use strokes that move in an outward direction from the umbra because this mimics the texture of a high-resolution sunspot photograph. Use a cotton bud to gently blend but don't try to totally erase the pencil strokes—this is a drawing and your visible pencil marks remind viewers that they're looking at a work of art, not a photograph. Don't forget to include any small, fainter regions around the outside of the main sunspot.

3. Repeat this step for the other three days, noting how the sunspot has evolved over time. Has it changed more or less than you expected? Do you think any of the changes in shape are due to foreshortening, when the sunspot is closer to the eastern or western limb? Next time, do you think it would be worth creating sketches more frequently than every other day?

Top Close-up details of the large sunspot group, drawn in pencil with close attention to the shape of the umbra and penumbra regions. Credit Mary McIntyre **Bottom** Comparing the four drawings made over the six-day period shows the evolution of the sunspot. Initially, it gets larger, divides into two distinct regions, then begins to shrink again. Credit Mary McIntyre

LUNAR PHOTOGRAPHY ART PROJECT

With nothing more than your unaided eye, you can observe the Moon changing day by day throughout its 29.5-day lunar phase cycle. You can then record these changes with a pencil and paper to create your own Moon phase graphic, so that's the focus of Part 1 of this project. The aim is to capture the Moon during each waxing and waning phase. You don't have to do this every day—the weather is unlikely to permit that anyway—but try to observe it as a Waxing Crescent, First Quarter, Waxing Gibbous, and Full Moon. Once you pass the full phase, attempt to capture a Waning Gibbous, Last Quarter, and Waning Crescent. We can't see the Moon when it's new. You'll need to check local rise and set times for the days you're observing—and some of the waning phases will require an early-morning session.

When observing with the naked eye you can't see a huge amount of detail but if you spend some time looking carefully you'll notice that certain features will be more prominent at certain times during the cycle. Most obvious will be the darker regions—the lunar maria. Just capturing the changing appearance of the maria as the Moon cycles through its phases is interesting, but careful study will reveal a little bit more surface detail.

Bottom left Pencil sketch of a 2.7 day old Waxing Crescent Moon, blending the shadow side with a cotton bud/Q-tip. Credit Mary McIntyre
Bottom right Pencil sketch of a 5-day old Waxing Crescent Moon. Lunar maria regions do not have harsh edges when viewed with the naked eye so the cotton bud/Q-tip used to blend the shadow side helps to draw the features. This gives a softer, subtler effect than drawing with pencil then blending. Credit Mary McIntyre

Part 1
Materials needed
- Piece of white paper
- HB pencil
- Cotton bud (Q-tip)
- Eraser
- Blending stump (optional)

1. Begin by drawing some circles on a piece of paper. Alternatively, you can create a digital observation sheet with circles on it and print it out. It's useful to have space under each circle to record how many days into the lunar cycle you are and what percentage of the Moon's surface is illuminated at the time you're creating the sketch; this information can be found on Stellarium or using a free smartphone app such as Lunescope.

2. Look at the shape of the illuminated part of the Moon and mark that on your first circle, then shade in the non-illuminated side and gently blend it a bit with a cotton bud.

3. Now it's time to move onto drawing your surface features. When creating astronomy sketches we're always looking for ways to mimic the textures and tones we see using drawing tools. While the lunar maria are darker than the surrounding areas, the contrast between them and the highland regions, when observed visually, is quite subtle and there are no harsh edges, so you may find it easier to put down your pencil and use the cotton bud you've just used for blending as a drawing tool instead. In the process of blending you will have picked up some excess pencil on the tip of the cotton bud; smudging this on the paper gives a really soft, blended stroke that's perfect for mimicking the texture of the maria.

4. Once the maria have been added, look to see if any additional features are visible and add those with the pencil or a small blending stump.

5. Repeat the above steps for each of the lunar phases. As the illumination increases, you may notice some more subtle differences in tone across the southern half of the Moon, and when you're close to Full Moon you may even spot some ejecta rays from the crater Tycho. If you see an area that needs to be brighter, you can use a clean corner of an eraser to remove a patch of smudged pencil. Cutting a thin slice off the end of the eraser will give you the clean, sharp edge you'll need for this.

Top right Nine pencil sketches showing different phases of a full lunar cycle as seen with the naked eye. Credit Mary McIntyre **Bottom** A full lunar cycle captured in sketches but this time using white and black pastel pencils on black paper. This medium gives high contrast and more closely resembles what we actually see when observing the Moon at night. Credit Mary McIntyre

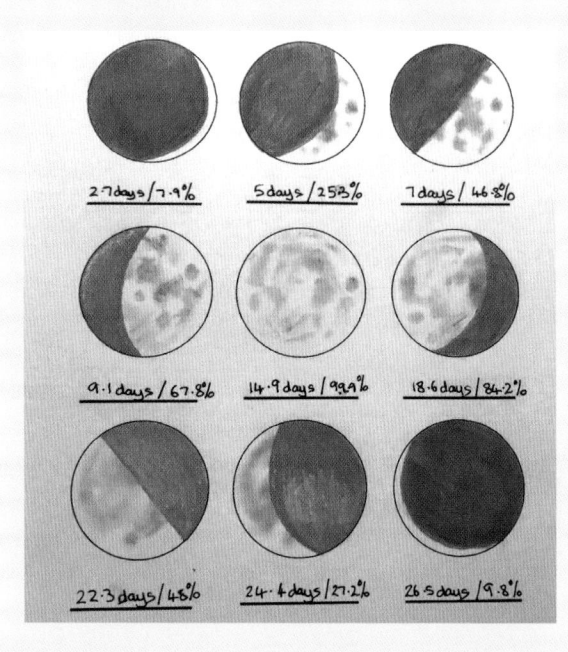

When you've completed your sketches, take some time to compare them. Which features changed the most during the cycle? Did you notice changes in the apparent position of the features due to libration? How can you explain the differences in orientation of the Moon at the time of drawing?

6. If you have a camera you can also take a photo of the Moon during each phase and use these to create a lunar phase graphic. It's always interesting to compare how much additional detail the camera captures compared to the naked-eye observations. And if you have pastels and black paper, you can also try creating a lunar phase graphic using those tools, following all the same principles of capturing texture and tone to mimic what you see.

Part 2

The lunar surface is an incredibly beautiful landscape. If you observe the regions close to the lunar terminator—the boundary between the illuminated and non-illuminated side of the Moon—you'll see prominent shadows and highlights. If you study this region repeatedly over a few hours you can see dramatic changes in the illumination of some areas. If the Moon is waxing, you can observe the Sun rising over an area that was previously in shadow, then watch the long shadows get shorter as the Sun altitude increases. If the Moon is waning, you can see the Sun setting over a crater, then watch the crater or mountain shadows get longer and longer until eventually the entire area falls into shadow.

If you have a telescope you can observe, sketch, or photograph these changes directly, and for this second part of the project we'll be photographing a lunar crater sunrise. If you don't have a telescope you can still complete this project, but rather than taking your own images you can use the NASA Scientific Visualization Studio (SVS) Moon Phase and Libration webpage. You simply enter a date and time to obtain a very detailed projected lunar image that can then be downloaded. We'll be using this tool for Part 3, but you can use it for Part 2 as well, if needed.

The crater chosen for our sunrise images was Copernicus—with a diameter of 57.8 miles (93 kilometers)—taken through a 71-mm refractor with an ASI120MC CMOS camera fitted with a 3x Barlow lens. You can use any telescope and camera combination to take your own photographs—even a smartphone camera attached to a telescope will work.

1. Look near the lunar terminator and pick a large crater that's just starting to catch the sunlight.

2. Take a photograph and note the time it was taken.

3. Repeat the process at regular intervals over the next few hours, then compare the images. (The photographs shown here were taken on February 10, 2023 at 15:50 UT, 19:11 UT, 21:00 UT, and 22:30 UT. You can see a dramatic change in illumination around Copernicus itself and also the surrounding area over this 6-hour-40-minute period.)

4. You can even align your images and create a timelapse video showing the Sun rising over the area.

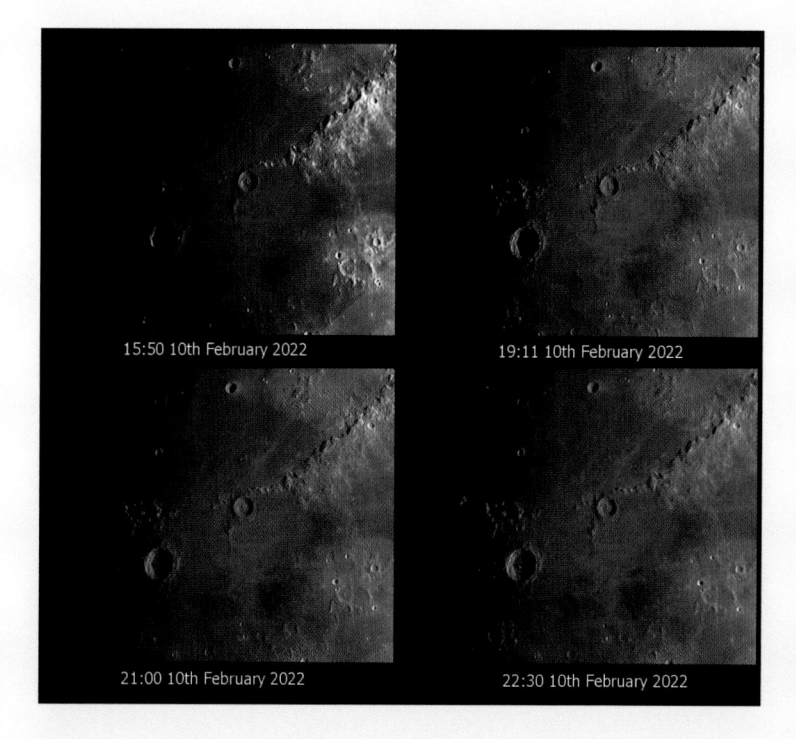

Left Four images showing the Sun rising over the lunar crater Copernicus and surrounding area. At 15:50 only the rim of the crater is visible and the highlights and shadows along the lunar Apennine Mountains to the upper right of the image are very prominent. At 19:11 more of the crater wall becomes visible, along with more high lying features near the lunar terminator. At 21:00 the left side of the crater floor of Copernicus is illuminated and the whole crater is casting a triangular shadow to the left. At 22:30 the central peak of Copernicus is visible and the crater shadow is shorter. The highlights and shadows on the lunar Apennine Mountains are less prominent. Credit: Mary McIntyre

Part 3

For this next stage we'll create a sequence of lunar sketches showing the Sun setting over the Ptolemaeus crater trio, when the Moon was near last quarter phase. This trio is comprised of Ptolemaeus (diameter 95.7 miles [154 kilometers]), Alphonsus (diameter 74.5 miles [120 kilometers]), and Arzachel (diameter 59.7 miles [96 kilometers]). You can create the sketches at the eyepiece, from your own photos, or you can download image projections from the NASA SVS page. The advantage of using the SVS page is that you aren't constrained by Moon rise and set times—you can download images covering a much longer time period. We saved four images, each one two hours after the previous one.

The example sketches were created using black and white pastel pencils on black sketching paper because pastels (or chalks) make it easier to mimic the starkly contrasting textures and tones we see that can be achieved with traditional pencil sketches, but you can still create stunning pencil sketches if that's all you have access to. By using light and shade in this way you can create a sketch that looks extremely three-dimensional, just by following some very simple steps.

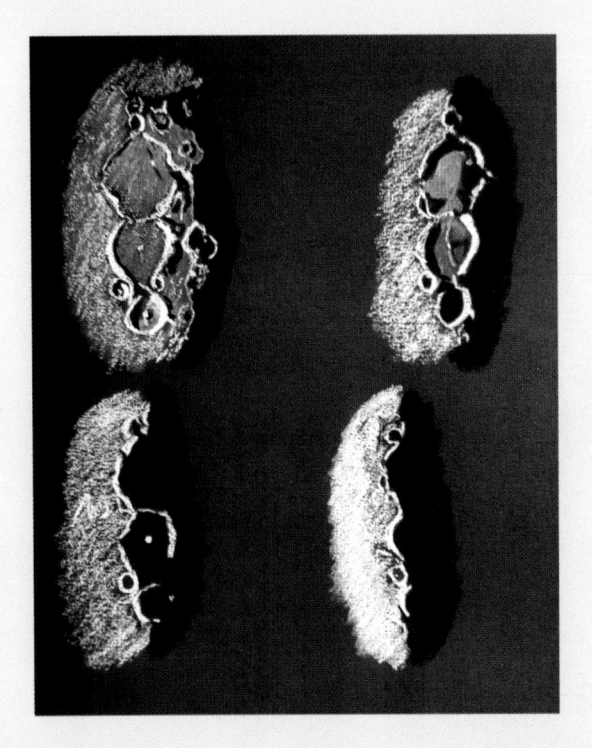

Materials needed
- Black and white pastel pencils
- Black sketching paper
- Cotton bud (Q-tip)

1. Begin by drawing the outline of the craters in white and mark out the parts that are catching the most sunlight.

2. Shade the crater floor and area around the outside of the crater and gently blend with the cotton bud, paying attention to the differences in tone. With this trio, the crater floor was slightly darker than the area surrounding the crater, so it was blended to a slightly darker tone.

3. Once the bright highlights and lighter shading are in place, go in with the black pencil and mark out the dark features around the crater, including any crater shadows and the inside edges of any smaller craters. Then add pastel strokes that mimic the textures seen in the area around the crater and on the crater floor, remembering that you can use your cotton bud as a drawing tool if a soft texture or subtle tonal difference is required.

4. Finally, finish the sketch by shading in the dark edge that is in shadow and blend it.

5. Repeat this step for the other images, paying particularly close attention to how the shadows and highlights within the crater and surrounding areas are changing. How does this series of lunar crater sunset sketches differ from the sunrise series you created? Did the shadows change faster or more slowly than you were expecting?

Left Four pastel sketches showing the Sun setting over the Ptolemaeus Trio on the Moon. The shadow terminator moves from right to left as the Moon wanes and the trio changes from being fully illuminated to almost completely in shadow.

Top left the outline of the crater trio is drawn on black paper using a white pastel pencil, and the brightest highlights marked onto it. **Top right** The crater floor and surrounding area are shaded and blended, paying attention to tonal differences, then the darker shadows are added to the crater trio with a black pastel pencil. **Middle left** The non-illuminated side of the crater trio is shaded in with the black pencil and blended to complete sketch number one. **Middle right** These steps are repeated for sketch number two. This time the shadow terminator has moved closer to the crater trio and the crater walls and peaks are casting deeper shadows. **Bottom left** Sketch number three shows further movement of the shadow terminator and now only the crater rims and highest peaks are catching any sunlight. **Bottom right** In sketch number four the shadow terminator has now almost completely covered the crater trio and only the left side crater rims are catching any sunlight. Credit: Mary McIntyre

CREATE YOUR OWN ASTRONOMICAL ART

There are two different ways of approaching astronomical art. One is to create an exact copy of a view seen through an eyepiece or captured photographically. For scientific purposes, accuracy is clearly important, and the act of drawing an object will really teach you about its structure and form. You can also approach space art in a looser way, by treating the same image as inspiration. If 20 people all created a drawing or painting of the same nebula or lunar crater, each one would be different, yet they would all still be recognizable for any astronomer who was familiar with the subject.

This is how it should be; your art is an expression of your own interpretation of the object you're seeing, and your brush or pencil strokes are completely unique to you. So many factors affect how a sketch comes together—how you hold the pencil, how much tension you have in your hand, even your mood; in fact, if you asked an astronomy sketcher to draw the same crater every day for a week, each sketch would be different. So when carrying out this project, if your art doesn't look *exactly* like the image you're copying, please remember that's completely okay!

A quick internet search for any astronomical object will turn up thousands of images that you can use for inspiration. The entire Hubble Space Telescope and James Webb Space Telescope (JWST) archives are a great place to start, but there are also a great many astrophotographers who are happy for others to use their images as inspiration—just make sure you ask for their permission first. For this project we're using a photograph of the Tarantula Nebula taken by the JWST, and we'll be working with colored pastels on black paper—the pastels blend beautifully and give a lovely texture for nebulae sketches. Choose paper that has a smooth surface because it will allow you to blend your colors more easily.

Pastel pencils have the advantage of being less messy than pastel sticks, but you'll need an old-style pencil sharpener to use with them because the blade in the small sharpeners designed for graphite pencils sits at the wrong angle for pastels and will just make them crumble. The steps that follow can also be applied to painting on canvas with acrylic or oil paints, but you'd need to paint the canvas black first.

Above James Webb Space Telescope NIRCam of The Tarantula Nebula (30 Doradus), an extremely luminous emission nebula located 160,000 light years from Earth, in the Large Magellanic Cloud. The James Webb Space Telescope reveals details of the structure and composition of the Tarantula Nebula, as well as dozens of background galaxies. Stellar nursery 30 Doradus gets its nickname of the Tarantula Nebula from its long, dusty filaments. Located in the Large Magellanic Cloud galaxy, it's the largest and brightest star-forming region near our own galaxy, plus home to the hottest, most massive stars known. The center of this image, taken by Webb's Near-Infrared Camera instrument (NIRCam), has been hollowed out by the radiation from young, massive stars (seen in sparkling pale blue).

Materials needed
- Pastel pencils or sticks in a variety of colors
- Black sketching paper
- Scissors
- Small, soft paintbrush (completely dry)
- Cotton buds (Q-tips)
- Fine-tip Posca acrylic paint pen (white)

1. Begin by lightly marking out the basic shape of the brighter parts of the nebula with a white pencil. You can also mark the brighter stars at this stage if it will help you to map out the shapes more accurately.

2. This next step involves using colored pastels in a way you may not be familiar with. To get a really soft blend, use the scissors to scratch some pastel dust onto your paper in the places you need it, then use the paintbrush to gently tap and blend it in. Work from the outside in, using the darker colors first, then go over them with some of the lighter colors to create some depth. This process lays down gorgeous, well-blended foundation layers for the nebula colors.

3. Once you're happy with most of the background, go in with your pencil and start to shade in the brighter parts of the nebula. Use the side of the pencil and blend with a cotton bud, then go over any sharper areas with the pencil tip. If you have pastel sticks you can use the corners for this step.

4. Keep checking your reference photo to make sure your shapes and colors are correct, and as you work, ensure that you look at your sketch from a small distance as well as up close so that you get a better idea of the overall effect, not just the details.

386 THE BACKYARD STARGAZER'S BIBLE

5. Once you have most of the brighter parts done, go in with dark blue and black pencils to enhance the parts of the sky that are darker and have less of the nebulosity in them, then blend these with a cotton bud. Adding these darker tones will really start to make the colored regions pop.

6. When you've reached a level of detail you're happy with, it's time to add stars. Nebulae look strange without stars present, so this step will immediately transform your work. Use the Posca paint pen to add the brighter, more prominent stars. Some regions have hundreds of stars in the background, so it would be very time-consuming to add every single dot; a much quicker method is to create the impression of lots of stars by using your scissors to scratch a bit of white pastel onto the page, then pressing it down gently with your finger. These stars might not land in exactly the right place, but as long as you've put the brighter stars in the correct position, the faint background stars won't matter too much.

7. If any of the stars have an obvious color, you can tap over these with a pastel pencil in an appropriate color.

8. Once finished, always sign your work!

You can use these same principles to create drawings of any astronomical object, or to create your own imaginary nebula. No matter what medium you're working with, if you have lots of blended colors on a black background, the second you add stars it will be immediately transformed into space art.

It's very common for people to dislike their own drawings, but standing it up and looking at what you've created from a bit of a distance will allow you to see it more clearly—and putting it in a frame will then give it a really polished look. Even if you're not happy with it at the time of completion, you'll probably find that it grows on you. Be proud of every piece of art you've created, and never throw away your work, even if you hate it, because you'll appreciate having a record of your art progress.

Opposite The basic outline shape of the inner edge of the nebula is drawn on black paper with a white pastel pencil. The shapes within the layers of the nebula are added by scratching the pastel pencil dust directly onto the page using various colors. Each layer of pastel dust is blended using a dry, fluffy brush. This gives a soft, cloud-like texture that mimics nebula clouds. More detail is added to the inner edge of the nebula using the white pastel pencil. **Above** Mary McIntyre's completed pastel sketch of the Tarantula Nebula (30 Doradus) based on James Webb Space Telescope NIRCam image of the nebula.

ADDITIONAL RESOURCES

RECORDED FOR FUTURE ASTRONAUTS
WHO MAY ONE DAY WALK ON MARS

I don't know why you're on Mars. Maybe we're on Mars because we recognized that if there are human communities in many worlds, the chances of us being rendered extinct by some catastrophe on one world is much worse. Or maybe we're on Mars because of the magnificent science that can be done there that the gates of the wonder world are opening in our time. Or maybe we're on Mars because we have to be because there is a deep nomadic impulse built into us by the evolutionary process. We come, after all, from hunter-gatherers. And for 99.99 percent of our tenure on Earth, we've been wanderers, and the next place to wander to is Mars. But whatever the reason you're on Mars is, I'm glad you're there, and I wish I was with you.

Carl Sagan,
Co-founder, Planetary Society (shared on Earth, 1996,
shared on the surface of Mars via the Phoenix lander, 2008)

Previous pages The Perseid meteor shower is a stunning site against the Milky Way, silhouetted in the foreground by a pine forest. **Opposite** The wonder and beauty of the night sky has captivated millions of people around the globe for centuries. **Page 393** Enhanced image of a lunar eclipse.

ACKNOWLEDGEMENTS

The contributors wish to thank ...

Mary McIntyre wishes to thank
Thank you to Caitlin Doyle and all at HarperCollins for the opportunity to be involved with writing this book. My thanks also extends to Ian Ridpath for such an enjoyable collaborative experience. Huge thanks to my husband Mark McIntyre for his endless support and help with proof-reading, and to Steve Warbis, Stuart Atkinson, Cath Adams, Rachel and Jonathon Wood, Olga Soby, Cathrin Machin, and the Global Meteor Network team for their support of this project. Last but by no means least, I am eternally grateful to Paul Money and Brian Jones because they are the reason I became a published astronomy writer.

Ian Ridpath wishes to thank
Those whose photographs are used in this book; pictures are credited individually where they appear. The following sources have been consulted extensively during the preparation of this book: the SIMBAD database; the Washington Double Star Catalogue; the General Catalogue of Variable Stars; and the International Variable Star Index of the AAVSO.

The photographic Moon maps are courtesy Mark Rosiek and the U.S. Geological Survey, Flagstaff, Arizona. Dimensions of formations are from the International Astronomical Union's Gazetteer of Planetary Nomenclature. Richard McKim of the British Astronomical Association gave valuable guidance on the Mars albedo maps.

Further details of the origin and mythology of the constellations may be found in *Star Tales* by Ian Ridpath (Lutterworth, Cambridge). For more on the meanings of individual star names, see *A Dictionary of Modern Star Names* by Paul Kunitzsch and Tim Smart (Sky Publishing, Cambridge, Mass.).

Rachel Federman wishes to thank
Many thanks to my ever-faithful editor Caitlin Doyle for her generous insight and good cheer. Mary McIntyre for your careful read. Sugar Bear, thanks for getting Wally to love space when he was younger. I miss those days of planet songs and constellation drawings. Hopefully he'll circle back. Thanks to Venus for keeping me company on the dark autumn mornings when I searched for signs of life far from home. Petra & Wally—you keep me pointed skyward.

The publisher wishes to thank
Michael Sand at Abrams for his enthusiastic collaboration on this beautiful series and Soyolmaa Lkhagvadorj at Abrams for her wonderful cooperation and keen eye; Ellie Ridsdale for her breathtaking design and hard work; Milena Harrison-Gray for always going above and beyond in her inspired picture research; Lynn Hatzius for her beautiful cover illustration; Angela Koo for her meticulous edit; Helena Caldon and Rachel Malig for their excellent editorial work; Geraldine Beare for the thorough index; Chris Wright for helping ensure the book made it to print; and Myles Archibald and Hazel Eriksson for their continued teamwork. Thank you to Stephen Maran for such inspiring words in the form of the Foreword. Immense thanks go to Mary McIntyre, Ian Ridpath, and Rachel Federman for their expertise and patience. What a joy it has been to spend months poring over stunning images of the incredible night sky, guided by your words. And finally, a big thank you to each and every contributor to this book.

FURTHER READING AND RESOURCES

Books

Collins Stars & Planets, 5th edition, by Ian Ridpath and Wil Tirion (William Collins, 2018)

The Backyard Astronomer's Guide, 4th edition, by Terence Dickinson and Alan Dyer (Firefly Books, 2021)

Constellations: The Story of Space Told Through the 88 Known Star Patterns in the Night Sky by Govert Schilling and Wil Tirion (Black Dog & Leventhal, 2019)

Dark Skies: A Practical Guide to Astrotourism by Valerie Stimac (Lonely Planet, 2019)

The End of Night: Searching for Natural Darkness in an Age of Artificial Light by Paul Bogard (Back Bay Books, 2014)

Expanding Universe: Photographs from the Hubble Space Telescope by Charles F. Bolden, Jr., John Mace Grunsfeld, Owen Edwards, and Zoltan Levay (Taschen, 2020)

The Night Sky: An Astronomer's Guide to the Night Sky and the Universe by Nigel Henbest (Cassell, 2023)

100 Things to See in the Night Sky, Expanded Edition: Your Illustrated Guide to the Planets, Satellites, Constellations, and More by Dean Regas (Adams Media, 2020)

Sharing the Skies: Navajo Astronomy by David Begay and Nancy C. Maryboy (Rio Nuevo Publishers, 2010)

The Sky Atlas: The Greatest Maps, Myths and Discoveries of the Universe by Edward Brooke-Hitching (Chronicle Books, 2020)

Skywatcher's Companion: Constellations and their Mythology by Stan Shadick (Strong Nations Publishing, 2007)

Stargazer's Atlas: The Ultimate Guide to the Night Sky (National Geographic, 2022)

The Stargazer's Guide: How to Read Our Night Sky by Emily Winterburn (Perennial, 2009)

To Infinity and Beyond: A Journey of Cosmic Discovery by Neil deGrasse Tyson and Lindsey Nyx Walker (National Geographic, 2023)

Turn Left at Orion: Hundreds of Night Sky Objects to See in a Home Telescope – and How to Find Them, 5th edition by Guy Consolmagno and Dan M. Davis (Cambridge University Press, 2019)

The Usborne Book of Astronomy & Space by Lisa Miles and Alastair Smith (Usborne, 2016)

Sites

Astronomy magazine
www.astronomy.com

Astronomy Picture of the Day
https://apod.nasa.gov/apod/astropix.html

BBC Sky at Night magazine
https://www.skyatnightmagazine.com

EarthSky
https://earthsky.org/earthsky

European Space Agency
https://www.esa.int

Go Stargazing
https://gostargazing.co.uk

HubbleSite
https://hubblesite.org

NASA
https://www.nasa.gov

National Park Service, Night Skies
https://www.nps.gov/subjects/nightskies/stargaze.htm

Night Sky Tourist
https://nightskytourist.com

The Planetary Society
https://www.planetary.org

Sky & Telescope: The Essential Guide to Astronomy
https://skyandtelescope.org

Space
https://www.space.com/skywatching

Articles

"Astronomy for Beginners" (The Planetary Society)
https://www.planetary.org/night-sky/astronomy-for-beginners

"How Reading the Night Sky Helped Black Americans Survive" by Heather Greenwood Davis (*National Geographic*)
https://www.nationalgeographic.com/travel/article/dark-sky-astronomy-african-american-history

"The Impact of Astronomy on Climate Change" (*Nature Astronomy*)
https://www.nature.com/collections/fhfcdebecc

"National Parks are Embracing Indigenous Astronomy" by Stephanie Vermillion (*Outside*) https://www.outsideonline.com/adventure-travel/national-parks/national-parks-indigenous-stars

"Why (and How) Astronomers Should Teach Climate Change" by Travis Rector (American Astronomical Society https://aas.org/posts/news/2021/03/why-and-how-astronomers-should-teach-climate-change

Apps
Google Sky
ISS Spotter
NASA app
Night Sky 11 — best for Apple device users
PhotoPills — best for astrophotographers
SkySafari 7 Pro — for novice and serious astronomers alike
Sky Tonight — best immersive app (and for AR)
SkyView
Star Chart — best free app
Star Tracker
Star Walk 2 — best for beginners
Stellarium Mobile Plus

FOR KIDS
Books
Astronomy Activity Book for Kids: 100+ Fun Ways to Learn About Space and Stargazing by Aurora Lipper and Victoria Stebleva (Z Kids, 2021)

A Child's Introduction to the Night Sky: The Story of the Stars, Planets and Constellations – and How You Can Find Them in the Sky by Michael Driscoll and Meredith Hamilton (Black Dog & Leventhal, 2019)

Constellations for Kids: An Easy Guide to Discovering the Stars by Kelsey Johnson (Rockridge Press, 2020)

Eyewitness Universe (Dorling Kindersley, 2023)

Find the Constellations by H. A. Rey (Clarion Book, 2016)

Glow: A Family Guide to the Night Sky by Noelia González and Sara Boccaccini Meadows (Harry N. Abrams, 2023)

The Mysteries of the Universe: Discover the Best-kept Secrets of Space by Will Gater (Dorling Kindersley, 2020)

Night Sky by Stephanie Warren Drimmer (National Geographic Kids, 2017)

Planets and the Solar System by Patricia Murphy (Smithsonian Kids; Cottage Door Press, 2019)

Solar System for Kids: A Junior Scientist's Guide to Planets, Dwarf Planets, and Everything Circling Our Sun by Hilary Statum (Rockridge Press, 2020)

Star Finder!: A Step-by-Step Guide to the Night Sky (Dorling Kindersley, 2017)

Stargazing for Kids: An Introduction to Astronomy by Jonathan Poppele (Adventure Publications, 2022)

What's Hidden in the Sky: Animal Constellations Around the World by Aina Bestard (Tra Publishing, 2022)

Sites
A Kid's Guide to Stargazing (American Museum of Natural History) https://www.amnh.org/explore/ology/astronomy/a-kids-guide-to-stargazing
Astronomy for Kids https://www.astronomy.com/tags/astronomy-for-kids/
NASA Kids' Club https://www.nasa.gov/learning-resources/nasa-kids-club/

Articles
"Stargazing with Kids: How to Inspire Young Astronomers" by Ruth Perkins (*BBC Sky at Night* magazine, 2023) https://www.skyatnightmagazine.com/advice/skills/stargazing-for-kids

"Stargazing With Kids Is Fun, Free, and Full of Possibility" by Rebecca Boyle (Atlas Obscura, 2023) https://www.atlasobscura.com/articles/stargazing-with-kids

"Stargazing with Kids" by Amity Hook-Sopko (*Green Child* magazine, 2022) https://www.greenchildmagazine.com/stargazing-with-kids-101/

GLOSSARY

A

aberration—Any defect that adversely affects the quality of the image in an optical system. Examples are astigmatism, chromatic aberration, and spherical aberration.

aberration of starlight—An effect caused by the Earth's motion across the path of incoming starlight that produces a slight change in the star's apparent position. Aberration of starlight resulting from the Earth's orbital motion is termed annual aberration; there is also a much smaller effect caused by the Earth's rotation known as diurnal aberration. Unlike parallax, aberration does not depend on the distance of an object.

absolute magnitude—The apparent magnitude that an object would have if moved to a standard distance of 10 parsecs (32.6 light years). Absolute magnitudes provide a way of comparing the luminosity of different objects.

absorption line—A dark line or band in an object's spectrum caused by the absorption of its light. The wavelength of an absorption line depends on the element that's causing it.

achromatic—A lens that does not suffer from chromatic aberration.

active galactic nucleus (AGN)—The central region of a galaxy that is producing enormous amounts of energy. The energy is thought to come from matter falling toward a black hole with a mass of up to a billion Suns, or more, at the galaxy's core.

albedo—A measure of the proportion of incoming radiation that's reflected from a surface. A surface with an albedo of 1 would reflect all incoming radiation and be perfectly white, while a surface with an albedo of 0 would absorb all radiation and be completely black. Planets covered in bright clouds have high albedos, while dark rocky bodies have low albedos.

alt-azimuth (Alt-AZ) mount—A mounting that allows a telescope to rotate freely in altitude (vertically) and in azimuth (horizontally). Such a mounting needs to be moved in both vertical and horizontal axes to track a celestial object as the Earth turns, but this can be accomplished automatically by computer control.

altitude—The angular distance of a given point above an observer's horizon. It's measured in degrees from 0° on the horizon to 90° directly overhead (the zenith).

annular eclipse—A solar eclipse in which the dark disc of the Moon is surrounding by a bright ring (an annulus) of sunlight. Such eclipses occur when the Moon is near its farthest point from the Earth, i.e. the apogee, and thus its apparent diameter is smaller than that of the Sun.

aperture—The diameter of the main lens or mirror in a telescope. Larger apertures admit more light (or radio waves, in the case of radio telescopes), allowing fainter objects and finer details to be discerned. Radio telescopes can combine the output from a series of smaller dishes to emulate the performance of a much larger instrument—a technique called aperture synthesis.

aphelion—The point in an elliptical orbit around the Sun that is farthest from the Sun.

apochromatic—A lens or optical system with a combination of elements designed to eliminate chromatic aberration completely.

apogee—The point in an elliptical orbit around the Earth that is farthest from the Earth.

apparent magnitude—The brightness an object appears as seen from the Earth. It's a product of the star's intrinsic luminosity and its distance from Earth.

asterism—A distinctive pattern formed by stars within one or more constellations. Examples are the Plough or Big Dipper, which is an asterism within Ursa Major, and the Square of Pegasus, which is an asterism formed from stars in Pegasus and Andromeda.

asteroid—A body smaller than a planet in orbit around the Sun; also known as a minor planet. Most asteroids lie in a belt between the orbits of Mars and Jupiter but others stray outside this belt and some cross the Earth's orbit. The so-called trojan asteroids lie at gravitationally stable points along the orbits of the major planets.

asteroid belt—The region in the Solar System between about 2 and 3.3 astronomical units from the Sun, where most asteroids lie. Asteroids in this region are known as main-belt asteroids.

astigmatism—An optical defect that causes point sources, such as stars, to appear elongated.

astrometry—The branch of astronomy concerned with the precise measurement of positions of celestial objects; also known as positional astronomy.

astronomical unit (au)—The average distance of the Earth from the Sun. Its value is defined by the International Astronomical Union (IAU) as 149,597,870,700 meters. Distances in the Solar System are often expressed in astronomical units.

astrophysics—The application of physics to the study of the Universe and the objects within it, notably stars, galaxies, and interstellar matter. Astrophysics began in the nineteenth century with the use of spectroscopy to study starlight. Astrophysics differs from astrometry and celestial mechanics, which are concerned with the positions and motions of objects.

aurora—An electrical effect in the Earth's upper atmosphere caused by particles from the solar wind, mostly electrons, being accelerated down the Earth's magnetic lines of force. The electrons excite molecules in the upper atmosphere, causing them to glow. Activity is confined to a ring around the magnetic poles known as the auroral oval. The width of the auroral oval varies according to the strength of the solar wind. Also known as the aurora borealis or northern lights in the northern hemisphere, and aurora australis or southern lights in the south.

azimuth—The angular distance of a given point from north, measured eastward (clockwise) around the horizon from 0° to 360°. It is the coordinate at right angles to the object's altitude.

B
barred spiral galaxy—A type of spiral galaxy in which the spiral arms extend from the ends of a central bar of stars and gas, rather than directly from the galaxy's nucleus. Many spiral galaxies have a bar at their centers, ours included.

Bayer letter—A letter of the Greek alphabet assigned to stars in approximate order of their brightness in a given constellation. The system is named after the German astronomer Johann Bayer, who introduced it in his star atlas *Uranometria*, published in 1603.

Big Bang—The explosive event that's thought to have marked the origin of the Universe. At the time of the Big Bang, all matter and space was concentrated into a superhot, superdense point, from which it has been expanding ever since. Current observations of the rate of expansion of the Universe suggest that the Big Bang occurred around 13.8 billion years ago.

binary star—A pair of stars that orbit their common center of mass. They are physically connected by gravity, unlike an optical double, in which the two stars simply lie in the same line of sight. The orbital periods of binaries range from minutes to many thousands of years.

black hole—An object with such a strong gravitational field that no light or any other radiation can escape from it, i.e. its escape velocity exceeds the speed of light. Stellar-mass black holes are formed by the collapse of a massive star at the end of its life; supermassive black holes with the mass of millions of stars lie at the centers of galaxies.

blueshift—The decrease (shortening) in wavelength of light or other radiation from a source due to its movement toward us. The opposite effect, caused by recession, is a redshift.

brown dwarf—A star with a mass less than 0.08 that of the Sun. In such stars, central temperatures are not high enough to start nuclear reactions; instead they radiate gravitational energy as they slowly shrink in size. Their surface temperatures are no more than 2,200 K, cooler than that of red dwarfs. As a result they're extremely faint optically and are best seen in the infrared.

C

Cassegrain—A type of reflecting telescope in which the focal point lies behind the main mirror. Incoming light is reflected by the main mirror on to a convex secondary, which directs it back through a central hole in the main mirror. The focal point behind the mirror is termed the Cassegrain focus. Advantages of Cassegrain reflectors are that they are compact in length, and that heavy instrumentation can conveniently be mounted at the Cassegrain focus. See also Schmidt–Cassegrain.

catadioptric—An optical system that uses both lenses and mirrors to form an image. A common variety is the Schmidt–Cassegrain.

CCD—Abbreviation for charge-coupled device, a type of sensor found in some dedicated astronomy imaging cameras.

celestial equator—The projection of the Earth's equator onto the celestial sphere, i.e. the circle of 0° declination.

celestial mechanics—The study of the motion and gravitational interaction of bodies in space, from artificial satellites to planets, stars, and galaxies.

celestial pole—Either of the two points on the celestial sphere directly above the Earth's geographical poles, 90° from the celestial equator. The celestial sphere appears to rotate around the celestial poles as the Earth turns on its axis.

celestial sphere—An imaginary sphere with the Earth at its center on which all celestial bodies appear to lie. The concept of the celestial sphere is useful for drawing up astronomical coordinate systems to specify the position of objects in the sky.

Cepheid variable—A type of variable star that pulsates with a time-period that's related to its luminosity (the absolute magnitude). This relationship between period and luminosity is named Leavitt's law after its discoverer, Henrietta Leavitt. Cepheids are highly luminous so can be seen over long distances, including in other galaxies, which has made them useful distance indicators. They are named after their prototype, Delta Cephei.

chromatic aberration—A defect in lenses that causes colored fringes around an image. It arises because light of different wavelengths is refracted by different amounts, so different colors are brought to a focus at different distances from the lens. The effect can be corrected by combining two or more lenses made of different types of glass.

chromosphere—The layer of gas immediately above the visible surface (the photosphere) of a star. In the case of the Sun, the chromosphere is about 6,200 miles (10,000 kilometers) deep. The name means "color sphere" and arises because it is visible as a ring of red light around the Sun at a total eclipse.

circumpolar—A celestial object that does not rise or set during the night as seen from a particular location, but circles around the celestial pole as the Earth rotates. For an object to be circumpolar, its angular distance from the pole must be less than the observer's latitude.

CMOS—Abbreviation for complementary metal-oxide semiconductor, a type of camera chip found in DSLR cameras, cell phone cameras, and dedicated astronomy imaging cameras.

comet—A small body consisting of ice and dust left over from the formation of the Solar System. Most comets reside in two main swarms beyond the planets called the Oort Cloud and the closer-in Kuiper Belt. The solid nuclei of comets can be perturbed from these swarms to approach the Sun, where they warm up and emit gas and dust. Comets that have been observed more than once are termed periodic comets. Long-period comets have orbits that may last many thousands or millions of years.

conjunction—An alignment of celestial bodies as seen from the Earth, such as two planets or a planet and the Sun. Mercury and Venus are said to be at inferior conjunction when they lie between the Earth and Sun, and at superior conjunction when they lie on the opposite side of the Sun. The outer planets are at conjunction when they lie on the far side of the Sun.

constellation—One of the 88 areas of sky as defined by the International Astronomical Union in 1930. The areas are bounded by arcs of right ascension and declination. Stars within each constellation can lie at vastly different distances and are not physically related. They serve as a convenient guide for identifying and naming celestial objects.

corona—The extremely hot but tenuous outer layer in the atmosphere of the Sun and other stars. The corona can be seen only when the main disc of the Sun is obscured at a total eclipse, or by using special instruments. Atomic particles from the corona stream outward to form the solar wind.

cosmic microwave background (CMB)—Radiation at microwave wavelengths left over from the Big Bang explosion that's thought to have marked the origin of the Universe. The radiation comes uniformly from all directions in space. It was discovered in 1964 and is one of the strongest pieces of evidence for the Big Bang theory.

culmination—The maximum altitude above the horizon that a celestial body reaches on a given date. It's usually the moment the object crosses the meridian, unless it moves appreciably during the night.

D

declination—The angular distance of an object north or south of the celestial equator, the equivalent of latitude on Earth. Declination runs from 0° to 90° and is positive if north of the celestial equator, or negative if south. It's one of the coordinates in the equatorial coordinate system, the other being right ascension.

deep-sky object—An object beyond the Solar System, such as a star cluster, nebula, or galaxy. The term is not usually applied to individual stars.

diffuse nebula—A luminous cloud of interstellar gas and dust that glows because of the effect of ultraviolet radiation from nearby stars; also known as an emission nebula.

distance modulus—The difference between the apparent magnitude and the absolute magnitude of a star, from which its distance can be calculated.

diurnal motion—An object's daily motion, such as the rotation of the celestial sphere.

Dobsonian mounting—A type of altazimuth mounting in which the telescope rests in a cradle that rotates on a flat base. Dobsonian mounts are used for reflecting telescopes of Newtonian design. They can be made from simple materials at low cost, although they lack the ability to track objects. The design was invented by the American amateur astronomer John Dobson, after whom it is named.

Doppler shift—The change in the observed frequency of waves received from an object as it moves toward or away from an observer. For a receding source, the frequency decreases and the wavelength becomes correspondingly longer (i.e. a redshift), while for an approaching source, the frequency increases and the wavelength becomes shorter (i.e. a blueshift).

double star—Two stars that appear close together in the sky. If the two stars are close enough to be linked by gravity they're known as a binary star. In other cases the stars are unconnected and simply lie in the same line of sight by chance, when they are known as an optical double. Most double stars are true binaries.

DSLR—Abbreviation for digital single-lens reflex camera. Such cameras have interchangeable lenses offering different focal lengths, and are widely used for astrophotography.

dwarf planet—A body orbiting the Sun that's midway in size between an asteroid and a true planet. Dwarf planets are sufficiently massive to have become spherical under the inward pull of their own gravity, but are not big enough to have cleared away other material around their orbits. The term was introduced by the International Astronomical Union in 2006 to accommodate bodies such as Pluto.

dwarf star—A star that lies on the main sequence of the Hertzsprung–Russell diagram. The term arose in the twentieth century when stars were classified as either giants or dwarfs. Most stars fall into this category, including the Sun.

E

eccentricity—A figure that indicates the ellipticity of an orbit, i.e. the extent to which it deviates from a circle. A circle has eccentricity 0, while a parabola has eccentricity 1.

eclipse—The passage of a celestial body into the shadow of another, as a result of which it is partially or totally darkened. For example, a lunar eclipse occurs when the Moon passes through the shadow of the Earth. The term is also applied to the passage of the Moon in front of the Sun, but this is strictly an occultation.

eclipsing binary—A pair of stars orbiting each other in which one passes in front of the other on each orbit, temporarily blocking off all or some of its companion's light from Earth. Hence the total light output of the pair appears to fade and recover on a regular basis.

ecliptic—The projection of the Earth's orbital plane onto the celestial sphere; it is the apparent path of the Sun in front of the background stars during the year. Because of the Earth's axial tilt, the ecliptic is inclined by about 23.5° to the celestial equator; this value is known as the obliquity of the ecliptic. The ecliptic and celestial equator cross at two points, known as the equinoxes.

elliptical galaxy—A galaxy without spiral arms, which appears rounded or lens-shaped. Elliptical galaxies consist mostly of older stars and contain little or no interstellar gas or dust to make new ones. They range widely in size, from dwarf ellipticals a few thousand light years in diameter, to giant ellipticals over 100,000 light years wide.

elongation—The angular separation of a planet (usually Mercury or Venus) from the Sun, or between a planet and its satellite, as seen from Earth. Greatest elongation is the maximum angular separation.

emission nebula—A cloud of gas that emits its own light because the atoms have been ionized by ultraviolet light from a hot star nearby.

equatorial mount—A telescope mounting in which one axis, known as the polar axis, points to the celestial pole. The other axis (the declination axis) is at right angles, parallel to the Earth's equator. The telescope is attached to the declination axis. The polar axis is turned to keep an object in the telescope's field of view as the Earth rotates.

equinox—One of two points on the celestial sphere at which the Sun's path, the ecliptic, crosses the celestial equator. The Sun crosses from south to north at the vernal (or spring) equinox on March 20; the autumnal equinox, when the Sun crosses from north to south, occurs on September 22 or 23. Around the equinoxes, day and night are equal in length everywhere in the world.

exoplanet—A planet orbiting a star other than the Sun. Also known as an extrasolar planet.

eyepiece—A lens or system of lenses for magnifying the image formed by a telescope. The magnification of an eyepiece depends on its focal length, with shorter focal lengths providing higher magnification.

F

fireball—A meteor brighter than about magnitude −5, i.e. brighter than Venus.

first point of Aries—Another name for the vernal (spring) equinox, the point where the Sun crosses the celestial equator from the southern into the northern celestial hemisphere. Due to the effect of precession it's no longer in Aries but now lies in neighboring Pisces.

Flamsteed number—The number given to each star listed in the catalog compiled by the first Astronomer Royal, John Flamsteed, published in 1725. The numbers were not actually allocated by Flamsteed himself but were added by later astronomers.

flare, solar—A localized outburst on the Sun caused by the release of energy in the intense magnetic fields around sunspots. Solar flares emit radiation over all wavelengths from gamma rays to radio waves, along with high-speed atomic particles. Flares typically last a few hours, and are more common near the maximum of the 11-year solar cycle.

focal length—The distance between the lens or mirror of an optical system and its focal point, where the image is formed.

focal ratio—The focal length of a lens or mirror divided by its aperture; also termed the f ratio.

Fraunhofer lines—The dark lines seen crossing the Sun's spectrum, caused by the absorption of light at particular wavelengths by gases in its atmosphere. The lines are named after the German physicist Joseph von Fraunhofer, who first identified them in 1814.

G

galaxy—An immense system of stars, dust, and gas bound together by its own gravity. Galaxies range in size from dwarfs with fewer than a million stars to supergiants containing billions of stars and hundreds of thousands of light years wide. They can be classified into four main shapes: spiral, barred spiral, elliptical, and irregular.

geocentric—A term meaning Earth-centered, for example as in the geocentric system of cosmology, or a geocentric orbit.

giant star—A star of similar mass to the Sun that has swollen in size as it reaches the end of its life. Giant stars are larger and more luminous than main-sequence stars of the same temperature.

globular cluster—A near-spherical group of stars in the halo around a galaxy. Globular clusters contain from tens of thousands to millions of mainly old stars and are typically 100 light years in diameter.

Greenwich Mean Time (GMT)—Local time as measured on the Greenwich meridian, the Earth's 0° line of longitude. It's now more usually known as Universal Time (UT).

H

heavy elements—In astronomy, any element heavier than hydrogen or helium is known as a heavy element. They are also termed metals, even though not all of them are metallic in the true sense.

heliocentric—A term meaning Sun-centered, for example as in the heliocentric system of cosmology, or a heliocentric orbit.

Hertzsprung–Russell diagram (HR diagram)—A graph on which the color or temperature of stars is plotted against their brightness. It was developed by the Danish astronomer Ejnar Hertzsprung and the American Henry Norris Russell. The position of a star on the H–R diagram indicates the stage it has reached in its evolution. Most stars lie in a diagonal band known as the main sequence. Above and to the right of this are stars in a later stage of evolution known as giants.

Hubble constant—A measure of the rate at which the Universe is expanding, as determined by observation. It's usually quoted in units of kilometers per second per million parsecs (km/s/Mpc).

Hubble law—The relationship which states that galaxies are receding from the Earth at speeds that increase in proportion to their distance, i.e. the farther away a galaxy is, the faster its recession. The law was originally named after Edwin Hubble who announced it in 1929, and is also known as the Hubble–Lemaître law, to acknowledge that the expansion of the Universe had been predicted by Georges Lemaître.

I

IC number—The number given to an object in one of the two Index Catalogues, which were supplements to the New General Catalogue of Nebulae and Clusters of Stars (NGC).

inclination—The angle between the orbital plane of a body and a reference plane, such as between the orbital plane of a planet and the ecliptic (i.e. the plane of the Earth's orbit). Axial inclination is the angle between the equatorial plane of a body and its orbital plane.

inferior conjunction—The alignment of the inner planets Mercury and Venus directly between the Earth and Sun.

K

Kuiper Belt—A region of the Solar System beyond Neptune consisting of millions of small icy remnants from the formation of the planets; its members are also known as trans-Neptunian objects (TNOs). Pluto is its largest known member.

L

light curve—A graph showing the variation in an object's brightness over time.

light pollution—Illumination of the night sky by artificial sources of light, such as street lights. Light pollution not only affects astronomical observation but also wildlife and human health.

light year—The distance traveled by light or any other form of electromagnetic radiation through a vacuum during one year (365.25 days). One light year equals 9.46 million million kilometers, or 63,241 astronomical units, or 0.3066 parsecs.

limiting magnitude—The faintest object that can be detected on a given night with either the naked eye or a particular instrument.

Local Group—The cluster of galaxies that includes the Milky Way. The group is about 10 million light years in diameter. Over 80 members are known, most of which are small and faint.

long-period variable—A variable star with a period of more than about 100 days. Most such stars are pulsating red giants or supergiants. They are also known as Mira stars, after their prototype.

luminosity class—A classification of stars according to their luminosity (i.e. total energy radiated per second). A star's luminosity class indicates whether it's a dwarf, giant, or supergiant.

M

magnitude—A measure of the brightness of a celestial object, either as seen from Earth (the apparent magnitude) or its actual light output (absolute magnitude). Naked-eye stars range from 1st to 6th magnitude. Fainter objects are given progressively larger positive magnitudes, while brighter objects are assigned negative magnitudes.

main sequence—A stage in the life of a star during which it creates energy by converting hydrogen into helium by nuclear reactions at its center. Stars spend most of their lives in this stage; the exact length of time depends on their mass.

mean sun—An imaginary body that moves along the celestial equator at uniform rate, unlike the real Sun. It represents the motion that the real Sun would have if the Earth's orbit were circular and its axis were upright and not inclined. The position of the mean sun is the basis of mean solar time.

meridian—In astronomy, the great circle on the celestial sphere that passes overhead and through the north and south points on the observer's horizon. On a planet, a meridian is a circle of longitude.

Messier object—An object in the catalog of non-stellar objects compiled by Charles Messier to distinguish them from comets, for which he was searching. His final list, with 103 entries, appeared in 1781 but with later additions by others the total number of objects with M designations is 110.

meteor—The streak of light produced when a small particle of interplanetary dust (a meteoroid) enters the Earth's atmosphere at high speed and burns up; also known as a shooting star. A handful of meteors, known as sporadics, can be seen every hour, but when the Earth encounters a stream of dust left along the orbit of a comet, a meteor shower occurs; at such times dozens of meteors an hour appear to emanate from an area of sky known as the radiant.

meteorite—A piece of interplanetary debris large enough to pass through the atmosphere and hit the surface of the Earth or other planetary body. Meteorites can be composed of rock or metal.

meteoroid—A small piece of debris in interplanetary space. Meteoroids that enter the Earth's atmosphere either burn up as meteors or, if they are large enough, fall to Earth as meteorites.

Milky Way—The faint band of light that consists of innumerable distant stars lying in the plane of our Galaxy; also, another name for the entire Galaxy to which the Sun belongs.

minor planet—Another name for an asteroid.

moon—Any object orbiting a larger body in the Solar System; also known as a natural satellite.

multiple star—A group of three or more stars bound together by their gravitational attraction.

N

nebula—A cloud of interstellar gas and dust in the spiral arms of a galaxy; plural nebulae. There are three main types: emission nebulae give out their own light; reflection nebulae reflect light from nearby stars; and dark nebulae are seen in silhouette against a brighter background.

neutron star—A very small, dense star in which the protons and electrons of the star's atoms have been compressed together to form neutrons. They are formed when massive stars explode as supernovae. Spinning neutron stars can be detected as pulsars.

Newtonian—A type of reflecting telescope in which incoming light is collected by a concave mirror at the base of the tube and reflected onto a flat mirror that diverts it into an eyepiece at the side of the tube. The design is named after Isaac Newton, who invented it in 1668.

NGC number—The number given to an object in the New General Catalogue of Nebulae and Clusters of Stars compiled by J. L. E. Dreyer, published in 1888.

northern lights—See aurora.

nova—A type of variable star that suddenly brightens by 10 magnitudes or more over a few days before slowly fading back to its original brightness. Novae are close binary systems in which matter flows from one star onto a white dwarf companion, igniting a nuclear explosion.

nucleus—In a comet, the nucleus is the solid body composed mostly of ice that gives off gas and dust to form the coma and tail. In a spiral or barred spiral galaxy, the nucleus is the central concentration of stars and gas, possibly also containing a supermassive black hole.

O

object glass—The main lens of a refracting telescope, which collects light and brings it to a focus; also known as the objective.

obliquity of the ecliptic—The angle between the Earth's equatorial plane and the ecliptic, resulting from the Earth's axial tilt. It is currently about 23.5°, but changes slightly over time.

occultation—The passage of one astronomical body in front of another, totally or partially obscuring it from view. Examples are the occultation of a star by the Moon, or of a moon by its parent planet. A solar eclipse, when the Moon moves in front of the Sun, is actually an occultation.

Oort Cloud—A swarm of icy bodies surrounding the Solar System extending out to about 100,000 astronomical units. Objects can be perturbed out of the cloud by passing stars and fall in towards the Sun, where they become visible as comets. The cloud is named after Jan Oort, who predicted its existence.

open cluster—A loosely packed group of stars with no overall shape, unlike the densely packed and rounded globular clusters. Open clusters lie in the spiral arms of galaxies and can contain from a few dozen to several hundred stars.

opposition—The alignment of a body in the Solar System in the opposite direction to the Sun, as seen from Earth. An object at opposition is due south at midnight and is visible all night. The outer planets are at their closest to Earth and appear at their brightest around opposition.

optical double—Two stars that appear close together in the sky but in fact lie at different distances from us in the same line of sight; hence they are unrelated, unlike the components of a true binary star.

orbit—The path followed by a celestial body as it moves around another. The time taken for the body to complete one circuit is known as its orbital period.

P

parallax—The angular distance that an object appears to move against a fixed background when viewed from two different points. In astronomy, the term is usually taken to mean the difference in position of a celestial object as seen from opposite sides of the Earth's orbit, and is used to find the distance to the nearest stars. By extension, the term can also be used as a synonym for distance.

parsec—A measure of astronomical distance equal to 3.2616 light years. It is the distance at which a star would show a parallax shift of 1 second of arc when observed from the ends of a baseline 1 astronomical unit long.

penumbra (1)—The lighter outer region of the shadow cast by a planet or moon. At a solar eclipse, an observer within the penumbra of the Moon sees the

Sun only partially covered. At a penumbral lunar eclipse, the Moon lies within the penumbra of the Earth's shadow.

penumbra (2)—The lighter, outer region of a sunspot, not as cool or dark as the central umbra.

perigee—The point in an elliptical orbit around the Earth that is nearest the Earth.

perihelion—The point in an elliptical orbit around the Sun that is nearest the Sun.

periodic comet—A comet that has been observed to orbit the Sun more than once. The names of periodic comets are prefixed with P/, as in P/Halley.

perturbation—A small disturbance to the motion of a body caused by the gravitational pull of another object, such as the effect of the planets on a comet.

photometer—A device for making accurate measurements of the brightness of objects. The application of such instruments is known as photometry.

photosphere—The visible surface of the Sun or other star, which emits most of its energy. The Sun's photosphere is a layer of gas about 300 miles (500 kilometers) deep with a temperature around 5,800 K. The name means "light sphere."

planet—A non-luminous body in orbit around a star that's sufficiently massive to have become rounded in shape by its own gravity and which has cleared its orbital neighborhood of smaller objects. Planets can consist of rock and metal, like our Earth, or of gas and liquid, as is Jupiter. The name dwarf planet is given to an object that's rounded in shape but which has not cleared the neighborhood around its orbit. An object with a mass more than about 10 times that of Jupiter is a brown dwarf.

planetary nebula—A shell of gas surrounding a hot central star, having been thrown off by the star during a late stage in its evolution. They are so named because of their resemblance to the disc of a planet when seen through a telescope.

population, stellar—Either of two main categories of star, based on age and composition. Stars of Population I are relatively young and have a high content of heavy elements, while Population II stars are old and contain few heavy elements. The Sun is a Population I star.

position angle—The relative positions of two objects, such as the components of a double star, measured as an angle from north through east.

precession—The gradual change in direction of the Earth's axis which causes it to trace out a circle on the celestial sphere every 25,800 years. The radius of the circle is about 23.5°, which is the tilt of the Earth's axis. As a result of precession, the position of the equinoxes move westward along the ecliptic by 1° every 72 years or so, leading to a gradual change in the right ascension and declination of stars.

prime focus—The point at which the primary mirror in a reflecting telescope brings light to a focus.

prominence—A cooler, denser cloud of gas in the Sun's inner corona, often seen on the Sun's limb at a total solar eclipse. When seen against the photosphere they appear dark and are termed filaments. Quiescent prominences can last for many weeks, while active prominences have lifetimes of only a few days or hours.

proper motion—The angular change in position of a star on the celestial sphere resulting from its motion through space relative to the Sun, expressed in seconds of arc per year. The shapes of the constellations are gradually changing over many thousands of years because of the proper motions of their stars.

protostar—A forming star in which central temperatures have not yet become high enough to trigger nuclear reactions.

pulsar—A radio source that emits short pulses of radiation at regular intervals, typically every second or so. They are thought to be rapidly rotating neutron stars emitting beams of radiation along their magnetic axes that sweep across the Earth every time they spin. Some pulsars have also been detected at optical wavelengths.

pulsating variable—A star that varies in brightness due to periodic changes in its size. Such pulsations can be regular, as in Cepheid variables, or irregular, as in giant and supergiant stars.

Q

quasar—A galaxy with a small, highly luminous core. Quasars have large redshifts and hence must be very far off in the Universe. The source of their energy is thought to be a disc of gas circling around a central black hole with a mass of tens of millions of Suns.

R

radial velocity—The velocity of an object in the line of sight, either toward or away from an observer. The radial velocity can be calculated from the redshift or blueshift of lines in an object's spectrum.

radiant—The region of sky from which members of a meteor shower appear to radiate as they burn up in the atmosphere. The shower is named after the constellation in which the radiant lies (or sometimes the nearest bright star); for example, the Perseid meteors radiate from the constellation Perseus.

radio galaxy—A galaxy that's an unusually strong source of radio waves, around a million times more powerful than a normal galaxy. Most are elliptical galaxies. They have a central radio source and a pair of radio-emitting lobes either side of the visible galaxy, thought to be jets of gas ejected from the active nucleus.

red dwarf—A small, cool star at the lower end of the main sequence with a surface temperature below about 5,000 K and a mass less than 0.8 that of the Sun. Red dwarfs are the most common type of star in our Galaxy but are difficult to see because of their faintness. Cooler stars still are known as brown dwarfs.

red giant—A large and highly luminous star with a cooler surface temperature than the Sun. Stars swell up into red giants toward the end of their lives when they have used up all the hydrogen in their cores and have switched to nuclear reactions involving helium and heavier elements. They are over 10 times larger than the Sun and hundreds of times brighter.

redshift—The increase (lengthening) in wavelength of light or other radiation from a source due to its movement away from us. For distant galaxies, the redshift in their light is caused by the expansion of the Universe.

reflection nebula—A bright nebula that shines by reflecting light from a nearby star. The dust in the nebula scatters the starlight so the nebula appears bluer than the star.

reflector—A design of telescope in which light is collected and focused by a concave main mirror. In the largest telescopes the main mirror is composed of segments.

refractor—A design of telescope in which light is collected and focused by a lens, also known as the object glass. The practical upper limit to the aperture of refractors is 3 feet (1 meter), beyond which the lens becomes too thick and heavy.

resolving power—The ability of a telescope or other instrument to detect fine detail; also known as resolution. Larger apertures have better resolving power.

retrograde motion—Movement or rotation from east to west, opposite to the normal west-to-east direction of motion in the Solar System, which is known as direct or prograde motion. The outer planets appear to move retrograde for a short time when the Earth catches up and overtakes them in its faster orbit. A body that orbits from east to west, such as some moons of the outer planets, are said to have retrograde orbits, while bodies that rotate from east to west are said to have retrograde rotation.

right ascension (RA)—The celestial equivalent of longitude on Earth. It's measured eastward (anticlockwise) along the celestial equator from the vernal equinox and is usually expressed in hours, minutes, and seconds of time from 0h to 24h. An hour of right ascension is the same as 15° of arc.

S

Schmidt–Cassegrain—A modified form of Cassegrain reflector in which the incoming light passes through a thin lens known as a corrector plate at the front of the tube before reaching the main mirror. Such telescopes are compact and portable, and are popular with amateur astronomers.

seeing—The unsteadiness in a telescopic image caused by air currents in the Earth's atmosphere. Such atmospheric turbulence causes stars to twinkle, particularly when close to the horizon.

sidereal period—The time taken for a planet or moon to go once around its orbit as measured against the star background. The sidereal period of a planet is said to be its year.

sidereal time—A time system based on the rotation of the Earth relative to the stars. The length of a sidereal day is 23 hours, 56 minutes, 4 seconds.

solar cycle—The periodic variation in activity on the Sun, such as the number of sunspots and flares. The average length of the solar cycle is about 11 years but it has varied from 9 to 14 years.

Solar System—The Sun and all the bodies that are gravitationally bound to it, such as planets and their moons, asteroids, and comets. By extension, the term is also applied to the planetary systems of other stars.

solar wind—The continuous outflow of atomic particles from the Sun, chiefly electrons and protons.

solstice—Either of the points on the celestial sphere at which the Sun reaches its most northerly or southerly declination (23.5° north or south). It is at its most northerly point on June 21 (summer solstice in the northern hemisphere, winter solstice in the south) and at its most southerly on December 21 or 22 (winter solstice in the northern hemisphere, summer solstice in the south).

spectral line—A dark or bright line in a spectrum caused by absorption or emission of light. The specific wavelengths of the spectral lines allow the chemical element causing the absorption or emission to be identified.

spectral type—A system of classification of stars on the basis of their spectra. The spectral type is an indication of the star's surface temperature, the hottest ones being the bluest and the coolest ones the reddest.

spectroscope—A device for splitting an object's light into a spectrum. The details of the spectrum can reveal the physical nature of the object, such as its temperature and composition. The application of such instruments is termed spectroscopy.

spectroscopic binary—A binary star in which the components are too close together to be seen individually, but the motion of one star around the other is revealed by the Doppler shift of lines in its spectrum.

spectrum—The entire range of electromagnetic radiation from gamma rays to radio waves. The term is often restricted to visible wavelengths only.

spherical aberration—An optical defect in which different parts of a lens or mirror have different focal lengths, thereby producing a blurred image.

spiral galaxy—A galaxy with arms of stars, gas, and dust spiralling outward from its nucleus. If the nucleus is noticeably elongated the galaxy is classified as a barred spiral.

star—A luminous ball of gas that shines from the energy released by internal nuclear reactions. The Sun is a typical star.

star cluster—A group of stars held together by their own mutual gravitational attraction. Open clusters are irregular in shape, lie in the spiral arms of our Galaxy, and contain up to a few hundred relatively young stars. Globular clusters are rounded balls containing thousands or even millions of old stars, are much larger than open clusters, and they lie in a halo around our Galaxy.

sunspot—A cooler area on the Sun's visible surface (the photosphere) that appears dark by contrast with its brighter surroundings. Sunspot numbers rise and fall with the 11-year solar cycle, being greatest at solar maximum. Similar markings also occur on other stars, when they are termed starspots.

supergiant—The largest and most luminous type of star, formed when stars more massive than the Sun evolve toward the ends of their lives. They are often variable due to fluctuations in size.

superior conjunction—The alignment of the inner planets Mercury and Venus directly behind the Sun, as seen from Earth.

superior planet—A planet that lies farther from the Sun than

Earth, namely Mars, Jupiter, Saturn, Uranus, and Neptune.

supernova—The violent explosion of a star at the end of its life. This can occur either when stars more massive than the Sun run out of nuclear fuel at their centers and collapse, or when matter falls onto a white dwarf in a binary system, causing it to ignite and explode.

synodic period—The time taken for an object to return to the same position in the sky as seen from the Earth, for example between two oppositions or two full moons. It differs from the sidereal period because of the Earth's own motion in orbit.

T
terrestrial planet—Any small, rocky planet with a solid surface. In the Solar System, the terrestrial planets are Mercury, Venus, Earth, and Mars.

transit (1)—The moment when a celestial body crosses the observer's meridian.

transit (2)—The passage of an object of small apparent size in front of a larger body, such as the planets Mercury or Venus across the face of the Sun, or a moon or a surface feature across the disc of its parent planet.

twilight—The period after sunset or before sunrise when the sky is not fully dark. Three types of twilight are defined: civil twilight, when the center of the Sun is less than 6° below the horizon; nautical twilight, when the centre of the Sun is between 6° and 12° below the horizon; and astronomical twilight, when the center of the Sun is between 12° and 18° below the horizon.

U
umbra (1)—The dark inner region of the shadow cast by a planet or moon. An observer within the umbral shadow of the Moon sees the Sun totally eclipsed.

umbra (2)—The dark central part of a sunspot.

Universal Time (UT)—A standard timescale used for scientific purposes, derived from the meridian transit of stars; also known as Greenwich Mean Time (GMT). Coordinated Universal Time (UTC) is the time given by atomic clocks, and differs by up to a second from UT because of irregularities in the Earth's rotation.

V
variable star—A star that varies in brightness, either because of changes in its actual light output, as happens in a pulsating variable, or because part of its light is temporarily blocked by another star, as in an eclipsing binary.

vernal equinox—The point on the celestial sphere where the Sun crosses the celestial equator from south to north each year, which currently occurs on March 20; also known as the spring equinox. It's the point from which the celestial coordinate right ascension is measured.

W
white dwarf—A small, hot star, the core of a former red giant that has lost its outer layers as a planetary nebula. The Sun will probably end up as a white dwarf.

Z
zenith—The point on the celestial sphere directly above an observer. It's the opposite point to the nadir.

zodiac—A band of sky extending for 8° either side of the ecliptic in which the Sun, Moon, and planets are found.

zodiacal light—A faint cone of light caused by sunlight being scattered off interplanetary dust. It extends from the horizon and along the ecliptic. It can only be observed from dark locations.

INDEX

absolute magnitude 39
accretion 222
Achernar 31, 135
Adams, Cath 373
Akatsuki spacecraft (Venus Climate Orbiter) 244, 246
al-Sūfī, Abd al-Ralman 58
Albion 276
Albireo (beta) Cygni 202, *300*, 301
Aldebaran 24, 26, 180, 197, 199, *302*, 303
Aldrin, Edwin "Buzz" 221
Algol (beta) Lyrae *206*
Algol (beta) Persei *205*
Almach 99
Almagest (*Mathematike Syntax*) (Ptolemy) 43, 56–7, *57*, 58
Alphard 141
Alphecca 125
Alpheratz (Alpha Andromeda) 63, 99
Alshain 103
Altair 103, 151, 299, *299*
altitude-azimuth (Alt-AZ) system 294–5, *296*, 314, 316
Amalthea 257
Andromeda Galaxy 11, 26, 39, 40, 50, 53, *64*, *98*, 99, 144, 162, 183, *302*, 303, 323, *365*
angular diameter 263
Antares 24, 174, 195, 197, 207
Antlia, the Air Pump 63, 100
aphelion 237
Apis, the Bee 155
Apollo missions 221
apparent magnitude 39
apps 305
Apus, the Bird of Paradise 101–2, 161
Aquarius, the Water Carrier 28, 102, 167
Aquila, the Eagle 103, 151, 171
Ara, the Altar 104, 149, 156
Aratus 57
Archer, Frederick Scott 359
Arcturus 24, 26, 29, 31, 107, 186, 199, 299, *299*
Arecibo Observatory, Puerto Rico 241
Argo Navis, the Ship of the Argonauts 26, 63, 115, 168
Ariel 264
Aries, the Ram 28, 105, 183
Armstrong, Neil 221
art 7, 12, 330–2

projects 376–82, 384, 387
artificial light at night (ALAN) 318, 319, 321
Asellus Australis 110
Asellus Borealis 110
asterisms 25
asteroids and comets 269, 272–3
  asteroid classification 274, 276
  asteroids and Jupiter *275*
  comets 8, 11–12, 44, 47, 48, 277–80, *279*, *280*, *281*, *304*, 337
  Kuiper Belt 276–7
  Oort cloud 276
astronomy
  background 7–8, 11–12
  clubs/organizations 325
  recording techniques 38–40
astrophotography 358–61, 364, 366–9, 371–2
see also photography
astrophysics 38, 50
Atlas Coelestis (Flamsteed) 62–3, *62*
Augsburg Book of Miracles *336*, 337
Auriga, the Charioteer 31, 63, 106, 150
Aurora Australis (Southern Lights) 306
Aurora Borealis (Northern Lights) *215*, 306, *350–1*
auroral ovals 215
auroras/aurorae 215, 306, *307*, 309, *374*
averted vision 292
axial tilt 240

Barlow lens 314, 323, 364
Barlow, Peter 314
Barnard's Star 165, 197
Bartsch, Jakob 109
Bayer, Johann 26, 60, 61
Bayeux Tapestry 8, *9*
Bell Telephone Laboratories, New Jersey (U.S.A.) 41, *41*, 54
Bennu 274
BepiColombo mission 242
Bessel, Friedrich William 37, 49, *49*
Betelgeuse 24, 31, 113, 197, 207
Bethe, Hans 54, *55*
Big Bang 40–1, 41, 53
Big Dipper *see* the Plough
binoculars 311
BL Lacertids 144
black holes 200, 210

Blaeu, Willem Janszoon 60
Blake, William 7
Blood Moon 270
Bode, Johann Elert 63, 65, 144, 169
Boeing 747 *362*
Bonestell, Chesley 12, 335
Book of the Fixed Stars (al-Sūfī) 58, *58*
Boötes, the Herdsman 31, 107, 113, 186, 299
Bortle, John E. 322
Bortle scale 322
Bouvard, Alexis 265
Bowie, David 16
Brahe, Tycho 35, 44, *44*, 45, 60, 116
British Association Catalogue 146
British Astronomical Association (BAA) 306, 309
Brocchi's Cluster, the Coat Hanger 193, *300*, 301
Bruce Peninsula National Park, Ontario (Canada) 7
Bunsch, Robert 38

Cacciatone, Niccolò "Nicolaus Venator" 131
Caelum, the Chisel 108
Caesar's Comet 278
calendars 8
California Nebula 163
Callisto 256–7, *257*
Camelopardalis, the Giraffe 60, 109, 109–10, 294
Cancer, the Crab 110–11, 114, 141
Canes Venatici, the Hunting Dogs 111
Canis Major, the Greater Dog 26, 31, 112–13, 147, 154
Canis Minor, the Lesser Dog 31, 113–14
Canopus 31, 165, 198
Capella 31, 106
Capricornus, the Sea Goat 102, 114
captured (synchronous rotation) 219
Carina, the Keel or Hull 31, 63, 115, 165, 195, 197
Carpenter, James 347
Cassini, J.D. 182
Cassini space probe 255, 260
Cassiopeia, the Queen 16, 25, *64*, 70, 116–17, 303
Castor 26, 137, 202
Catalogus Stellarum Fixartum (Hevelius) 62

celestial globe 57, *60*
celestial sphere 20–1
Cellarius, Andreus 33, 34, 342
 chart of constellations *343*
Centaurus A *211*
Centaurus, the Centaur 31, 117–18, 121, 128, 149, 174, 197, 202
Cepheids 39, 40, 51, 52, *205*
Cepheus 28, 118–19
Cerberus, triple-headed monster 62
Ceres 266, *272*, 273, 276
Cetus, the Whale 102, 119–20, 207
Chacornac, Jean 349
Chamaeleon, the Chameleon 120–1
chaos terrain 257
Charon 266
Chiron 276
chondrites 286, *286*
chromosphere 216
Church, Frederic Edwin 337, 350
Circinus, the Compasses 121–2, 184
Circlet 166
circumpolar 21
Claws of the Scorpion 148
clusters *see* star clusters
CN Leonis, the Wolf 145
Coalsack Nebula 155
Columba, the Dove 60, 122
Coma Berenices, Berenice's Hair 123, 123–4, 190
Coma Star Cluster 123
comets *see* asteroids and comets
constellations 16, 21, 25, *30*, 47, 48, 59, *61*, 62
 charts 71, 98–171
 names 66–7, 294
Copernican system 34–5, *34*
Copernicus, Nicolaus 8, 34–5, 44, *44*, 52
Corona Australis, the Southern Crown 124, 182
Corona Borealis, the Northern Crown 107, 124, 125–6, 294, 332
coronal holes 215
coronal mass ejection (CME) 215
Corvus, the Crow 126–7
CP Lacertae 144
Crab Nebula 11, 26, 48, 50, 180, 200
Crab Pulsar 200
Crater, the Cup 127–8
Crescent Moon 361
Critchfield, Charles 55
Crux, the Southern Cross 11, 25, *25*, 117, 128–9, 161, 174, 195
Curiosity Rover 251
Cygnus Rift, the Northern

Coalsack 129
Cygnus, the Swan 28, *30*, 49, 57, 103, 129–30, 144, 151, 171, 193, 200, 301

Daguerre, Louis-Jacques-Mandé 359
Dark and Quiet Skies Global Outreach Project 324
dark sky sites 324–5
dark-adapted eyes 292
DART mission 275, 287
data gathering/recording 38–41, 44, 47, 48, 49, 50, 51
daylight-saving time (DST) 71
De revolutionibus orbium coelestium ("On the Revolutions of the Heavenly Spheres") (Copernicus) 44, 245
declination (DEC) 21, 28, 295
deep-sky objects 71
Deimos *253*, 273
Délimitation scientifique des constellations (Delporte) 65
Delphinus, the Dolphin 131–3
Delporte, Eugène 65
Delta Aquariids 114
Delta Cephei *206*
Delta Pegasi 63
Deltoton 183
Dendera, Egypt 332
Deneb 27, 103, 129, 151, 199, 299, *299*, 301
Didymos-Dimphros asteroid system 275
Dione 261
Divine Couple Ring *333*
Dobsonian mounts 312
Dorado, the Goldfish 132, 165, 192, 195
Double Asteroid Redirection Test (DART) 274
Double Cluster 163
Double-Double (Epsilon Lyrae) 202
Draco, the Dragon 133–4
Draper, John Williams 39
Dreyer, J.E. 26
Dubhe 186
Dunhuang, China 21, 65
Dürer, Albrecht 58, 59, 334
dwarf planets 266

Earth 21, 23–4, 27, 35, 40, 47, 54, 112, 114, 133, 269, *272*, 338
 rotation 297
 wobble (precession) 28–9
earth-grazing meteors 285

earthshine 219, 361
eclipses and occultations 268–9
 lunar eclipse 270–1, *271*
 Saros cycles 271
 solar eclipse *268*, 270, *270*, *290*
eclipsing binaries 202
Eddington, Arthur 52, *52*
Edgeworth, Kenneth 276
Eimmart, Georg Christoph 345
Eimmart, Maria Clara 345
Einstein, Albert 51, *51*, 338
elliptic 21
elliptical orbits 35
Enceladus 261–2, *262*
epicycle 34
equant 34
equatorial (EQ) coordinate system 294, 295, *296*
equinox 21, 24, 28
Equuleus, the Little Horse 134–5
Equuleus Pictorium 165
Eridanus, the River 31, 108, 135–6, 142, 164
Eris 266
ESA Mars Express 250
Eta Aquariids 114, 281
Eudoxus 33
Europa 256–7, *257*
Europa Clipper mission 238
European Space Agency (ESA) 242

False Cross 115, 189
Faltkalendar mit Moratsbildern (folding calendar) *331*
Farnese Atlas (sculpture) 57, *57*
Farnese, Cardinal Alessandro 57
Felix the Cat 63
filaments 216
Firmamentum Sobiescianum (Hevelius) 62, *62*
First Point of Aries 105
First Point of Libra 148
Flamamsteed numbers 26
Flamsteed, John 26, 62–3
Fludd, Robert 340, 341, *341*
Fomalhaut 102
Fornax, the Furnace 63, 135, 136–7
Foucault, Léon 349
Fraunhofer, Joseph von 38
Fraunhofer lines 38
French Royal Academy of Sciences 62
Friedrichs Ehre, Frederick's Glory 144
Fromalhaut 166

Gaia (spacecraft) 28, 29
galaxies 209–10

active 210
barred spiral 210
elliptical 209–10
irregular 210
radio 210
redshift in light *40*
spiral 210
Galaxy 29, 39–40, 50, 54, 123, 151, 158, 159, 172, 195, 200, *208, 210*
Galilean satellites *257*
Galileo Galilei 12, 35, 36, 45, *45*, 240, 245, 256–7, 338, *338*
Galileo probe 255, 257
Gamma Aurigae (Beta Tauri) 62, 63
Gamma Virginis *203*
Ganymede 238, 256–7, *257*
Gaposchkin, Sergei 54
Gassendi, Pierre 341, 344
Geisler, Christoph 319
Gemini, the Twins *13*, 26, 137–9, 159
Geminids 114, 137, 283
General Catalogue of Nebulae and Clusters of Stars (JFW Herschel) 49
geocentric (Earth-centered) model 33–4, 43, 44, 240, 334
Giant Impact Theory 221
Gilbert, William 338
Giotto di Bondone 337
Global Meteor Network (GMN) 283, 285
globular clusters 209
Globus Aerostaticus, the Hot Air Balloon 63
Goddard Space Flight Center 213, 305
Gonggong 266
Goodricke, John 118
gravity 37, 46
Great Bear *see* Ursa Major
Great Comets (1680 & 1811) 278
Great Square of Pegasus 11, 99, 162
the Greater Dog *see* Canis Major
greenhouse effect 246
Greenwich Mean Time (GMT) 23
Greenwich Observatory 62
Grus, the Crane 138
Guardians of the Pole 188
guest star 8, 11
Gum, Colin S. 189

Hadley, John 157
Hadley Rille 222
Hale-Bopp 279, *280*
Halley, Edmond 9, 37, 47, *47*
Halley's Comet 9, 12, 37, 47, 159, 277, 337
Hardy, David A. 335, 355

Saturn from Enceladus 355
Harmonia Macrocosmica ("Cosmic Harmony") (Cellarius) 33, *34*, 342
Harriot, Thomas 338
Harvard College Observatory, Massachusetts (U.S.A.) 39
Haumea 266
Hawking, Stephen 338
Heart Nebula *370*
Heinfogel, Conrad 58
heliocentric (Sun-centered) model 34–5, 43, 44, 45
hemisphere charts 71, *72–3*
Hercules 139, 139–40
Herschel, Caroline 12, 48
Herschel Crater *262*
Herschel, John Frederick William 48, 49, *49*
Herschel, William 12, 36, 37–8, 48, *48*, 49, 261
Hertzsprung, Einar 198
Hertzsprung-Russell diagram 198, *198*, 199
Hevelius, Johannes 60, 61, 62, 63, 111, 144, 146, 150, 177, 179, 193, 345
Hind's Variable Nebula 180
Hipparchus 26, 33, 43, 134
Hipparcos Catalogue 117
Hipparcos (spacecraft) 28, 29
Holst, Gustav 7
Horologium, the Pendulum Clock 140–1
Houtman, Frederick de 60, 101, 120, 132, 138, 143, 155, 161, 164, 184, 185, 192
Hoyle, Fred 53
Hubble constant 53
Hubble, Edwin 40, 41, 53, *53*
Hubble Space Telescope 12, 40, 53, 248, 254, 256, 257, 259, 260, 263, 264, 265, 275, 314
Hubble's Law 40
Huggins, Margaret Lindsey 50
Huggins, William 39, 50, *50*
Huygens, Christian 245, 260
Huygens probe 261
Hyades 180
Hydra, the Water Snake 141–2
Hydrus, the Lesser Water Snake 135, 142–3
Hygiea 273
Hyginus 57

Index Catalogues (IC) 26
Indus, the Indian 143–4
infrared radiation 48

Ingenuity helicopter 251–2, *251*
Inner Solar System *275*
inter-crater plains 242
International Astronomical Union (IAU) 26, 56, 63, 65, 266, 324
International Dark Sky Places programme 324
International Dark-Sky Association 324
International Date Line 23
International Space Station 305
interplanetary dust 280
Io 256–7, *257*

James Webb telescope 12, 258, 264, 265
Jansky, Karl 54, *54*
Japanese Aerospace Exploration Agency 242
JAXA Hayabusa2 spacecraft 274
Job's Coffin 131
Johnson Space Center, Houston, Texas (U.S.A.) 221
Journal of Astronomical History and Heritage 349
Journal of the British Astronomical Association 276
Juno spacecraft 255
Jupiter 8, 24, 33, 35, 46, 254, *254*, *255*, *256*, *258*, 269, 273, 274, *275*, 276, 293, 309
  cloud belts 254–5
  composition 256
  core 256
  moons and rings 256–8
  Shoemaker-Levy 9 impact 256

Kant, Immanuel 38
Kapteyn's Star 165
Kepler, Johannes 35, 44, 45, *45*, 200, 341
Kepler's Star 45, 158, 200
Keyser, Pieter Dirkszoon 60, 101, 120, 132, 138, 143, 155, 161, 164, 184, 185, 192
Keystone of Hercules 126, 139
Kirchhoff, Gustav 38
Kirkwood, Daniel 273
Kirkwood gaps 273
Kochab 188
Kometenbuch ("The Comet Book") 285
Kreutz Group 280
Kuiper Belt 276, 277
Kuiper Belt Object (KBO) 276, *276*, 277
Kuiper, Gerard 276

Lacaille, Nicolas Louis de *47*, 62, 63, 100, 108, 115, 121, 136, 140, 152, 153, 156, 157, 165, 168, 169, 170, 176, 182, 184, 189
Lacerta, the Lizard 144–5
Lagoon Nebula *194*
Laille, Nicolas Louis de *47*, 63
Lalande, Joseph Jérome de 62, 63, 186
Lalande (red dwarf) 186
Lapetus 261
Large Magellanic Cloud (LMC) 132, 152, 165, 170, 195, 200, 210
  see also Magellanic Clouds
Lascaux Cave, Montignac (France) 294, 332
Leavitt, Henrietta 39, 51, *51*, 52
LED lights 319, 321
Lemaître, Georges 40–1, 53, *53*
Leo, the Lion *30*, 145, 145–6
Leo Minor, the Lesser Lion 146–7
Leonard, Frederick C. 276
Leonids 284
Lepus, the Hare 147–8
Levy, David 256
Libra, the Scales 148–9
libration 219
libration in latitude 219
light
  frames 366
  pollution 319, 321, 324
  speed of 39–41
  years 27–8
light pillars *319*
light pollution *322*, *323*
light spill *320*
limb darkening 213
Limón, Ada 238
Liperhey, Hans 35
literature 7
the Little Bear see Ursa Minor
Local Group 145, 176, 183, 209
Lochium Funis, the Ship's Log and Line 63, 169
Lowell, Percival 247, 251
Lupus, the Wolf 149–50, 156, 174, 195
Lynx, the Lynx 25, 150
Lyra, the Lyre 28, 31, 103, 151–2, 299, 301
Lyrid meteor shower 151

M numbers 26, 38, 40, 48, 52
M77 (Messier 77) in Cetus *211*
M87 (Messier 87) in Virgo *211*
Machin, Cathrin 354
  The Passage 356
  When Two Souls Meet 356
Machina Electrica, the Electrostatic Generator 63
McIntyre, Mary 269, 284, 297, 300, 301, 305, 306, 309, 320, 363, 367, 376–83
Magellan probe 246
Magellanic Clouds 11, 142
  see also Large Magellanic Cloud
Makemake 266
Malevich, Kazimir 341
Malus, the mast of Argo 169
Mariner probes 241, 244
Mars 8, 24, 33, 246, *247*, 248, *250*, 253, 269, *269*, 273, 286, 293, 309
  atmosphere 250–1
  canals 246, 248
  canyons 250
  craters 250
  life on 252
  moons of 252
  volcanoes 248
  water on 251–2
Mars Global Surveyor 251–2
Mars Reconnaissance Orbiter 253
May, Brian 12
Mellan, Claude 344, *344*
Mensa, the Table Mountain 63, 152–3
Merak 186
Mercury 8, 33, *214*, 241–2, *241*, *243*, 269, 293, 309, *345*
MErcury Surface Space ENvironment, GEochemistry and Ranging (MESSENGER) probe 241, 242
Messier, Charles 26, 38, 48, *48*, 172
meteors and meteorites 282–3
  meteor cameras 285
  meteor shower 11, 107, 114, 283–4
  meteorites 8, 285–6, *286*
  notable meteorite falls 287
  random/sporadic 11
Microscopium, the Microscope 153–4
Midnight Sun *23*, 24
Milk Dipper 172
Milky Way 7, *7*, 11, 29, 37, 54, 57, *68–9*, 71, 103, 104, 115, 116, 117, 128, 129, 131, 132, 136, 149, 158, 163, 168, 171, 172, 176, 177, 183, 189, 208–9, 301, *305*, 306, *325*, *326*, *327*, 362, *374*
Miller, William Allen 50
Mimas 261, 262, *262*
Mira, Omicron Ceti *206*
Mirach 99, *302*, 303

Miranda 264
Miró, Joan 354
  Ciphers and Constellations in Love with a Woman 354
Monoceros, the Unicorn 60, 154–5, 294
Mons Maenalus 62
monthly charts 71, *74–97*
Moon 8, 12, 33, *36*, 39, 45, 47, 147, 218–19, *218*, 269, *272*, 286, 298, 309, 338, *339*, *344*, 346, *347*, *361*, *365*, *381*, *382*, *383*
  Apollo missions 221
  eclipses 270–1
  lunar surface 219–20
  motions 219
  origin/evolution 221–2
  phases 222, *223*, 291, *308*, *362*, *379*, *380*
  pink supermoon *318*
  seas, mountains, craters 220
  space-probe exploration 220–1
The Moon Considered as a Planet, a World and a Satellite (Nasmyth & Carpenter) 335, *347*
Moon features
  Abulfeda 233
  Agrippa 227
  Albategnius 233
  Aliacensis 233
  Alpetragius 233
  Alphonsus 233
  Archimedes 227
  Aristarchus 225
  Aristillus 227
  Aristoteles 227
  Arzachel 233
  Atlas 229
  Autolycus 227
  Barocius 233
  Birt 233
  Blacanus 233
  Bulliadus 231
  Burckhardt 229
  Bürg 229
  Capella 235
  Cassini 227
  Catharina 235
  Clavius 233
  Cleomedes 229
  Copernicus 225
  Cyrillus 235
  Delambre 233
  Encke 225
  Endymion 229
  Eratosthenes 227

Eudoxos 227
Euler 225
Flamsteed 231
Fra Mauro 233
Franklin 229
Frascastorius 235
Gassendi 231
Geminus 229
Godin 227
Grimaldi 231
Hainzel 231
Harpsius 225
Heraclitus 233
Hercules 229
Herodotus 225
Herschel 233
Hevelius 225
Hippaius 231
Hipparchus 233
Hyginus 227
Janssen 235
Kepler 225
Lambert 227
Langrenus 235
Lansberg 231
Le Monnier 229
Letronne 231
Linné 227
Longomontanus 233
Mädler 235
Maginus 233
Mairan 225
Manilius 227
Mare Crisium 229
Mare Fecunditatis 235
Mare Humorum 231
Mare Imbrium 227
Mare Nectaris 235
Mare Nubium 233
Mare Serenitatis 227
Mare Tranquillitatis 229
Marius 225
Maurolycus 233
Messier and Messier A 235
Moretus 233
Oceanus Procellarum 225
Palitsch 235
Petavius 235
Piccolomini 235
Pilatus 233
Plato 227
Plinius 229
Posidonius 229
Prinz 225
Proclus 229
Ptolemaeus 233

Purbach 233
Pythagoras 225
Pythea 227
Regiomontanus 233
Reiner 225
Reinhold 225
Rümker 225
Scheiner 233
Schickard 231
Schiller 231
Sinus Iridum 225
Sirsalis and Sirsalis A 231
Snellius 235
Stadius 227
Stöfler 233
Taruntius 229
Thales 229
Thebit 233
Theophilus 235
Timocharis 227
Triesnecker 227
Tycho 233
Vallis Alpes (Alpine Valley) 227
Vallis Rheitax (Rheita Valley) 235
Walther 233
Werner 233
Moon maps
 North Central section *226*, 227
 Northeast Section *228*, 229
 Northwest section *224*, 225
 South Central section *232*, 233
 Southeast section *234*, 235
 Southwest section *230*, 231
Mount Wilson Observatory, California (U.S.A.) 39–40, 41, 52
Musca, the Fly 155–6
Museo Galileo, Florence (Italy) 36
music 7, 12, 16

NASA 213, 305, 357
Nasmyth, James 347, *347*
near-Earth objects (NEOs) 273, 277
Nebra sky disk *330*
nebulas/nebulae 40, 50, 196, 197, *348*
Neowise Comet *304*
Neptune 265, *265*, 276, 293
neutron star 200
New General Catalogue of Nebulae and Clusters of Stars (NGC) 26, 49
New Horizon mission 266
Newton, Isaac 35, 37, 46, *46*, 312
NGC/IC numbers 26
Norma, the Set Square 121, 156–7, 184
the Northern Coalsack, Cygnus Rift 129
Northern Cross 129

Northern Crown, Corona Boralis 124, 125–6
Northern Lights *see* Aurora Borealis
Nova Persei 163
novae 207
Nuremberg Chronicle (Wolgemut) 334, *335*, 341

Oberon 264
observations and findings 294–5, 297
 naked-eye observation 306, 309
observing diary 292
occultations *see* eclipses and occultations
Octans, the Octant 157, 157–8
Officina Typographica, the Printing Shop 63
O'Keefe, Georgia 354
Oort Cloud 277
Ophiuchus, the Serpent Holder 158–9, 165, 178, 182, *203*
orbital eccentricity 240
orbital inclination 240
Orion, the Hunter 13, 16, 21, 25, 26, 31, 50, 57, 62, *62*, 113, 147, 154, 159–61, 298, 301, *302*
Orion Nebula 39, 159, 195, *195*, 196, *365*, 367
Orionids 159, 281
Orionis, Trapezium 195, *195*
Orion's Belt 301, 303
OSIRIS-REx spacecraft 274
Outer Solar System Object Positions 276
Ovid 57

Palermo Catalogue 131
Pallas 273
Pandora 260
Papin, Denis 100
Paracelsus 341
parallax 28
pareidolia 342
Paris Observatory 349
Parsons, William *see* Rosse, William Parsons, 3rd Earl of
Pavo, the Peacock 161–2
Payne-Gaposchkin, Cecilia 54, *54*
Pegasus, the Winged Horse 16, 63, 162–3
penumbra 214
Penzias, Arno 41
perihelion 237
Perseids 11, 114, *282*, 283, 284, *284*, *287*, 306
Perseus 57, 58, *58*, 163–4
Perseverance rover ("Percy") 251–2,

251, 253
Pether, Henry 346
  The Thames and Greenwich Hospital by Moonlight 346
Pherkad 188
Philosophiae naturalis principia mathematica ("Mathematical Principles of Natural Philosophy") (Newton) 37, 46
Phobos *253*, 273
Phoenix, the Phoenix 161, 164–5
Pholus 276
photography 39, 49, 50
  basics 359
  early photography 359
  focal lengths/ISO settings 360–1
  image processing 372
  lunar art project 379–82
  photographing the Sun 367
  solar art project 376–8
  stacking 323, 367–9, 371
  through a telescope 364, 366
  timelapse 374–5
  see also astrophotography
photosphere 213
Piazzi, Giuseppe 131
Picasso, Pablo 354
Pico del Castillo, Spain 294, 332
Pictor, the Painter's Easel 63, 165–6
Pisces, the Fish 28, 102, 166–7
Piscis Austrinus, the Southern Fish 120, 167–8
Plancius, Petrus 60, 109, 122, 154, 161
planetary nebula 199
planets 8, 11, 24, 33–4, 62, 236–7, 293–4
  distance from the Sun *237*
  orbits 237–40, *238*, *239*
  rotational/orbital periods *240*
  size to scale *238*
  see also named planets
planisphere 293
Plaskett, John Stanley 154
Plato 33, 266
Pleiades, Seven Sisters 180, 195, *302*, *368*, *372*
the Plough (the Big Dipper) 11, 16, 17, 25, *30*, 65, *65*, 107, 186, *298*, 298, 299, *299*
Pluto 266, *267*, 276
Pogson, Norman 26
polar night 24
Polaris 28, 186, 188, 199, 297, *298*
Pole Star 28, 116
potentially hazardous asteroids (PHAs) 273

Praesepe, the Manger 110
precession (earth wobble) 28–9, *29*
preferential scattering 251
Procyon 31, 49, 113, 197
Prometheus 260
prominences 216
  eruptive/quiescent 216
Proxima Centauri 27, 28, 117, 202
Ptolemaic system 33–4, *33*, 334
Ptolemy, Claudius 33–4, *42*, 43, 56–8, 63, 134
pulsars 200
Puppis, the Stern 26, 63, 115, 168–9, 189
Pyxis, the Compass 169, 169–70

Quadrant Muralis, the Mural Quadrant 114
Quadrantids 114
Quaor 266
quasars 210

radio galaxies 210
Rappenglueck, Dr Michael 332
Ravidat, Marcel 332
Reber, Grote 54
red dwarf 186, 202
red giant 198–9, 207, *299*, *302*
Red Rocks, Sedona, Arizona (U.S.A.) 287
red supergiant 207
redshift 40
Reflector magazine 325
reflectors *41*, 312, 314
refractors 311–12
Reticulum, the Net 170–1
retrograde motion 33, 34
Rhea 261
Rigel 31, 159, 199
right ascension (RA) *21*, 28, 295
Rigil Kentaurus 31
rilles 222
Ritchey-Chrétien telescope 314
Roberts, Isaac 39
Rømer, Ole 46, *46*
Rosetta spacecraft 279
Rosse, William Parsons, 3rd Earl of 38, 39, 50, *50*, 111, 349, *349*, 353
Rotanev 131
Royer, Augustin 144
Rudaux, Lucien 335
Rudolphine Tables 45
Russell, Henry Norris 198

Sagan, Carl 16
Sagitta, the Arrow 171–2

Sagittarius, the Archer 40, 114, 172–3, 182, 298
St. Elmo's Fire 137
Samarkand, Uzbekistan 43
Saros cycles 271
Saturn 8, 24, 33, 259–60, *259*, *260*, 276, 293, 309, 335, *355*
  moons 261–2
  rings 260–1
Schedel, Hartmann 334
Schiaparelli, Giovanni 241, 247
Schiller, Julius 342
Schmidt-Cassegrain (SCT) telescope 314
Scientific Visualization Studio (NASA) 305
Scorpius, the Scorpion 148, 149, 156, 174–5, 182
Scorpius-Centaurus (or Sco-Cen) 174, 195
Sculptor Dwarf 176
Sculptor, the Sculptor 176
Scutum, the Shield 62, 177
seasons 21, 23–4
Sedna 266
Serpens, the Serpent 178
Sextans, the Sextant 145, 179
Seyfert, Carl 210
Seyfert Galaxy 211
Shapley, Harlow 39–40, 52, *52*
Shoemaker, Carolyn and Eugene 256
Shoemaker-Levy 9 256
Sickle of Leo 25
sidereal day 240
sidereal month 219
Siqueiros, David Alfato 357
Sirius (Dog Star) 26, 29, 31, 37, 49, 50, 112, 113, 195, 197
sky
  changing appearance *22*
  rotation 297–8
Sky & Telescope magazine 322
Small Magellanic Cloud 39, 51, 185
Soby, Olga 357
  "Dancing Universe" series *357*
Society for Popular Astronomy (SPA) 306, 309
Solar Dynamics Observatory (SDO) 376
solar nebula 237
Solar System 40, 46, 222, 237–40, 246, 247, 248, 256–7, 263, 269, 274, 276–80, 286, 293, 295, 306
solar wind 216, 279
solstices 21, 24
Southern Cross, Crux 11, 25, 60,

115, 117, 128–9, 155, 161
Southern Crown, Corona Australis 124
southern sky 60
Southern Triangle, Triangulum Australe 156
SpaceX 357
spectroscopic binary 202
spectroscopy 38–9
Spica 197
Square of Pegasus 25
Stabius, Johannes 58
stacking 367–9, *367*, *371*, *372*
Star of Bethlehem 11, 12
star charts *63*, *64*, *65*, 69–70, *70*
    constellations 71
    hemisphere 71
    monthly 71
star clusters
    globular 195
    open or galactic 195
    stellar associations 195
star hopping 299, 301, 303
stargazing 8, 9, 16–17
    basics 290–5, 297–9, 301, 303
    equipment 311–14, 316–17
    experience 323
    for free 304–6, 309
stars 55, 194–5, *297*
    binary system 202, 203
    brightness (or magnitude) 26–7, 31, 39
    composition 38
    death 199
    distances 27–8
    double and multiple 202
    evolving 198–9
    luminosity 52, 201
    naming 26, 60
    parallax 27
    as pointers 299, 301, 303
    positions 28–9
    proper motions 29, *30*
    size and lifetimes 24, 197
    spectral types 201
    temperature and color 24, 197–8
    variable 204, 207
    velocity 52
    visibility 24
Stellarium app 302, 305
Strong, Jim 305
Sualocin 131
suburban skies 318–19, 321–3
Summer Triangle 11, 103, 129, 151, 299, *299*, 332
Sun 8, 27, 28, 29, 33, 35, 38, 47, 52, 55, 105, 110, 114, 137, 151, 166, 195, 198, 212–17
    corona and solar wind 216, *217*
    eclipses 270
    flares, CMEs, aurorae *212*, 215
    observing *213*, 217
    outer layers 216
    photographing *217*, 363
    planetary orbits and distance 237–40
    solar cycle 215
    spectrum *38*
    sunspots and solar rotation 214
sungrazers 280
sunspots 213, 214, 215, *363*, *376*, *377*, *378*
supernova 45, 200
synodic month (or lunation) 219

Table Mountain, South Africa 63
Tarantula Nebula 195, *384–7*
Tarazed 103
Taurids 180
Taurus, the Bull 57, 62, *62*, 106, 110, 135, 180, 200, 303, *308*
Teapot of Sagittarius 25
telescopes 35, *36*, 39–41, 45, *46*, 47, 48, 50, 240, 298, *310*, 311, *313*, *315*, *316*, *317*, 349, *364*, *366*
    compound 314
    mounts 314, 316
    photography through *364*, *366*
    smart *372*
Telescopium, the Telescope 25, 182
Tempel-Tuttle 145
terminator 219
terminology 293–4
Tethys 261
Thales 188
time 23–4, 298, 338
Time magazine 335
Titan 261
Titania 264
Tobin, William 349
Trapezium *see* Orionis
Triangulum Australe, Southern Triangle 184
Triangulum Minus 62
Triangulum, the Triangle 183, *302*, *373*
Triton *236*
Tropic of Cancer 24, 110, 114
Tucana, the Toucan 161, 185
Tycho's Star 116

UFOs 8, 11, 244
Ulugh Beg 43, *43*
umbra 214

Umbriel 264
Universal Time (UT) 23
Universe 27, 33–5, 37–41, 44, 47, 48, 53, 54, 338
Uranographia (Bode) 63, *64*
Uranometria (Bayer) 26, 60, *61*
Uranus 12, 37, 263–4, *263*, *264*, 276, 293
urban skies 318–19, 321–3
Ursa Major, the Great Bear 8, 16, 25, 107, 111, 116, 146, 150, 186–7, *203*, 298, *298*, 299, *299*
Ursa Minor, the Little Bear 11, 57, 188, *298*, 299, *305*
US Aeronautical Chart and Information Center 247

Van Gogh, Vincent 7, 12, 349, 350, 353, *353*
    In the Starry Night 352
    Starry Night over the Rhone 328–9
Vega 28, 29, 31, 39, 103, 151, 299, *299*, 301
Veil Nebula 129, *196*, 200
Vela, the Sails 26, 63, 115, 189
Venera probe 246
Venus 8, 11, 12, 24, 33, 244, *244*, 245, 246, 269, 293, 309, 338
Venus Express probe 246
Vera C. Rubin Observatory 266, 277
vernal (or spring) equinox 28, 105
Vesta 273
Viking 1 orbiter 250
Virgo Cluster 123, 190
Virgo, the Maiden 113, 190–1, 197
Volans, the Flying Fish 192
Vopel, Caspar 123
Voyager probe 256, 259
Vulpecula, the Fox 171, 193

Whirlpool nebula *38*, 50, 349
white dwarf 199
Wilson, Robert 41
Wolgemut, Michael 334, *334*
Wood, Rachel and Jonathan 371
Woolsthorpe Manor, Lincolnshire (U.K.) 37

zenith 21
zenith hourly rate (ZHR) 283, 284
Zeta Herculis *203*
zodiac 25, *59*, 294
Zubenelgenubi, southern claw 148
Zubeneschamali, northern claw 148

PICTURE CREDITS

All reasonable efforts have been made by the authors and publishers to trace the copyright owners of the material quoted in this book and of any images reproduced in this book. In the event that the authors or publishers are notified of any mistakes or omissions by copyright owners after publication, the authors and publishers will endeavor to rectify the position accordingly for any subsequent printing.

Key: t: top, b: below, m: middle, l: left, r: right

Alamy: 33, 36b, 38tl, 41b, 42, 47r, 49l, 50l, 51l, 52tl, 52r, 53tl, 54tr, 55tl, 275, 334, 336, 338, 341, 345t, 345b, 346, 348, 349, 354
© 2015 AstroArt by David A. Hardy: 355
Bibliothèque de l'Observatoire de Paris: 63
Bibliothèque nationale de France, département Cartes et plans, GE DD-1439 (RES): 62r
© Bodleian Libraries, University of Oxford, CC-BY-NC 4.0.: 58
© Cathrin Machin: 356t, 356b
© Cath Adams: 373
Getty Images: 36tl, 53tr
© Ian Ridpath: 64b, 195, 196l, 196r, 198, 211t, 211b, 213, 223
© Ian Ridpath/Wil Tirion: 203
Illustrations by Lynn Nordbeck Hatzius: 312t, 312b, 313t, 315t, 315m, 315b
Library of Congress: 57l
Library of Congress: 339t
© Mark Radice, Refreshing Views Observatory: 248t
© Mary McIntyre: 214l, 214r, 239, 268, 269, 270, 271t, 282, 284, 286, 294, 296t, 296b, 296l, 296r, 297, 298, 299t, 299b, 300t, 300b, 301, 302t, 302m, 302b, 307t, 307b, 308t, 308b, 313bl, 313r, 316l, 316r, 317, 320t, 320b, 362t, 362b, 363, 364, 365tl, 365r, 365b, 366, 367, 368, 372l, 372r
The Metropolitan Museum of Art, Harris Brisbane Dick Fund, 1951: 56, 59
The Metropolitan Museum of Art, Purchase, Friends of European Sculpture and Decorative Arts Gifts, 1990: 60
NASA/JPL-Caltech/ASU, additional processing by Mary McIntyre: 253bl, 253r
NASA/JPL-Caltech/SwRI/MSSS additional processing by Mary McIntyre: 255tl
NASA/JPL-Caltech/SwRI/MSSS/Stuart Atkinson – Public Domain: 257b
NASA/Goddard Space Flight Center Scientific Visualization Studio: 352b
The Nuremberg Chronicle, World Digital Library: 335t, 335bl, 335r
© Olga Soby: 357t, 357b
The OMNIKA Foundation: 333
© Rachael and Jonathan Wood: 370
© Dr Rafael Ball, ETH-Bibliothek: 39, 61t, 61b, 62l, 64t
© Sheelagh Evans: 215
Shutterstock: 4, 6, 7, 9, 10, 12, 13t, 13b, 14, 17, 18 & 19 (dps), 20, 32, 43, 44tl, 44r, 45r, 55tr, 68 & 69 (dps), 70, 98, 194, 208, 217, 281b, 287, 288 & 289 (dps), 290, 291, 292, 304, 305, 310, 318, 323, 325, 326t, 326b, 327t, 327b, 358, 361, 374, 388 & 389 (dps), 390, 393, 416
Sir William Huggins. Process print after Walker & Cockerell. Wellcome Collection. Public Domain Mark. Source: Wellcome Collection: 50r
Smithsonian Institution Archives: 54tl
© Successio Miro / ADAGP, Paris and DACS London 2024: 354
Wikimedia Commons - 23, 36tr, 37, 38br, 41t, 45l, 46l, 46r, 47l, 48tl, 48r, 49r, 51r, 57r, 65, 212, 236, 241, 243tl, 243r, 243b, 244, 245tl 245r, 245b, 247, 248b, 250l, 250t, 250b, 250r, 251t, 251b, 253t, 254, 255tr, 255b, 256, 257t, 258, 259, 260t, 260bl, 260t, 260br, 262t, 262b, 263, 264, 265, 267t, 267b, 271b, 272, 275b, 276, 279, 280, 281t, 319, 322, 328, 330, 331, 339b, 340tl, 340r, 340tr, 340r, 340b, 342 & 343 (dps), 344t, 344bl, 344r, 347t, 347b, 350 & 351 (dps), 352t, 353
© Wil Tirion: 21, 22tl, 22r, 22br, 25, 27, 29, 30, 40, 72–97, 99–193, 205t, 205b, 206l, 206r, 210, 218, 224, 226, 228, 230, 232, 234, 238

CONTRIBUTOR BIOGRAPHIES

Dr. Stephen Maran is a Senior Advisor with the American Astronomical Society and a Fellow of both the American Association for the Advancement of Science and the Royal Astronomical Society. He is an astronomer and author with 35 years' experience at NASA, during which he was the Assistant Director of Space Science for Information and Outreach at the Goddard Space Flight Center.

Dr. Maran is the author and editor of 12 books and over 100 popular articles on astronomy and space exploration, including *Astronomy for Dummies* (now in its fifth edition). Dr. Maran has been recognized with the NASA Medal for Exceptional Achievement and by the International Astronomical Union, which named Minor Planet 9768 *Stephenmaran* in his honor. He received the Astronomical Society of the Pacific's Klumpke-Roberts Award (1999) for outstanding contributions to the public understanding and appreciation of astronomy. And in 2008, the American Astronomical Society awarded him the George Van Biesbroeck Prize, which "honours a living individual for long-term extraordinary or unselfish service to astronomy."

Ian Ridpath is an internationally renowned writer on astronomy and space. He is editor of the authoritative *Oxford Dictionary of Astronomy* and author of a widely used series of night sky guides for beginners, including *The Collins Stars and Planets Guide* and *Collins Gem Stars*. Ian is the recipient of the Astronomical Society of the Pacific's Klumpke-Roberts Award for outstanding contributions to the public understanding and appreciation of astronomy. Ian is based in London, U.K.

Mary McIntyre is an astronomer, astrophotographer, astronomy and space artist, astronomy communicator, and author. With a life-long interest in astronomy, Mary is passionate about astronomy outreach and ensuring inclusion for all in the joy of stargazing. She has appeared on BBC TV's *The Sky at Night* and BBC Radio, is a co-host of *Comet Watch* on Astronomy FM radio and when time allows appears on the Space Oddities Live YouTube panel show. Mary is a Fellow of the Royal Astronomical Society, a member of the British Astronomical Association, is on the council of the Society for Popular Astronomy, and is involved with the Global Meteor Network. In 2021, she was awarded the Sir Patrick Moore Prize by the British Astronomical Association for contributions to astronomy outreach.

Rachel Federman is a writer and nonprofit researcher who has written about mermaids, llamas, herbs, writing, quiet, exploring the outdoors, teacher retention, and mental wellness, among other topics. Her recent books include *Mer-mania* (HarperCollins, 2019) and *The Backyard Chicken-keeper's Bible* (HarperCollins 2023). Rachel lives in New York City with her family. During her work on this project she saw Venus out the window many mornings in the cold, dark sky.